D0812345

Chicago Public Library
C
P
L
REFERENCE

Welding Handbook

Sixth Edition

SECTION 5

Applications of Welding

Edited by Charlotte Weisman
Published in 1973 by AMERICAN WELDING SOCIETY
2501 N.W. 7th Street
Miami, Florida 33125

Welding Handbook

IN FIVE SECTIONS

Prepared under the direction of

THE WELDING HANDBOOK COMMITTEE

Chairman	S. WEISS	*University of Wisconsin*
Editor	C. WEISMAN	*American Welding Society*
	S. D. REYNOLDS, JR.	*Westinghouse Electric Corp.*
	R. T. TELFORD	*Union Carbide Corp.*
	I. G. BETZ	*Frankford Arsenal*
	A. W. PENSE	*Lehigh University*

Library of Congress Catalog Card Number: 74-346
International Standard Book Number: 0-87171-100-1

Printed in the United States of America

Preface

With the publication of Section 5, the Sixth Edition of the WELDING HANDBOOK is now complete. For this final volume, the writers of the Handbook–the members of the various Handbook chapter committees–have carefully evaluated the material from the Fifth Edition of Section 5 and have preserved or updated the information which they felt was essential to the Handbook user. To this they have added important new data on welding applications which have developed since the previous volume appeared six years ago.

The chapters which have been substantially revised for this edition include Chapters 81, Buildings, and 82, Bridges, which reflect the work recently completed by the AWS Structural Welding Committee. In addition, Chapter 87, Nuclear Power, has appropriately undergone an extensive revision. Chapter 89, Railroads, also contains much new material relating to the design and construction of freight cars and other railroad facilities.

It is important to note here, however, that "revised" does not necessarily mean "expanded." In examining the preceding Section 5, the Welding Handbook Committee (the administrative body for the Handbook) decided that the two chapters covering clad metals, weld cladding and applied liners could be effectively combined into one. The new chapter is now Chapter 93, Clad Steel and Applied Liners.

Another decision made by the Welding Handbook Committee concerns Chapters 91, Aircraft, and 92, Launch Vehicles. The developments in the aircraft and aerospace industries of recent years have lead to a period of austerity for the technical personnel working in these fields, making it difficult for them to participate in such extracurricular endeavors as Handbook revisions. Since it would have been extremely difficult, if not impossible, to establish committees for these two chapters, the Welding Handbook Committee reviewed the existing chapters and concluded that they still contain vital information on all types of aerospace and launch vehicles. The two chapters are thus reprinted here essentially as they appeared in the last edition with minor editorial changes made by the AWS Headquarters staff.

The publication of this Handbook would not have been possible without the Handbook chapter committees and the chapter chairmen who unselfishly devoted their time, knowledge and resources to the revision of their assigned chapters. The project would also have been impossible without the expert guidance of the Welding Handbook Committee which formulated Handbook policies and painstakingly reviewed all chapter drafts. To all of them, we offer our sincere appreciation for their contribution to Section 5.

Charlotte Weisman, *Editor*

Contents

CHAPTER 81

BUILDINGS

PREPARED BY A COMMITTEE CONSISTING OF:

S. A. GREENBERG, *Chairman*
Welding Consultant

J. T. BISKUP
Canadian Welding Bureau

O. W. BLODGETT
The Lincoln Electric Co.

A. L. COLLIN
Kaiser Steel Company

W. A. MILEK, JR.
American Institute of Steel Construction

CHAPTER 81

BUILDINGS

INTRODUCTION

This chapter covers those structures which are to be subjected to static loading or a low frequency of dynamic stresses, as would normally be encountered in building construction. For those buildings, portions of buildings or particular members such as crane runway girders, machinery supports or for members of structures such as bridges, subjected to fluctuations in loading, the recommendations in the chapter on bridges should be followed.

Welding has become a major method of making joints in building structures. The advantages of welding are now generally recognized by designers and fabricators alike. This chapter, therefore, is primarily concerned with providing information that will permit the use of welding in the most economical manner, while at the same time, assuring that the required quality will be obtained.

GOVERNING CODES AND SPECIFICATIONS

The present state of welding would be chaotic without standard procedures and design criteria. Existing regulatory building requirements must be kept under continuous study in order to keep up with technical developments.

Present-day steel construction regulations are the result of many years of shop and field experience, and are affected by the extensive research carried out in recent years. The introduction of higher strength steels and corresponding modifications in welding specifications are notable examples. These are covered in the Structural Welding Code, AWS D1.1-72, and the AISC Specifications for the Design, Fabrication and Erection of Structural Steel for Buildings (February

12, 1969) and Supplement to the Specification (November 1, 1970). The documents issued by AWS and AISC are in general agreement. The AWS Structural Welding Code states in the footnote to Section 8 that, in the absence of any locally applicable building law, specification or regulation, it is recommended that the construction comply with the AISC Specification for the Design, Fabrication and Erection of Structural Steel for Buildings. The AISC Specification cites the AWS Code in several sections; however, the two documents are not consistent in every detail.

The following are significant areas where differences exist: both documents permit high-strength bolts in friction-type joints installed prior to welding to be considered as sharing loads with welds. Neither document permits the use of A307 bolts or A325 bolts in bearing joints in combination with welds. In the case of welded alterations to existing structures, the AISC Specification permits existing rivets or properly tightened high-strength bolts to be considered as carrying dead loads. The welding then need be adequate only to carry all additional stresses.

The AWS Code specifies that the effective throat thickness of fillet welds shall be taken as the shortest distance from the root to the face of the diagrammatic weld with all welding processes. For fillet welds made by processes other than submerged arc welding, the two documents are consistent. However, the AISC Specification recognizes the increased penetration that results from submerged arc welding and defines the effective throat thickness for fillet welds made by this process as equal to the leg size for 3/8 in. and smaller fillet welds, and equal to the theoretical throat plus 0.11 in. for fillet welds over 3/8 in.

Another difference exists in the requirements for intermittent fillet welds. In this instance, the AISC Specification states that these welds may be used to transfer calculated stress across a joint or faying surfaces when the strength required is less than that developed by a continuous fillet weld of the smallest permitted size, and to join components of built-up members. The AWS Code states that intermittent fillet welds may be used to carry calculated stress. In either case it is required that the effective length of any segment of intermittent fillet weld should not be made less than four times the weld size, with a minimum of 1 1/2 in.

The AISC Specification requires that, in the absence of more restrictive tolerances called for in bid documents, dimensional tolerances of welded built-up cross sections be in accordance with the applicable tolerances for comparable hot-rolled cross sections specified in ASTM A6. The AWS Code in Article 3.5 specifies certain special tolerances that for comparable items are essentially in agreement with ASTM A6, and therefore consistent with the AISC Specification requirements. However, the AISC Specification takes exception to the AWS tolerances for flatness of welded girder webs (not included in ASTM A6) on the basis that since out-of-flatness of webs has no effect on the static load-carrying capacity of such members, no limitation can rationally be established, nor is a limitation needed except when dictated by architectural considerations, as when girders are exposed.

Heretofore, differences in the AWS and the AISC documents existed between the joint forms that were recognized as prequalified for manual shielded metal-arc welded butt joints with partial penetration. These differences were eliminated in the 1969 AWS Code and the Seventh Edition of the AISC Manual of Steel Construction. A review of other building construction regulations indicates a general agreement or direct citation of the AWS Code and the AISC

Specification. Such general agreement or citation will be found in the regulations issued by the four regional code authorities: *National Building Code* (American Insurance Association, formerly National Board of Fire Underwriters), *Basic Building Code* (Building Officials Conference of America), *Southern Building Code* (Southern Building Conference) and the *Uniform Building Code* (International Conference of Building Officials). Federal agencies–Army Corps of Engineers, Navy Bureau of Yards and Docks, General Services Administration and the National Aeronautics and Space Agency also generally accept the AWS and AISC standards for welding.

METALS AND PROCESSES

BASE METALS

The AWS Structural Welding Code permits the use of steels conforming to any one of the following ASTM Specifications[1]:

A36 –Structural Steel
A53 (Grade B)–Steel Pipe
A242–High-Strength Low-Alloy Structural Steel[2]
A375–High-Strength Low-Alloy Hot-Rolled Steel Sheets and Strip
A441–High-Strength Low-Alloy Structural Manganese Vanadium Steel
A500–Cold-Formed Welded and Seamless Carbon Steel Structural Tubing in Rounds and Shapes
A501–Hot-Formed Welded and Seamless Carbon Steel Structural Tubing
A514–High-Yield Strength, Quenched and Tempered Alloy Steel Plate, Suitable for Welding
A529–Structural Steel with 42,000 psi Minimum Yield Point (½ in. maximum thickness)
A570 (Grades D and E)–Hot-Rolled Carbon Steel Sheets and Strip, Structural Quality
A572–High-Strength Low-Alloy Columbium-Vanadium Steels of Structural Quality
A588–High-Strength Low-Alloy Structural Steel with 50,000 psi Minimum Yield Point to 4 In. Thick
A618–Hot-Formed Welded and Seamless High-Strength Low-Alloy Structural Tubing[2]

The gamut of chemistries, mechanical properties and the thickness range covered by the full spectrum of these specifications is extremely wide. It encompasses steels having tensile stengths from about 58,000 to 135,000 psi. The thickness range is from that of thin-walled tubing and sheet to sections with an unlimited thickness. (Plate 6 in. and more in thickness is being welded regularly. Present-day jumbo column sections weighing over 700 lb per ft and having flanges 5 in. thick are listed as standard). It is reasonable to expect the designer and fabricator to recognize that these steels encompass a very wide range of properties and quality. Selection of the steels to use for any given project requires careful consideration of the welding characteristics and quality as well as the mechanical properties of the steels considered.

---

[1]ASTM Specifications A7 and A373 have been withdrawn and replaced by A36. They should not be specified in Job Specifications.

[2]ASTM A242 and A618 Gr. I must have properties suitable for welding.

Brittle fracture can occur in a structural member that is subjected to service temperatures lower than the "transition temperature" of its metal, that is, the temperature below which the behavior of the metal becomes brittle rather than ductile. This temperature should be determined by a test that reflects the velocity of local deformation resulting from the service loading of the member, as well as the concentration of stress caused by any notch-like details that are involved, and other characteristics of the stress. Also, the degree of constraint against ductile behavior resulting from the thickness of the base metal and geometric shape should be reflected. If such a test has not been developed, experience may provide judgment for applying a temperature correction factor to transition temperatures determined by a standard test such as the Charpy V-notch test.

While impact properties are not a part of steel specifications today, they may have to be added as a job requirement or at least evaluated in terms of the welded joints obtained using a given steel. This concept is contained in the filler metal requirements for submerged arc, gas metal-arc, flux cored arc, electroslag and electrogas welding in the AWS Code and should be applied in other cases as service conditions may warrant.

Welding Processes

The AWS Code provides for use of seven welding processes: shielded metal-arc, submerged arc, gas metal-arc, flux cored arc, electroslag, electrogas and stud welding. The first four of these processes can be used without welding procedure qualification testing under the Code provided the welding procedures conform to certain specific limitations specified in the Code. Use of electroslag, electrogas and stud welding is permitted if the procedures are qualified as prescribed in the Code.

Information is available and in many welding shops experience exists to permit welding of these base metals by the processes cited. However this information and experience are not universally available. Moreover, the Code requirements are prescribed as the minimum representing the usual type of structure as to size and service conditions.

To assure an economical and safe structure, the final responsibility rests with the designer as to the selection of the steels to be used and the welding processes and procedures that are to join them. The contractor, whether fabricator or erector, shares this responsibility and must determine that the requirements for the structure will be met under the specific capabilities and facilities of his shop.

In practical terms, this means that welding should be done using written welding procedures that have been established by tests. The tests can be of prior origin, having been used for previous jobs. In any event, the testing should be sufficiently extensive so that the required service properties are obtained from the steels, welding processes and methods, facilities and abilities to be used for the given job.

In most cases, the extent of testing prescribed in the Code will be adequate. However, this should not be taken for granted and test requirements should be modified as necessary–adding or even omitting tests–based on the job at hand.

Another consideration is base metal quality. Welding of edges having extensive inclusions, voids or laminations will lead to unsound joints; the unsound base metal must be removed prior to welding. The Code now provides limits on acceptability and the repair of defects in cut edges of plate. For very critical work, precise limitations–even ultrasonic examination at the mill–have been specified. Some shops have adopted the practice of ultrasonically inspecting base metal over some specified thickness (for example, 2 in.) themselves before doing any work on it.

One test sometimes used to check a questionable edge is to heat the section to about 200 F, allow it to cool and then magnetic particle or dye penetrant inspect it. The heat will tend to open up any void extending to the surface. Oxygen cutting will also reveal discontinuities. However, whatever the precautions are, the edges of base metal should be examined at least visually before they are welded.

FILLER METAL

Currently the five AWS Specifications listed below provide for filler metals used for shielded metal-arc, submerged arc, gas metal-arc and flux cored arc welding. Again, as with the base metal specifications, there are numerous classifications of filler metals and combinations of flux and electrode that can be used. Selection must be made with due regard to the specific requirements for the job at hand.

A5.1 –Mild Steel Covered Arc-Welding Electrodes. (For shielded metal-arc welding.)
A5.5 –Low-Alloy Steel Covered Arc-Welding Electrodes. (For shielded metal-arc welding.)
A5.17–Bare Mild Steel Electrodes and Fluxes for Submerged-Arc Welding
A5.18–Mild Steel Electrodes for Gas Metal-Arc Welding
A5.20–Mild Steel Electrodes for Flux-Cored Arc Welding

With the diversity of steels available for structural use, it is common to use more than one steel for a job, or even for different parts of the same fabricated member. The extent of such practice should be governed by economic considerations. The important thing to recognize is that proper identification of the different steels used is necessary to avoid a mix-up. Color coding is being used and is satisfactory if personnel are familiar with the colors and the steels they identify. Some objection to this practice has been raised because of the possible incidence of color blindness in some people. As an alternative, painting of the ASTM specification number on each section has been used. The argument against this is the cost of doing it. Whatever the means used, positive identification and separation of the steels are a must.

In the same way, welding materials–electrodes, rods, fluxes and shielding gases–must be clean, properly stored and positively identified. They should be carefully controlled to prevent use of wrong or unsuitable material.

It should be only a matter of economics whether field welding or some other means of joining is used. However, control of the steels and materials used must

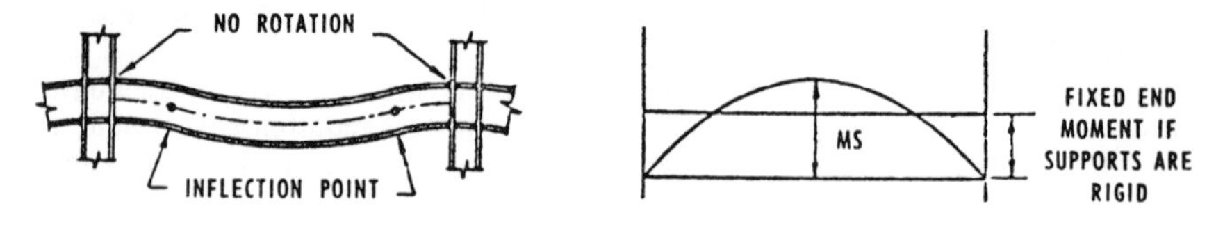

Fig. 81.1.–Rigid framing designated Type 1.

be even tighter in the field, since the likelihood of error and contamination is greater.

Strengthening and repair of existing structures by welding require knowledge of the properties of the steel to be welded. If this information is unavailable from records, then an investigation is warranted. Depending on the nature of the work and the severity of the service, it may be desirable to make chemical and mechanical tests of samples of the existing steel. In any event, welding should be preceded by full removal of paint, plaster, cement, scale and other foreign material from around the area of the joint.

DESIGN OF STRUCTURAL ELEMENTS

While the AWS Code usually governs materials, workmanship, weld details, welding processes and techniques and inspection, the design of structural elements comes under the jurisdiction of the AISC Specifications.

TYPES OF STRUCTURAL ELEMENTS

Structural elements may be classified by the type of stress[3] they resist, as follows: tension members–tie-rods, hangers, truss members; compression members–columns, struts, pedastals, truss members and shear members–shafts, shear keys, studs, lugs.

In the design of even a simple structural element, many types of stresses may have to be accounted for. Structural elements may have to be designed to resist a combination of stresses such as a column carrying an axial load and resisting a bending moment at a beam connected to it.

Structural elements may also be classified by the types of loads or forces they are designed to resist such as main member, secondary member, bracing member, wind member, crane runway support, etc.

TYPES OF CONSTRUCTION

The AISC defines the following three basic types of construction, based on the amount of rigidity by which the structural elements are connected to each other.

Type 1:–*rigid frame (continuous frame) construction*, which assumes that the members are connected with sufficient rigidity to maintain the original angle between them under the assumed loading (Fig. 81.1).

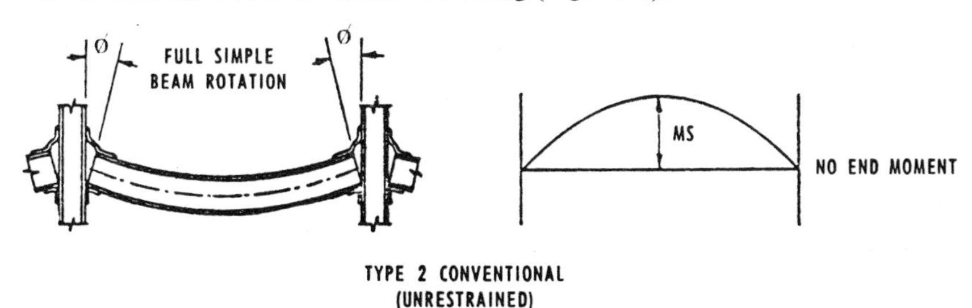

Fig. 81.2.–Simple framing designated Type 2.

[3]Flexural members, such as beams and girders, are variously required to resist any of the appropriate types of stress.

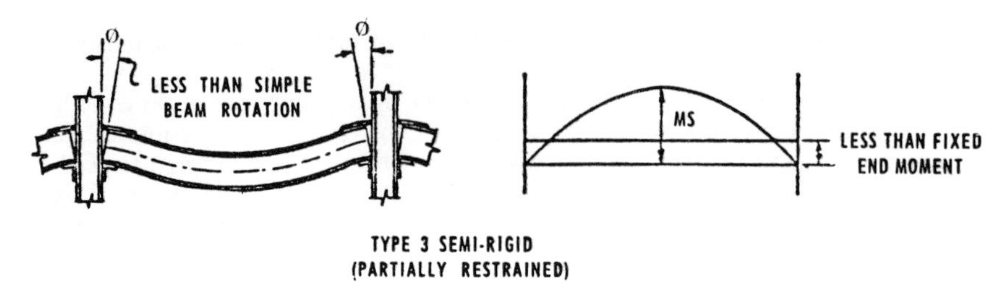

Fig. 81.3.–Semirigid framing designated Type 3.

Type 2:–*simple (unrestrained or free-ended) framing*, which assumes that the ends of the members are connected for shear only and are free to rotate under gravity loads (Fig. 81.2).
Type 3:–*semi-rigid (partially-restrained) framing*, which assumes that the connections of members have a known and predictable moment capacity somewhere between that of Type 1 and Type 2 (Fig. 81.3).

LOADS AND FORCES

Structural elements in building construction must be designed for a number of possible types of combinations of loads and forces such as dead loads, impact, seismic tremors, vibration, live loads, wind forces and fatigue.

Minimum unit loads are usually set by the applicable local building code, or by one of the Model Codes. In the absence of any local requirements, design loads should not be less than those recommended by Minimum Design Loads in Buildings and Other Structures, ANSI A58.1-1972.

ALLOWABLE STRESSES

Allowable stresses in tension, compression, shear, bearing and bending for members are prescribed in the AISC Specifications as functions of the yield strength and ultimate strength of the steel used. For axially-loaded compression members, the slenderness ratio (Kl/r) is a factor. Appendix A of the AISC Specification contains tables of allowable compressive stresses for a wide range of effective slenderness ratios for steels of different yield strengths.

Design requirements for structural elements subjected to combined stresses and other requirements such as fatigue, stability, stiffness, wind and seismic forces, deflection and vibration are described in the AISC Specification.

Allowable stresses for welds in structural joints are prescribed by the AWS Code and included in the AISC Specification.

GROOVE WELDS

For groove welds, stresses of whatever kind are ascribed the same maximum allowable values as for the members or elements of members that they join. The values are a function of the base metal, welding process and filler metal used.

FILLET WELDS

The standards for judging the strength of fillet welds differ from those for groove welds. Groove welds are usually a continuation of the base metal. However, the joint configuration of fillet welds is more complex, and

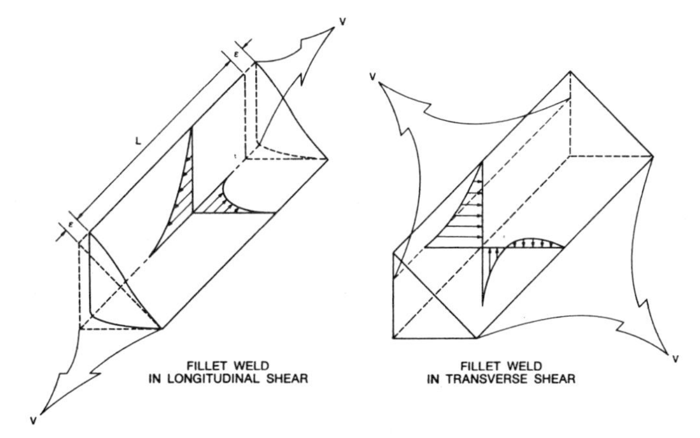

Fig. 81.4.–Distribution of stresses in fillet welds.

deformation relates to the direction of the loads. As shown in Fig. 81.4, both bending and shearing stresses are present in fillet welds. Designers concerned with transferring forces from one plane to another need not confront such an indeterminate problem, for recently completed tests assure the transferring of forces with adequate safety factors.

The AISC Specification and AWS Code define the allowable force a weld can resist as the throat area, multiplied by an allowable shearing stress. The throat dimension is the same for both longitudinally- and transversely-stressed joints. In reality, a fillet weld stressed transversely to its axis can sustain much higher loads than if it were stressed longitudinally.

The joint AWS-AISC advisory group on fillet welds conducted a series of tests to determine the ultimate strength of fillet welds subjected to longitudinal and transverse shear. Test specimens were prepared by both eastern and western fabricators, under actual shop conditions. The steels tested ranged from A-36 with a yield strength of 36 ksi, to A-514 which has a yield strength of 100 ksi. Electrode strengths ranged from 60 to 100 ksi, and the weld sizes were 1/4, 3/8 and 1/2 in. All tests were conducted by an independent testing laboratory.

Under previous welding specifications, two values for allowable shear stresses in fillet welds were permitted: 13.6 ksi were allowed for E-60XX manual electrodes on all weldable grades of steels and 15.8 ksi for E-70XX electrodes used with A36, A242 and A441 steel. When these values were applied to joints used in the test program to determine allowable loads, the loads represented but a small fraction of the actual strength of the joints. For example, welds stressed longitudinally exhibited average safety factors of 4.8 for A36 steel welded with E-60XX electrodes, while for A514 steel welded with E-110XX electrodes, the average safety factor was 6.6.

Welds stressed transversely exhibited safety factors much higher than those stressed longitudinally. The safety factors ranged from 8.2 for A36 steel welded with E-70XX electrodes to 11.5 for A514 steel welded with E-110XX electrodes.

It was also noted that the failure plane for welds stressed transversely was not the typical throat dimension that has historically been accepted as the weakest plane, but a plane at a relatively flat angle from the horizontal.

Although the Code does not recognize higher values for fillet welds stressed transversely, the designer can achieve greater strength by laying out the joint to

Table 81.1—Fillet weld strength with allowable loads in kips per inch of fillet weld

Matching ASTM Base Metals	Filler Metal (Electrode and flux-electrode combinations)	Allowable Unit Shear Stress (ksi)	Process	Fillet Weld Size									
				1/8	3/16	1/4	5/16	3/8	1/2	5/8	3/4	7/8	1
A500 Grd. A A570 Grd. D	AWS A5.1E60XX AWS A5.17 F6X-EXXX AWS A5.20E60T-X	18.0	Manual submerged arc	1.6 2.3	2.4 3.4	3.2 4.5	4.0 5.6	4.8 6.8	6.4 8.3	8.0 9.9	9.5 11.5	11.1 13.1	12.7 14.7
A36, A53 Gr. B, A242, A375, A441, A500 Gr. B, A501, A529, A570 Gr. E, A572 Gr. 42 to 60, A588, and A618	AWS A5.1 or A5.5 E70XX AWS A5.17 F7X-EXXX AWS A5.18 E70S-X or E70U-1 AWS A5.20 E70T-X	21.0	Manual submerged arc	1.9 2.6	2.8 3.9	3.7 5.3	4.6 6.6	5.6 7.9	7.4 9.7	9.3 11.6	11.2 13.4	13.0 15.3	14.8 17.2
A572 Gr. 65	AWS A5.5 E80XX Gr. 80 submerged arc, gas metal-arc, or flux cored arc weld	24.0	Manual submerged arc,	2.1 3.0	3.2 4.5	4.2 6.0	5.3 7.5	6.4 9.0	8.5 11.1	10.6 13.2	12.7 15.4	14.8 17.5	17.0 19.6
A514 over 2 1/2 in. thick	AWS A5.5 E90XX Gr. 90 submerged arc, gas metal-arc, or flux cored arc weld	27.0	Manual submerged arc	2.4 3.4	3.6 5.1	4.8 6.8	6.0 8.4	7.2 10.1	9.6 12.5	11.9 14.9	14.3 17.3	16.7 19.7	19.1 22.1
A514 over 2 1/2 in. thick	AWS A5.5 E100XX Gr. 100 submerged arc, gas metal-arc, or flux cored arc weld	30.0	Manual submerged arc	2.7 3.8	4.0 5.6	5.3 7.5	6.6 9.4	8.0 11.3	10.6 13.9	13.3 16.6	15.9 19.2	18.6 21.9	21.2 24.5
A514, 2 1/2 in. thick and under	AWS A5.5 E110XX Grd. 110 submerged arc, gas metal arc, or flux cored arc weld	33.0	Manual submerged arc	2.9 4.1	4.4 6.2	5.8 8.3	7.3 10.3	8.7 12.4	11.7 15.3	14.6 18.2	17.5 21.1	20.4 24.0	23.3 27.0

use transverse fillet welds. The allowable stresses for transversely stressed fillet welds will almost certainly be increased when additional test information is available.

From this study and other continuing welding research, the AWS and AISC concluded that a substantial increase in allowable stresses for fillet welds was justified, and inclusion of high-strength steel and electrodes was necessary. Allowable stresses are shown in Table 81.1, which also lists the strength of different sizes of fillet welds for each combination of base and filler metal.

The table recognizes the deeper penetration obtained by submerged arc welding. The additional penetration, conveniently expressed in greater equivalent throat thickness, accounts for the higher strength capacities for such fillet welds as shown in Table 81.1.

MINIMUM FILLET WELD SIZE

The minimum size fillet weld as related to plate thickness is specified by both AISC and AWS as listed in Table 81.2. This requirement has long existed to prevent cracking, and is related to the heat flow characteristics and the increased restraint to which welds in thicker metal are subjected. The many parameters of this phenomenon are under investigation and include the following:

1. Total heat input from sources such as the welding process, preheating, interpass heat, ambient temperatures, etc.
2. Heat sink characteristics due to metal mass (rate of cooling).
3. Metallurgical composition.

COLUMN BASE PLATES

Base plates are required on the ends of columns to distribute the concentrated compressive load of the column over a much larger area of the material that is to

Table 81.2–Minimum fillet weld size

Material Thickness of Thicker Part Joined (in.)	Minimum Size* of Fillet Weld (in.)
1/4 or less	1/8
Over 1/4 to 1/2	3/16
Over 1/2 to 3/4	1/4
Over 3/4 to 1 1/2	5/16
Over 1 1/2 to 2 1/4	3/8
Over 2 1/4 to 6	1/2
Over 6	5/8

*Weld size need not exceed the thickness of the thinner part joined. For material 1/4 in. or more in thickness, the maximum fillet weld size permitted is 1/16 in. less than the thickness of the material, unless the weld is designated on the drawings to be built out to obtain full throat thickness.

support the column. The overhanging portion of the base plate is assumed to be a cantilever beam, with a fixed end just inside of the column and uniformly loaded with the bearing pressure of the supporting material.

AISC suggests the method shown in Fig. 81.5 to determine the required thickness of bearing plate, using a maximum bending stress of $F_\sigma = 0.75F_y$.

The procedure shown in Fig. 81.5 results in the formula:

$$t = \sqrt{\frac{3\rho n^2}{\sigma}}$$

where:

ρ = Allowable bearing pressure on support, psi

σ = Allowable bending stress in plate, psi = $0.75F_y$ (F_y = minimum yield strength of steel)

t = Thickness of plate, in.

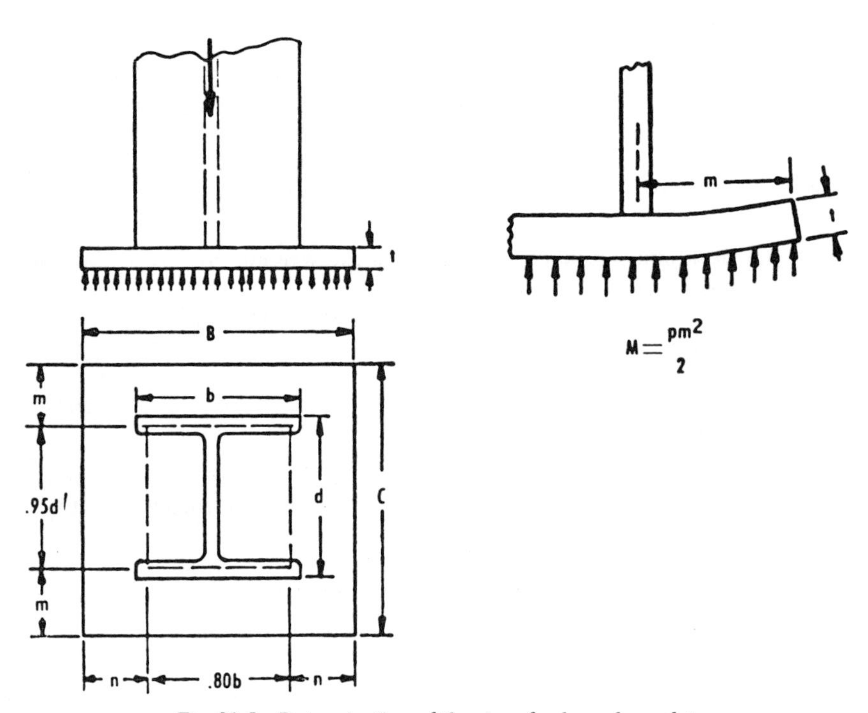

Fig. 81.5.–Determination of the size of column base plate.

Where compression members have full-milled bearing on base plates, AISC specifications require sufficient welding to hold the parts securely in place. These must be proportioned to resist any tension developed by specified wind forces acting with 75% of the calculated dead load stress and no live load, if this condition will produce more tension than with full dead load and live load applied.

Some examples of welded column base plates are shown in Fig. 81.6. If there is a moment to be transferred at the base, gusset plates may be added.

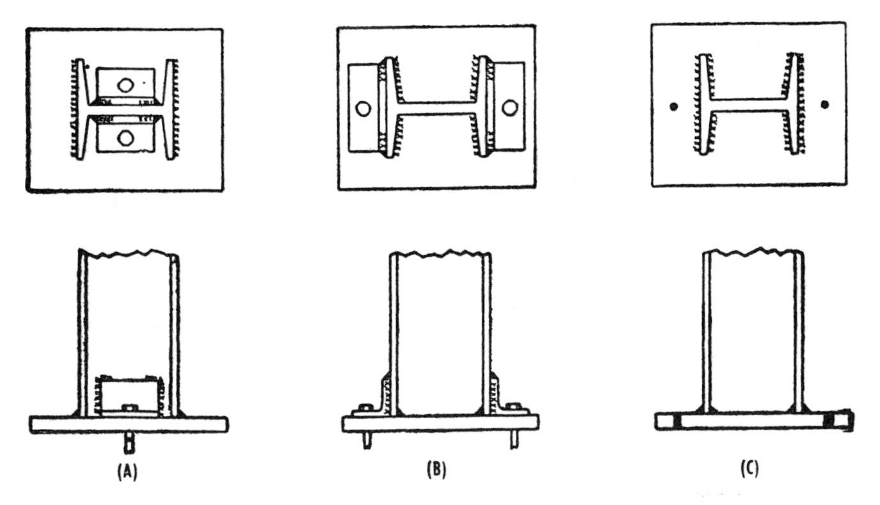

Fig. 81.6.–Examples of welded column base plates.

COLUMN SPLICES

The requirements for the splicing of full-milled tier building columns are the same as for column base plates.

Several methods of making column splices without punching the main columns are shown in Fig. 81.7.

Beam Seats and Framed Connections

AISC Manual of Design includes the methods of designing both stiffened and unstiffened seats as well as tables for proportioning these types of connections.

Similar data are included in the AISC Manual for framed angle connections.

RIGID CONNECTIONS

There are two basic types of rigid connections used at the present time: (1) the directly connected beam in which the flanges are groove welded to the supporting member without any connecting plates (2) the beam cut short and attached with additional connecting plates.

AISC Specifications allow a 10% increase in the allowable bending stress or $0.66\sigma_y$ for a compact section. In addition, in the negative moment region at the support 90% of the end moment may be used. This reduction does not apply to a cantilever beam. The section modulus used at this point must not be less than that required for the positive moments in the same beam, and the compression flange shall be regarded as unsupported from the support to the point of contraflexure. These values also apply to the connecting welds. The design of a connecting weld should take into account the combined effect of stresses from moment and shear and the reduction in available section because of snipes. The local increase in unit stresses may be reduced by adjusting the size of the reinforcing welds on the complete penetration welds at the toes.

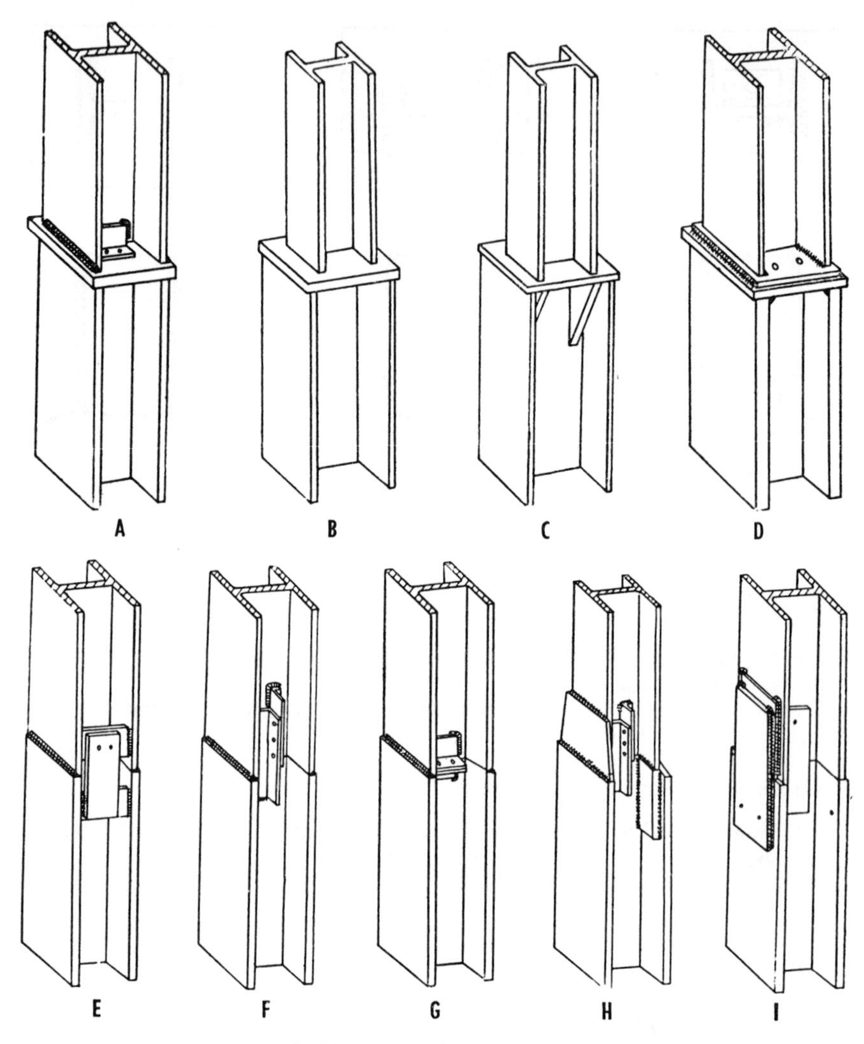

Fig. 81.7.–Typical column splices.

SEMIRIGID CONNECTIONS

The design of a semirigid connection is similar to that for rigid connections except that the degree of restraint produced by the connection has been established by test for the particular type of joint under consideration. The only type of semirigid connection in which the actual restraint can be predicted with any degree of accuracy uses a top connecting plate. Here the tensile force in the top flange is found from the end moment, and the cross-section area and length of this attaching plate is computed so as to provide the proper amount of elastic elongation for the amount of rotation required.

WIND BRACING REQUIREMENTS

For wind bracing connections in a simply supported structure, the top plate is designed to carry the calculated wind moment at a one-third higher stress. The welds connecting the top plate to column and beam are designed at normal unit stress[4] to develop the value of the top plate at the increased allowable stress for wind loads. The connection to the seat is designed for the same moment couple at corresponding values for the base metal and welds. It is assumed that any combination of gravity and wind loading that would produce a total stress in the connection exceeding the yield point would be relieved by plastic flow in the unwelded area in the top plate. Since the moment due to gravity load in the simple frame is not reversible, no provision need be made in the seat for plastic behavior.

The effect of a reversal of the direction of beam end rotation upon the top plate after it has gone through a cycle of loading producing plastic flow is to reduce the tension from yield point stress to some lower value. This may result in appreciable compression in the top plate. Where such a condition might occur, the top plate should be made thick enough to prevent buckling under compression. This may be accomplished by making the thickness not less than 1/24 of the unwelded distance. Repeated cycles of tension and compression in building construction would not occur often enough to consider the endurance limit of the metal.

TRUSSES

In trusses of arc-welded design, gusset plates are generally eliminated or reduced to the minimum requirement for the transfer of forces. Tension members in a welded design are lighter because the entire section is effective, and the amount of extraneous detail metal is reduced to a minimum. Gusset plates may be used at some panel points on heavy trusses to reduce the criticality of fit-up, a measure required if all connections are butt welded. In the case of heavy trusses using rolled sections, gussets may be used to provide for smoother transition in the joint. Consequently, a substantial reduction of severe stress concentrations which accompany any sudden change of cross section (Fig. 81.8a) will also be obtained. Similarly, several gusset plates may be individually used in one joint for exactly the same purpose (Fig. 81.8b).

Welded trusses may be designed in various ways, using T and H shapes for chords. The web members are generally angles or channels.

The simplest and most common type of truss construction is made of H or I and angle shapes. In this type truss, the bottom and top chords are made of H or I sections, with angle sections for the web members. This truss is easy to fabricate and weld, as the sections lap each other and fillet welds are used.

Some trusses make use of T sections for their diagonal members. However, it has to be borne in mind that if a T section is cut from a WF section, a straightening operation will most likely be needed; secondly, the rather thin thicknesses of stems will limit the size of fillets, which in turn may result in length of welds longer than the stems can accommodate.

---

[4]This is not clear in the AISC Specification which states " . . . to avoid overstress of the welds" but fails to specify the applicable allowable stress from the four possibilities: (a) steady allowable (b) 1.33 higher (c) 1.0% higher or (d) yield strength of the weld.

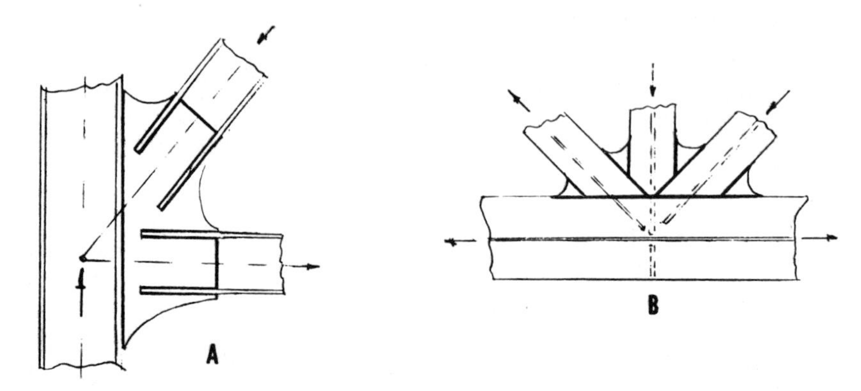

Fig. 81.8.–Typical truss connections.

The chord sections must be reinforced by stiffeners where the diagonals and post intersect the chord. This is necessary if the web of the chord is to participate in the carrying of the applied loads. This type of truss, with members having high resistance to bending, invites secondary stresses, often of an appreciable magnitude.

Heavier trusses may be made of wide flange beams and H sections, with the web of the top and bottom chords in the horizontal position. The welding of these members would consist mainly of flanges that are groove welded together. Under severe loading, gusset plates may be added between the flange connections to strengthen the joint and reduce the possibility of concentrated stresses.

The selection of "web vertical" versus "flange vertical" orientation of H section trusses may present a problem if one or the other is not predetermined by other pertinent conditions. The following factors are offered here for consideration:

1. In terms of fit-up, the "web vertical" orientation is better since all web members frame into one flange of the chord, easily accommodating the full geometry of the connecting members. The "flange vertical" orientation creates many more planes of fitting (one for each flange and one or more for the web). This, combined with variations permitted by mill tolerances, results in a very acute problem in matching and fitting of joints.
2. From the point of view of web member design (verticals and diagonals), the "flange vertical" orientation restricts the designer to the use of only one family of beam sizes, and thus his main concern becomes geometry rather than stresses. On the other hand, the "web vertical" trusses allow any beam size to be used, and proportioning for stresses is entirely feasible.
3. The "flange vertical" orientation can transfer high forces very easily with simple flange groove welds without adding stiffeners.
4. Since on the average, beam sections are deeper than they are wide, the "web vertical" alternative will provide for more open joints and hence better access for welding.
5. In addition, the "web vertical" truss members have inherently greater flexural capacities with their stronger axes in the plane of bending. Secondary moments should be investigated in the design.

DESIGN OF JOINTS

In order to properly define a welded joint, Fig. 81.9 will be useful. Notice that the designations *joint* and *weld* are independent of each other. We may have a butt joint, but this in itself does not indicate the type of weld used. For example, it could be welded using a square groove, a bevel groove, a Vee-, J- or U-groove weld. The tee joint may be fillet-welded or groove-welded. The lap joint could be fillet-welded, plug-welded or slot-welded. Figure 81.10 shows typical welded joints combining the basic types of welds and joints.

The AWS Structural Welding Code has set up certain prequalified joints for manual shielded metal-arc, submerged arc, gas metal-arc and flux cored arc welding for both the complete joint penetration groove welds and the partial joint penetration groove welds.

GROOVE WELDS

There are several items that make up a given welded joint; these include type of joint, included angle, root opening and root face. Each of these items is important and will greatly influence the quality of the welded joint.

TYPES of JOINTS

Butt	(B)
Tee	(T)
Corner	(C)
Lap	(L)
Edge	(E)

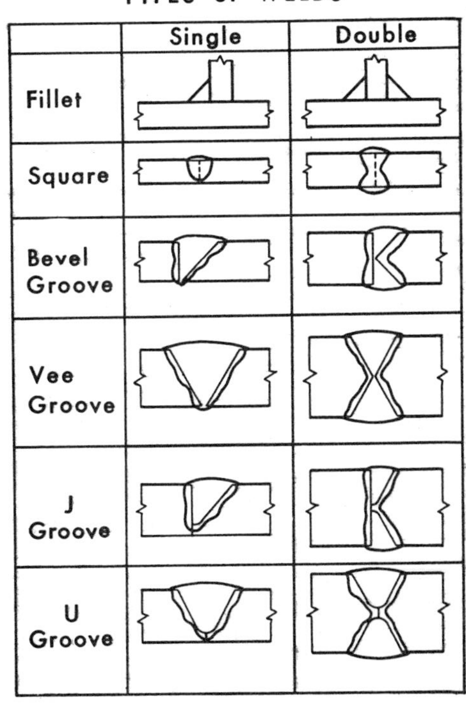

Fig. 81.9.–Basic types of joints and welds.

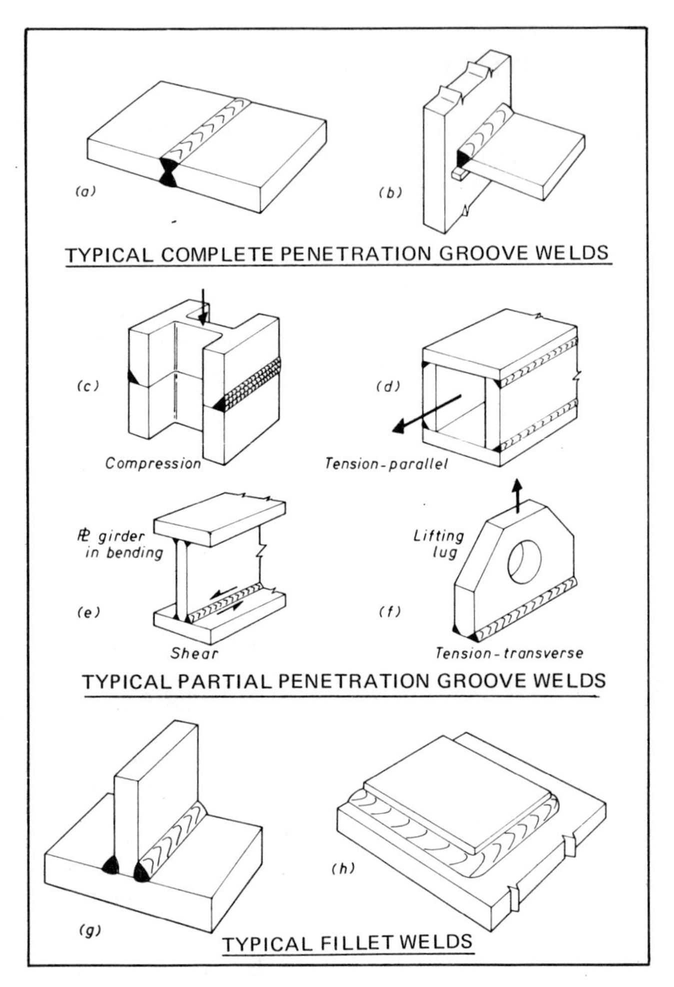

Fig. 81.10.–Configurations of welded joints: (a) butt joint (b) tee joint (c) butt joint (d) corner joint (e) tee joint (f) tee joint (g) tee joint (h) lap joint.

Figure 81.11 indicates that the root opening, R, is the separation between the members to be joined. A root opening is needed to provide electrode accessibility to the root of the joint. The smaller the angle of the bevel, the larger the root opening must be for good fusion and penetration at the root. If the root opening is too small, root fusion is difficult to obtain and smaller electrodes must be used, thus slowing down the welding speed. If the root opening is too large, more weld metal is required, thus increasing both the cost and the weld shrinkage or distortion.

Figure 81.12 indicates how the root opening must be increased as the groove angle is decreased. Larger root openings are used where welds are to be made against a backing. All of these preparations are acceptable as they are conducive to good welding procedures and good weld quality. Selection is usually based upon cost.

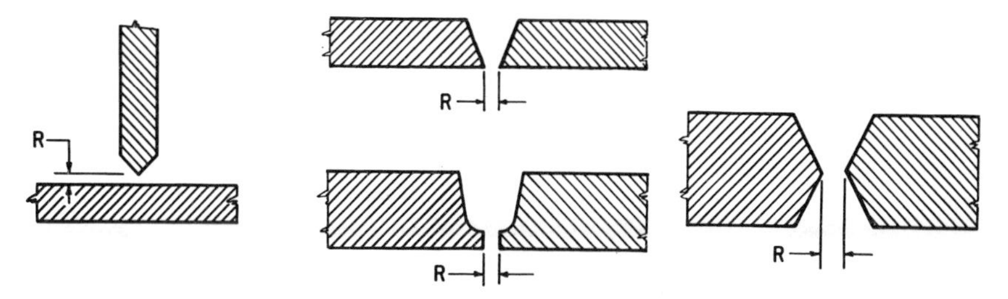

Fig. 81.11.–Root openings of typical groove welds.

In Fig. 81.13, if the bevel or root opening, or both, are too small, the weld will bridge the gap leaving slag at the root or producing incomplete penetration. Excessive back gouging is then required. Figure 81.13b shows how proper joint preparation and fit-up will produce good root fusion and minimize back gouging. In Fig. 81.13c, too large a root opening will result in burn-thru.

Backings are commonly used when welding is done from one side, or when the root opening is excessive. Backings, shown in Fig. 81.14, are generally left in place in building structures and become an integral part of the joint. Spacers are used, especially in the case of deep double-Vee joints, to prevent burn-thru where the root openings are large. The spacer shown in Fig. 81.14d, must be gouged out before welding the second side of the joint.

Backing metal should be similar to the base metal, or be of a steel permitted by the AWS Structural Welding Code. Feather edges of the plate are preferred when using a backing. Short, intermittent tack welds should be used to hold the backing tightly in place, and these should preferably be staggered rather than positioned opposite each other to reduce any initial restraint of the joint. The backing must be in close contact with both plate edges to avoid trapping slag at the root.

For a butt joint, a nominal weld reinforcement to prevent an undersized weld is all that is required. The AWS Structural Welding Code permits a 1/8 in. reinforcement as a maximum. Additional buildup serves no useful purpose and will increase the weld cost. Care should be taken to keep both the width and the height of the reinforcement to a minimum.

The main purpose of a root face, as opposed to a feather edge, is to provide an additional thickness of metal to minimize any burn-thru tendency. A feather edge preparation is more prone to burn-thru than a joint with a root face, especially if the root opening is too large, or when no backing is used.

A feather edge is usually obtained by one cut with a torch, while a root face may require two cuts or possibly a torch cut plus grinding. A root face may require back gouging if a complete penetration weld is required. A root face is

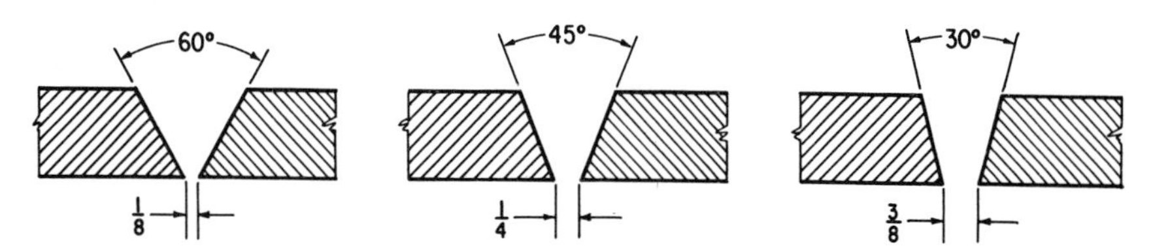

Fig. 81.12.–Variations in root openings for different groove welds.

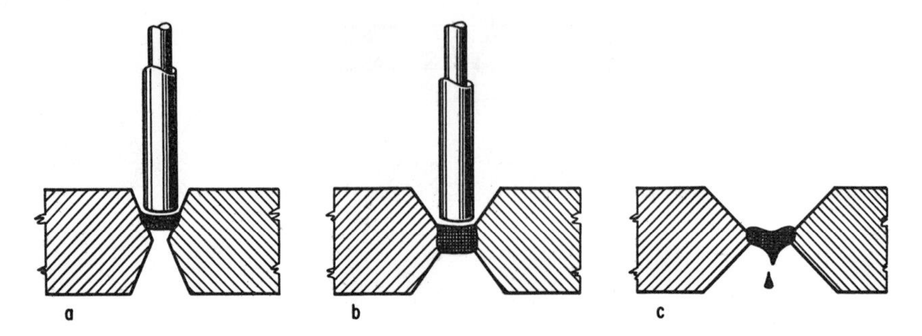

Fig. 81.13.–Good and bad joint fit-up.

not permitted when welding onto a backing, since it may be difficult to reach the bottom of the joint.

To obtain complete joint penetration when welding without a backing, back gouging is required by the AWS Structural Welding Code and the AISC Specification. This may be done by any convenient means–grinding, chipping or air carbon-arc gouging. The last method is generally the most economical and leaves an ideal contour for subsequent welding. Without back gouging, penetration is considered to be incomplete for welds made by shielded metal-arc, gas metal-arc, and flux cored arc welding. For the other processes, the specific welding procedures will determine the need for back gouging. Proper back gouging should be deep enough to expose sound weld metal, and the groove contour should permit complete accessibility for welding.

In the past, the prequalified manual shielded metal-arc welded single- and double-bevel and single- and double-Vee joints, such as illustrated in Fig. 81.15, had no root face, but merely a feathered edge and a 1/8 in. root opening. Such preparation leads to difficulties when the fit-up is less than perfect along the length of the joint. Frequently, the member has a slight bow, resulting in a root opening in excess of 1/8 in. along some portion of the length. Burn-thru along the feathered edge could then become a problem. Figures 2.9.1a and 2.9.1b of the AWS Structural Welding Code permit the root opening to be detailed from

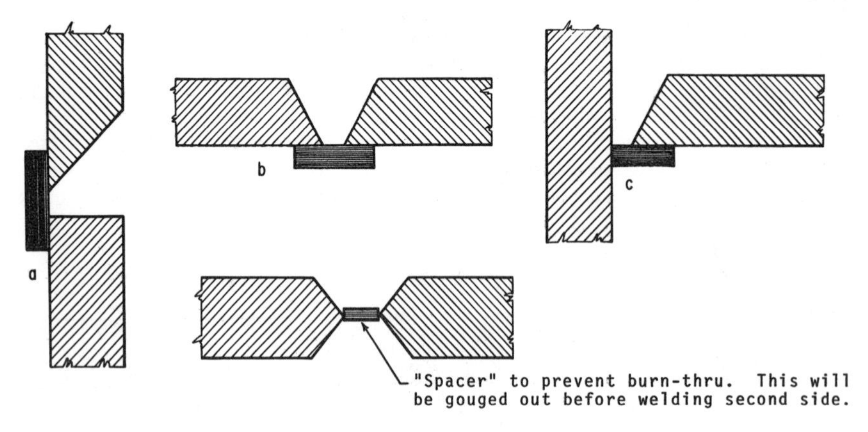

Fig. 81.14.–Use of backings for welds made from one side, or when the root opening is excessive, for welds made from both sides.

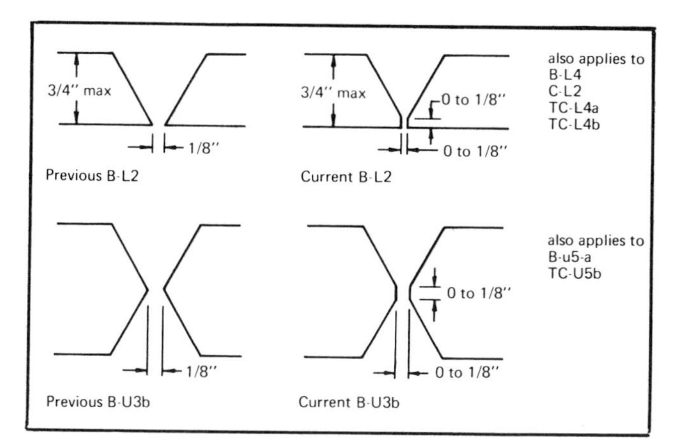

Fig. 81.15.–Revised root opening and root face for shielded metal-arc welds.

0 to 1/8 in. together with a root face of 0 to 1/8 in. The option, as illustrated in the right hand sketches of Fig. 81.15, now permits a joint to be detailed with a 1/8 in. root face and no root opening. If there should be poor fit-up along the length, the weld can still be made without burning through.

An option is also provided for prequalified submerged arc complete joint penetration groove joints shown in Fig. 81.16. Joints B-L2b-S and B-L3-S (Fig. 2.11.1 in the AWS Code) are prequalified submerged arc complete joint penetration groove joints for plate thicknesses over 1/2 in. and up to and including 1 1/2 in. With plates having thicknesses exceeding 1 1/2 in., it was formerly necessary to use one of the following: a joint (B-U2-S) that required a backing and a 5/8 in. root opening, a joint (B-U3a-S) that required a 5/8 in. x

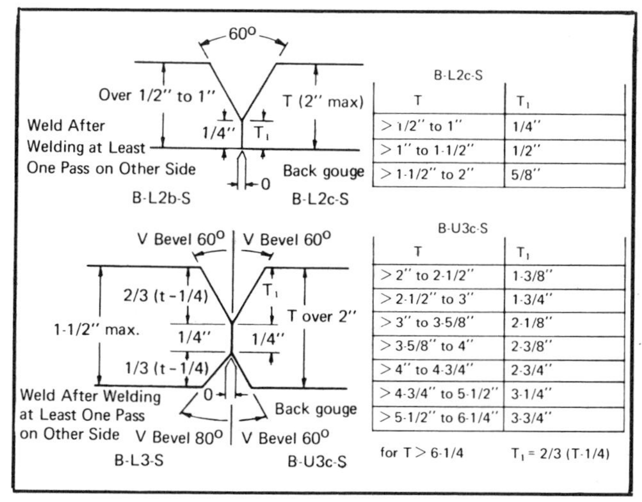

B-L2c-S

T	T_1
> 1/2" to 1"	1/4"
> 1" to 1-1/2"	1/2"
> 1-1/2" to 2"	5/8"

B-U3c-S

T	T_1
> 2" to 2-1/2"	1-3/8"
> 2-1/2" to 3"	1-3/4"
> 3" to 3-5/8"	2-1/8"
> 3-5/8" to 4"	2-3/8"
> 4" to 4-3/4"	2-3/4"
> 4-3/4" to 5-1/2"	3-1/4"
> 5-1/2" to 6-1/4"	3-3/4"
for T > 6-1/4	T_1 = 2/3 (T-1/4)

Fig. 81.16.–Change in submerged arc prequalified complete penetration groove joints.

1/4 in. spacer, a joint (B-U3b-S) which required that a portion of the joint be made with manual shielded metal-arc welding or a joint (B-U7-S) that required U-gouging. Now, B-L2c-S and B-U3c-S with root back gouging may be used for the thicker plates instead of the more costly alternatives. See the joints in the right-hand portions of Fig. 81.16 in which each joint sketch represents two different joints. The left half represents the previous joint with its limitation in thickness; the right half is the current joint with the permitted increase in thickness.

FILLET WELDS

The fillet weld is one of the most commonly used welds. It requires no plate preparation, and less accessibility is necessary to make the weld. Because of the configuration of tee and corner joints, there should be no problem with burn-thru. The strength of the fillet weld is in its throat. For the conventional equal leg weld, this throat is 0.707 of the leg size. However, for fillet welds made by submerged arc welding, the deeper throat penetration is recognized by ascribing higher allowable stresses for such welds as shown in Table 81.1.

PARTIAL JOINT PENETRATION GROOVE WELDS

If a standard Vee-, J- or U-groove is used for manual shielded metal-arc welding, submerged arc welding, gas metal-arc welding or flux cored arc welding, it is assumed that the bottom of the groove can easily be reached for welding. Thus, the effective throat of the weld (t_e) is equal to the actual depth of the prepared groove (D). In the case of a bevel joint in which the included angle is 45 deg or less, the effective throat is 1/8 in. less than the depth of preparation for manual shielded metal-arc welding in any position and for gas metal-arc and flux cored arc welding in the vertical and overhead positions.

Just as fillet welds have a minimum size for thick plates because of fast cooling and greater restraint, so partial penetration groove welds are required to have a minimum throat (t_e) of

$$t_e = \sqrt{\frac{t_p}{6}}$$

where t_p is the thickness of the thinner plate.

Partial joint penetration groove welds are allowed in building construction. They have many applications, primarily in splices of columns, built-up columns, built-up box sections for truss chords, etc. The AWS Structural Welding Code prohibits their use under cyclic or fatigue loading if they are subject to tension transverse to the axis of the weld. Joints containing such welds, made from one side only, shall be restrained to prevent rotation. The AISC Specification for the Design, Fabrication, and Erection of Structural Steel for Buildings prohibits their use in butt splices in plate girders and beams.

Partial joint penetration groove welds are similar to fillet welds in that the root of the joint may be unwelded. They are treated in a manner similar to fillet welds. They both have an unwelded root and their throat is the shortest distance from the root to the face (see Fig. 81.16). For the same throat (strength), the partial joint penetration groove weld requires less weld metal than the fillet weld. In fact, for the 45 deg included angle single-bevel joint, the partial joint penetration groove weld has just one-half of the weld metal volume that would be required for the same strength of fillet weld (see Fig. 81.17). Costs of fillet and partial joint penetration groove welds are plotted in Fig. 81.18.

WORKMANSHIP

GENERAL REQUIREMENTS

Among the factors that affect the performance of a welded joint in service, quality of workmanship in all phases of fabrication and erection is the most influential. Fine cutting operations, accurate assembly and alignment provide production conditions conducive to sound and efficient execution of welding. But for such conditions to be truly reflected in better workmanship, attention must be given to a number of other requirements.

Welding procedures must have their parameters individually adjusted to yield optimum properties in the joint under anticipated service loads. Such procedures should be in writing and they should be followed in construction.

The influence of notch effects, especially in the case of structures subjected to relatively low temperatures or to fatigue conditions in service, must be translated into joint configuration, geometry of details, degree of finish and extent of internal weld soundness. These will assure uniform transfer of stress. The avoidance of discontinuities, re-entrant corners and abrupt changes in section is imperative in the design of dynamically-loaded structures such as bridges, crane girders and crane runways. These factors should also be considered in planning statically-loaded structures.

In view of the high local stress concentrations created by notches, it is essential that both the base metal and the welded joints have the capacity to undergo substantial deformation without fracture. This must be considered the primary objective of a sound welding procedure. The designer must be fully aware of stresses introduced by fabrication and welding. There should be strict

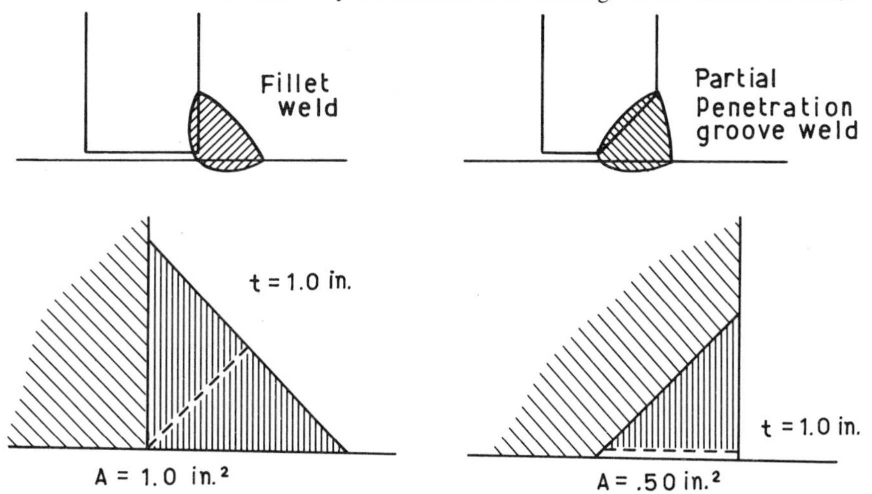

Fig. 81.17.–Root penetration of fillet and partial penetration groove welds.

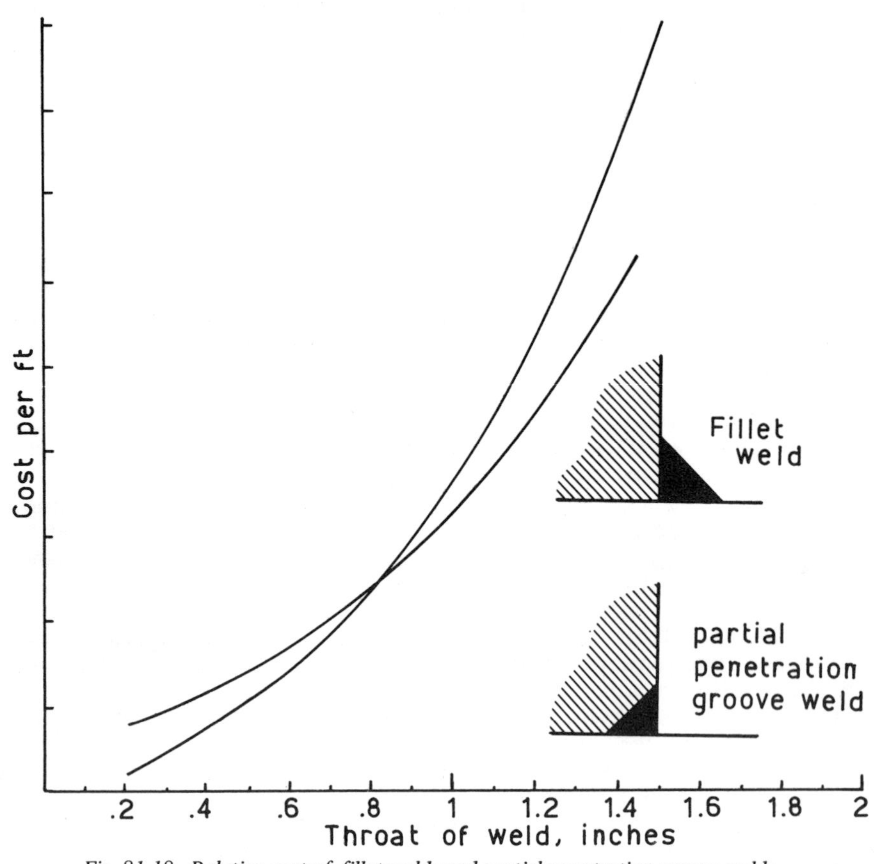

Fig. 81.18–Relative cost of fillet welds and partial penetration groove welds.

adherence to the AWS Code tolerances for finished products and the specific job requirements. Due consideration should also be given to good preparation, sound welding procedures and proper geometrical form. These are all intrinsic factors necessary to produce a sound welded structure.

PREPARATION OF MATERIAL

There are several established methods for cutting, assembly and alignment used in the preparation of edges of plates or rolled structural sections for welding. These are mechanical cutting and machining (shearing, machine planing), oxygen cutting and gouging and air carbon-arc and plasma-arc cutting. The choice will depend on the class of work, on the thickness of metal and the type of joint. Shearing is usually confined to the preparation of straight edges in metals up to 3/4 in. in thickness.

The most prevalent and economical method of preparation is oxygen cutting. This method can be used manually for short lengths, but its versatility and potential are best demonstrated in machine cutting operations. This type of operation extends from an adjustable portable moving track arrangement to automatically-controlled straight line and shape cutting machines with single and multiple torches using a variety of guidance systems to ensure accurate operation.

Recent improvements in speed and quality of cut have made plasma arc cutting economically competitive for the preparation of mild steel.

Air carbon-arc gouging has established itself as an excellent tool for the preparation of grooves and at the same time for providing sound weld metal conditions for immediate welding from the "other side" of the joint.

Surfaces to be welded should be smooth, uniform and free from scale, rust, grease or paint. For most applications, vigorous wire brushing will be sufficient to remove any objectionable solid material. For girders, all mill scale shall be removed from the surfaces on which flange-to-web welds are to be made by submerged arc welding or by shielded metal-arc welding with low-hydrogen electrodes.

In the assembly of metal to be welded, utmost attention must be paid to the proper and correct alignment of the parts. Joint configuration and fit-up within tolerances as stipulated in the AWS Code should be maintained. Excessive deviations in this respect would constitute grounds for qualification testing of joints normally considered prequalified or, more likely, rejection of the work. Also, additional costly penalties may be incurred either in poor access, involving a higher risk of defects, or in an excessive volume of weld metal, reflected in higher deposition costs and greater distortion.

WELDING PROCEDURES

The primary objectives of a welding procedure are to produce a welded joint satisfying the requirements of soundness in the most economic way, while retaining the mechanical properties of the base metal. There is general agreement within the applicable codes and specifications concerning the essential constituents of any welding procedure which include the following:

- Designation of metal or alloy
- Sketch of joint showing the geometry of preparation and fit-up
- Welding symbols
- Position of welding
- Welding data (usually presented in a tabular form): size of fillet welds or thicknesses of metal for groove welds
- Electrode classification and size
- Amperage and voltage
- Sequence and number of passes
- Preparation at the root before welding from the other side
- Preheat, interpass temperature and postweld heat-treatment requirements.

In any automatic or semiautomatic process, these additional factors should be given:

- Travel speed
- Type of shielding (gas or gas mixture, flux and the AWS or trade designations)
- Rate of gas flow (if used)
- The official classification of the "filler metal/shielding medium" combination.

Many other detailed points may warrant inclusion. Their extent and presentation will vary with specific processes and jobs. For gas metal-arc welding (GMAW), the mode of metal transfer across the arc must be identified. In the case of electroslag (EW) and electrogas, a number of additional parameters characteristic to these processes should be included. Among them may be details of consumable guide, traverse length of oscillation, dwell time, and type of molding shoes.

In the case of submerged arc welding, restrictions have been placed by the AWS Code on the maximum amperage permitted for all passes that have fusion to both faces of the joint, except in the case of the final layer. Similarly, limitations have been put on the depth and width of a deposited weld that must not exceed the width at the face of the same weld (see Fig. 81.19). Both limitations have been imposed to prevent use of a weld configuration recognized to be prone to hot cracking.

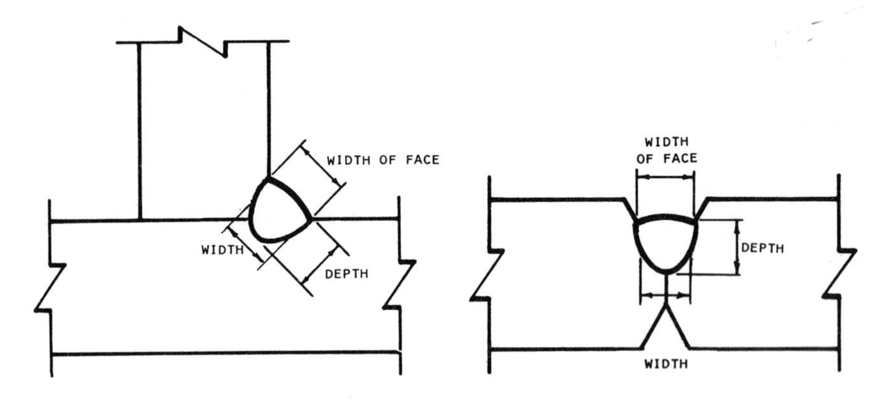

Fig. 81.19.–Schematic of weld pass showing that the depth and width shall not exceed the width at the face for fillet weld (top) and groove weld (bottom).

Welding procedures may be prequalified as to joint geometry and fit-up. However, the concept of "prequalification" is not limited exclusively to joint preparation and fit-up, but extends to include a number of procedural rulings within each individual welding process. These rulings are based on sound and proven welding practices. Among others, they impose limitations on the maximum size of electrodes, maximum thicknesses of root passes and intermediate layers for different types of welds, processes and positions of welding.

A cursory glance at the list of essential constituents of a welding procedure just given will indicate that factors such as chemical composition (designation of steel), thickness of material, electrode class and size, speed of travel, amperage and voltage and preheat temperature are all mutually interdependent. The proper combination of all these variable factors is the primary requisite of a welding procedure. Soundness must depend on the attainment of appropriate optimum cooling rates.

Similarly, notice should be taken of the fundamental difference in the heat input requirements for non-heat-treated carbon steels, as compared with those for heat-treated quenched and tempered steels. In the first case, the objective is to generate sufficiently high heat input so as to effect slow cooling rates while in the case of quenched and tempered steels, the foremost concern is to restrict heat input below a maximum value in order to maintain fast cooling rates.

Generally, the basic premise in welding procedures is to provide filler metal matching the base metal in terms of mechanical properties. However, in the case of exposed applications of ASTM A242 and A588 steels, where weld metal with atmospheric corrosion resistance and coloring characteristics similar to that of base metal is required, the emphasis shifts towards chemical compatability. The electrodes satisfying this condition may have higher yield and ultimate tensile strengths.

For unpainted applications of the same steels and single-pass welding where exact color match is not essential, the required atmospheric corrosion properties as provided by the high degree of dilution (characteristic of a single-pass weld) are considered acceptable. The AWS Code contains provisions for color and atmospheric corrosion matching of base and weld metals.

Preheat

Preheat is one of the most conveniently adjustable parameters of a welding procedure. The present AWS Code continues, however, to base its preheat and interpass temperature recommendations as functions of chemistry (steel designation–carbon equivalent), thickness of material and type of filler metal (low hydrogen, etc.). It entirely disregards the very potent parameters of heat input of the welding process (joules (J)/in.), manner of dissipation of heat from the joint and type of welding process.

The present position of the Code is to maintain the somewhat arbitrary preheat recommendations until sufficient work has been done to justify their more realistic and more scientific evaluation.

Preheat is usually a separate operation preceding actual welding. The simplicity or sophistication in the setups used depends largely on the thickness of metal, amount of preheat required and the type and length of the joint to be preheated. However, in some special applications, a group of three welding arcs in one moving system has been used with the leading arc some 12 in. ahead of the joint, and with the two trailing arcs (tandem submerged arc welding) assigned to substantially remelt and anneal the first pass and deposit the required weld. (The leading arc may be submerged arc or flux cored arc welded with CO_2 shielding.) This application to automatic fillet welding of heavy column sections produced most satisfactory results at substantial savings due to the elimination of a separate preheat operation. Hence, in automatic multiple-arc welding, the heat input by the leading electrode can, under certain circumstances, be used to serve as a preheat.

Care and Storage of Electrodes and Fluxes

A quality factor of increasing importance is proper care and storage of electrodes and fluxes. This is especially true of the low-hydrogen types, in view of their greater susceptibility to moisture absorption.

Moisture is a potential source of hydrogen. Hydrogen, when entrapped in the weld metal, can cause cracks to form. The unique ability of low-hydrogen electrodes to prevent underbead cracking is overwhelmingly dependent on a minimum moisture content in their coating. The extent of strict control and enforcement of these provisions, as well as their severity, increases with the mechanical properties (and hence classification) of the low-hydrogen electrodes used. These factors will assume crucial importance for electrodes that are to be used on quenched and tempered steels, and when welding under severe restraint conditions.

Limits on exposure time for given humidity levels, comprehensive reconditioning procedures in terms of temperature and soaking time must be clearly stated. Appropriate rules governing care, storage and reconditioning of low-hydrogen electrodes, as well as fluxes for submerged arc welding, are included in the AWS Code. These should be strictly followed.

Run-off Bars

In flange splices of beams or plate girders and in groove welded beam-to-column connections, the welding procedure should provide for mandatory use of run-off bars. These effectively maintain the proper throat thickness at the start and end of the joint, and at the same time continue any starting porosity or terminal defects outside the effective length of the joint. Such run-off bars need not be removed in statically-loaded building structures unless required by the engineer, usually for architectural reasons.

Distortion Control

The use of welding without proper control in the joining of some structural units into composite members may result in distortions that are costly to correct, but which are predictable to a certain degree and hence can be controlled or eliminated. There are several stages where care and knowledge may be applied to achieve the desired result. If the designer selects the parts of a composite member so that the amount of welding on each side of each axis is balanced, the welding will be less prone to produce distortion.

When two or more elements are assembled into a composite member, distortions may result that require expensive shop straightening before shipment. This is an important item when coupled with the present tendency in architectural practice requiring exposed structural members at the sides and eaves of industrial buildings to provide almost perfect alignment for continuous steel sash, glass block construction and finished cave lines.

Where an angle is riveted to the flange of a channel, the punching of holes in the channel tends to increase the length of the outer edges of the flange with respect to the back of the channel producing a bow. Subsequent riveting may accentuate this effect. Attaching a masonry plate or combinations of plate and angle to a beam has the same effect where there is a much larger number of punched holes on one side of the center line. Welding such assemblies instead of riveting produces similar camber in the opposite direction. Riveted buildings and ships tend to grow and result in greater length than the sum of the units; welded structures tend to shrink, resulting in a shorter finished structure unless allowances are made for the effect of the joining methods employed.

The explanation for the welding shrinkage depends on the type of welds involved. Where isolated spots are heated by welding to a plastic condition, the restraining action of the surrounding cold metal increases the thickness in the unrestrained direction. Solidifying in that state during cooling retains the upset. While the original volume is regained, the increased thickness due to the upset reduces the area and results in a residual tension that must be balanced by an area of residual compression causing a reduction in the length of the member. The application of heat in this manner may be employed to take out buckles in plates or to straighten a curved structural member.

A somewhat similar effect is obtained when a longitudinal weld is employed. The shrinkage of the weld metal results in a residual tension in the weld, balanced by areas of residual compression in the surrounding base metal and a reduction in length or local buckling. If the weld is on one side only of the neutral axis, a curved member results. When fully aware of the effects of welding, the designer may arrange the parts of a composite member to eliminate or greatly reduce this condition. For example, where a masonry plate is attached to a beam to carry masonry on only one side of the center line of the beam, increasing the width of plate so that the connecting welds may be made on both edges of the flange will aid in balancing the welds and reducing the distortion.

When equal amounts of welding are not possible on both sides of the center line, there are several other effects which, if controlled, will assist in producing the desired result. Since continuous welds result in greater shrinkage than intermittent welds, carrying the lengths and sizes of the welds to offset varying distances from the neutral axis may assist in balancing the shrinkage. When equal size welds are placed on opposite sides of the center line and one of these welds is completed before the other is started, the distortion effect of the second weld probably will not equal that of the first weld. Even when the welds are balanced, distortion from this effect may result. This is best controlled by making the two welds simultaneously, employing two qualified welders. This may also be accomplished by alternately welding short lengths on either side in a staggered sequence.

Another cause of excessive distortion is the tendency of some designer and fabrication personnel, unfamiliar with welding, to use an excess of welding. Such overwelding adds greatly to the cost, exaggerates the distortion problem and requires more drastic straightening operations.

It is also possible for the excess heat input resulting from overwelding to sufficiently raise the temperature of the base metal at some points to reduce the ability of the material to resist residual stress introduced in the steel mill during the operations of rolling, cooling and cold straightening. This results in distortion. Since the existence and location of such residual stress are impossible to predict, the distortion produced by excessive welding is unpredictable

When the designer is familiar with the problems of welded design, the distortions which may result and how they may be overcome in design and detail, it is desirable to show on the design drawings the results of this experience. When this is unknown, it would be preferable for the designer to outline the main considerations only, and leave the detailing and fabrication to the competent fabricator.

Shop personnel learn by experience how to overcome the problems of distortion during fabrication. Under some conditions, variation in procedure can result in complete elimination of distortion. The degree and direction of distortion can be anticipated only after considerable experience with the same or similar assemblies. The amount of experience and general ability of the shop organization will be reflected in the fit and straightness of the finished structure and, to a degree, in fabrication costs.

Effective control of distortion contributes in a significant manner to the efficiency, economy and soundness of welded construction. The fabricator must be aware of the distortion-producing factors and have reasonable insight into the magnitude of their individual contribution to the final effect.

Although some empirical guidance is provided to assist in predicting the response of steel to certain fabrication measures, individual experience, continually enriched by a statistical control of all the pertinent factors, is the final criterion.

The following are generally considered to minimize distortion due to shrinkage:

In preparation:

1. Minimum root opening and minimum included angle for a given size of electrode.
2. Double-sided preparation (Vee, U), resulting in the least volume of weld metal, balanced symmetrically about the center of the thickness.
3. Minimum size fillet welds to satisfy strength or other requirements.
4. Intermittent fillets, where permissible, with emphasis on small size rather than short length (however, the minimum size must be in agreement with the AWS Code).

In fit-up and welding:

1. Proper fit-up.
2. No overwelding.
3. Balancing of weld passes on both sides of double-welded butt joints.
4. Use of either high deposition rate electrodes or automatic welding at optimum currents for maximum speed and adequate penetration.
5. Positioning of work for downhand and horizontal welding.
6. Balancing of welds about the neutral axis if member is symmetrical.
7. Welding away from the restrained parts of the member.
8. First welding those joints which will produce the greatest distortion.
9. Use of presetting or prebending before welding when necessary.
10. Use of mechanical restraint by clamping during welding or joining similar sections back to back.
11. Use of jigs and fixtures to maintain alignment and proper fit-up during welding (however, these should allow freedom of end movement).
12. Provisions permitting contraction in the member.
13. Division of work into subassemblies.
14. Making of closing welds preferentially in the compression element in rigid assemblies where high residual stresses are likely.

Distortion considerations may in some way dictate a certain mode in shop fabrication as well as a certain manner of field erection. This is duly recognized in the operations or decisions listed below as they bear on welding of rigid moment connections in a multistory building:

1. Verification of any preconceived ideas of the mode of fabrication (single beams and columns or standardized preassemblies welded in the shop for a minimum of welding in the field).
2. Establishment of additional shop welding procedures if scope and type of work are not adequately covered by existing standards, or if any specific features or nature of the job warrant additional elaboration.
3. Decision by the erection department on a definite order of erection based on the speed, efficiency and structural stability requirements at any stage.
4. Development of field welding procedures.

ERECTION

Welding and workmanship in general are no different for field welding during erection than for shop fabrication. All that has been said about welding procedures, distortion control and weld quality apply. Welding should be done on the ground to the fullest extent possible. This makes positioning for flat and horizontal welding possible and hence is more economical. This particular advantage is further enhanced by the greater ease of welding and the lesser probability of defect inclusion.

While clamps, clamping devices and jigs are used rather extensively on some classes of shop welding to hold the parts in line or in contact, they are not in such general use in the field welding of framed structures. The use of clamps and hitches of various sorts is quite common to plate and tank work, and in ship work, where the lining up of contiguous parts of the large plate justifies such equipment.

The more usual practice in framed structures is to use ordinary rough bolts in punched holes to hold the members during plumbing and welding. Generally, two bolts are so arranged that the erection loads and the service loads are carried on shop-welded seats. Where erection loads are carried directly by the bolts, it may be necessary to increase the number. Holes are usually restricted to detail material. A seat angle (in the case of shallower beams, 12 in. or less) or a one-sided angle or plate for beams of greater depth is generally used for directly welded beam-to-column connections. This is done with a sufficient number of bolts appropriately spread to provide a certain amount of restraint, but is mainly intended to carry vertical erection loads. Where possible, location material is so designed as to avoid holes in the main material.

A minimum of two bolts is desirable even for minor bracing since the erector can pin through one hole with the spud end of his wrench, prying the holes into line until he can place a bolt in the second hole. Depending on the location and importance of the joint, a second bolt is then placed in the other hole and both bolts turned up tight, or a drift pin is placed in the second hole and driven up tight. In some cases, the second hole is left open. In any case, after the completion of field welding, the erection bolts are usually permitted to remain in place since the cost of removal usually exceeds the value of the bolts. Drift pins may be removed since one or two sharp raps of a hammer will remove them.

For some forms of rigid frame where the joints are left exposed and a more pleasing appearance results from smooth areas clear of all bolts and nuts, the use of welded hitch plates or angles is common. All the holes are confined to the hitch material, which is cut off with an oxygen-cutting torch when the welding has proceeded to a point where the hitch is no longer required. The surface is then chipped or ground before painting.

In an otherwise all-welded structure, it is considered good practice to use rough bolted connections, where specifications permit, for such items as roof purlins, roof bracing, girts and secondary beams. Since two bolts would generally be required for erection purposes alone, and few or no additional bolts to carry the service load, substantial savings result if the joint is designed for field bolting.

For a series of bents, particularly rigid-frame bents with rather large knee and ridge joints, some erectors prefer to do the welding on the ground and raise the bent into position in one piece. This practice is limited by the relative size and weight of bent and the available hoisting equipment (and the personal preference of the erector). It is advisable to determine whether the joints are to be welded on the ground or in the air before designing the details of the various joints.

For some types of structures, the accuracy in fabrication is not much different from that required for riveted construction. This would be true of simple beam and column construction where the gravity loads are carried on shop-welded seats with two bolts to hold the beam during plumbing. Loose top angles or top plates are field welded in position and the beams are cut to usual mill-cutting tolerances. However, if erection costs are not to be greatly increased, the fabrication must be more carefully done and to closer tolerances for structures employing connections by butt joints. Where the abutting edges of joints are too close to permit penetration at the root, field cutting or gouging must be used. If the root opening is too wide, the volume of welding is greatly increased. If root openings are exceptionally wide, the joints must be "buttered" before welding.

Rigid frame field splices should be located in the narrower portions of the rafter beam, preferably at or near the point of inflection so that the field welding is minimized. The highly stressed knee joint may be completely shop welded where the control of the joint clearances and the welding process may be more efficiently accomplished.

In the specific case of erection of welded multistory structures using direct beam-to-column connections, the following operations normally take place:

1. Use of a definite order of erection on the premise of speed, efficiency and structural stability at any state.
2. Use of a welding sequence which prescribes the sequence of welding operations as well as their pace so as to parallel to a predetermined degree the progress of erection. This sequence should consider the effect of cumulative shrinkage due to welding on the plumbness of the structure, and provide recommendations for preventive or corrective action. The sequence of welding of individual rigid connections should be stipulated so as to provide a uniform pattern of shrinkage stresses, and effect their most proportionate distribution.
3. The preheat recommendations should be included with detailed information as to when, where, how, how much and for how long.
4. Provision for repairs should be included in the procedures. Consideration should be given to the relative cost of using electric welding machines as compared with gasoline-driven machines for providing electrical power to the welding sites (as at a high story).

The above pertinent information may appear either on erection drawings, or may be issued as a separate document, together with supplementary data identifying typical joints and their location in the structure. The most effective means of communicating this information to the welders are through the welding supervisor. He should inform welding personnel prior to commencement of work, and at appropriate time intervals thereafter.

The assignment of joints to individual welders should be accompanied by full instruction and explanation of rules governing the execution of welding of a given or a number of given joints in the structure. Through this practice, welders will give their full attention to the production of sound joints and will also understand the overall objectives of shrinkage and distortion control in the structure.

INSPECTION

Inspection is a necessary and important part of the entire program of operations required to obtain sound welding. Properly done, it is reasonable in cost and can contribute to cost savings by eliminating the necessity for repairs to completed welds. Improperly performed, inspection can be harmful by providing a false sense of security.

Inspection is not a substitute for supervision of the welding operations or for proper welding procedures, welding equipment and materials. The purpose of good inspection is to assure that all of the requirements applicable to a job are being followed, and that the resultant welds are as specified on design drawings and in job specifications. Proper inspection requires the performance of checks on the work as it progresses–not just the inspection of completed work, or, worse yet, work which was completed in the shop and has already been erected.

The nature and extent of inspection will vary with the criticality of the work and its complexity. The methods of examination to be employed should be established in advance of the actual work, and made known to the fabricator and erector. The steps to be followed in the course of inspection should be instituted from the start of the work, not long after the work has been erected. Deviations found at the start can be corrected on the first members produced, and it is reasonable to expect that the same deviations will not recur. If deviations are found after extensive work has been done, the corrections may then be extensive and costly. If the steel has already been erected, it may even be necessary to dismantle part of the structure or make modifications to correct deficiencies

Inspection should never be made part of the fabrication or erection contract. Inspection should be contracted for and paid directly by the engineer or other representative of the owner. Inspectors assigned to a job will then be responsible directly to the engineer, and will not be called upon to pass judgment on the work of their customers.

Regardless of the inspection employed for acceptance, a competent fabricator or erector will make his own inspections during the progress of the work. Such inspection serves to assure that the work will be acceptable when inspected by the representative of the purchaser. It is even more valuable in revealing basic difficulties at the start of work. These can then be corrected before too much of the work has been done, reducing the amount of repairs necessary. Such self-inspection or quality control has been found to more than pay for itself.

Every examination method used for welding has been used for inspection of structural welding. These include visual, magnetic particle, liquid penetrant, ultrasonic and radiographic examinations. For a given job, the method to use and the extent of the examination will vary with the nature of the work and the criticality of certain joints. The actual extent of inspection must be decided by the engineer, based on his knowledge of the nature of the work and the possible difficulties that can be expected under given circumstances.

Whether or not any other method of examination is used, all welding operations should be subject at least to visual examination. Even if additional examinations are to be employed, visual examination should not be omitted. Visual examination should encompass examination of the weld joint before welding, ascertaining that the fit-up and joint geometry are correct and that the work is clean. During welding, visual examination should include seeing that the proper electrodes are used and that only qualified welding personnel are making the welds. The finished root pass and other passes selected at random should be examined for the presence of cracks, slag, porosity or excessive undercut. In addition, check should be made that no welds have been omitted or that no unspecified welds have been made.

Magnetic particle examination can readily be done without any additional cost if full-time inspection is used. Even if the completed weld is to be radiographed or examined ultrasonically, magnetic particle examination of the root passes of welds will reveal any root cracks, slag or excessive porosity. These can then be removed without the need for removing all of the weld. Liquid penetrant examination is used in the same way as magnetic particle examination when there is no convenient current source, or when the number of welds to be examined is so small as to make the use of magnetic particle examination equipment unfeasible. Liquid penetrant examination reveals only surface defects, while magnetic particle examination can be expected to extend up to 1/8 in. below the surface

Radiography is used in building structures for critical welds, for welds that are difficult to make or as a spot check. Using radiography early in the work will reveal any existing basic defects. A spot check will determine whether further radiography is necessary. Complete radiography of structural welds is never necessary. Radiographic technique and standards of acceptance should conform to the AWS Code unless local requirements or some special circumstances of the work requires otherwise.

Ultrasonic examination is now an everyday practice in structural work. Technique and standards of acceptance are included in the AWS Code The required level of quality for acceptance is equal to that obtained by other methods of nondestructive examination.

Use of ultrasonic examination requires personnel competent to use the equipment. Fortunately, the training programs offered by the manufacturers of ultrasonic equipment are producing an increasing number of such people.

The speed of examination by ultrasonic means has been a great attraction. Many welds can be examined in a day and the results are available immediately. These attractive features, however, have served to encourage even more untimely inspection. When the work being done is not inspected until it is ready for shipping, or worse yet, until it has been erected, the delays and costs of labor for repairs of defects found by this practice are often serious.

If ultrasonic examination is to be a boon to structural welding quality, it must be used properly. Untimely inspection has been the cause of concern to many people in the structural field.

One practice that has been used effectively on a limited basis is to specify in the job contract that any weld repairs necessitated by untimely inspection will impose an additional cost on the owner. The cost of making the repair is increased due to handling and the more difficult welding conditions under which the repair might have to be made. Such a requirement recognizes the responsibility of the contractor to produce welds meeting the quality requirements of the AWS Code (or Job Specifications), and also recognizes the responsibility of the owner (or the Engineer as his representative) to make his inspections for acceptance timely.

BIBLIOGRAPHY

Design of Welded Structures, Omer W. Blodgett, J. F. Lincoln Arc Welding Foundation (1966).

Structural Welding Code, D1.1-72, American Welding Society.

General Specification for Welding of Steel Structures, CSA Standard W59.1, Canadian Standards Association.

"Structural Welding," Lambert Tall, *Welding Design and Fabrication* (January through December 1964).

Specification for Design, Fabrication and Erection of Structural Steels for Building, American Institute of Steel Construction.

Standard Specifications for Welding of Steel Structures, CSA Standard W59, Canadian Standards Association.

Ultrasonic Testing Inspection for Butt Welds in Highway & Railway Bridges, Bureau of Public Roads (1968). Available from Superintendent of Documents, U. S. Govt. Printing Office.

Steel Construction Manual, Seventh Edition, American Institute of Steel Construction (1970).

Welding Fundamental Principles and Practices, Canadian Welding Bureau.

Welded Structural Design, Canadian Welding Bureau.

CHAPTER **82**

BRIDGES

PREPARED BY A COMMITTEE CONSISTING OF:

F. H. RAY, *Chairman*
Ohio State Dept. of Highways

P. A. BARNES
N. Y. S. Dept. of Public Works

J. L. BEATON
California Div. of Highways

O. W. BLODGETT
The Lincoln Electric Co.

M. F. COUCH
Bethlehem Steel Co.

S. A. GREENBERG
Welding Consultant

A. W. MOON
New York State Dept. of Transportation

CHAPTER 82

BRIDGES

INTRODUCTION

In most situations, bridges are subject to dynamic and repetitive live loads, and buildings are subject to static live loads. This difference in loading constitutes the principal difference other than geometry in the structural aspects of bridges and buildings. This requires bridge designers to consider material properties and design constraints that are in addition to those required for building design. A second and perhaps equally important difference between bridges and buildings is the differing exposure to weather of the main structural elements. The structural elements of bridges are usually exposed while the structural elements of buildings are usually protected from such exposure. The more conservative allowable unit stresses established for bridges are largely justified by this difference.

This chapter includes basic information about design specifications, metals for fabrication and erection of welded steel bridges. The departure from the preceding chapter on buildings is basically due to the above noted differences.

GOVERNING CODES AND SPECIFICATIONS

The American Welding Society prepares and publishes a Structural Welding Code. It covers the general area of welding only; therefore it must be used in conjunction with the prescribed standard specifications for the specific type of structure.

Highway bridges in all fifty states are designed according to the Standard Specifications for Highway Bridges as adopted by the American Association of State Highway Officials (AASHO). Yearly revisions to the AASHO Standard

Specifications are published as Interim Specifications which bear the year of their issue. These specifications are frequently supplemented by special provisions formulated by the individual State Highway Departments. The supplements usually formalize a design procedure or constraint not mentioned in the Standard Specifications but which reflect the policy of the state highway department issuing it.

Railway bridges are designed in accordance with the American Railway Engineering Association *Manual for Railway Engineering.*

When a reference is made in this chapter to one of the above specifications, such reference is to the most recent one published by the issuing agency; the reader is cautioned to ascertain that he is referring to the latest copy.

BASE METALS

The present AWS Structural Welding Code provides specifically for the use of five ASTM steels for plates and/or shapes and three ASTM steels for tubular members. These are ASTM A36, A441, A514, A572 Grades 42, 45 and 50, and A588 for plates and/or shapes and ASTM A500, A501 and A618 except Grade I for tubular members. ASTM A242 and A618, Grade I steel are also permitted when subject to special investigation concerning their weldability.

The yield and tensile strengths of the above steels are shown in Table 82.1. Chemical requirements for each of the steels may be obtained from the ASTM specifications. The many types and grades of some of the steels, especially for A514 and A588, make a listing of these requirements cumbersome.

Although these steels can be welded by any of the arc welding processes commonly used in structural fabrication, consideration must be given to the need for preheat and special techniques in relation to material, thickness, restraint and joint design. For specific recommendations on welding procedure, reference should be made to the AWS Code.

Electrodes and fluxes conforming to the specifications listed in Table 82.2 are considered to be satisfactory for welding any of the previously mentioned steels, although presently only A5.5 covers filler metal satisfactory for A514 steels.

Steels conforming to ASTM A514 are especially useful in welded bridge construction when the deadweight reduction achieved through use of a high-strength material provides an economic advantage. When fabricating these quenched and tempered steels, welding procedures must be used which limit heat input to the steel manufacturers' recommendations and which utilize the preheat and interpass temperatures at or above the AWS minumum specified. A514 steels have been successfully welded by the submerged arc, gas metal-arc and shielded metal-arc welding processes. The electroslag and electrogas processes, because of their high heat input characteristics, are not permitted by the AWS Code, for A514 steels. E110XX electrodes conforming to AWS Specification A5.5 are appropriate for the shielded metal-arc process, and Grades F110, E110S and E110T are satisfactory with submerged arc, gas metal-arc and flux cored arc welding, respectively. These last three grades are not yet covered by AWS filler metal specifications, but have mechanical and nominal chemical properties defined by the AWS Structural Welding Code.

The A242 and A588 high-strength, low-alloy steels are used for welded structures where increased resistance to atmospheric corrosion is essential or desirable. These steels provide increased strengths as well. Proprietary steels in

Table 82.1—Tensile properties of nine ASTM structural steels used in bridge construction

ASTM Specification		Yield Pt. ksi (min)	Tensile Strength, ksi (F_u)
A36 (Structural Steel)		36	58–80
A441 (High-Strength Low-Alloy Structural Manganese Vanadium Steel)		50 to 3/4 in. 46 to 1 1/2 in. 42 to 4 in.	70 min 67 min 63 min
A500 (Cold-Formed Welded and Seamless Carbon Steel Structural Tubing in Rounds and Shapes)		*Round Structural Tubing*	
		Gr. A 33	45 min
		Gr. B 42	58 min
		Shaped Structural Tubing	
		Gr. A 39	45 min
		Gr. B 46	58 min
A501 (Hot-Formed Welded and Seamless Carbon Steel Structural Tubing)		36	58 min
A514 (High-Yield Strength Quenched and Tempered Alloy Steel Plate, Suitable for Welding)		100 to 2 1/2 in. 90 to 4 in.	115–135 105–135
	Grade		
A572 (High-Strength Low-Alloy Columbium-Vanadium Steel of Structural Quality)	42	42	60 min
	45	45	60 min
	50	50	65 min
A588 (High-Strength Low-Alloy Structural Steel with 50 Ksi Minimum Yield Point to 4 in. thick)		50	70 min
A242* (High-Strength Low-Alloy Structural Steel)		50 to 3/4 in. 46 to 1 1/2 in. 42 to 4 in.	70 min 67 min 63 min
	Grade		
A618 (Hot-Formed Welded and Seamless High-Strength Low-Alloy Structural Tubing)	I*	50	70 min
	II	50	70 min
	III	50	65 min

*Weldability to be established by test.

this category provide atmospheric corrosion resistance ranging from four to eight times that of carbon steel without copper. In the unpainted condition, these steels are capable of developing a tightly adhering oxide coating on the surface that acts as a barrier to moisture and oxygen, thus preventing continued corrosion. This natural coating requires no further maintenance and is represented as being esthetically pleasing. Such steels have had limited use in bridge structures in the unpainted condition but are becoming increasingly

Table 82.2—Filler metal specifications for various welding processes

Welding Process	Filler Metal Specification
Manual shielded metal-arc welding	AWS A5.1 & A5.5
Submerged arc welding	AWS A5.17
Gas metal-arc welding	AWS A5.18
Flux cored arc welding	AWS A5.20
Electroslag welding	None
Electrogas welding	None

attractive economically as the cost for providing and maintaining painted surfaces increases. The environment of such unpainted structures must not provide contaminants that will adversely affect the formation and effectiveness of the oxide coating.

It is essential that filler metal in finishing passes of multipass welds in bridge structures of ASTM A242 and/or A588 steels to be unpainted have a chemical composition that will provide a resistance to corrosion similar to the base metal. Where an early color match of welds and base metal is desired, careful selection of the filler metal used for the finishing passes must be made. E80XX electrodes having a composition similar to the steel to be welded and conforming to AWS Specification A5.5 are specified by the AWS Structural Welding Code for the shielded metal-arc process; only mechanical and chemical properties are specified for filler metals used with the submerged arc, gas metal-arc or flux cored arc welding processes. The Structural Welding Code recognizes the effects of dilution by permitting some latitude of filler metal selection for single-pass welds not requiring an exact color match.

DESIGN

HIGHWAY BRIDGE LOADINGS

Live loads used for the design of highway bridges are systematized in the H and HS standard truck and lane loadings of the AASHO, Standard Specifications for Highway Bridges.

H loadings consist of a truck as illustrated in Fig. 82.1 or a corresponding lane load as shown in Fig. 82.2.

HS loadings illustrated in Fig. 82.3 consist of a tractor and semi-trailer or the corresponding lane load of Fig. 82.4.

The number following the H or HS of the load designation is respectively the gross weight of the truck or the tractor exclusive of the semi-trailer. The year of adoption to the AASHO Standard Specification is indicated by the last number of the load designation i.e., HS20-44.

The lane load formulated for mathematical tractability approximates the effect of an H load preceded and followed by trucks like the H load except that each has a gross load and axle loads equal to 3/4 of the H load values. All vehicles in the train are spaced at intervals of 30 ft. Each lane loading, as shown in Fig. 82.2 consists of a uniform load per foot of traffic lane combined with one or two concentrated loads. Two concentrated loads are used only when the spans are continuous, and then only for calculating maximum negative moments. Both the uniform and concentrated load(s) are placed to produce the maximum

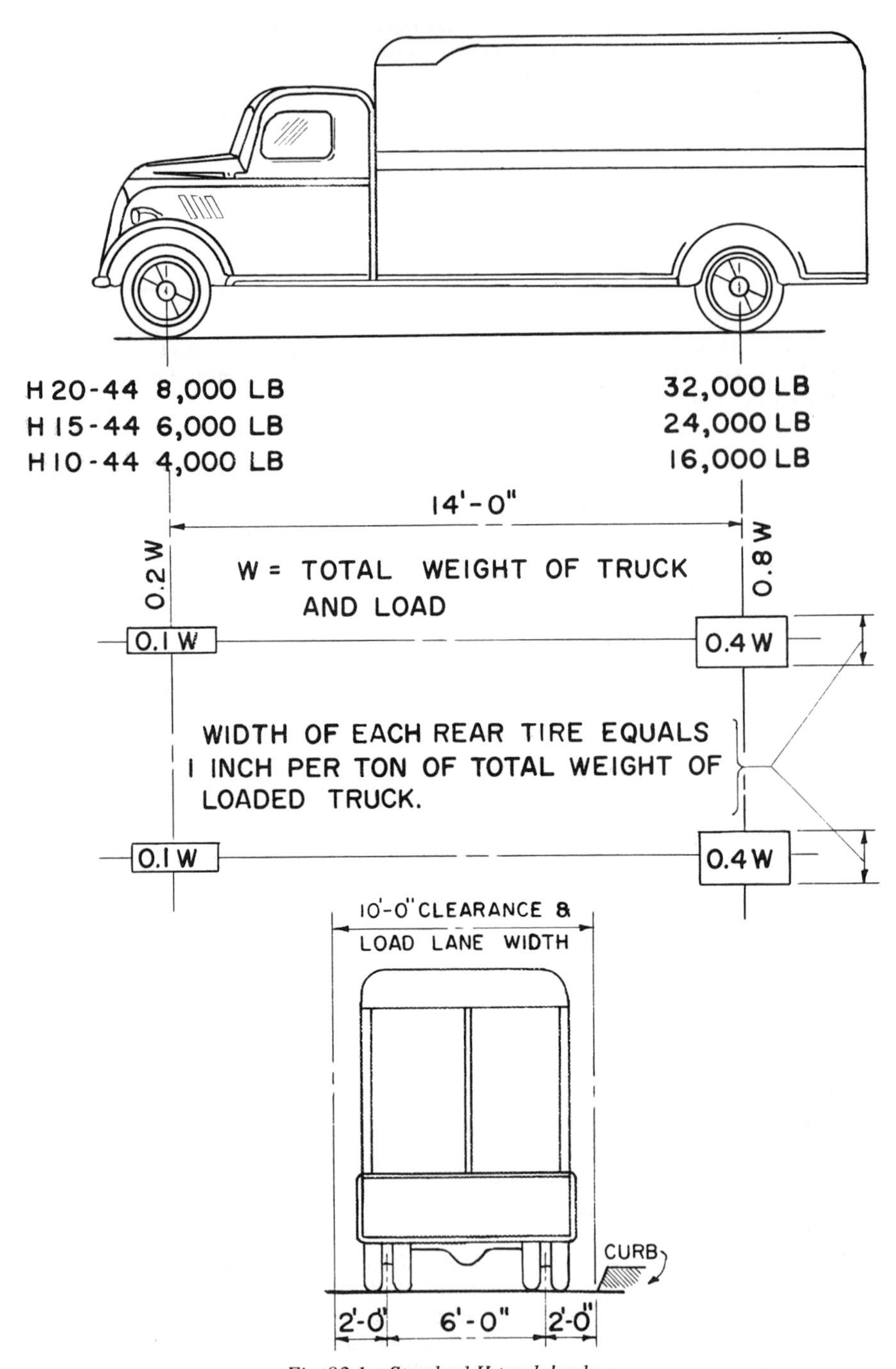

Fig. 82.1.–Standard H truck loads.

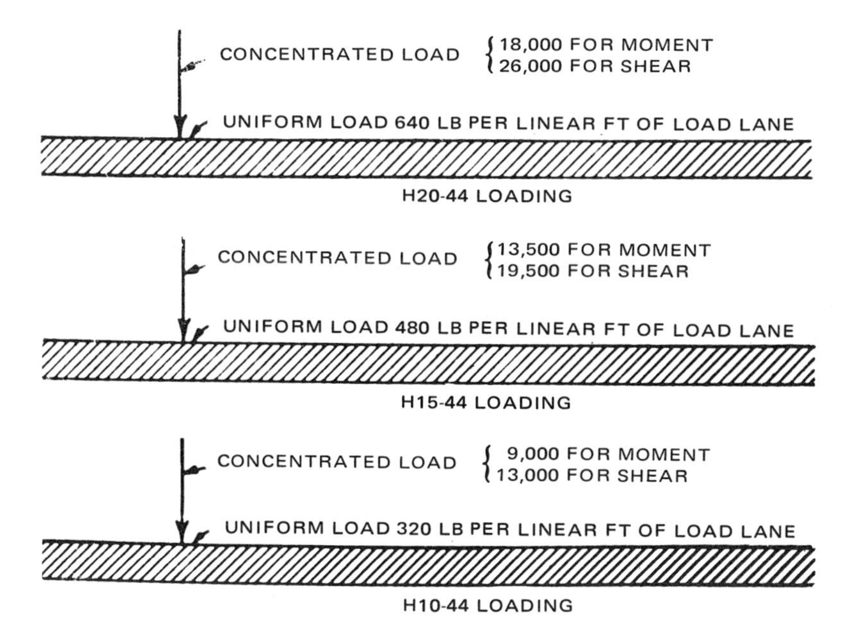

Fig. 82.2.–H lane loadings: H20-44 loading (top); H15-44 loading (middle); H10-44 loading (bottom).

stress at the point being investigated. The concentrated load, like the uniform load, is considered to be uniformly distributed over a 10-ft width on a line normal to the centerline of the lane. Different concentrated loads are used as indicated in Figs. 82.2 and 82.4 for calculating moments and shears. The lighter concentrated load is used for calculating moments and the heavier concentrated load is used for calculating shears.

In calculating the maximum stress in a member, either the lane load or truck load is used, whichever produces the maximum stress.

The HS15-44 loading is usually the lightest load used for bridges which are expected to carry heavy trucks, and HS20-44 loading should be used where the volume of heavy trucks is expected to be relatively high.

Traffic lanes, usually 12 ft wide, used in loading the structure should always be a whole number. The 10-ft wide truck or lane loading may occupy any position within the traffic lane that will produce the maximum stress. Reductions for improbable coincidental live loading are allowed when more than two lanes are loaded simultaneously.

Distribution of the loads for designing the structural elements of the superstructure and substructure is defined by empirical formulas included in the AASHO Standard Specifications.

Impact, the dynamic effect of the live load, is assumed to be a percentage of the direct vertical effect. The amount is determined by an empirical formula and is limited to a maximum of 30%.

Other forces on the structure including those caused by wind, traction, braking, roadway curvature, temperature changes, stream flow, buoyancy, earth pressure and earthquakes must also be considered, when appropriate, in the design of a bridge. The magnitude of these forces and the procedures for their combination with live load and structure weight forces are also detailed by the AASHO Standard Specifications.

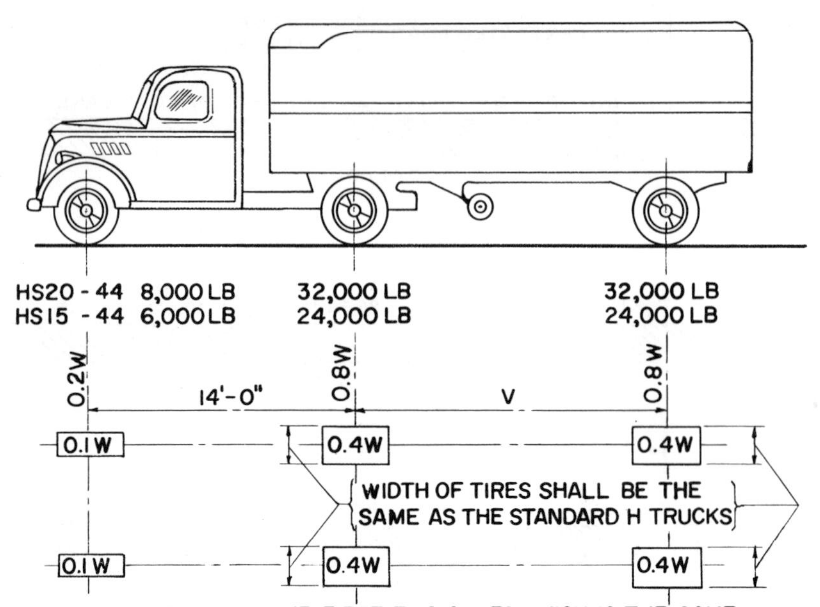

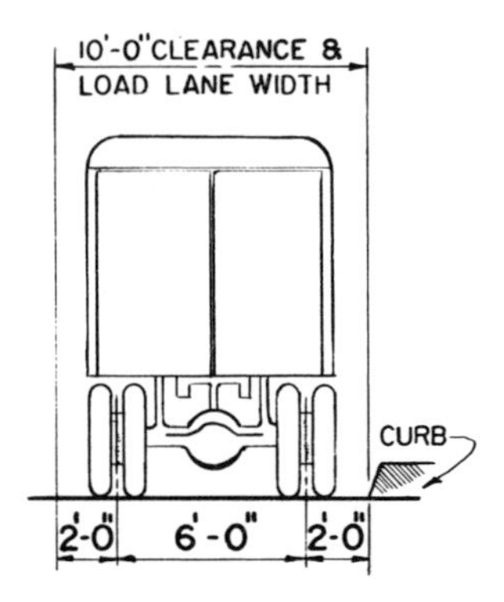

Fig. 82.3.–Standard HS trucks.

Dead load or self weight of the structure should be calculated by use of the weights per unit volume listed in the AASHO Standard Specifications. Future additions to the dead load, such as new wearing surface, should also be considered.

RAILWAY BRIDGE LOADINGS

The live loading used for railroad bridge design is usually the Cooper E system. It represents two steam locomotives with tenders pulling a train having a

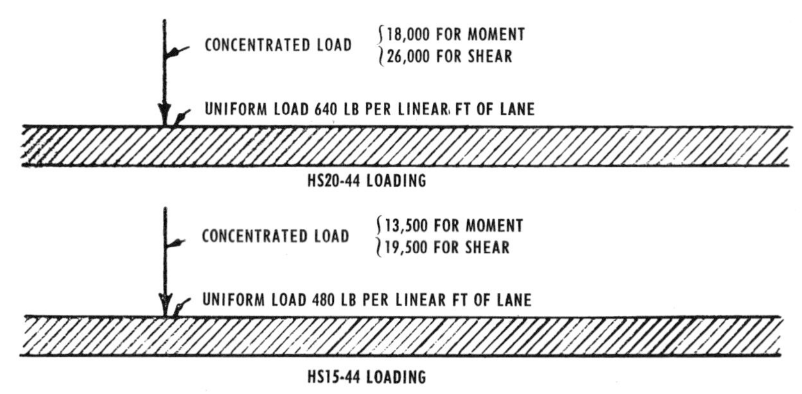

Fig. 82.4.–Standard HS loadings: HS20-44 loading (top); HS15-44 loadings (bottom).

uniform weight per unit of length. Although diesel locomotives are used almost exclusively, the Cooper loading is still a practical and convenient system. The American Railway Engineering Association's Manual for Railway Engineering recommends the Cooper E-80 live load for each track. This loading is shown in Fig. 82.5. The number associated with the loading is the axle load of the locomotive driving wheels in thousands of pounds as well as the weight of the uniform train in hundreds of pounds per foot. All other axle loads have a fixed proportional relationship to these loads.

Tables and equivalent uniform weight charts to facilitate the calculation of end shears, moments and floor beam reactions of simple spans due to the Cooper loading were included in many older textbooks dealing with simple structures. Electronic computers have diminished the utility of these tables but where hand computation is to be used, they still represent a valuable aid. Stress and load data are readily converted from one load to another by direct proportion, for example an E-80 loading is $\frac{80}{72}$ of an E-72 loading.

Impact is assumed, as for highway bridges, to be a percentage of the direct vertical effects except as it is adjusted when more than one track is loaded. In this latter case, the impact percentage is adjusted to reflect the improbability of coincidental effect.

A special effect of the railway live load that must be considered is nosing. This is the lateral force produced on the rail by the locomotive weaving between the rails. It is specified to be a single moving force of 20,000 lb applied at the top of the rail in either direction. It is to be applied to one track only, regardless of the number of tracks carried by the bridge.

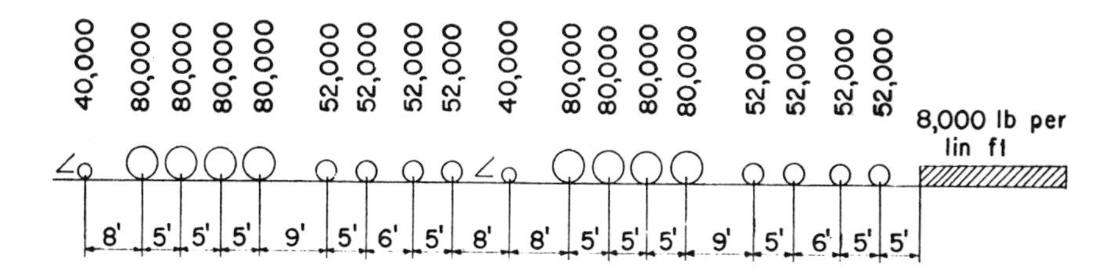

Fig. 82.5.–Cooper E-80 railroad loading.

Table 82.3—Filler metal requirements for complete joint penetration butt welds

Base Metal[1]	Welding Process[2 3]			
	Shielded Metal-Arc	Submerged Arc	Gas Metal-Arc	Flux Cored Arc
ASTM A36[4], A53 Gr. B, A106, A131, A139, A375, A381 Gr. Y35, A500, A501, A516, A524, A529, A570 Gr. D and E, A573; API 5L Gr. B; ABS Gr. A, B, C, CS, D, E, R	AWS A5.1 or A5.5, E60XX or E70XX	AWS A5.17 F6X or F7X-EXXX	AWS A5.18 E70S-X or E70U-1	AWS A5.20 E60T-X or E70T-X (Except -2 & -3)
ASTM A242, A441, A537 Gr. A, A572 Gr. 42 through 60, A588, A618; API 5LX Gr. 42; ABS Gr. AH, DH, EH	AWS A5.1 or A.5.5 E70XX[5]	AWS A5.17 F7X-EXXX	AWS A5.18 E70S-X or E70U-1	AWS A5.20 E70T-X (Except -2 & -3)
ASTM A572 Gr. 65, A537, Gr. B	AWS A5.5 E80XX[5]	Grade F80	Grade E80S	Grade E80T
ASTM A514, A517	AWS A5.5 E110XX[5]	Grade F110	Grade E110S	Grade E110T

Use of same type filler metal having next higher mechanical properties as listed in AWS Specification is permitted.

[1] In joints involving base metals of different yield points or strengths, filler metals applicable to the lower strength base metal may be used.

[2] When welds are to be stress relieved the deposited weld metal shall not exceed 0.05 percent vanadium.

[3] Does not cover electroslag and electrogas weld metal requirements.

[4] Only low-hydrogen electrodes shall be used for welding A36 steel more than 1 in. thick for bridges.

[5] Low-hydrogen classifications.

Distribution of the loads laterally and longitudinally for the design of the superstructure elements is detailed by the Manual for Railway Engineering which also defines the magnitude and condition for determining the effects of centrifugal forces, wind on the loaded and unloaded bridge and longitudinal forces from braking and traction.

Dead load of the structure should be determined by use of the weight per unit volumes listed in the Manual for Railway Engineering. The weight of the rails inside guard rails and fastenings is assumed to be 200 pounds per foot for each track.

WELD DESIGN

Welded joints in members subject to static loading or infrequent variation in loading should be designed using the unit stresses as outlined in Article 9.3 of the Structural Welding Code, AWS 1.1-72, which follows:

9.3 Basic Unit Stresses[27]

Note: The application of these stresses is modified by the requirements of 9.4.[1]

9.3.1 The permissible unit stresses, applicable to the effective throat area as defined in 2.3, for complete joint penetration groove welds made with the electrode or electrode-flux combinations specified in Table 4.1.1 (Table 82.3) shall be those allowed for the same kind of stress for the base metal.

[27] Unless specified in the general specifications, it is recommended that the basic unit shear stress in the net section be 72% of the basic allowable stress in tension.

9.3.2 Complete joint penetration groove welds between girder web and flange and all partial joint penetration groove welds may be used to transfer stress subject to the following limitations:

9.3.2.1 Welds subject to substantial compression stress applied normal to their longitudinal axis shall be made with the appropriate electrode or electrode-flux combinations as specified in Table 4.1.1 (Table 82.3). The permissible stress for such welds, applicable to the effective throat area, shall be that allowed for compression on the base metal.

9.3.2.2 Welds required to transmit shear, and/or tension in a direction other than parallel to the longitudinal axis, shall be proportioned on the basis of the permissible stress shown in Table 9.3.2.1 (Table 82.4), applicable to the effective throat, but shall not exceed the allowable shear stress of the base metal. Lower strength weld metal than shown on the same line in Table 9.3.2.1 (Table 82.4) for a base metal may be used.

9.3.3 Plug and slot welds required to transmit shear shall be proportioned in accordance with the provisions of 9.3.2.2.

9.3.4 The permissible shear stress for fillet welds, applicable to their effective throat area, shall be those listed in Table 9.3.2.1 (Table 82.4). Stress on fillet welds is considered as shear regardless of the direction of application.

9.3.5 Groove and fillet welds parallel to the longitudinal axis of tension and/or compression members are not considered as transferring stress and hence may take the same stress as that in the base metal regardless of electrode (filler metal) classification.

---

[1] Fatigue stress provisions.

Table 82.4—Permissible shear stresses

Base Metal	Corresponding Electrode or Electrode-Flux Combinations	Permissible Stress on Effective Throat Area (1)
A500 Grade A	AWS A5.1, E60XX electrodes; AWS A5.17, F6X-EXXX flux-electrode combination; or AWS A5.20, E60T-X electrodes.	16,500 psi
A36, A441, A500 Grade B, A501, A572 Grades 42 through 50, A588 and A618	AWS A5.1 or A5.5, E70XX electrodes, AWS A5.17, F7X-EXXX flux-electrode combination; AWS A5.18, E70S-X or E70U-1 electrodes; or AWS A5.20, E70T-X electrodes.	19,000 psi
	AWS A5.5, E80XX electrodes; Grade 80 submerged arc, gas metal-arc or flux cored arc weld metal.	22,000 psi
A514 over 2 1/2 in. thick	AWS A5.5, E90XX electrodes; Grade 90 submerged arc, gas metal-arc or flux cored arc weld metal.	24,500 psi
A514 over 2 1/2 in. thick	AWS A5.5, E100XX electrodes; Grade 100 submerged arc, gas metal-arc or flux cored arc weld metal.	27,000 psi
A514 2 1/2 in. or less in thickness	AWS A5.5, E110XX electrodes; Grade 110 submerged arc, gas metal-arc or flux cored arc weld metal.	30,000 psi

(1) Permissible stresses are applicable with base metal and electrodes or electrode-flux combinations listed on the same line. Electrodes and electrode-flux combinations may be used, but at their tabulated stress value, with stronger base metals appearing on lower lines. They may also be used with a base metal listed one line above, but at the permissible stress given on the line on which they are listed.

Table 82.5—Allowable fatigue design stress parameters

Category	Type and Location of Material	Type of Maximum Stress	100,000 Cycles			500,000 Cycles			2,000,000 Cycles		
			fro, psi	α	k_2	*fro*, psi	α	k_2	*fro*, psi	α	k_2
A	Base Metal (Plates and Rolled Sections)	Tension or Reversal	60,000	0	1.00	36,000	0	1.00	24,000	0	1.00
B	Weld Metal or Base Metal Adjacent to Butt Weld[1]	Tension	20,500	0.65	0.55	17,200	0.23	0.62	15,000	0	0.67
		Compression[2]	13,300	0.65	–	10,600	0.23	–	9,000	0	–
C	Flanges with Stud Shear Connectors	Tension	20,500	1.06	0.55	16,500	0	0.65	11,500	0	0.75
D	Base Metal at End of Partial Length Cover Plate	Tension or Compression	21,000	0	1.00	12,500	0	1.00	8,000	0	1.00
E	Base Metal Adjacent to or Connected by Fillet[3] or Plug Weld	Tension or Compression[4]	15,000	0	0.70	10,000	0	0.80	7,000	0	0.83
F	Weld Metal	Shear	12,000	0.78	0.50	10,800	0.36	0.55	9,000	0	0.62
G	Base Metal of Girder Web or Flange Adjacent to Stiffener Weld	Tension	20,500	0	0.55	17,200	0	0.62	15,000	0	0.67
H	Weld Metal or Base Metal Adjacent to: Continuous Flange-Web Weld[5] and Butt Weld Conforming to 9.4.5	Tension or Compression	45,000	0	1.00	27,500	0	1.00	18,000	0	1.00

NOTES:

[1] Longitudinal joints and joints with welds ground flush conforming to criteria set forth under WELD DESIGN shall be governed by Category H criteria.

[2] Use the Formula: $Fr = \dfrac{0.55Fy}{1-\left(\dfrac{0.55Fy}{k_1 f_{ro}}-1\right)R}$; Fy – specified minimum yield point or yield strength of the type of steel being used (psi)

[3] The usual continuous fillet-welded flange-web connections and similar connections shall be governed by Category H criteria.

[4] Applies to compression stress only where the weld transmits the full repeated load from one member to another, but not to a compression flange to which fillet-welded shear connectors are attached. Brackets, clips, gussets and other detail material shall not be welded to members or parts subjected to tensile stress unless the maximum stress at the point of attachment does not exceed F_r, Category E.

[5] Category F, Weld Metal, does not apply in this case. Where the shear stress in the weld exceeds 15,000 psi, $F_r^2 \geqslant F_b^2 + 3F_v^2$, in which F_b and F_v are the maximum bending and shear stresses in the weld and F_r is the allowable fatigue design stress for Category H, Weld Metal or Base Metal Adjacent to Continuous Flange-Web Weld.

Table 82.6—Cycles of maximum stress for design of highway and railway bridges

	Highway Bridges		*Railway Bridges*
Number of Cycles of Maximum Stress	Type of Road	Type of Load Producing Maximum Stress	Loading Condition
100,000 or less	1. Freeways 2. Expressways 3. Major highways and streets	Lane Loading	Not Applicable
	Other highways and streets not included in above	HS or Lane Loading	
500,000 or less	1. Freeways 2. Expressways 3. Major highways and streets	HS Loading	Loaded length of single track is over 100 ft or for two or more tracks of any length
	Other highways and streets not included in above	H Loading	
2,000,000 or less	1. Freeways 2. Expressways 3. Major highways and streets	H Loading	Loaded length of single track is 100 ft or less

[1]Where HS and H loadings produce the same maximum stress, use the number of cycles specified for H loadings.

Welded joints subject to repeated variations or reversal of stress should be designed so that the maximum stress does not exceed the basic unit stresses of the applicable general specifications; the permissible basic unit stresses (see 9.3) nor the allowable fatigue stress, F_r, given by the formula:

$$F_r = \frac{k_1 f_{ro}}{1 - k_2 R} \quad \text{in which}$$

R = algebraic ratio of the minimum stress to the maximum stress and

$$k_1 = 1.0 + \alpha\left[\frac{F_u}{58{,}000} - 1\right], \text{ but not less than } 1.0;$$

f_{ro}, k_2 and α = values of coefficients given in Table 82.5

F_u = minimum specified tensile strength in psi. See Table 82.1.

The number of cycles of maximum stress used should be according to the general specification or to Table 82.6 if none is listed in the general specification.

Grinding the surface of butt welds flush with the base metal improves their fatigue performance with the mild steels showing the greatest improvement. *Category H* of Table 82.5 may be used to determine the allowable unit stresses for weld metal or base metal adjacent to a butt weld when the surface of the weld is ground flush, the parts are of equal thickness and equal width or transitioned according to Fig. 82.6(a) and weld soundness is established by radiographic or ultrasonic testing.

Partial joint penetration groove welds are permitted only between components of built-up main members designed primarily for axial stress such as in corner and tee joints parallel to the direction of computed stress, in secondary or non-stress carrying members and in shoes, etc.

Intermittent fillet welds are permitted only to connect stiffeners to beam and girder webs because when they are used elsewhere, they very seriously impair the fatigue life. Intermittent groove welds are prohibited.

If two or more of the general types of welds (groove, fillet, plug, slot) are combined in a single joint, the allowable capacity of each weld should be separately computed with reference to the axis of the group in order to determine the allowable capacity of the combination.

Tee and corner joints that are to be subjected to bending about an axis parallel to the joint must have their welds arranged to avoid concentration of tensile stress at the root of any weld.

DETAILS

Details are defined as portions that can be considered independently. The definition is valid for structural details, but it would not be prudent for the designer to do so. Each detail for a welded joint, when not expressly defined by the specification, should be selected after considering the basic service requirements of the joint, the kind and thickness of steel, the welding process and then their effect on the fabrication and erection of the entire structure.

Shop welding can usually be done more economically than field welding. Horizontal and flat position welding for all processes except electroslag and electrogas are usually preferable to vertical and overhead welding because they are not only economical, but are easier for the welder; acceptable weld quality is obtained more consistently. Details can often be selected to avoid turning a

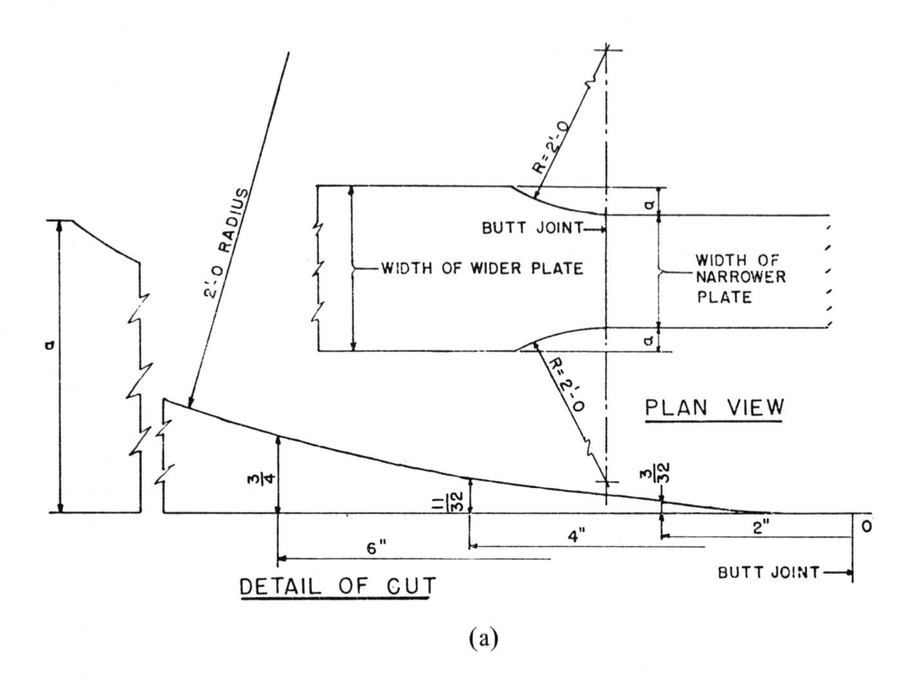

(a)

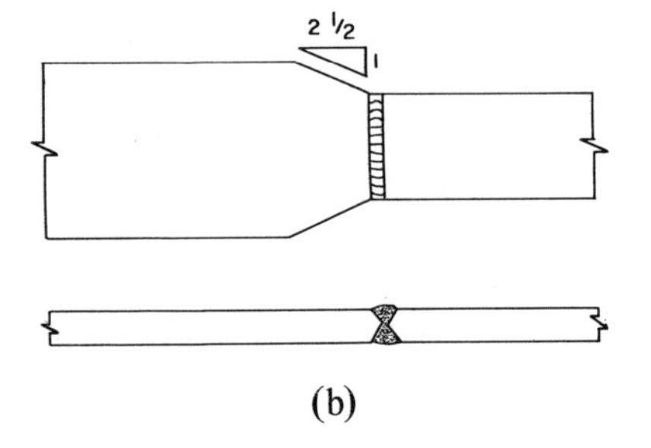

(b)

Fig. 82.6.–Transition of plate width.

piece in the shop. These are but a few of the many items which influence the overall cost of a structure.

The designer should specify the type of weld required: fillet, groove, plug or slot. For fillet, plug and slot welds, the size, spacing and other details of the welds should be clearly shown. The groove depth should be specified for partial joint penetration groove welds.

For complete joint penetration groove welds, the designer should let the fabricator select and use the AWS prequalified joint or fabricator-qualified joint detail best suited, unless he has a specific reason for showing a particular joint preparation. The joint selected should be reviewed and approved by the designer.

Refer to Fig. 81.9 in Chapter 81 for an illustration of the basic types of joints and welds. Figure 82.7 shows single-bevel groove welds which may be used in a butt, tee or corner joint.

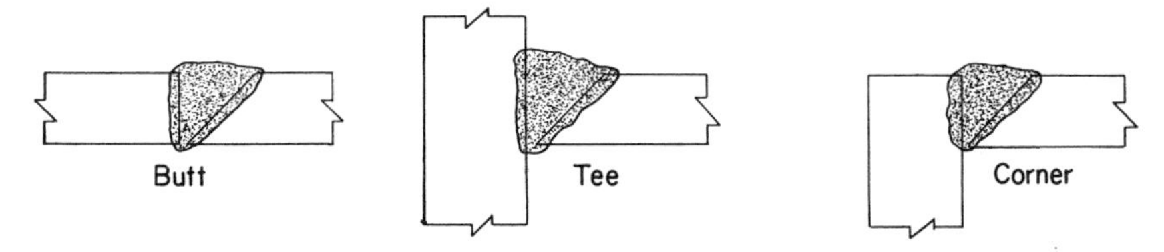

Fig. 82.7.–Example of single-bevel groove welds.

The AWS Structural Welding Code has established certain joints for manual shielded metal-arc, submerged arc, gas metal-arc and flux cored arc welding as being prequalified. They are regarded as prequalified because:

1. Over a period of years, these joints used with conventional procedures have established a record of satisfactory service performance that is supported by experience and many tests which preclude the need for further testing.
2. These tests have shown that procedures required for welding filler metals conforming to a given AWS classification made by one manufacturer produce welds which have properties similar to those made with welding filler metals conforming to the same classification manufactured by another.

Unless otherwise instructed, the prequalified joints shown in the AWS Code may be used without performing joint welding procedure qualification tests providing:

1. The weld metal conforms to one of the AWS filler metal specifications (see Table 82.3).
2. Preheat and interpass temperatures are held to at least specified minimums.
3. Joints are welded in a position to which their prequalification is restricted.
4. The AWS Code workmanship requirements are met.
5. The techniques required by the AWS Code are employed.

The geometric factors that must be defined for a given joint when developing a welding procedure include type of joint preparation, groove angle, root opening and root face, joint thickness and dimensions of backing, if used. Each of these items will greatly influence the procedure for welding the joint.

The root opening is the separation between the members to be joined (R in Fig. 81.11, Chapter 81) and is used for electrode accessibility to the root of the joint. The smaller the angle of the bevel, the larger the root opening must be to get good fusion at the root. If the root opening is small, smaller electrodes must be used, and this slows down the welding operation. If the root opening is too large, weld quality does not suffer unless burn-thru is a problem; however, more weld metal is required, which again slows the welding and increases the cost. Large root openings also tend to increase distortion.

Figure 82.8 indicates how the root opening must be increased as the included angle is decreased. Larger root openings are used for joints made with backings.

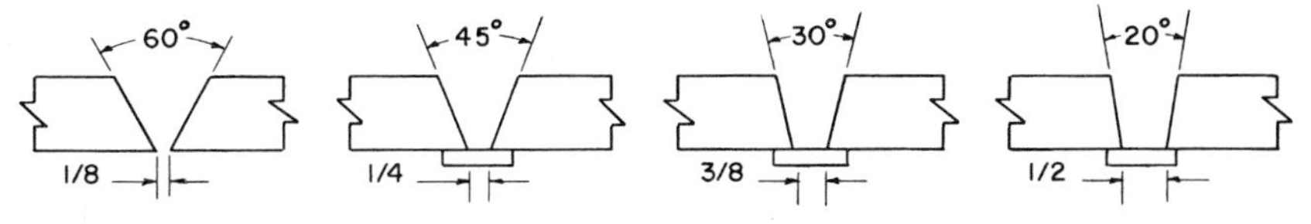

Fig. 82.8.–Root opening vs. included angle (manual shielded metal-arc welding examples).

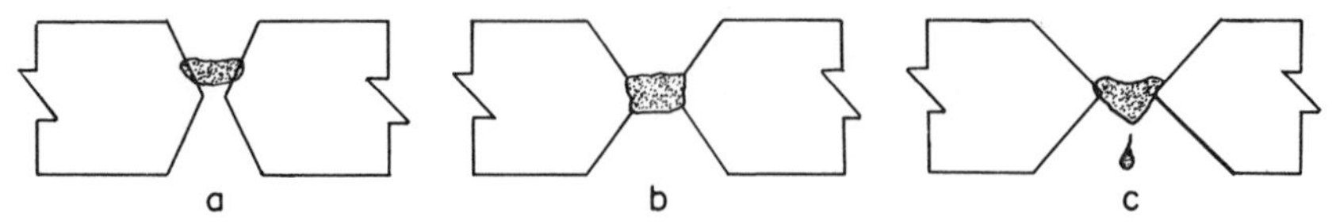

Fig. 82.9.–Root openings: (a) too close (b) proper (c) too wide.

All of these preparations are acceptable; all can be used to obtain good weld quality. Selection, therefore, is usually based upon cost and the practice of a particular shop.

If the bevel or root opening or both are too small, the weld may bridge the gap and leave slag entrapped at the root (Fig. 82.9a). Excessive back gouging is then required. Figure 82.9b shows how proper joint preparation and procedure will produce good root fusion and will minimize back gouging. In Fig. 82.9c, a large root opening results in a burn-thru.

Backing material is commonly used when all welding must be done from one side, or it may be used when the root opening is excessive. Backing shown in Fig. 82.10 is generally removed on bridge structures where it is accessible, since it reduces the fatigue strength of the joint. A spacer bar, Fig. 82.10c, may be used in double-Vee joints to prevent burn-thru; it must be gouged out before welding the second side of the joint.

Material for backings should conform to any one of the specifications for base metal. Feather edges of the plate (zero root face) are recommended when using backing. Short intermittent tack welds should be used to hold the backing in place, and these should be staggered to reduce any initial restraint of the joint. They should not be directly opposite one another. The backing must be in close contact with both plate edges to avoid trapped slag at the root (Fig. 82.11).

On a butt joint, a nominal weld reinforcement, approximately 1/16 in. above surface is all that is necessary although the AWS Code will permit up to 1/8 in. Additional buildup may reduce the fatigue life and will increase the weld cost (Fig. 82.12). Care should be taken to keep both the width and the height of the reinforcement to a minimum.

The main purpose of a root face (Fig. 82.13) is to provide an additional thickness of metal at the root of the joint, as opposed to a feather edge, in order to minimize any burn-thru tendency.

A feather edge preparation is more prone to burn-thru than a joint with a root face, especially if the gap gets a little too large (Fig. 82.14).

A root face is not as easily obtained as a feather edge which is usually a matter of one cut with a torch; a root face usually requires two cuts or possibly a torch

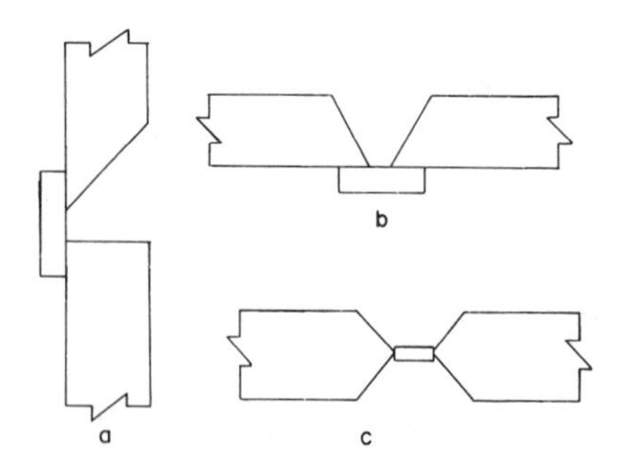

Fig. 82.10.–Examples of backing and spacer bar.

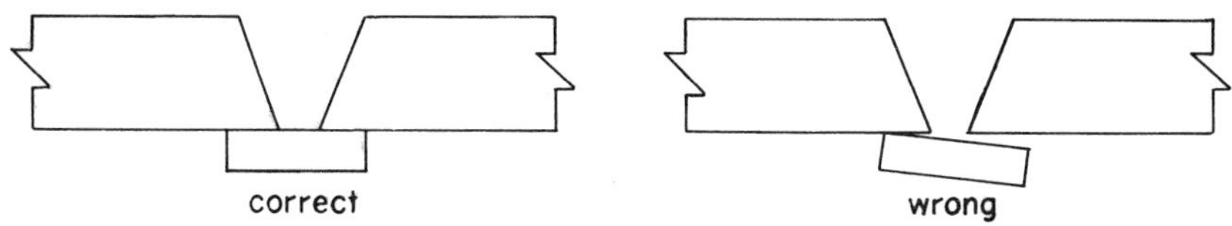

Fig. 82.11.–Use of backing.

cut plus grinding. Back gouging is usually required for groove welds having a root face when a 100% penetration weld is to be made with manual shielded metal-arc welding. When backing is used, a root face is not recommended because of the difficulty in reaching the bottom of the joint.

To consistently eliminate fusion defects at the root of groove welds when using manual shielded metal-arc welding without a backing, the back side of the joint must be removed to sound metal. This may be done by any convenient means: grinding, chipping or air carbon-arc gouging. The last method is generally the most economical and leaves an ideal contour for subsequent weld beads. Gouging produces a good groove for welding the second side; for double groove welds only the first side needs to be prepared. Without back gouging, inadequate joint penetration or other defects are very likely to occur. Proper back gouging must be deep enough to expose sound weld metal, and the contour should permit accessibility for the electrode (Fig. 82.15).

The AWS Code requires that splices in beams and girders or any tension or compression member be made with complete penetration groove welds.

Partial joint penetration groove welds are permitted as indicated in the weld design shown in Fig. 82.16.

Partial joint penetration groove welds are similar to fillet welds in that they both have unwelded roots. When stressed parallel to the axis they are given the same allowable stress as that which is permitted for the base metal. Partial joint penetration groove welds can be used where it would be possible to use fillet welds. Where fillet welds would require one plate to extend out slightly in a corner joint, partial penetration groove welds allow the joint to be flush. This flush surface provides smoother appearance and facilitates painting and maintenance of the structure. In general they require about half of the weld metal as the same strength fillet weld. There is the added cost of some extra preparation in providing the bevel, and initially the deposition rate of the weld metal may be lower because of the welding procedure used in order to reach the root of the joint. However, as the required weld size increases, there will be more of an advantage in using the partial joint penetration groove welds.

Butt joints between material having equal width but unequal thickness should be transitioned as shown in Fig. 82.17. The specified 1 to 2 1/2 slope is a maximum; when the surface offsets are such that the maximum is not required, the weld surface should be sloped to produce a smooth transition from the surface of the thinner part to the surface of the thicker part.

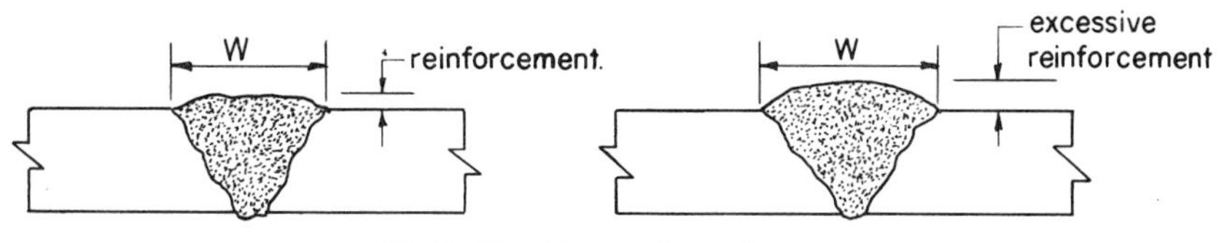

Fig. 82.12.–Butt weld reinforcement.

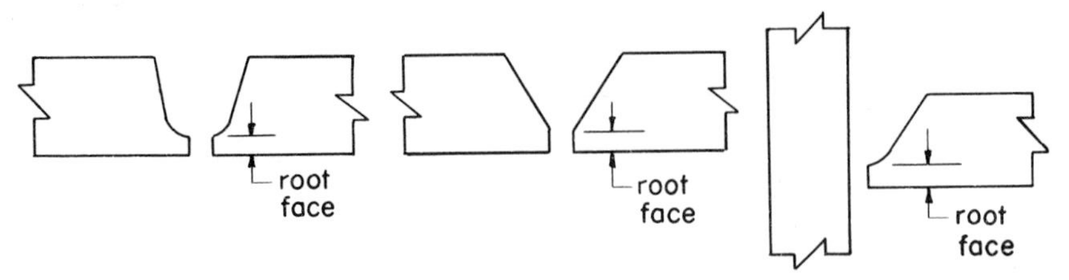

Fig. 82.13.–Root faces.

Butt joints between material having equal thickness but unequal width should be transitioned as shown in Fig. 82.6(b) or as shown in Fig. 82.6(a) if Category H (Table 82.5) is used as the basis for design of certain butt welds.

WELDING PROCESSES

The Structural Welding Code provides for the use of the following welding processes:

1. Manual shielded metal-arc welding
2. Submerged arc welding
3. Gas metal-arc welding
4. Flux cored arc welding
5. Electrogas welding
6. Electroslag welding
7. Stud welding

When the first four processes are used in conformance with the applicable sections of the Code, they are deemed to be prequalified and are therefore exempt from procedure qualification tests.

Electroslag and electrogas welding are not prequalified and require that the contractor qualify the procedure specification by tests in accordance with the terms of the Code.

Stud welding is the process used for welding steel studs to steel for connection of members and connection of devices to concrete such as shear connectors of composite construction. A separate section of the Code deals with the qualification, attachment and inspection of welded studs.

The detailed requirements under which these processes may be used are prescribed in the Code. All requirements of these sections are based on experience with use of the processes in bridge construction, and when properly followed, using the prequalified joints listed in the Code for the first four processes, sound and adequate welds can be obtained.

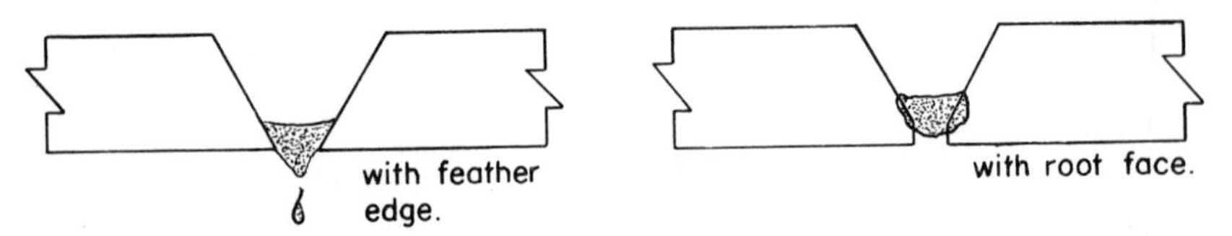

Fig. 82.14.–Use of root faced joint to prevent excessive melt-thru.

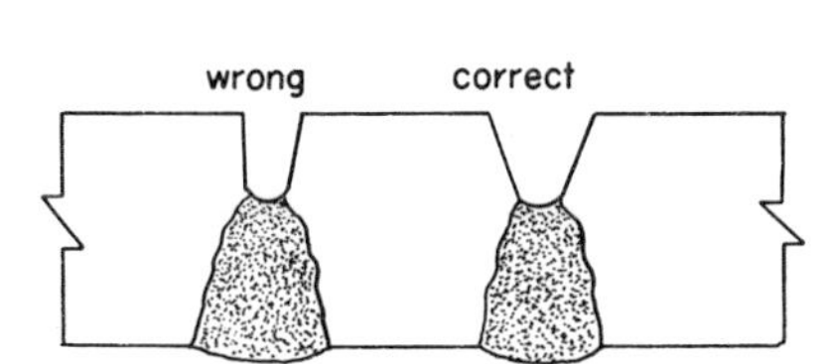

Fig. 82.15.–Back gouging.

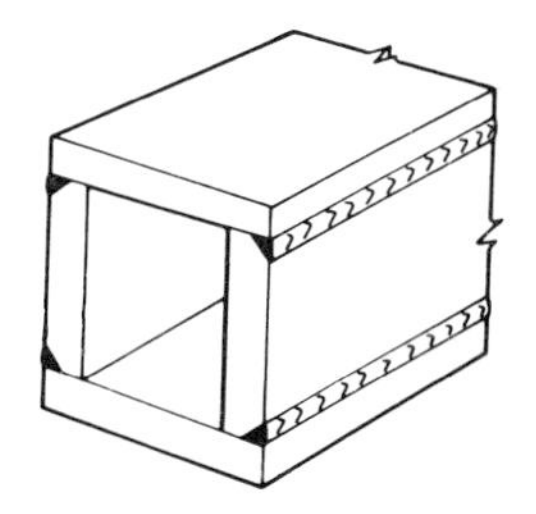

Fig. 82.16.–Partial penetration groove welds on box sections.

Detailed descriptions of the welding processes are included in Sections 2, 3A and 3B of the Welding Handbook.

It is important to note that when a steel not listed in the Code for bridge construction, or a steel specified by the Code as subject to a special investigation as to weldability is to be used, it is the responsibility of the Engineer to see that all welding procedures are suitable for it.

WORKMANSHIP

GENERAL

Workmanship requirements apply to all welding, whether done in the shop or field.

The responsibility for good workmanship on welded bridges falls on many shoulders. The welding engineer, designer, fabricator, welder and erector should all recognize each other's problems. Complete cooperation among these people can result in strong, durable and economically-welded bridge structures.

The inevitable variations from theoretical dimensions of rolled structural shapes lead to increased costs of fabrication and erection. Dimensional tolerances, straightness and flatness of structural shapes and plates are prescribed in ASTM A6 and the AWS Structural Welding Code. Welded connections should be designed to accommodate variations to the maximum permitted tolerance, or methods for their compensation should be decided upon in advance.

The avoidance of severe notch effects, abrupt discontinuities or changes in section and other forms of severe geometrical constraint is especially important in the design of dynamically-loaded structures.

Avoidance of notches, etc., is even more critical if the structures are to be subjected to relatively low temperatures. A notch, a sharp re-entrant corner, or any abrupt change in section may cause a high local concentration of stress. The notch may also cause deformation to be concentrated within a very small volume of metal so that the capacity of the metal to deform without fracture is reached prematurely at that point.

Snipes or copes should be used at points of severe restraint or where welds meet at a single point. For example, it is common practice to cope the corners of girder stiffeners if they are to be welded to a flange. The size of cope should be at least two times the fillet weld size.

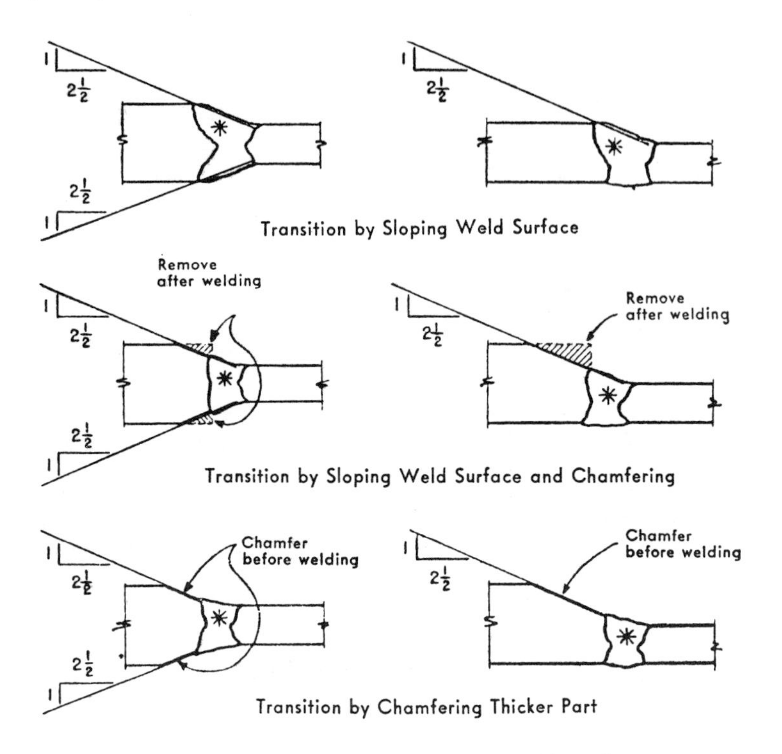

Fig. 82.17.–Transition of butt joints in parts of unequal thickness.

PREPARATION OF MATERIAL

Surfaces to be welded should be smooth, uniform and free from scale, rust, grease or paint. For most applications, vigorous wire brushing will be sufficient to remove any objectionable solid material. However, the AWS Code requires that all mill scale be removed from the surfaces on which flange-to-web welds are to be made by submerged arc welding or by shielded metal-arc welding with low-hydrogen electrodes.

For edge preparation prior to welding, several established methods are commonly used in the fabricating shop. These are mechanical cutting and machining (shearing, machine planing), oxygen (flame) cutting and gouging and arc and plasma-arc cutting. The choice of any of these will depend on the class of work, on the thickness of material and the type of joint. Oxygen cutting is most frequently used with emphasis on machine cutting for important work. However, air carbon-arc cutting is slowly gaining ground in the preparation of certain types of grooves.

Surface roughness requirements for oxygen-cut edges are established in the AWS Code to assure that good shop practices are followed.

ASSEMBLY

In the assembly of material to be welded, utmost attention must be paid to the correct alignment of the parts. Joint configuration and fit-up within tolerances as stipulated in the AWS Code must be maintained. Excessive deviations in this respect would constitute grounds for qualification of joints, normally considered prequalified, or rejection of the work. Also, additional costly penalties may be incurred either in poor access, involving a higher risk of defects, or in an excessive volume of weld metal, reflected in higher deposition costs and greater distortion.

In establishing tolerances, consideration should be given to the permissible mill tolerances as they relate to the end product and its use. Tolerance should be kept as liberal as possible consistent with engineering requirements, since working to unrealistic tolerances only adds to costs.

After the material is shaped, cut and prepared for assembly, it should be moved to the fitting and welding location. At this point, some shops combine fit-up and welding in a common jig, while others employ one jig for fitting and tack welding, and another fixture for the final welding operation.

When a considerable amount of duplication of work is involved, it is more economical to use jigs and fixtures for fitting and welding. This practice facilitates accuracy and helps ensure proper fit-up during erection.

Fillet and groove welding on bridge girders is now most commonly done by automatic or semiautomatic welding using either submerged arc or gas metal-arc/CO_2, or a combination of these processes. However, the use of other processes is also possible.

To facilitate the use of automatic or semiautomatic welding, there has been an increase in the use of material-positioning devices. Some shops utilize manipulators or side-beam or track-mounted carriages to position the welding equipment. Both single and multiple electrode equipment is used. Multiple electrode welding equipment permits greater welding speeds. However, good material handling must be associated with high welding speeds for maximum economy. Set-up and handling time is often much greater than actual welding time. Therefore, high welding speeds are only part of the answer to obtaining greater production rates and lower costs.

DISTORTION CONTROL

The reader is referred to Chapter 81 in which material on distortion control, pertinent to this chapter, is presented in detail.

ERECTION

Welded bridge erection is geared toward handling the largest possible lengths of section in the field. Shop assembly rather than field assembly is the most economical approach. Consequently, welding in the field should be held to a minimum. Various types of fit-up clamps and clips are often used instead of erection bolts for the field splices to help in the fitting of the girders before welding.

Field splices as well as shop splices of built-up plate girders are best made in one plane. When the splice must be made in position in the field, most of the

5. Examination of finished welding
   a. Examine every welded joint for compliance with the working drawings with respect to size, length and location of welds, and see that no welds have been omitted or added.
   b. Examine the welding for compliance with standards of workmanship with respect to the contour of the weld surface, surface defects, craters, undercutting, overlapped edges of welds, cracks, etc.
   c. Require corrective measures to be taken wherever necessary, in compliance with the methods prescribed by the specifications.

Nondestructive Testing

Nondestructive testing procedures of bridge welding are well-established. The AWS Structural Welding Code now includes requirements for technique and standards of acceptance for the following types of examination: magnetic particle, liquid penetrant, radiography with X-rays or gamma rays and, most recently, ultrasonic examination.

Destructive Testing

This type of testing is normally used to determine the mechanical and metallurgical qualities of welds and joints obtained with the proposed welding procedures. Such tests would include those for tensile strength, yield point, cold bend properties, shear strength and impact and fatigue properties.

Destructive testing would be employed in the course of fabrication or erection only if there was serious reason to question the quality of the welds or material, as evidenced by information obtained from nondestructive examination, or when it becomes necessary to requalify a welding procedure.

Everything that has been said about examination in Chapter 81, Buildings, applies equally to bridge welding. For more detailed information on specific methods of inspection see also Chapter 6 in Section 1 of this Handbook and AWS WELDING INSPECTION.

BIBLIOGRAPHY

Design of Welded Structures, O. W. Blodgett, The James F. Lincoln Welding Foundation (1966).

Structural Welding Code, D1.1, American Welding Society (1972).

"Specifications for Steel Railway Bridges," American Railway Engineering Association (1971).

Fatigue of Welded Steel Structures, W. H. Munse and L. Grover (ed.), Welding Research Council (1964).

Fatigue in Welded Beams and Girders, W. H. Munse and J. E. Stallmeyer, Highway Research Board (1961).

"Standard Specifications for Highway Bridges," The American Association of State Highway Officials (1969).

"Construction Inspection of Welded Steel Bridges," Texas Highway Dept. Bridge Division (July, 1956).

Weldability of Steels, R. D. Stout and W. D. Doty, Welding Research Council (1971).

"Manual of Design for Arc Welded Steel Structures," L. Grover, Airco (1946).

Ultrasonic Testing Inspection for Butt Welds in Highway and Railway Bridges, U. S. Bureau of Public Roads (1968). (For Sale by Superintendent of Documents, U. S. Government Printing Office).

CHAPTER **83**

FIELD-WELDED STORAGE TANKS

PREPARED BY A COMMITTEE CONSISTING OF:

L.J. CHRISTENSEN, *Chairman*
Chicago Bridge & Iron Co.

J. H. ADAMS
Pittsburgh-Des Moines Steel Company

T. W. HOWLETT
Graver Tank & Manufacturing Co.

CHAPTER **83**

FIELD-WELDED STORAGE TANKS

The growth of cities and towns has increased the demand for water, oil, natural gas and propane storage. The increased use of automobiles, trucks and aircraft has created greater demand for storage of petroleum products. In addition, industrial and scientific developments–basic oxygen steelmaking processes and rocket propulsion systems, respectively–have emphasized the need for volume storage of cryogenic liquids such as oxygen, nitrogen and hydrogen. Expansion of the fertilizer industry through economical storage of refrigerated ammonia has also created a need for large volume field-erected storage tanks.

GENERAL CHARACTERISTICS

This discussion of field-welded storage tanks will be confined to bulk storage tanks too large for shop assembly, thus necessitating field erection. Such tanks, usually having diameters greater than 12 ft, are generally erected vertically. These include spheres, spherical tanks and cylindrical steel tanks with vertical axes, designed to contain liquid products. They generally operate at atmospheric pressure or at not over 15 psi gage pressure measured above the liquid level. Oil-storage tanks of various types, standpipes, reservoirs and elevated tanks for the storage of water and tanks for storing miscellaneous liquid and industrial products are included in this category.

GOVERNING CODES AND SPECIFICATIONS

Field-welded storage tanks are usually constructed in accordance with codes or specifications prepared to meet the specific needs of a particular industry. Such codes and specifications for one industry should not be indiscriminately applied to other industries without a full understanding of the design and construction concepts used.

METALS

Modern industrial requirements have created a demand for tanks of such large volumes that they must of necessity be erected in the field. In the majority of cases, they remain uncovered or unprotected, and are therefore subjected to rapid changes and extremes of weather conditions.

Safe, dependable tankage presently in service is constructed of practically every available metal. Formal rules, such as codes or specifications, have been developed to govern the design, selection of metals and construction methods for many of these structures. Unfortunately, such codes can only specify minimum requirements for reliability under average conditions. In addition to adhering to such rules, the designer must apply sound engineering judgment in order to produce a reliable structure with maximum economy.

The great majority of field-erected storage tanks have been built of plain carbon steels such as ASTM A283. Engineers have recognized that critical combinations of stress and temperature can cause brittle fracture of steel resulting in catastrophic failure. Although no single test will predict performance of all metals under all conditions, industry has utilized the Charpy V-notch impact test to qualify metals for low and cryogenic temperatures. Authors of various codes such as API, ASME and AWWA are now including the use of notch-tough steels for areas of low mean temperatures, and for storage of liquids and gases at sub-atmospheric temperatures.

Case histories indicate that tanks and vessels constructed using plain carbon steel can fail in brittle fracture at ambient temperatures when sufficient stresses are present. Notch toughness of carbon steel can be improved by variations in steelmaking practices such as deoxidization, small additions of manganese and other elements for controlling grain size. Post-manufacture heat treatment, such as normalizing or quenching and tempering the steel, will further improve the notch toughness. Additions of nickel up to 9% can reduce the transition temperature to below -325 F.

Field-welded storage tanks designed to hold petroleum products or other liquids at atmospheric pressure are usually constructed of one of the metals listed below. Specifications for pipe, bolting, castings and forgings are designed to adhere to the same general quality and service severity requirements as those specified for plates and structural shapes.

PLATES

Steel plates used are limited to open hearth, electric furnace or basic oxygen process steels conforming to ASTM specifications. Designations used are: A36, A131, A283, A284, A285, A353, A442, A514, A516, A537, A553 and A573. Limitations involving thickness, impact testing, heat treatment and operating temperatures are usually governed by applicable codes. Refer to these codes for specific information.

STRUCTURAL

Structural steel shapes are open hearth, electric furnace or basic oxygen process steels conforming to ASTM specifications A36 or A131.

COPPER-BEARING STEELS

Copper-bearing steels, containing approximately 0.20% copper, may be specified where a slightly greater resistance to atmospheric corrosion is desirable.

OTHER METALS

Wrought iron product forms used in tank construction are covered by ASTM specifications A42, A189 and A207. See Chapter 61, Section 4, for welding procedures.

Aluminum and its alloys in various product forms conform to ASTM specifications B26, B108, B209, B211, B221, B241, B247 and B308. See Chapter 69, Section 4.

Austenitic stainless steel and its product forms are covered by the following ASTM specifications: A182, A240, A269 and A312. See Chapter 65, Section 4.

DESIGN

TYPES OF TANKS

The most common shape for storage tanks is the vertical tank with a cone roof and flat bottom. Figures 83.1 and 83.2 show the more common types. The need to prevent excessive evaporation losses of volatile oils has led to the development

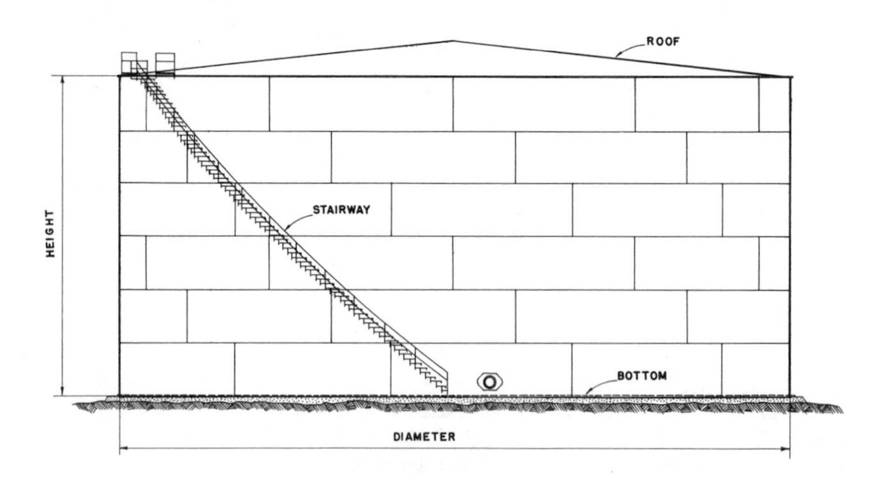

Fig. 83.1.–Cylindrical oil-storage tank with cone roof.

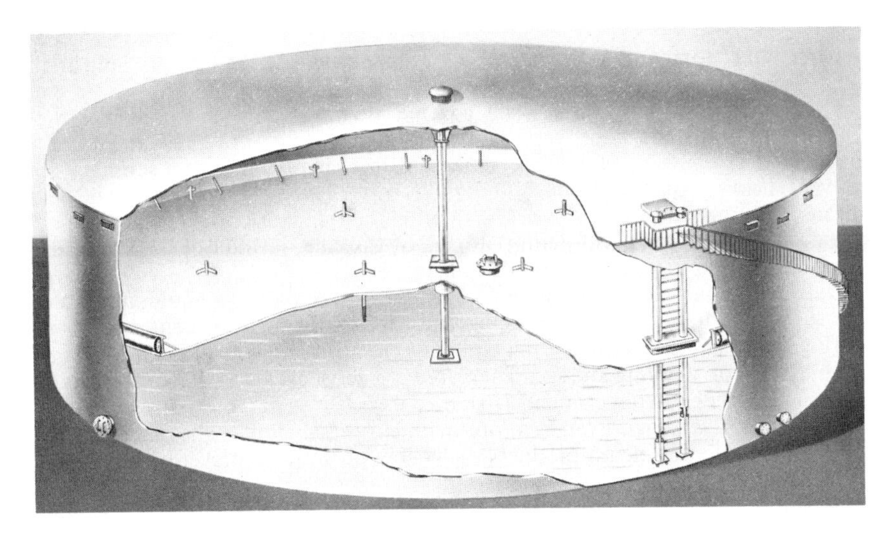

Fig. 83.2.–Covered floating roof tank.

of special types of oil-storage tanks differing in various details. In the case of water-storage tanks, it is sometimes necessary to elevate the tank on a tower to obtain a suitable gravity head for distribution purposes.

Each type of tank requires special accessories, such as foundations, vents, gages, drains, overflows or manholes depending upon the content, local conditions and regulations. The design, construction and application of these accessories should be handled by personnel experienced in this field.

Cylindrical storage tank with cone roof and flat bottom.–This type is used for

Fig. 83.3.–Oil-storage tank with floating roof.

general bulk storage of oil products, tar, alcohol and miscellaneous industrial liquids (Fig. 83.1).

Covered floating roof tank.—The covered floating roof tank combines the benefits of both the cone roof tank and the floating roof. The fixed roof provides a cover to exclude rain, snow and dirt. It also acts as a sunshade to prevent heating of the product under the roof. The floating roof provides a means of minimizing vapor losses, oxidation, contamination or degradation of the product. This tank will permit storage of gasoline, jet fuel or high porosity chemicals without loss or contamination (Fig. 83.2).

Oil-storage tank with floating roof.—The roof floats upon the liquid, insulating the surface of the liquid and reducing vapor losses. These roofs conserve vapors during filling and emptying of the tank, but are not as tight against vapor loss as lifter roofs for standing storage. They are very serviceable for active storage of gasoline or other volatile products (Fig. 83.3).

Domed-roof tank with flat reinforced or curved bottom.—These are usually used for storing petroleum products under low pressure to prevent evaporation (Fig. 83.4).

Spheroid for storing petroleum products under low pressure.—Pressure should not exceed 15 psi, which permits a pressure above atmospheric to be applied to the surface of the liquid with nearly uniform stresses throughout the shell. These vessels may be plain or noded. Figure 83.5 shows a noded spheroid.

Sphere for storing gas or petroleum products under pressure.—These tanks permit a pressure above atmospheric, with nearly uniform stresses throughout the shell. Such tanks are frequently built for a pressure range about 15 psi gage. In such cases, they are considered pressure vessels (Fig. 83.6).

Elevated water-storage tank.—These tanks are used to store water and, by elevation, to provide sufficient pressure for water distribution systems or fire protection (Figs. 83.7 and 83.8).

Fig. 83.4.—Dome roof tanks with curved and flat reinforced bottoms.

Fig. 83.5.–Noded spheroid for storing products under low pressure.

Cylindrical water standpipe with domed roof.–This tank is used for bulk storage of water for domestic service. Pressure is obtained by locating the tank on ground elevations or by pumping. The domed roof is self-supporting (Fig. 83.9).

Fig. 83.6.–Sphere for storing gas or petroleum products under pressure.

Fig. 83.8.–Modern design elevated water-storage tank.

Fig. 83.7.–Conventional elevated water-storage tank.

Fig. 83.9.–Cylindrical water standpipe with domed roof.

Cylindrical water standpipe with conical roof.—This tank is used for bulk storage of water for domestic service. Pressure is obtained by locating the tank on ground elevations or by pumping (Fig. 83.10). The roof is supported by rafters, and in turn by columns which extend to the bottom of the tank.

Process tank with cylindrical shell, elliptical roof and conical bottom on structural supports.—This tank is used in many industrial plants for storing and handling liquids or dry products during manufacture (Fig. 83.11).

Cryogenic storage tank.— The needs of the U.S.A. space program led to the development of double-shelled insulated flat bottom tanks for the storage of large quantities of liquid oxygen, nitrogen and hydrogen. These liquefied gases are stored at essentially atmospheric pressure, which means that their temperatures are -297 F, -320 F and -423 F respectively. This type of tank has become very common for the storage of liquefied natural gas (liquid at one atmosphere and -260 F) to meet peak winter demands (Fig. 83.12).

Underground storage tank.—This tank is usually used for storing finished petroleum products. It is used chiefly where tanks located above ground would

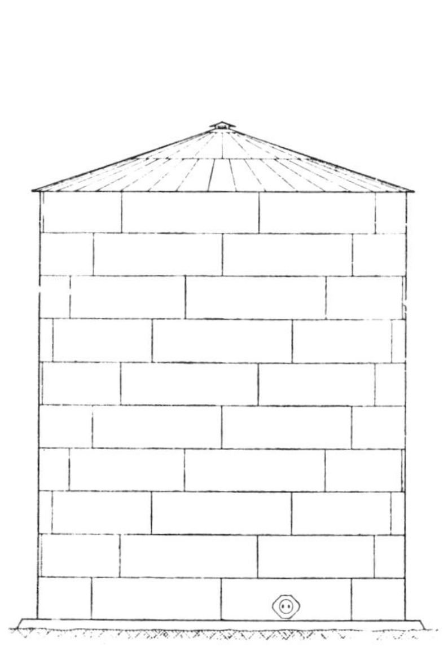

Fig. 83.10.—Cylindrical water standpipe with conical roof.

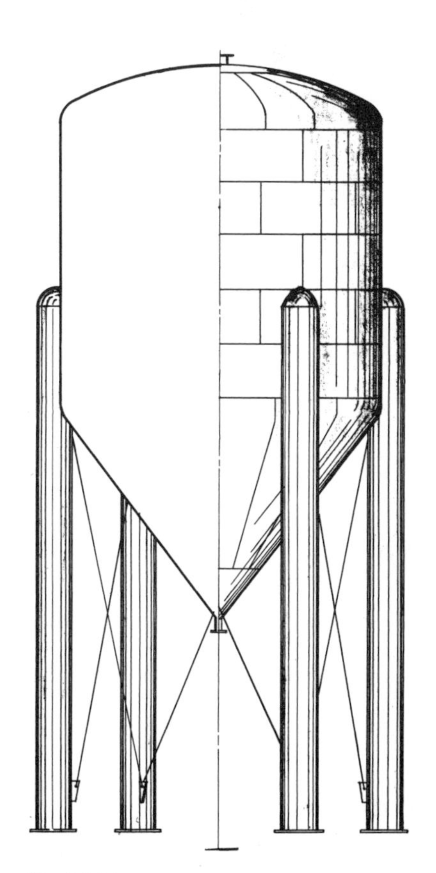

Fig. 83.11.—Process tank with cylindrical shell, elliptical roof and conical bottom.

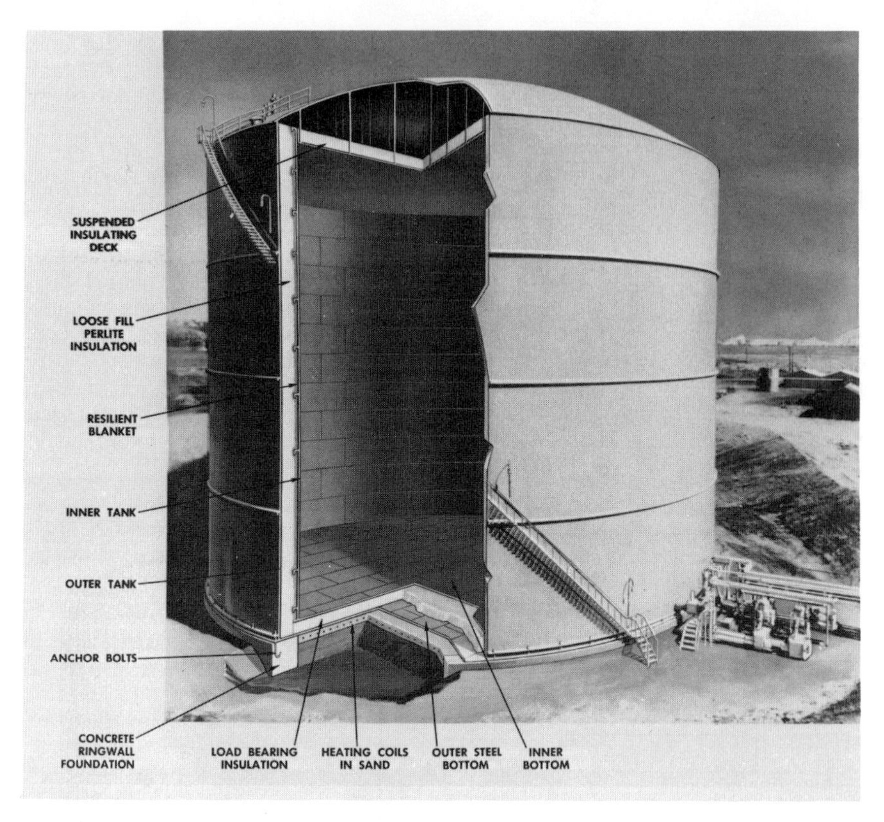

Fig. 83.12.–Cryogenic storage tank.

be considered objectionable because of their use for military purposes or because of zoning requirements.

UNIT STRESSES

Steel Plates

Industry usually specifies the unit stress requirements for the steel plating to be used in the construction of storage tanks. AWWA provides a 15,000 psi stress for certain lower strength carbon steels used in water-storage tanks. Stresses vary with conditions up to the 32,000 psi permitted by API 650, Appendix G when using a quenched and tempered steel such as A537, Grade B.

Allowable stresses specified by any individual code may incorporate unstated factors of design and application that may not safely be carried over to another industry. For instance, API 650 allows 21,000 psi on the basis that the petroleum product to be stored has a specific gravity of approximately 0.85. A water test (specific gravity 1.0), therefore, provides a pressure above the expected working stress. Hence, storage of heavier products, in effect, lowers the factor of safety. For that reason, specifications mentioned in the bibliography should be referred to for unit stresses and other pertinent information.

Welds

Butt joints welded from both sides with complete penetration, and double lap joints fully fillet-welded will usually show 100% strength values when tested with the conventional test coupons. Wide unwelded plates and wide welded plates will generally show lower tensile values than narrow test coupons. It is therefore advisable to assign efficiency values for welded joints to provide for the foregoing condition and for deficiencies in the welded joints.

The following joint efficiencies are used primarily in the design of field-welded storage tanks: 85% efficiency for butt joints welded from both sides in tank shell plating based upon the unit stress used for the plate material when spot radiography is used; 100% efficiency when 100% radiography of seam welds is specified; 70 to 75% efficiency based upon the unit stress when double lap joints in tank shell plating, full fillet-welded, are used for the plate material.

TYPICAL WELDING DETAILS

The plate seams, which sustain primary stress from the weight or pressure of tank contents, require welded joints with strength equal to the base plate material. These are the vertical joints in cylindrical tank shells, joints below the point of support in tanks with suspended bottoms and all the joints in low-pressure tanks. These joints should be welded to obtain complete penetration and fusion throughout the thickness of the plate. They are usually welded from both sides, using square-groove butt welds for plate 5/16 in. or less in thickness, and other groove preparations for heavier plate.

Seams subject to secondary stress, such as the circumferential seams of cylindrical tank shells, do not require full strength joints. They are usually welded from both sides to obtain at least two-thirds penetration. The unwelded portion, not over one-third of the thickness of the plate, is located substantially at the center of the joint. If single-Vee groove welds are used, it is customary to require complete penetration.

The joint connecting the shell to a flat bottom on vertical cylindrical tanks is usually made with a continuous fillet weld on both sides of the shell plate. It is of a size equal to the thickness of the bottom plate or the shell plate whichever is smaller, with a maximum fillet size of 1/2 in.

Flat bottom and roof plates on vertical cylindrical tanks, if the thickness does not exceed 1/2 in., are usually connected with fillet welds applied to the top side only. Heavier plates are usually groove-welded. Dished bottoms and dome and cone roofs on tanks subjected to low internal pressures must be designed with a connecting weld capable of carrying the resulting stresses.

WORKMANSHIP

WELDER AND WELDING PROCEDURE QUALIFICATIONS

To ensure that only skilled men shall be employed, all welders and welding processes must be properly qualified. The welding procedure to be used for each type of joint, in each welding position, should also be properly qualified. The qualification should be performed according to the requirements of the AWS Standard Qualification Procedure, or some other applicable governing code or specification as agreed upon by the supplier and the purchaser.

WELDING PROCESSES

Welding may be performed by any of the conventional fusion processes. In addition to the widely used manual shielded metal-arc welding (SMAW) process, several semiautomatic and automatic processes have come into use. These reduce costs and can increase the reliability of the welded joints by eliminating the human factor as much as possible.

One such widely used process is submerged arc welding of the girth seams in the "3 o'clock" position. Joints may be made in plate thicknesses up to 3/8 in. without beveling the plates. Heavier plates may be welded using a double-bevel groove, welding from one side at a time or welding from both sides simultaneously. Back gouging is generally not required with this process.

Another process recently introduced is the electrogas welding process. Water-cooled copper shoes mold the deposited weld metal to a desired contour until solidification occurs.

Additional methods that may be used include several semiautomatic processes, such as those using composite electrode (flux cored) wires, with or without external shielding gases; and solid bare wire with external gas shielding. These welding processes are particularly applicable for lap jointed bottoms and roofs, and fillet welds.

Surface Preparation

Surfaces to be welded must be free from loose scale, slag, heavy rust, grease, paint and any other foreign material. Tightly adherent mill scale can be tolerated. A light film of linseed oil or spatter film compound may be disregarded. Such surfaces should be smooth, uniform and free from fins, tears and other defects that adversely affect welding. A fine film of rust adhering on cut or sheared edges after wire-brushing need not be removed.

Weather Conditions

Welding should not be performed when the temperature of the base metal is below 0 F; when surfaces are wet from condensation, rain, snow or ice; when rain or snow is falling on the surfaces to be welded; during periods of high wind, unless the welder and the welding set-up are properly protected. At temperatures between 0 and 32 F, the surfaces of all areas within 3 in. of the point where the weld is to be started should be heated to a temperature warm to the hand before welding is begun.

Cleaning Between Passes

Each pass of a multiple-layer weld should be cleaned of slag and other loose deposits before depositing the next pass.

Treatment of Weld Root

The root of the first pass of double-welded butt joints should be chipped, ground, oxygen or air carbon-arc gouged to eliminate all unsound metal. This requirement is not intended to apply to any process of welding where proper fusion and penetration are otherwise obtained, or where the root of the weld remains free from impurities.

INSPECTION AND TESTING

INSPECTION

It is advisable to have an experienced inspector check all construction details. The inspector should not be responsible to, or in the employ of, the superintendent or foreman in charge of construction. He should be given full authority to insist upon compliance with the applicable codes or specifications.

The purchaser may, if he so desires, have the steel inspected at the rolling mills, and the fabrication inspected in the fabricating shop, to be assured that materials and workmanship are acceptable. Field inspection of the erection may also be performed by the purchaser if desired. The important welded joints should be checked by suitable methods.

Aside from visual examination, important welded joints may be examined by sectioning methods, and by liquid penetrant, magnetic particle, radiography or ultrasonic testing. Sectioning methods are generally losing favor as an examination tool in modern construction. This is due to the uncertainty regarding the soundness of the metal deposited to fill the void created in removing the original test section. Sectioning is most frequently replaced by radiography or ultrasonic testing. Magnetic particle or liquid penetrant examination is often employed to check root passes and back-gouged grooves of butt joints to assure freedom from rejectable defects.

Increased popularity of radiographic examination is associated with its inherent advantage of being a nondestructive method that can be applied by means of lightweight portable apparatus to either partial or complete examination of main vessel seams. Being nondestructive, radiographic examination suffers none of the disadvantages of sectioning, while providing a permanent record of joint quality. Ultrasonic examination is increasing in popularity and will undoubtedly continue to do so.

For detailed rules regarding any examination method, reference should be made to various codes listed in the bibliography or to job specifications.

TESTING

Flat Tank Bottoms and Shells

Joints, including root pass and back-gouged grooves, in tank bottoms resting on grades or foundations, may be tested by the magnetic particle method, or by leak testing. The joints are coated with a suitable material such as linseed oil, and air pressure or vacuum is applied which will disclose leaks by the appearance of bubbles.

Whenever possible, the storage tank should be tested by filling with water.

BIBLIOGRAPHY

Standard for Steel Tanks, Standpipes, Reservoirs and Elevated Tanks for Water Storage (D100*, AWS D5.2*) American Water Works Association-American Welding Society.

Water Tanks for Private Fire Protection, NFPA No. 22.

Welded Steel Tanks for Oil Storage, API Standard 650. "An Investigation of the Impact Properties of Vessel Steels." F. B. Hamel, Paper presented at 23rd Meeting of the API Division of Refining (May 1958).

Control of Steel Construction to Avoid Brittle Failure, Welding Research Council (1957).

Recommended Rules for Design of Large, Welded, Low Pressure Storage Tanks, API 620*.

WELDING INSPECTION, Second Edition, American Welding Society (1968).

---

*Refer to latest edition.

CHAPTER 84

PRESSURE VESSELS AND BOILERS

PREPARED BY A COMMITTEE CONSISTING OF:

D. C. BERTOSSA, *Chairman*
General Electric Co.

L. J. CHRISTENSEN
Chicago Bridge & Iron Co.

W. J. ERICHSEN
Westinghouse Electric Corp.

H. L. HELMBRECHT
Babcock & Wilcox

L. K. KEAY
Lukens Steel Co.

G. A. LECLAIR
Foster Wheeler Corp.

R. E. LORENTZ, JR.
Combustion Engineering, Inc.

CHAPTER 84

PRESSURE VESSELS AND BOILERS

INTRODUCTION

It is essential that all those involved in the field of pressure vessels and boilers keep abreast of the evolution in materials, design and fabrication techniques. These new techniques in welded construction permit weight reductions, design freedoms and higher strength and pressure levels. Such advances will ultimately lead to improved quality, safety, longer service capability and economy of construction.

Emphasis on welding quality assurance programs and new inspection techniques supplies a greater level of safety and demonstrates the high quality of welding which is now demanded by governing codes and regulations. The use of nondestructive inspection methods such as radiography, ultrasonics, magnetic particle, liquid penetrants and mass spectrometer leak tests offers means of developing the integrity and workmanship characteristics of vessels to a degree previously unavailable.

Those involved in the fabrication of welded pressure vessels now have years of experience to draw upon and data from the supporting disciplines of metallurgical research and fracture mechanics. These supporting disciplines are dealt with in depth in other chapters of the Handbook. The welding viewpoint of pressure vessel and boiler construction will be stressed in this chapter.

CODES, STANDARDS AND REGULATIONS

Metallic vessels under pressure could constitute a substantial hazard if catastrophic failures occur. Therefore, safety codes, standards and regulations have been prepared to cover the construction of steam-generating equipment and pressure vessels. The following is a list of generally accepted codes, standards and regulations (latest editions as applicable) covering the construction and maintenance of boilers and pressure vessels fabricated by welding:

1. ASME Boiler and Pressure Vessel Code (referred to herein as ASME Code)
   Section 1–Power Boilers
   Section II–Material Specifications
      Part A–Ferrous
      Part B–Nonferrous
      Part C–Welding Rods, Electrodes and Filler Metals
   Section III–Nuclear Power Plant Components
   Section IV–Heating Boilers
   Section V–Nondestructive Examination
   Section VI–Care and Operation of Heating Boilers
   Section VII–Care of Power Boilers
   Section VIII–Pressure Vessels
      Division 1–Rules for Construction of
      Division 2–Alternative Rules for Construction of
   Section IX–Welding Qualifications
   Section X–Fiberglass-reinforced Plastic Pressure Vessels
   Section XI–Inservice Inspection of Nuclear Reactor Coolant Systems
   Case Interpretations and Addenda
2. U. S. Navy
   MIL-STD-00248–Welding and Brazing Procedure and Performance Qualification
   MIL-STD-278–Fabrication Welding and Inspection; and Casting Inspection and Repair for Machinery, Piping and Pressure Vessels in Ships of the United States Navy
   NAVSHIPS 250-1500-1–Welding Standard
3. Marine Engineering Regulations and Material Specifications; United States Coast Guard
4. ABS Rules for Building and Classing Steel Vessels; American Bureau of Shipping
5. Standards; Tubular Exchanger Manufacturers Association, Inc.
6. Lloyd's Rules and Regulations; Lloyd's Register of Shipping
7. CSA-STANDARD-B51–Code for the Construction and Inspection of Boilers and Pressure Vessels–Canadian Standards Association
8. Various State and City Codes or Regulations
9. Various Foreign Codes or Regulations

These documents detail permissible materials, sizes, shapes, design, service limitations, heat treatments, inspection and testing requirements and, in addition, contain requirements for the qualification of welding procedures, equipment and welders and welding operators.

The ASME Boiler and Pressure Vessel Code has been adopted in its entirety or in part with certain modifications for boilers and pressure vessels used for stationary service by most states and municipalities of the United States and provinces of Canada.

The extent to which the Code has been adopted by the various jurisdictions and the department or division having control is outlined in a guide "Synopsis of Boiler and Pressure Vessel Law, Rules and Regulations" obtainable from the Uniform Boiler and Pressure Vessel Laws Society, 57 Pratt Street, Hartford, Connecticut 06103.

The various codes and standards listed are subjected to modifications and revisions for the inclusion of new methods of construction, materials, interpretations and such additional modifications for safe construction and operation as dictated by service experience. When applying the rules for construction and welding of pressure vessels, the designer or fabricator should use the latest issue, case interpretation, addenda or amendment of the code, standard or regulation specified. If there is any question about which code or standard is to be observed, the local authorities at the installation site should be contacted for clarification.

METALS

The first welded boiler drums and pressure vessels were operated under conditions where plain carbon steels of moderate tensile strength could give satisfactory and economical construction. Drum plate metal was confined to conventional flange and firebox grades of 55,000 or 60,000 psi minimum tensile strength steel. This type of steel, for many years the outstanding metal for riveted construction, was suitable for arc and gas welding. This same general class of metal was used for other component parts of pressure vessels and auxiliary equipment such as headers, piping, boiler tubes and superheater and economizer tubular elements.

Welded pressure vessels of today are designed to operate under conditions which are much more varied and usually more stringent than in the past. Pressures up to 3,500 psi, temperatures from 1200 F down to –423 F, pressure and temperature cycling, environments containing hydrogen and corrosive agents and neutron irradiation have brought into common use a variety of alloys and filler metals, heavier sections and new welding processes. The diversity of service conditions and materials together with increasing sophistication of pressure vessel design require welding of matching complexity and suitability for service.

The following properties of metals are important in determining the most suitable and economical metal for a particular application:

1. Short time tensile strength at room and at maximum operating temperature.
2. Short time ductility at room and at maximum operating temperature.
3. Short time yield strength at room and at maximum operating temperature.
4. Resistance to deterioration by corrosion, corrosion fatigue, stress corrosion.
5. Notch toughness and impact resistance, particularly for low temperatures.
6. Creep strength (for service temperatures in plastic flow range).
7. Rupture strength and ductility (for service temperatures in plastic flow range).
8. Structural stability under intended service conditions.
9. Age-hardening and embrittlement under intended service conditions.
10. Weldability and need for special preheat and postweld heat-treatments.
11. Work-hardening characteristics.
12. Thermal conductivity, thermal expansion and thermal shock resistance.
13. Resistance to repeated applications of load (fatigue).

The metals generally employed in boilers and pressure vessels may be grouped as follows:

Plain carbon steels–rimmed, semikilled and killed
Low-alloy steels
Quenched and tempered steels
Chromium steels
Austenitic steels
Clad steels
Nonferrous metals

The most widely used pressure vessel metals are listed in Section II, Material Specifications, of the ASME Boiler and Pressure Vessel Code.

PLAIN CARBON STEELS

Specifications for rimmed or semikilled carbon steel plate were based on riveted and forge-welded construction. These steels contain approximately 0.15 to 0.30% carbon and, although the welding quality of these steels has been improved, the specifications themselves are very little different from what they were twenty years ago.

Nevertheless even today an appreciable proportion of plain carbon pressure vessels is fabricated from plain carbon steels–particularly in lighter thicknesses; occasionally, semikilled, or more rarely, rimmed steels have been used.

All carbon steel plates more than 2 in. thick and the higher strength 65,000 and 70,000 psi tensile strength grades are silicon-killed. Some specifications for carbon steel pipe permit the use of rimmed steel, although most seamless piping is made of killed steel. The ASME Code recognizes two grades of killed steel pipe, one in which the silicon content is less than 0.10% and the other in which it is 0.10% minimum. The ASME Code has also established allowable working stresses at elevated temperatures for each grade.

Steels of relatively low carbon and high manganese content made to fine grain practice are used for their good notch toughness, particularly in the normalized or quenched and tempered condition.

LOW-ALLOY STEELS

As compared with plain carbon steels, there are a number of low-alloy steels which possess one or more of the following properties:

1. Improved ambient temperature strength.
2. Improved high-temperature strength.
3. Greater resistance to impact at low temperatures.
4. Improved corrosion resistance.
5. Improved oxidation resistance and stability at high temperatures.
6. Improved depth of hardening in thick sections.

Chromium-manganese-silicon steel possesses the first of these characteristics and also resistance to atmospheric corrosion; nickel steel possesses the first and third; carbon-molybdenum steel possesses the second; 2% chromium-1/2% molybdenum, 2 1/4% chromium-1% molybdenum, 5% chromium-1/2% molybdenum steels and steels of similar analyses possess the second, fourth and fifth of these characteristics. Because of its high strength at elevated temperatures and its resistance to sulfur corrosion, 5% chromium-1/2% molybdenum steel is used mainly in the oil refining industry. Steels in this group require greater care in welding than plain carbon steels and may require thermal stress relief after welding.

In some pressure vessel steels, the primary function of the alloying elements is to increase response to heat treatment, so that the advantages in strength and toughness of tempered martensitic and bainitic microstructures may be realized—even in heavy sections.

QUENCHED AND TEMPERED STEELS

A quenching and tempering heat treatment is often used for plain carbon and low-alloy pressure vessel steels as a substitute for normalizing and as a means of maintaining notch toughness in heavy thicknesses.

In addition, there are many pressure vessel steels for which quenching and tempering form a part of the specification and are used either to enhance strength or notch toughness, or both. Familiar examples are Mn-Mo nuclear reactor vessel steel, 9% nickel steel and the various proprietary high-yield-strength low-alloy steels.

Quenching and tempering per se do not greatly affect weldability; however, the improvement in properties usually effected by this heat treatment does impose requirements on the welding procedure necessary to maintain the advantages in the joint. In particular, welding energy input and level of preheat should be balanced to avoid degradation in the heat-affected zone. (See ASME Code, Section VIII, Division 1, Part UHT.)

CHROMIUM STAINLESS STEELS

Both ferritic and martensitic chromium iron alloys are used for pressure vessel construction. In practice, both types will often be duplex, thus largely avoiding grain growth and brittleness problems associated with fully ferritic alloys, or the intense heat-affected zone hardening of fully martensitic alloys. (Refer to ASTM-ASME A(SA)240.)

AUSTENITIC STAINLESS STEELS

The most common of the austenitic steels are AISI Types 304, 304L, 316 and 316L. They are used for both pressure parts and internal linings and are suitable for very low-temperature service and maintaining good impact resistance below -300 F; they possess corrosion resistance and substantial high-temperature creep and rupture strength. In this group, the ASME Code contains material specifications for chromium-nickel steels which are modifications of AISI Types 304, 316, 321 and 347 and rules for their use in construction of pressure vessels. These steels are covered by ASME Material Specification SA-240. Design and fabrication rules are found in Sections I and VIII, Part UHA.

CLAD STEELS

Carbon or low-alloy steel plates with an integral cladding of stainless steel or a nonferrous metal are widely used for pressure vessels where corrosion is a design factor. ASME material specifications for clad plate in which the cladding thickness may be used in the calculation of the vessel wall thickness are:

SA-263: Straight Chromium Stainless Steel Clad on ASME Carbon and Low-Alloy Steels

SA-264: Chromium-Nickel Stainless Steel Clad on ASME Carbon and Low-Alloy Steels

SA-265: Nickel and Nickel Alloys Clad on ASME Carbon and Low-Alloy Steels

Cladding thickness is usually treated as corrosion allowance rather than as part of the pressure design thickness.

NONFERROUS MATERIALS

The ASME Code covers many nonferrous materials applicable to arc-welded pressure vessels: aluminum, aluminum alloys, nickel, nickel alloys, copper, copper alloys and titanium. Section 4 of the WELDING HANDBOOK contains details of weldability on the individual metals. Nonferrous plate metals specifications for use in pressure vessels are covered in ASME Material Specifications:

SB-209: Aluminum and Its Alloys
SB 162, 127, 168, 333, 334, 424: Nickel and Its Alloys
SB 11, 152, 402, 169, 96, 98: Copper and Its Alloys
SB 265: Titanium

ARC WELDING FILLER METALS AND FLUXES

The pressure vessel user, designer or builder must select electrodes, shielding fluxes or gases and welding procedures which will provide mechanical and chemical properties appropriate to the vessel service conditions. Generally, codes and regulatory bodies specify room temperature strength and ductility properties and low-temperature toughness requirements where applicable, but matching of elevated temperature properties and chemical analysis of weld metal to service conditions are left largely to the vessel owner. Manufacturers of welding filler metals usually have information on the properties and applications as well as the operating characteristics of their products to assist the vessel builder in the proper selection of these.

As new welding materials and processes come into wide usage, standards are written to categorize the weld deposits in terms of chemical and mechanical properties, as well as other characteristics. The American Welding Society and the American Society of Mechanical Engineers jointly issue comprehensive specifications for welding rods, electrodes and filler metals.

DESIGN

Application of modern welding techniques to pressure vessels and boilers has made possible improved design approaching uniform strength throughout each structure. Together with advances in mechanics and metallurgy, this has led to more economical use of metals. Overall economic considerations usually dictate design for complete rather than average safety. However, damages associated with even a small failure could offset material and fabrication savings. Accordingly, safety codes and designers proceed cautiously in considering safety factors. Proven experience or appropriate testing should determine each advance step.

Safety codes referred to throughout this chapter are the minimum required for safety. Unusual service requirements or particular hazards not covered should be evaluated carefully. Corrosion and other deteriorating circumstances can appreciably alter the characteristics of existing structures. Consequently, code rules are intended to apply to new construction only.

Provisions against deterioration and frequency of service inspection are matters of user and enforcement agency experience. Unfortunately, some forms of deterioration such as intergranular corrosion, stress corrosion and corrosion fatigue do not involve visibly discernible loss of weight or pitting. Where suspected, all possible means of detection should be employed.

Pressure vessel shapes include cylinders, spheres, ellipsoids, tori and cones in many combinations. In addition to axially symmetrical vessels, occasional use is made of flat sides or elliptical or obround shells, with or without stiffeners or staybolts. The considerations that affect shape selection include economics (material and fabrication), space limitations and service requirements. Conventional construction employs cylindrical shell courses with either ellipsoidal or two radii torispherical heads, commonly called dished or basket heads.

DESIGN LOADING CRITERIA

Pressure vessels are used to contain liquids, vapors or finely-divided solid materials, singly or in combination, under pressure and at widely variant temperatures. Protection against excess pressure is obtained by designing for the maximum pressure that can be generated in the system, or, by pressure relief equipment.

Operating pressures are usually 5 to 25% below the maximum pressure permitted. In code terminology the design pressure is the *maximum allowable working pressure*, and relief devices must reach full rated discharge capacity at 10% over this pressure (6% in the case of boilers). The maximum temperature the pressure shell will attain is designated the *maximum working temperature*. The maximum working pressure and temperature governing the design are required to be stamped on the vessel (or on an appurtenance in some cases).

Pressure and temperature, while principal factors, are not the only design criteria. Other loadings must be considered even though some codes contain no specific rules for handling them. Such loadings include the following:

Dynamic shock; fluctuating and repeated pressure loading

Thermal stresses; heat transfer; differential expansion; fluctuating and repeated temperature; thermal shock; start-up and shutdown effects

Superimposed loadings; piping reactions; reactions of platforms and attached equipment

Gradients such as varying load of contents with depth, or varying temperature in a fractionating tower

Wind and earthquake loading

Dead loads such as vessel and contents

Effects of supports such as lugs, saddles and skirts.

Since all loading does not occur simultaneously, parallel design conditions must sometimes be considered (erection, start-up, normal operation, shutdown, emergency or operation above design conditions). For example, loading under erection is usually taken at stripped weight (without insulation, piping, equipment, etc.) with full effects of wind and earthquake; loading start-up usually assumes vessel is cold but under pressure, pipe connecting lines are hot and assumes full effects of wind and earthquake.

SAFE LIMITS FOR LOADING

The safety of pressure equipment involves evaluation of service, materials, design, fabrication and inspection. Since these factors individually are not susceptible to complete quantitative assessment, the only presently reliable measure of safety is experience.

Many factors of safety derive quantitative significance only in relation to simplified tensile tests and do not extend beyond single applications of load. At elevated temperatures, where creep rate is appreciable, they are meaningless. With the variety of design and fabrication details permitted, the true safety factor is dependent on the magnitude of localized stress around openings, intersections, undercuts or at defects in welds, etc., in combination with the resistance of the material to sustained shock or cycle loading.

BASIS FOR ESTABLISHING STRESS VALUES FOR FERROUS METALS

The ASME Code specifies maximum allowable stresses in each metal based upon tensile strength, stress to produce rupture in a specified time and creep strength. The design criteria, based upon these properties, vary somewhat between Sections I, III and VIII of the Code (for a complete description of these criteria see Paragraph A-150 of Section I, Appendix II of Section III and Appendix P of Section VIII). Codes other than the ASME Code specify other design criteria which result in higher allowable stresses than those in the ASME and some result in lower allowable stresses. Figure 84.1 shows curves of the ASME Code Section I, allowable stresses together with typical tensile, creep and stress-rupture curves for a killed carbon steel.

DESIGN FORMULAS

Formulas based on the theory of elasticity are used for the entire temperature range; at high temperatures, however, plastic flow or creep proportional to stress tends to equalize stress distribution. The margin is in the direction of safety with higher temperatures. The simple stress theory is employed, ignoring interaction of strains along other axes, which is probably the best compromise since the strain energy, maximum strain or other criteria for predicting yield are as yet insufficiently explored for correlation with high-temperature behavior.

For shells and spheres, approximations of Lamé Formulas are used in the Codes, and for cones a similar formula is used. For ellipsoidal or dished heads the Membrane Formula is used directly or with test and experience data, to arrive at factors to be used in simplified formulas. In this connection, except for a few pressed heads, all heads are two radii (torispherical) in shape with high stresses at the torus-sphere intersection, which are lowered as yield under

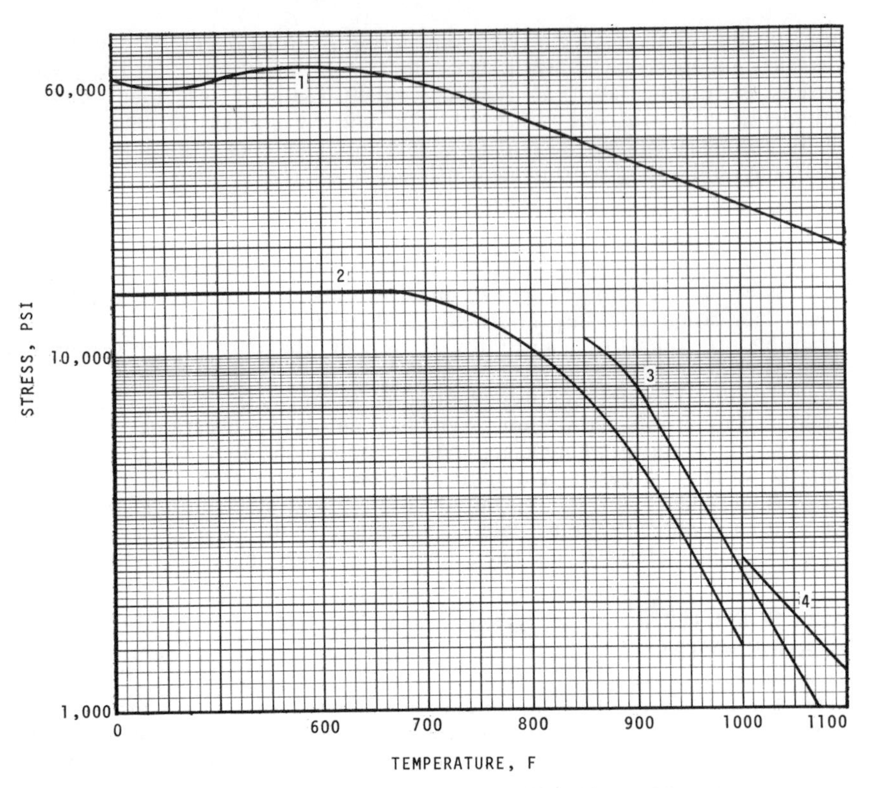

Fig. 84.1.–ASME allowable stress curves and typical tensile, creep and creep and stress rupture curves for a killed carbon steel.

hydrostatic test pressure will permit an ellipsoidal contour to be approached.

For flat covers, the thin plate formula is used with a different constant to take care of edge fixation and the particular details used; in the case of bolted covers with inside gaskets, there is an added factor for the edge moment due to the gasket reaction. Bolted flange rules (inside gasket facing) that evaluate the distribution of moment between the flange ring and the hub are given in organized form, which assumes that the bolt load is that required for a tight joint. It is increased by one-half the strength of any excess bolt area, as a protection against overtightening. For simplicity, this formula ignores direct and discontinuous pressure effects. It also provides for longitudinal bending stresses 50% above the nominal allowable design stress, provided the average of the flange and hub stresses is within the Code value. With the additional stress imposed by direct end pressure, the total-hub stress may reach twice the nominal design value. Localized stress conditions of this and higher orders, for example, are regularly encountered in vessels, in the knuckles of minimum Code heads or around openings. The averaging of flange and hub stresses is entirely justified since the combined structure will not begin to yield until both are overstressed.

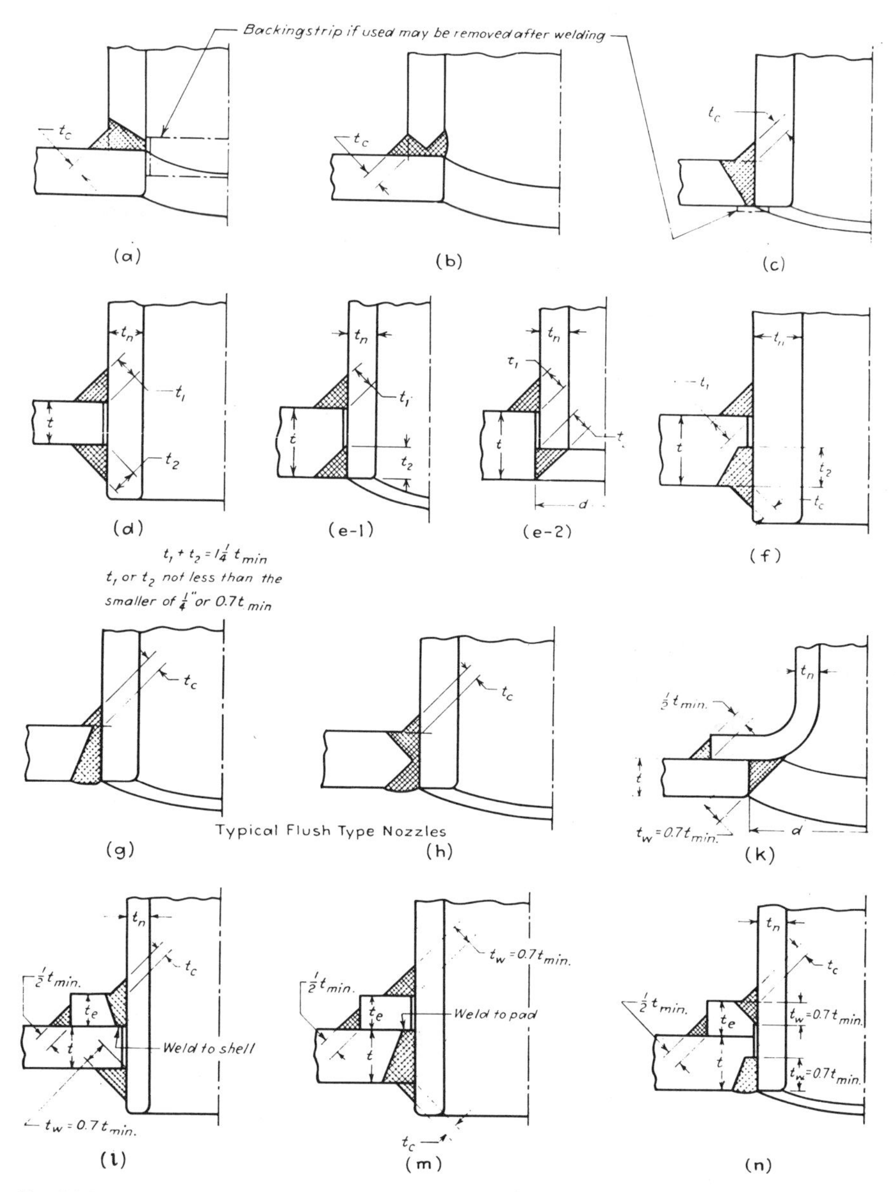

Fig. 84.2a.–Some acceptable types of welded nozzles and other connections to shells, drums and headers. From ASME Boiler and Pressure Vessel Code, Section VIII, Division 2.

Code rules permit specified small openings without checking for strength, on the basis that the coupling or nipple and weld will always provide sufficient strength for the weakened area to prevent excessive stress intensification. Elsewhere, reinforcements within a prescribed area must essentially replace the

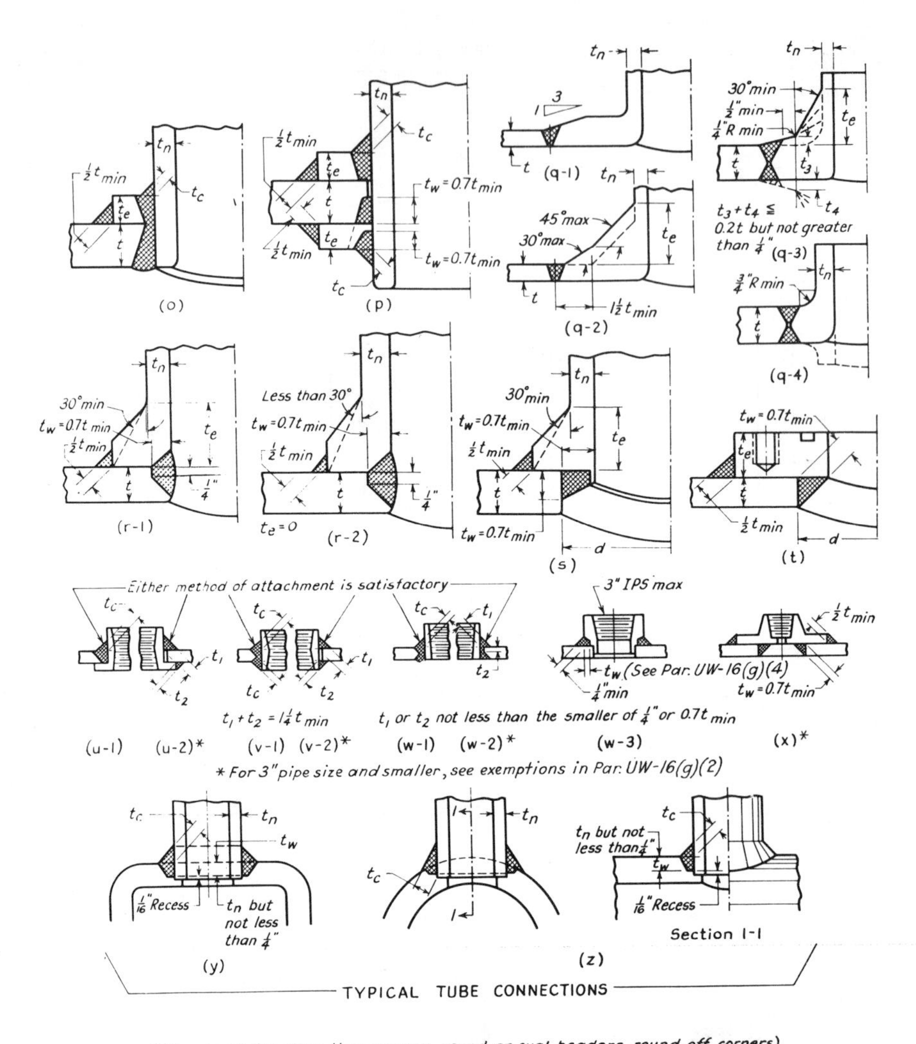

Fig. 84.2b.–Some acceptable types of welded nozzles and other connections to shells, drums and headers. From ASME Boiler and Pressure Vessel Code, Section VIII, Division 1.

metal removed, and welds must provide comparable strength. Analysis and tests are under consideration to determine the relation between location and effectiveness of reinforcements and attachment welds, and the amount required to prescribe stress intensification. More attention should be paid to normal and bending stresses imposed on nozzles by piping, and reinforcement details best suited to such loading; also to multiple of numerous openings where effects overlap, to openings large in proportion to the vessel diameter, and to openings located in areas of contour changes such as in the knuckle area of heads.

Some acceptable types of arc- and gas-welded nozzles and other connections to pressure vessels, reproduced from the ASME Boiler and Pressure Vessel Code, are shown in Fig. 84.2. Radiographic examination of most of these welds can be made with specialized techniques. The single-welded butt joint type shown in Fig. 84.3 for a forging inserted into a vessel gives the easiest detail for radiographic inspection, but these designs are more expensive than other methods of attachment. Where economy is required, they are not used.

All of the foregoing is concerned with principal direct or bending stresses that result directly from pressure, bolt loading or other external sources. The single exception is the external pressure vessel where secondary stresses result from the stiffener resisting inward movement of the shell. Similar axially symmetrical discontinuity is encountered in any vessel where members or sections are joined, or at intersections of shapes of revolution having unequal membrane displacement. Sharp intersections, such as between a shell and cone are an example. The unbalanced radial component of the longitudinal membrane stress in the cone sets up a ring loading at the intersection, causing localized direct and bending stresses that may be excessive. Nonsymmetrical bending stresses (in so far as the vessel axis is concerned) may appear around nozzles, due to the difference in longitudinal and circumferential stress, and also to direct or bending loads, around lugs and saddles, around ties or staybolts and, in general, at all localized loads.

There is need for a recognized limit for such stress in comparison with the so-called allowable stress–used in Code rules to govern principal stresses only. This limit would necessarily differentiate between those cases where yield produces stronger or weaker contours, between internal and external loading, and between axial stress bending effects. The limit would give due significance to location and extent of secondary stress.

WELD JOINTS

Butt joints welded from both sides, or from one side but made equivalent to a double welded butt joint, offer greatest assurance of uniform or complete fusion. They should be used where maximum strength is required or where shock, fatigue or plastic deformation such as creep may occur. Equally satisfactory is the single-groove weld made against a backing, which is subsequently removed and the root surface repaired if necessary. Where the backing is not removed, an unknown root condition exists that could initiate failure under repeated loading, particularly on longitudinal welds. However, such welds are permitted under some Code applications.

Efficiencies for the joints are given in the Code rules. Other types of joints are the single-welded butt joint without the use of backing strip, the double-fillet lap joint, single-fillet lap joints with plug welds and single-fillet lap joints without

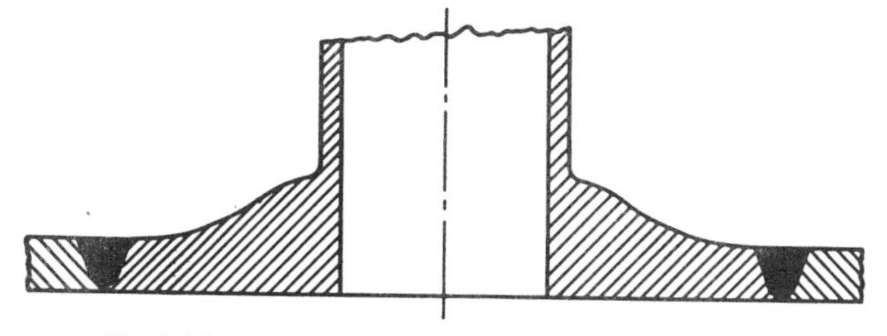

Fig. 84.3.–Forged nozzle welded into a pressure vessel.

plug welds. Their assigned joint efficiencies in proportion to butt joints welded from both sides can be justified only on the basis of static, non-fatigue comparison, since inadequate fusion and unknown root conditions are inherent to groove welds made from one side only, and to fillet welds. Vessels fabricated by such welding are usually for less critical service.

SUPPORTS

Vessels can be supported in many ways, but all introduce direct, bending or secondary stresses that should be maintained at a safe level.

Direct loading equally distributed around the circumference is most favorable. It is similar to skirt supports, which when fillet-welded to the shell introduce bending approximating a lap girth joint. When welded to the knuckle portion of the head, they introduce bending in proportion to load component normal to the surface at the point of attachment. Such bending stresses are at a minimum when a generous knuckle radius is used, and when the mean diameter of the vessel and skirt coincide.

Favorable design, using ring girders or band reinforcings to spread the concentrated support reactions, minimize their effect on the remainder of the vessel. Next favorable would be infinitely rigid legs or combined lugs and supports, which would take the total bending moment caused by the vessel weight without rotation. In practice, most legs or the beams of supporting structures are much less rigid than the vessel shell, which, therefore, carries essentially the entire bending effect.

Spreading the area of attachment and local reinforcement reduces induced stresses. For horizontal vessels, lugs and hangers or legs are sometimes used. More often, saddles, usually free but sometimes integral with the shell, are employed.

The bending moment from lugs and legs can be minimized by locating the centerline of the support to coincide with the mean radius of the shell at the point of tangency and by locating the supports at the head or other point of enhanced circumferential stiffness. Properly fitted saddles cause longitudinal bending along the curved edges, and circumferential bending at the ends. The former is controlled by saddle area and arc; the latter by saddle width and arc (principally the arc) so that a 120 deg minimum angle is usually necessary unless shell thickness in excess of pressure requirements is available.

When supporting vessels contain fluids at high temperatures, adherence to surfaces of revolution is effective in minimizing temperature differences introduced by heat loss at the supports. Due to the continuous metal structure, insulation is not sufficient to control temperature difference, because it equalizes conditions around the circumference. Accordingly, skirts are the preferred construction, with the advantage of permitting control of temperature at the steel or concrete supporting structure. Surfaces of revolution can absorb axial temperature gradients without stress except for end restraints, which are proportional to the gradient; flat plates are completely restrained with maximum stress. This accounts for distortion and cracking in lugs and their attachment welds, so that even local attachments are preferably made of pipe. For service at high temperatures, horizontal vessels should be avoided or, when necessary, rollers, hangers or similar constructions permitting free axial expansion should be used.

FABRICATION AND SHOP PRACTICES

PLATE QUALITY AND INSPECTION

Boiler and pressure vessel steel plates are routinely inspected at the mill during and after manufacture. The inspection and mill testing vary with the quality of plate specified. ASTM specifications (or for Code work, the equivalent ASME specification) are generally employed for purchasing tehse steels, and the plate qualities available are structural and pressure vessel.

Structural Quality

Structural quality steel is produced under normal manufacturing conditions and is applicable to various classes of structures specified by ASTM. ASTM Specification A6 covers the structural quality steels. An example of this group of steels is ASTM A283 which only requires one set of mechanical tests per heat of steel. Limitations on the chemical analysis of structural quality steels are specified for only three elements: phosphorus, sulfur and copper. Limitations on strength are provided that exert some control over the amount of carbon, manganese and other elements commonly found in plain carbon steel.

Pressure Vessel Quality

Prior to 1970, the terms "flange quality" and "firebox quality" were used to describe grades of steel of better quality than the structural quality in the preceding paragraph. These terms are no longer employed and the grades of steel previously described in this manner are now known as pressure vessel steels. This term includes steel specially manufactured for application in pressure vessels either exposed to or not exposed to fire or radiant heat. Special testing and marking is required for this steel as described in ASTM A20. An example of an A20 standard specification for carbon steel plates for pressure vessels for moderate and lower temperature service would be A516. Limitations on ladle analysis are prescribed for carbon, manganese, phosphorus, sulfur and silicon. One tension and one bend test is taken from opposite ends of each as-rolled plate. Plates furnished under this specification are to be free from injurious defects and are to have a workmanlike finish.

In addition to mill inspection of plates, it is advisable to carefully inspect the edges and surfaces of all plates in the shop before starting fabrication. If a plate defect is encountered at the later stages of fabrication the cost involved in discarding the plate can be appreciable. Sand blasting, shot blasting or pickling aid in surface inspection because they remove the covering mill scale which hides imperfections. It should be understood that unless the purchase order to the mill specifically states that the material is to be inspected after cleaning, the rejection of a plate due to surface imperfections can seldom be made.

Edge inspection during burning, machining and welding operations can reveal harmful laminations. It is desirable to inspect plate edges at an early stage of fabrication for the same reason surfaces are inspected.

In general, plate defects may be divided into two main categories: surface and internal. The surface defects can be further divided into ingot types (metallurgical) and mechanical types (Table 84.1). Besides showing surface flaws by visual inspection and presenting a good surface for coatings and layout markings, the

Table 84.1—Steel plate defects

	SURFACE	
INTERNAL	Metallurgical	Mechanical
Off Analysis	Blow Holes	Tears
Segregations	Seams	Causes
Porosity	Slivers	Low Rolling Temperature
Pipe	Alligator Hide	Coarse Dendritic Ingot
Center–Concentrated	Ingot Cracks	Nonuniform Heat
Center–Scattered	Snakes	Hot Shortness
Flakes	Steeples	Burned Steel
Nonmetallics	Breaks	Clamp Marks
Foreign Inclusions	Scabs	Nipped Corners
Laminations	Bootleg (Double Skin)	Washed Ingot
	Blurbs	Bottom Scabs
	Folds	Rolled In Scale
	Hanger Cracks	Rough Surface
	Cold Shuts	Laps
		Under-Fills and Over-Fills
		Split Ends
		Collar Marks
		Guide Marks
		Scarting Laps
		Fish Tail

removal of mill scale also presents a suitable surface in special cases where ultrasonic inspection is employed. This method of inspecting for subsurface imperfections is increasingly used for quality evaluation of plate material.

The removal of mill scale obviously decreases the measured plate thickness. This must be considered in calculating design thickness along with thinning as a result of forming or other heat treatments causing scaling. The thickness at the mill is measured with the scale included.

PLATE LAYOUT AND FORMING PREPARATIONS

The mill is capable of supplying plate to a reasonable degree of rectangularity, but layout in the shop of vessel shell plates is generally required in the flat before edge preparation and forming. The longitudinal edges of shell plates, which after forming are circumferential edges, are customarily beveled in the flat, but greater accuracy in assembly of heavy plates may be obtained in most shops if planing or burning is performed after forming. Heavy plate ends also are more accurately prepared after forming, but can be prepared before forming where the shop experience and equipment can produce the desired tolerances. It is considered poor practice to cold form thick plates in which edge tears or notches from shearing or burning have not been removed. Where alloy steels or higher carbon steels are involved, it is also desirable to remove surfaces hardened by oxygen cutting before extensive cold forming. The cold forming technique involved (dishing, rolling or pressing) must also be considered when edges are prepared.

In shell plates requiring press forming, one layout technique involves centerlines laid out at the ends of the plate. From these centers, equally spaced registration points are laid out extending to the longitudinal edges of the plate. These registration points are numbered at each end so the markings correspond, thus facilitating lining up the plate in a press die. To facilitate forming accuracy, the layout department may furnish templates of wood or metal cut to the desired curvature.

FORMING STEEL PLATES

Two main techniques are employed for forming pressure vessel plates: bending in pinch rolls or offset rolls in unidirectional manner, and press forming with dies. These operations are preferably performed cold but hot forming is also common, particularly in press-forming operations involving thick plates or sharp bending angles.

Prior to roll-forming plates, it is usually necessary to set the plate ends to the desired curvature with a press brake. Offset rolls or pinch rolls are incapable of properly bending the extreme ends of a plate. By presetting the ends it is often practical to roll a ring section with sufficient exactness to allow weld beveling of the ring joint before rolling and tack welding the joint while it is in the roll. The weld is made after the shell section is transported to automatic welding equipment.

Forming double curvature sections or very thick plate sections, such as transition sections (knuckle plates), spheroidal head sections, etc. is performed primarily by large hydraulic presses. Explosive forming has been developed for certain special plate applications, but it is not competitive when hydraulic presses and dies are available. To avoid overstressing the outer fibers of the plate during cold bending, it is sometimes necessary to give the plate an intermediate stress-relief heat treatment. The allowable strain on outer fibers varies depending upon the inherent plate ductility. Although the ductility of a plate has been demonstrated by laboratory test samples to withstand a 180 degree cold bend, the limit to which a large plate can be cold formed without cracking is far less. A safe limit, depending upon the plate material, will normally range from 3 to 4% elongation in the outer surface. However, the maximum allowable elongation is not fixed and it may even reach 5% when the fabricator has a great deal of experience.

Edge conditioning is of prime importance in cold forming operations. Notches act as stress raisers to initiate cracks which in some cases, such as during cold weather, may propagate entirely across a plate. In shops that are not heated appreciably, it is possible that the plate temperature may be below the ductile-brittle transition temperature for the metal involved. In these conditions, it may be profitable to warm the plates in addition to removing edge notches and flame-hardened surfaces.

Hot pressing minimizes some of the problems associated with cold forming– such as edge conditioning and outer fiber stress build-up. It also serves to refine the grain size if the temperature is not excessive. However, the handling problems and required manhours for forming encourage the use of cold forming and stress relief when practical. In many instances, cold sizing after hot forming is required.

Presses are available that can cold form plates five to seven inches thick. Figure 84.4 illustrates a press forming operation on thick plate. Both cold forming and hot forming are performed with this press, the decision resting upon plate thickness, plate strength, press capacity and width involved. In hot forming operations, plates to approximately twelve inches in thickness may be pressed at shops designed for heavy wall vessel fabrication (Fig. 84.5).

Fig. 84.4.–Setting ends of a thick plate prior to forming a cylinder.

Fig. 84.5.–Press forming a thick shell ring.

Boiler drums and pressure vessels are designed to meet individual requirements of purchasers. There can therefore be little standardization in the pressing operation due to the variety of diameters, plate thicknesses and curvatures. Press dies for each diameter and plate thickness are expensive. Excessive die costs are avoided by pressing with relatively narrow dies across a ring plate width. A single die may be used for curvatures close to the original diameter by using shims. The curvature is checked by a template supplied by the layout department. For head plates of varying diameters, a stockpile of dies that can be modified and shimmed is generally acquired.

PREPARATION FOR WELDING

Edge preparations for welding differ because of plate thickness, welding process employed, position of welding and individual shop preferences. Whenever possible, edges are prepared in the flat on side planers, gantry machine cutting equipment or on other similar devices. Thin plates in particular are generally edged in the flat. Due to thinning and stretching when formed, the edges of thick plates may require postforming preparation to obtain the desired fit with adjoining sections. To avoid exceeding the specified diameter tolerances in thick ring sections, the weld grooves normally are prepared after forming. In boiler drums designed for a thick tube sheet and thinner wrapper sheet, this method is also recommended.

Some conventional types of welding grooves are presented in Fig. 84.6A. The single-U and Vee-type grooves are perhaps the most commonly used. The single-U has a 1/8 in. root face at the bottom, curving with a 1/4 to 5/16 in. radius to a surface wall, and sloping approximately 1/8 in. per in. to the outside surface of the plate. Steps in welding with a single-U groove are illustrated in Fig. 84.6B. The single-Vee groove bevel has straight sloping sides, a root opening and is generally backed with a bar for welding. It has the advantage of being easily made by oxygen-cutting techniques when machining equipment is not available. This is also true of the double-Vee groove which finds extensive use in field welding operations.

The single-U and single-Vee grooves lend themselves to automatic submerged arc welding from one side, as well as manual welding. For relatively small

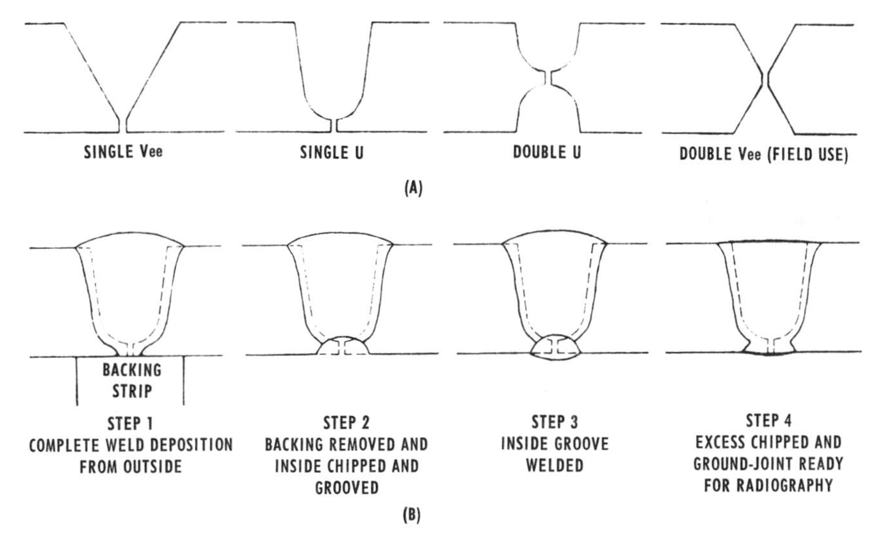

Fig. 84.6.–(A) Some conventional types of welding grooves. (B) Steps in welding a single-U groove.

diameter vessels, where preheat for welding is required, the best practice is to plan the groove so that welding is accomplished from the outside.

PLATE ASSEMBLY TECHNIQUES

In the assembly of cylindrical vessel or boiler courses, the mating edges are brought together and held together with lugs and bars and are frequently tack welded on the weld root side. The abutting beveled edges may be either touching or gapped to a predetermined distance for improved weld metal penetration. In heavy wall construction, tack welding is insufficient and key plates or bars are used to hold the cylinders in position for handling and welding. According to Section VIII, Par. UW-31(c) of the ASME Code (1971 Edition), tack welds may be employed in pressure vessels for fit-up, subject to the restrictions stated in respect to removal, preparation, examination and other requirements.

The stress due to weld contraction along and across a welded joint may be of yield point magnitude. The pull thus created tends to distort the plate locally, particularly in the circumferential direction on vertical welds. The normal effect of this weld contraction is to depress the cylinder at the joints so that the out-of-round tolerance allowed may be exceeded. In light plate the out-of-roundness can be corrected by rerolling or jacking before radiographic inspection.

In heavy plate fabrication, correction of departure from circularity is generally obtained by means of bevel selection, struts and jacks. A tolerance of 1% of the diameter may be allowed. It is highly desirable to try to prevent this distortion in heavy plate by pressing allowances during forming, by opening the chord distance of each element and by placing supporting struts at the cylinder during assembly and welding. Double-bevel joints instead of single bevel will also minimize distortion.

WELDING

METHODS

The common processes utilized in welding pressure vessels include manual shielded metal-arc, submerged arc, gas metal-arc, and gas tungsten-arc processes. Electroslag welding, electrogas welding and high heat input variations of the gas metal-arc and submerged arc processes are also used. Other processes used include pressure welding (particularly for butt welds in tubing and piping), resistance, friction, thermit, plasma-arc and electron beam welding.

Many of the welding processes include variations which may consist of separately added consumable filler metal, braided electrodes or cored electrodes with powdered flux or metallic alloy additions in the core. Gas welding also is used. In all the processes, the choice of filler metal and/or flux is predicated upon providing adequate mechanical strength and corrosion resistance for the particular service encountered.

For manual shielded metal-arc welding, depending upon the type of joint and fit-up, electrodes as large as 5/16 in. diameter may be employed. In cases where only a thin root face is provided or where a root opening exists (no backup bar), it may be necessary to utilize smaller electrodes at the start until a suitable thickness is developed. Subsequently, larger electrodes are employed as the thickness of the deposit increases.

Multilayer welds generally are built up in relatively thin layers, approximately 6 to 10 layers per inch of thickness. The beads may be deposited either in straight line progress (stringer passes), or the electrode may be oscillated from side to side to cover the full width of the welding groove (weaved). In either case the edges of the bead should fuse evenly into the walls of the joint or into the surface of the previously deposited weld metal.

Sharp depressions or heavy shoulders at the edges of the bead, caused by excessive convexity due to insufficient welding heat or ineffective side-wall fusion, are the primary cause of entrapped slag. These inclusions are difficult to melt out by the succeeding pass, and have to be removed before depositing subsequent passes. Improper side-wall fusion such as undercut and "rolled" beads are caused by improper distance between the electrode and the side wall of the joint.

The relatively thin layers or beads of weld metal permit a partial progressive grain refinement of preceding layers of ferritic weld metal and associated base metal HAZ. This minimizes the columnar structure characteristics of single beads of weld metal as deposited, and improves ductility and impact resistance of both weld and base metal HAZ.

Satsifactory welding results may be obtained with either direct (dc) or alternating current (ac). However, welding of magnetic materials with a-c transformers has a distinct operating advantage over d-c welding in that occasional troublesome arc blow is effectively minimized by ac.

In rigid types of structures–joints in heavy wall vessels or nozzles attached to thick shells or heads–it is advisable to weld continuously until the joint is finished, to avoid the tendency for cracks to occur in partially welded joints. When this is not possible, it may be necessary to eliminate stress raisers from backing rings or side wall undercut, or to utilize an intermediate stress-relief heat treatment to prevent cracking during or after cooling. It is usually better to maintain a constant and uniform preheat until welding is completed, than to perform heat treatment on a partially completed nozzle weld, unless the rate of heating is restricted.

Manual welding is also performed using the semiautomatic processes of submerged arc and gas metal-arc welding. The principles discussed above apply in general to these applications.

SUBMERGED ARC WELDING

The widespread use of submerged arc welding stems from its ability to produce satisfactory welds at high rates of deposition. In this method single or multiple electrodes (usually two) are automatically fed through a powdered mass of special flux covering the arc. In the case of single electrodes, the source of power may be either ac or dc. In the case of dual wires, combinations of the two types of current may be employed, or if preferred, both may be powered from a-c sources or both from d-c sources. Generally speaking, independent control of the welding conditions for each electrode is provided.

In the multiple wire system the electrodes are inclined slightly toward each other in the same plane so that both operate in the same pool of molten metal. With independent control of welding current and voltage for each electrode, a wide variety is available for controlling bead width and shape. In the case of single electrode submerged arc welding, an upper limit of travel speed and welding current exists beyond which poor bead shape, cracking or porosity result. In effect, multiple electrode submerged arc welding increases the limit of travel speed before improper bead shape takes over. A saving of weld time is therefore achieved.

One of the disadvantages of this process is the welding operator's inability to view the deposition of the filler metal. Some means for automatically guiding the electrodes may be acquired through mechanical or optical devices. Since inaccurate placement of the molten metal may easily result in a convex bead shape, with the resultant entrapment of highly refractory slag, it is virtually impossible to melt out the affected area by means of a subsequent pass. The removal of such a defective area becomes mandatory. Generally speaking, multipass weld joints, when executed satisfactorily, will result in slag formation that is self-peeling as it cools. This enables the welding operator to weld over extended periods of time without interruption.

In the case of high amperage welding, the maximum electrode usually utilized is 5/16 in. diameter. The limiting thickness of plate to which this type of welding can be adapted will be dictated by the alloy composition and the diameter of the vessel. As the wall thickness increases, the welding current should be increased, thus increasing the volume of molten metal. As this volume increases, it becomes more and more difficult to keep the metal from running downward and producing a razor-back type of surface. The razor-back or ridge, if formed, must be removed in order to comply with weld reinforcement requirements, as well as to provide an adequate surface for radiography. Therefore, as the diameter of the vessel decreases, there is a maximum thickness to which this type of welding can be applied. When this condition exists, multipass welding should be employed.

It should be pointed out that high amperage welding results in a very coarse columnar structure in the weld metal with an enlarged heat-affected zone in the base metal. This is important in connection with vessels that will operate at low temperature. The coarse columnar structure of the high amperage weld usually results in lower impact resistance than multipass welds. Therefore, particular consideration should be given to the welding procedure in terms of the service requirements involved.

In the case of low-alloy steels, the introduction of alloying elements may be made in one of two ways: (1) through the filler metal or (2) through the use of enriched fluxes containing alloying elements. In the first case, variations in welding conditions (such as current, voltage and travel speed) do not appreciably affect the composition of the deposited metal. In the second case, the final composition is dependent upon the volume of flux melted. Therefore, depending upon the welding condition employed, considerable variations in composition can be produced. For this reason, it is extremely important that once welding procedures have been established they are maintained throughout the fabrication of the vessel. Unless such precise control over welding conditions can be guaranteed, serious consideration should be given to the use of alloying elements introduced into the weld metal through the filler metal.

The process is also widely used for weld overlay cladding of pressure vessels to impart a corrosion-resistant sirface. Direct-arc or -arcs, series-arc with and without separate filler metal additions and other process variations are used.

GAS-SHIELDED ARC WELDING

Gas Metal-Arc Welding

This is a semiautomatic or automatic process in which bare filler metal as part of the electric circuit is deposited in a welding groove under an atmosphere of either inert gas(es) or carbon dioxide. It has been widely used both in the manufacture of nonferrous pressure vessels, and for ferrous pressure vessels such as carbon steel, low-alloy steel and stainless steel. It has an advantage over submerged arc welding in applications where a visible arc is desirable.

Electrodes in coils of many types, diameters, and chemical compositions are readily available. This, combined with innovations in specialized equipment and power sources, has resulted in greatly increased usage.

The process is also used in "vertical-up" welding, using high heat input similar to electroslag welding. It is particularly applicable to plates thinner than those used with electroslag welding. It is also used for weld overlay cladding of pressure vessels to impart a corrosion-resistant surface.

Other modes of gas metal-arc welding include the shorting-arc technique and the pulsing-arc technique. Both of these methods provide improved out-of-position metal deposition due to their reduced heat input. The shorting-arc mode is the lowest heat input method, and as a result has a tendency to produce cold laps. This is particularly true when it is not applied by a fully automatic method.

Gas Tungsten-Arc Welding

In this process, the heat source is provided by an arc between the base metal and a tungsten electrode. The filler metal that is added is not part of the electrical circuit. It has found wide application in the piping industry where full fusion welding, made from one side without the use of permanent backing rings, is required.

The process is widely used for girth butt welds in tubing where a smooth internal contour is desired. The first layer weld may be deposited by several different techniques, dependent upon factors that include metal combinations and possible fit-up tolerances. The tube ends may be abutted and the abutting ends simply fused; the tube ends may be abutted or separated slightly and filler metal separately added into the arc area; or the tube ends may be abutted against an intervening filler ring which is then completely fused.

ELECTROSLAG WELDING

This process is being used increasingly because of its high deposition rates and economic advantages. Its high heat input, however, can have an adverse affect on the base metal heat-affected zone properties. This may necessitate a postweld heat treatment of a specific type dependent upon the base metal. The process is particularly applicable to relatively thick materials. It is also applicable for large fillet welds and weld buildups.

GAS WELDING

Gas welding may be applied to the construction of pressure vessels of limited thicknesses. It is particularly suitable for the groove welding of small diameter pipe and tubing, and with skilled welders and proper joints, complete penetration may be obtained at the root without a backing ring.

It should be realized that since this process involves heating above the critical temperature for a prolonged period of time, the structure of deposited metal and the heat-affected zone is very coarse. This coarse structure results in severely lowered impact resistance, which should be considered where vessels are to be operated at low temperatures. In those cases, the adequacy of the joint must be proved.

METALLURGY

The basic principles involved in the successful welding of any pressure vessel steel include, but are not limited to, the following metallurgical aspects:

1. Base metal soundness and general cleanliness, chemical composition, including presence of possible deleterious trace elements and general chemical segregations; metallurgical structure including effects of previous working in imparting anisotropy; mechanical properties including resistance to brittle fracture, preweld and postweld heat treatments.
2. Weld metal chemical composition including hydrogen, general cleanliness, weld deposit metallurgical transformation structures and mechanical properties of these structures; and welding condition as it affects (a) weld deposit thicknesses, (b) heat input, (c) cooling rates of the weld deposit and (d) postweld heat treatments.
3. Heat-affected zone structure as affected by heat input in regulating width of the zone, variations of grain structure within the zones as influenced by varying cooling rates and resultant mechanical properties as additionally affected by possible postweld heat treatments.
4. Corrosion resistance of weld metal and heat-affected zone, as influenced by choice of filler metal, welding technique and postweld heat treatment.

PREHEATING

Heating before and during the welding of relatively heavy shell plate may prevent the formation of stress cracks and fissures in the weld. This is particularly important in welding the higher carbon high-tensile steels and the low-alloy steels, having a tendency toward air-quench hardening, which are now being used in shell construction. The preheat temperature required depends on the mass and rigidity of the joint, and the type of plate metal used. In general, preheat temperature need not exceed 300 F. The ASME Code, in its various sections, gives some mandatory and nonmandatory requirements for preheating.

HEADS–ASSEMLBY AND WELDING

In intermediate and heavy shells, the ends of the cylinders may be prepared for welding by cutting the welding groove in a boring mill or lathe, the same type of groove as is used for the shell. Oxygen-cutting, gouging and plasma-arc cutting are also used. To eliminate troublesome fitting-up and welding problems and for the sake of appearance, the circumferential travel (measurement) of heads and shell should be as close as good shop practice permits.

Heads are assembled to the shells with or without the backing strip or ring, depending upon welding process. Welding procedure for head joints is the same as for shell longitudinal joints except that for automatic welding–and for convenient manual welding–the shell is placed on power-driven turners so that it rotates under the stationary electrode. As in longitudinal joints, the inside, depending upon welding process, may be chipped and grooved to sound metal and welded.

NOZZLE WELDING

It is advisable to weld all required part attachments, such as nozzles and pads, to heads before the heads are assembled to the shell. At this stage, the heads may be conveniently handled and all welding may be done in approximately the flat position. It may be more convenient, however, to weld nozzles, pads and other external connections to the shell after the heads have been welded and after radiographic inspection, because these may interfere with both processes.

Nozzles and internal or external connections to pressure vessels may be welded by manual, semiautomatic or automatic methods. All welded connections are attached before postweld heat treatment of the vessels. Some rules or specifications about attaching nozzles require complete penetration in welds, thus making the nozzle an integral part of the structure. In some designs the complete penetration is through the nozzle wall. In other designs, the wall of the shell or head is Vee-grooved to permit complete penetration and, as in main seam welding, it is advisable to back-gouge the weld before completing the joint to remove defective material, or to remove undesirable notches or irregularities in the first metal deposited. In the event that a reinforcing ring around the nozzle is required, it also is welded the full thickness of the reinforcement, and the depth of the weld becomes the combined thickness of the vessel wall and reinforcing ring. The weld around the nozzle is finished off as a fillet weld on the outside of the vessel and also on the inside, if the nozzle has an inside projection. Nozzle reinforcing pads may be single rings or, in order to obtain a more uniform stress distribution for large openings and high pressures, may be built up of two or more rings, each ring above the first being smaller in area than the one below. Each ring, besides being welded to the nozzle neck, is fillet-welded at the periphery to the structure below it.

Other rules permit nozzles to be inserted in an opening with slight clearance, and attached by fillet welds of predetermined size, both on the inside and on the outside of the vessel wall. Here also, nozzle-opening reinforcing rings are used if stress calculations show them to be necessary. Inasmuch as all nozzles are part of the pressure unit, they require welding of the first quality; and the same care should be used as in welding the longitudinal and circumferential joints.

For some services it is mandatory to have nozzle welds radiographed. Under these circumstances it is desirable to have the reinforcement integral with the neck of the fitting. In this particular case the attachment groove weld between the shell and the fitting is sufficiently beyond the flange to permit radiographic inspection. Some fittings are not amenable to radiographic inspection since their attachment to the shell is such that the outside flange dictates an angular exposure to the radiation beam. Such exposure, together with lack of uniform thickness at the weld, results in distortion so that difficulty in film interpretation develops.

POSTWELD HEAT TREATMENT

Most boiler construction codes require that the residual-stress condition caused by the welding operation be minimized by subjecting the welded vessel to a heat treatment at elevated temperatures. Codes covering the construction of other classes of pressure vessels may or may not require such thermal treatment depending on the nature of the service, the thickness of the vessel and the desired welded joint efficiency. The postweld heat treatment for carbon steels usually consists of heating the welded vessel at a uniform rate to a temperature of 1100 to 1200 F, holding at that temperature for a period of 1 hour per inch of thickness and then cooling uniformly. While some codes permit local heat treatment of girth and nozzle joints by heating circumferential bands around the entire part, the general practice on pressure vessels is to heat the vessel as a unit.

The usual postweld heat treatments do not produce any appreciable changes in the microstructure of either the plate or weld metal. Some slight spheroidization of the carbide constituent will occur in the plate metal which will result in a slight reduction in its yield point and tensile strength, and in a slight increase in its ductility. The average reduction in the tensile strength of carbon steels will be about 5%. Postweld heat treatment also results in softening and increased ductility of the weld metal and heat-affected zones.

Such changes in mechanical properties of both base and weld metals are only incidental. The purpose of the heat treatment, in addition to removing stresses, also includes other aspects such as removal of cold work, imparting of dimensional stability, and control of the properties of the weld metal, base metal and heat-affected zones.

The properties necessary to be controlled vary with the metals and intended service. In addition to tensile and yield strengths, they may include toughness for resistance to brittle fracture, improvement of corrosion resistance and metallurgical stability for high-temperature creep and stress rupture uses. During fabrication, weldments may require heat treatment to eliminate cracking. Some weldments may require thermal treatment before cooling from preheat temperature. This treatment, which may be at relatively low temperature (400 to 1000 F) for a short time, helps to eliminate cracking. Necessary for certain weldments, it is generally termed an "interstage" thermal treatment because it is applied to individual parts of an assembly or after partial welding of subassemblies.

The principles of thermal stress relief are based on the fact that the elastic properties of a steel are so materially reduced at the temperatures specified, that a plastic readjustment occurs at the stressed regions. Followed by uniform cooling, the residual stress in the finished structure can be no higher than the yield point of the steel at the postweld heat-treatment temperature.

In view of the marked softening and reduced strength at 1100 F to 1200 F (or higher), it is necessary to provide suitable support in the furnace to very long heavy vessels, or for vessels having a large diameter-plate thickness ratio. Sagging and distortion through plastic flow induced by the weight of the vessel can be prevented by properly spaced saddle supports, which provide a large bearing surface, and by internal struts across the diameter.

So called "stress relieving" by vibration at the resonant frequency of the component being treated may be effective in stabilizing residual stresses, but it is not approved as a substitute for the thermal treatments specified by the ASME Code.

ECONOMIZERS, SUPERHEATERS AND HEAT EXCHANGERS

These units generally involve parallel tubular elements extending between two or more headers or drums. The arrangement will vary for individual jobs.

Figure 84.7 is a view looking upwardly into the furnace of a utility fossil-fueled boiler. The walls of the furnace are enclosed by many miles of waterwall tubes. These walls are joined to each other by longitudinal welds so as to form a pressure containment. A cross section of a portion of this wall is shown in Fig. 84.8.

At the top of Fig. 84.7 can be seen superheater loops hanging downwardly

Fig. 84.7.–Upward view internally into the furnace of a fossil-fueled boiler showing waterwall tubes.

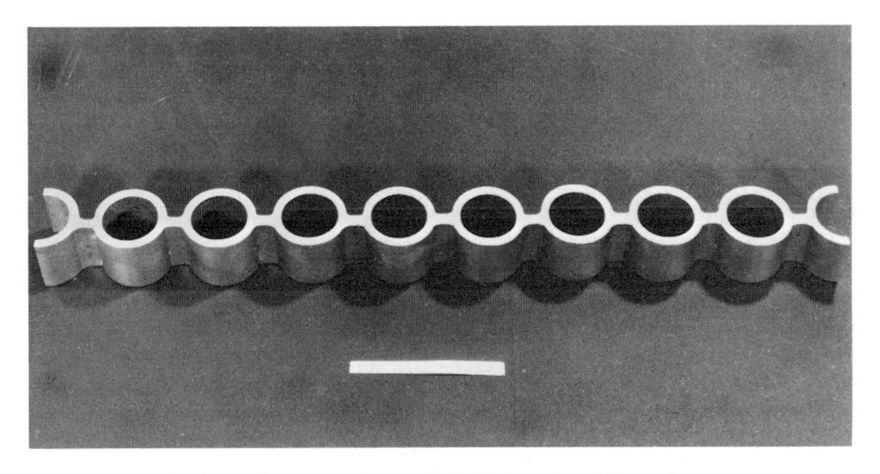

Fig. 84.8.–Wall cross section typical of the waterwall type furnace.

into the furnace enclosure. Also visible is one of the burner panels through which coal, oil, gas or other fossil fuel reaches the furnace enclosure.

Figure 84.9 shows a typical boiler "doghouse" (located above the furnace roof) of the type also shown schematically at the top left of Fig. 84.10. The welds in tubing are mostly girth butt welds made in shops by automatic gas metal-arc, induction, pressure, flash, or, in the shop and the field, by gas tungsten-arc for the first pass and shielded metal-arc welding for succeeding passes.

Economizers, as shown in Fig. 84.11, are placed toward the low-temperature end of the passage of the heating gases, and are used for heating the feed water before entering the steam-generating portion of the boiler unit. Since the temperatures attained in service of economizer parts are relatively low, plain

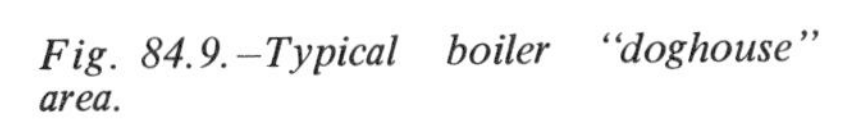

Fig. 84.9.–Typical boiler "doghouse" area.

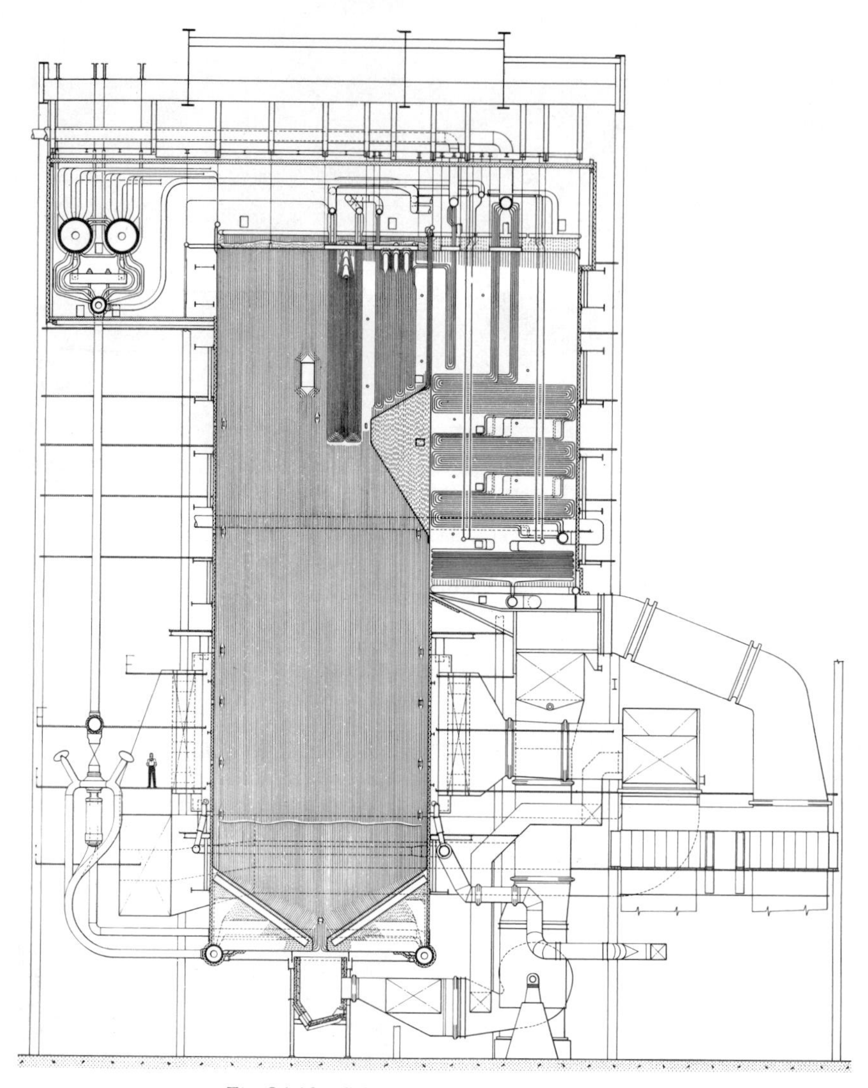

Fig. 84.10.–Schematic of a fossil-fueled boiler.

carbon steel is usually sufficient to meet the service temperature condition. *Superheater* units are usually installed toward the high-temperature end of the heating gas flow. As their name denotes, superheaters raise the temperature of the steam delivered from the primary steam-generating portion of the boiler unit. Superheater parts often require the superior oxidation resistance and creep resistance of alloy steels. This is also true of *heat exchangers*. Their function is the transfer of heat from one phase of the operation to another, and conditions may be such that compositions for high-temperature service are required. Steels containing chromium, silicon, aluminum, or combinations of these elements are preferred for increased oxidation resistance; molybdenum or tungsten additions for increased creep resistance. The quantity of these elements present will vary,

Fig. 84.11.–Internal view of an economizer as applied to an industrial boiler.

depending upon the maximum anticipated temperature at various portions of the superheater or heat exchanger.

Economizer and superheater headers are generally fabricated of tubular pipe stock or forged seamless drums, but may also be fabricated of rolled and welded cylinders. The same is true of heat exchanger shells. Welding is involved in all cases, either in the closure of the ends or in the attachment of nozzles to shells or heads. In the case of very long headers, the required length may be obtained by butt welding together shorter lengths.

The tubular elements required for superheater, economizer and heat exchanger units may be longer than the mill lengths normally obtainable. In such case, the required length is obtained by butt welding mill lengths of tubing.

INSPECTION

Inspection, as applied to the fabrication of pressure vessels, is fundamentally to ensure that the customer receives a quality product built to his requirements, that meets minimum safety requirements. From the fabricator's viewpoint, a well-coordinated program of inspection can reduce costly rework.

The most common methods of examination used in pressure vessel fabrication are visual, radiographic, magnetic particle, fluorescent and liquid penetrants, ultrasonic, sectioning and etching. The choice of method and extent of its use for inspection depends primarily on the intended service of the vessel, Code requirements, and the fabricator's own experience in determining risks involved with undetected flaws.

It is common practice in the fabrication of pressure vessels to complete radiographic and other nondestructive quality examinations before a vessel is given the final postweld heat treatment and pressure tests. This approach is favored for economic reasons, since it costs less to repair a defective weld before postweld heat treatment or hydrostatic testing. For the majority of metals, these do not introduce new defects or unduly propagate existing ones. For the highest degree of assurance, some purchasers will require radiographic examination after final postweld heat treatment.

VISUAL EXAMINATION

Visual examination is probably the most widely used of all examination methods because of its simplicity, low cost and minimum equipment. To be effective, visual examination should be made during all stages of fabrication, particularly for joint edge preparation, surface cleanliness and welding fitup. Welding procedures and welder qualifications should be checked, and the welding operation carefully observed for compliance with the welding procedure.

During welding it may be necessary to check significant variables of the welding procedure: preheat, amperage, volts, rate of welding progression, etc. Root passes contribute to final weld soundness. Cracks in root passes, if undetected, may propagate and extend to subsequent layers. Cracks, incomplete fusion and porosity frequently can be detected during welding and remedied while they are still accessible.

The finished weld is nearly always first visually examined for appearance, contour, and surface flaws such as cracks, undercut, overlap, porosity and other defects. Correct interpretation and evaluation of results are essential in visual inspection. The inspector must have a sound knowledge of the welding process involved, the Code requirements, the intended service requirements of the weldments, together with the experience and judgment needed to evaluate visually the quality of a weld.

Obviously more than visual examination is required to assure soundness of welds, particularly in pressure vessel fabrication; but if properly applied, visual examination can increase the probability of the joint being satisfactory and efficient in service.

RADIOGRAPHIC EXAMINATION

Radiographic examination uses radiation from X-ray or gamma-ray sources to detect internal defects in welds. By means of photographic films (and to some extent by fluoroscopic techniques), an accurate image of the defect can be obtained permitting determination of its size, shape and location within the weld. While radiography still remains by far the most effective method for determining weld soundness, it has certain limitations. It is not sufficiently sensitive for detecting microcracks in weld deposits. Also, tightly compressed separation in any orientation or cracks lying in a plane perpendicular to the rays may escape detection. Therefore, great skill and experience is required in the interpretation of radiographic data and their relation to the applicable specification.

The degree of sensitivity as required by the ASME Code is that which will reveal a penetrameter having a thickness ranging from 1% to 2% of the plate thickness being examined. This penetrameter is placed on the side of the weld opposite the film. The penetrameter consists of a thin strip of metal equal in density to the weld metal, and containing three small holes for comparing image sharpness and contrast of the film exposure. The greater the sharpness and contrast of the image, the greater the probability of crack detection.

X-ray machines up to 25 million volts are available and permit examining welds to thicknesses of 20 in. in steel; however, 200,000 to 400,000 volt machines are the most common. Gamma rays are obtained from the radiation of radium and several commercially available radioisotopes: cobalt 60, iridium 192 and cesium 137. It is difficult to compare the merits of X-rays and gamma-rays, since both have optimum fields of application. Sources of gamma-rays have the advantage of being portable and of being capable of greater accessibility for certain shapes and locations than standard X-ray equipment. For general pressure vessel applications, particularly in thicknesses up to 3 in., X-rays are preferred because of sharper definition, greater contrast, quicker examination and lower costs.

The most common types of defects found by radiography are cracks, inadequate joint penetration or incomplete fusion, slag and porosity. For pressure vessels, the ASME Code governs the requirements of radiographic examination and outlines minimum acceptance standards for welds in vessels thus inspected.

Radiographic examination of full penetration welded joints of pressure vessels is applied in accordance with the service requirements and the joint efficiency desired. Highest joint efficiency is permitted if both postweld heat treatment and radiographic examination are used. As an aid to the interpretation of radiographs, it is necessary that reinforcement on both surfaces of welds be smoothed to remove weld bead irregularities which may interfere with proper interpretation.

MAGNETIC PARTICLE EXAMINATION

This method is normally used for detecting surface and relatively shallow subsurface defects in magnetic materials only. Minute surface cracks in welds which may not be detected by radiography can be detected by this method. Subsurface defects may also be located, depending on their orientation, size and shape, and distance below the surface.

The principle of this method involves the action of a magnetic field which, when induced in the metal, will be disturbed by any discontinuity present. An application of a fine dust of magnetic particles will be attracted at the discontinuity and outline the pattern on the surface. A study of the various patterns is required to enable proper identification of the defects. Various defects such as cracks, seams, laminations, inclusions, segregations, porosity and incomplete fusion can be detected in welds and base metals. For magnetization, dc is generally preferred for use in detecting subsurface defects in pressure vessel applications. The dry magnetic powder method is considered superior to the wet method for subsurface flaws and for rough surfaces.

In pressure vessel applications, the magnetic particle method is often used for branch connection, manway, tube stub and nozzle welds which cannot be radiographed; for checking the root of a weld before applying subsequent layers; or after backchipping or checking repair cavities for removal of all defects before rewelding. It is also used to detect laminations or other defects in the beveled edge of weld joints in heavier plate thicknesses. For nuclear vessels, the ASME Code Section III outlines the requirements governing magnetic particle examination and acceptance standards.

Certain limitations exist with this method since irrelevant and false indications are often produced. Abrupt changes in physical contour such as sharp corners or edges, changes in magnetic properties of structures such as a heat-affected zone, and fusion zones in dissimilar metal (magnetic to nonmagnetic) welds will often give the same indication as flaws. Therefore, properly developed techniques and skillful interpretation are essential to an effective application of this method of examination.

LIQUID PENETRANT EXAMINATION

Penetrants–either dye or fluorescent types–can be used on any material for detection of minute surface defects not usually visible to the naked eye. Due to its low surface tension, the penetrant is drawn readily into extremely small surface openings by capillary action. Prior to applying the penetrant, the surface to be examined must be cleaned of foreign materials, especially grease and oil, and free of excessive roughness. After application of the penetrant and removal of excess, the surface is examined under ultraviolet light or by the application of a developer, producing a color contrast, making the indication of the defect visible.

Penetrant examination is particularly suited for detection of microcracks, porosity, seams or laps and cold shuts in nonmagnetic weldments and metals. For magnetic metals, the magnetic particle method of examination is usually preferred. Typical applications on pressure vessels include weld metal overlays, applied liners for leaks and tube-to-tubesheet welds. Special skills in the techniques involved and sound judgment and experience for interpretation of results are required with this examination method. For nuclear vessels, the ASME Code Section III stipulates the requirements governing liquid penetrant examination procedures and acceptance standards.

ULTRASONIC EXAMINATION

Although comparatively recent in the pressure vessel industry, ultrasonic examination is becoming widely used because of its rapid and accurate method of detecting surface and subsurface defects. Techniques have been devised for detecting cracks, porosity and incomplete fusion in weldments as well as laminations, inclusion and segregation in base metals.

The examination is accomplished by directing a beam of high frequency sound waves into the metal being inspected by means of a transducer–usually manually guided–which scans the surface of the metal. That portion of the signal which is reflected back by any discontinuity or flaw is displayed on an oscilloscope screen where the type, location and size of the defect can be interpreted. However, considerable skill and experience are mandatory for proper calibration of equipment and interpretation of results.

Various commercially available equipment will produce a range of ultrasonic frequencies from 200 kHz to 25 mHz with transducers designed to produce shear (angle) waves, or longitudinal (straight) waves. Shear waves are normally used for examination of welded joints in plate or piping; the longitudinal waves are used for detecting laminations or other defects on edges of weld joints before welding. Low frequencies are considered best for detecting deep defects in heavy sections.

Ultrasonic examination can be adjusted to be more sensitive for flaw detection than the other common methods of examination: it can detect flaws 0.5% of the plate thickness in the plates from 1/4 in. to about 12 in. This high sensitivity is not a disadvantage because the ASME Code allows equipment operation at an adjustable sensitivity. In pressure vessel fabrication, ultrasonic examination is commonly used for flaw detection in tubesheet forgings, weld metal overlays, plate and weld joint preparations, electroslag welded seams (ASME Code Case requirement) and specially critical welds. In nuclear boilers, ultrasonic examination is widely used for comparisons of weldments prior to service and after periods of service.

OTHER METHODS OF INSPECTION

Chemical Analysis

Samples for chemical analysis may be taken to verify whether the base metal and weld metal meet codes and specifications for composition requirements. Acceptability of weld metal overlays is usually checked by chemical analysis. This method is also helpful in determining the cause of certain welding difficulties such as cracking, excessive porosity (due to sulfur) or segregation of harmful elements in the base metal. Samples for chemical analysis must be representative of the area being checked.

A simple chemical spot check may also be used for identifying types of base metals. Usually a few drops of chemical solution applied to the properly prepared surface will result in a color distinctly characteristic for a particular metal.

Metallography

Metallographic tests are often required and frequently used for examination of weld surfaces to detect weld outlines and location of defects, or for examination of weld cross sections to determine structure and geometry. Samples are usually prepared for examination by grinding or polishing (if greater detail is desired) and then etched with a corrosive chemical solution. The etched specimen can then be examined visually with the naked eye or a low magnification with magnifying glass or binocular-type microscopes (20X).

Sampling by Sectioning

Removal of samples by sectioning is an accepted method of weld inspection approved by various pressure vessel codes. While the cavity must later be repaired by rewelding, the method is nevertheless considered a nondestructive test. Samples may be taken for chemical analysis, etch tests, subsize tension tests or impact tests. Method of removal of samples will depend on the size of specimen desired; however, a hole saw, trepanning tool, cold chisel or boat-cutter–a special power tool using a hemispherical saw–are common tools used for sample removal. As with chemical analysis, the samples must be carefully selected to be representative of the weld or base metal being checked.

TESTING

The practice of subjecting a completed vessel to a test which will produce stress somewhat above the design maximum is a long established safety measure. The principal objective aside from the detection of leaks is to give some assurance of the absence of serious defects in workmanship, materials and design. The test does not guarantee that the vessel will withstand subsequent applications of the same pressure or that periodic applications of the lower service pressure might not produce failure. For reasonable assurance of safe construction, other inspection for the detection of flaws is essential.

HYDROSTATIC TESTING

Practically all vessels are given their final test by means of hydrostatic internal pressure. The test for riveted vessels has long been standardized at 50% greater than the maximum allowable design pressure. When welded vessels were first sanctioned by the ASME Boiler Code, the test pressure was established at two times the maximum allowable working pressure (equivalent cold pressure in the case of a high-temperature vessel); this higher value was selected to increase the probability of bringing out defects and to demonstrate the safety of welded construction. The present ASME Code with few exceptions (Code Paragraph UG-99) requires that "vessels designed for internal pressure shall be subjected to a hydrostatic test pressure which at every point in the vessel is at least equal to one and one half times the maximum design pressure to be marked on the vessel multiplied by the lowest ratio (for the materials of which the vessel is constructed) of the stress value S for the test temperature on the vessel to the stress value S for the design temperature. All loadings that may exist during this test shall be given consideration."

Hydrostatic tests of pressure vessels are normally performed after the vessel has been completely radiographed and postweld heat-treated. The pressure to which the vessels are subjected in this test will be determined by the governing Code.

PNEUMATIC TESTING

Vessels which are not designed for liquid contents sometimes cannot support a hydrostatic test load and are therefore given a pneumatic test. In recognition of the increased hazard involved in subjecting a vessel to over-pressure in this manner, the codes permit testing at a lower pressure. The ASME Code requires a test of 1 1/4 times design pressure multiplied by the lowest ratio (for the materials of which the vessel is constructed) of the stress value S for the test temperature of the vessel to the stress value S for the design temperature. In no case should the pressure exceed 1 1/4 times the maximum allowable working pressure at the test temperature.

In some cases it is desirable to test vessels when partly filled with liquids. For such vessels a combined hydrostatic and pneumatic test may be used as an alternative to the pneumatic test; the liquid level must be set so that the maximum stress, including the stress produced by pneumatic pressure at any point in the vessel or in the support attachments, does not exceed 1 1/2 times the allowable stress value of the material multiplied by the applicable joint efficiency. After setting the liquid level to meet this condition, the pneumatic pressure is applied until equal to 1 1/4 times the design pressure multiplied by the ratio of test temperature allowable stress to the design temperature allowable stress.

In view of the measurably greater energy stored, pneumatic tests are potentially hazardous and should be used only where hydrostatic testing is not practicable. In applying this type of test, all persons should be withdrawn from the vicinity while the pressure is being increased. Pressure should be raised in easy stages with careful inspection at each stage. Practice permits approaching the vessel for inspection while under pressure if the pressure has remained constant for some minutes, or, preferably, has been reduced somewhat below the last maximum applied.

Under no circumstances should a vessel under air or gas test pressures be subjected to shock loading.

PROOF TESTING

Vessels or vessel portions, the strength of which cannot be accurately calculated, must pass suitable proof tests. The ASME Code requires that the maximum allowable working pressure be determined by yielding or burst tests. The allowable working pressure is established at a suitable proportion of the test value. (Refer to ASME Code Section VIII Par. UG-101.)

ROUTINE TESTING

The ASME Code does not cover testing in service, leaving such matters to users and enforcement authorities. Periodic testing at not less than 1 1/2 times the maximum working pressure is the generally accepted practice, particularly if the metal suffers deterioration in service due to corrosion or other reasons.

MECHANICAL TESTING

Welding procedure qualification test plates, as well as vessel test plates, when required by the various codes usually must be welded in accordance with the same rules specified for fabrication of the vessel, i.e., thermal treatment and radiographic or other examinations must be performed when required.

From these test plates the various specimens for tension, bend, or impact tests are removed; these must then exhibit certain minimum test values which are based on the properties of the metal used in fabrication, and in conformance with the applicable Code. These mechanical tests are usually witnessed by an authorized Code inspector to ensure full compliance of the test weld properties with minimum Code requirements.

SPECIAL VESSELS

FIELD-ASSEMBLED VESSELS

It is necessary to field-assemble vessels that are too large or heavy for complete shop fabrication and shipping conditions. Field practices often deviate from shop practices due to special conditions encountered in field work. The conditions handicapping field operations are as follows:

1. Welding must be performed in difficult positions and at elevated locations. This limits the use of automatic equipment, and in some instances welding efficiency.
2. Weather conditions present special problems to schedules, procedures and quality.
3. Local workmen must often be employed. These men are generally less effective than those in a standing organization accustomed to working as a team, even though their welding ability may be equal or superior.

4. Forming equipment to correct improperly fitting parts is not usually available.
5. Field work requires special versatile equipment, careful planning of material requirements and delivery coordination. There is no stockroom to draw upon for parts.

For these reasons a maximum amount of forming and fabrication is performed in well-equipped shops. Overhead, vertical and other out-of-position manual welding is minimized in shops through the use of welding positioners. In many instances, field crews may also be equipped with automatic horizontal, flat and vertical welding equipment. This allows the use of high deposition rate automatic welding methods to a maximum extent.

Shop-formed components and subassemblies often require prefitting and piece marking in the shop for ease of reassembly in the field. The lack of good forming facilities in the field makes it very difficult to correct poor fit-up, dimensional errors, etc.; hence, careful inspection and control of shop forming and prefitting are of great importance. However, complete trial fitting at the shop is unnecessary and wasteful if the vessel is simple in design or of a standard type. Complete trial fitting and match marking are advisable to assure proper field assembly of an unusually complicated vessel or a vessel type unfamiliar to the fabricator. An example of such an instance is illustrated in Fig. 84.12. This trifurcation for a hydropower plant was completely prefitted and match-marked in the shop prior to disassembly and shipment.

Problems of weather and obtaining skilled workmen require utmost vigilance on the part of the contractor and inspectors if satisfactory field welding is to be obtained. Welding on wet surfaces should be avoided (see API Specifications, Section 650). This is particularly true for higher strength steels which require low-hydrogen conditions to avoid loss of weld ductility and the occurrence of underbead cracking. Satisfactory platforms should be provided and shelters or wind guards may also be required.

Under field conditions improper electrode care is often the source of welding quality problems. Careful packaging, storage and handling of welding filler metals are necessary to prevent excessive moisture absorption. Portable electrode baking and holding ovens are commonly used when the higher strength steels are field-welded. The presence of hydrogen in detrimental amounts does not always evidence itself as a crack. It can severely reduce weld metal ductility. In tensile

Fig. 84.12.–Trial fitting a steel "trifurcation" before movement to a hydroelectric power project. The structure was field-welded and attached to the end of a steel penstock to conduct water in three directions to generators. After field welding, this entire structure was postweld heat-treated (stress-relieved) in the field at 1100 F. A furnace was erected around the structure to perform this heat treatment.

tests its presence is made apparent by the appearance of "fish eyes" or "flakes" in the fracture texture, in addition to the loss of elongation produced.

Edge preparations for field welding differ mainly in the method of application and the position demanded. Plates for main joints generally are "scarfed" to single-Vee or double-Vee grooves. Occasionally U-grooves or combination grooves are also used. Most fabricators prefer double-Vee grooves for the main joints of plates over 3/4 in. thick, which allows balanced welding and better control of distortion. For circumferential joints in vertical vessels, a single-bevel or double-bevel groove with a 45 deg bevel angle is sometimes used for examination (manual) welds.

The use of automatic field girth welding equipment leads to combination bevel angles. The increased use of radiography for examination led to the combination bevel technique illustrated in Fig. 84.13. This gap design is for plates over 3/4 in. thick. The code requirement of 2/3 fusion (minimum) can be met by the technique in Fig. 84.13A, but for better quality joints it is best to upgrade the bevel design to produce 100% fusion welds (Fig. 84.13B). This method of obtaining 100% fusion is preferable to opening the bevel angle.

Preheating is employed with increasing frequency as field operations encompass the more hardenable low-alloy steels, and thicker plates of low-carbon and low-alloy steels. Operations are not terminated during cold weather so plates must be warmed before welding to avoid abnormal cooling rates in the weld metal and plate heat-affected zone. Requirements for warming in cold weather differ depending on material classification and the various codes by which vessels are constructed. API Specification, Section 620 (low-pressure storage tanks) specifies that there should be no welding when the metal is below 0 F. In temperatures between 0 and 32 F, welding can be performed by preheating the metal so that it is warm to the hand. Also, when the plate is over 1 1/4 in. in thickness, it must be warm to the hand. Methods of warming include resistance strip heaters and gas torches.

For welding in other than the flat position, successful use has been made of the backstep sequence to control shrinkage. In this technique successive sections of a weld are deposited opposite to the direction of progression. The use of backing strips has also been found advantageous for complete penetration welding in areas that are inaccessible or difficult to weld from the back side. If the back-up bars are not removed, it is imperative that the butt weld splices

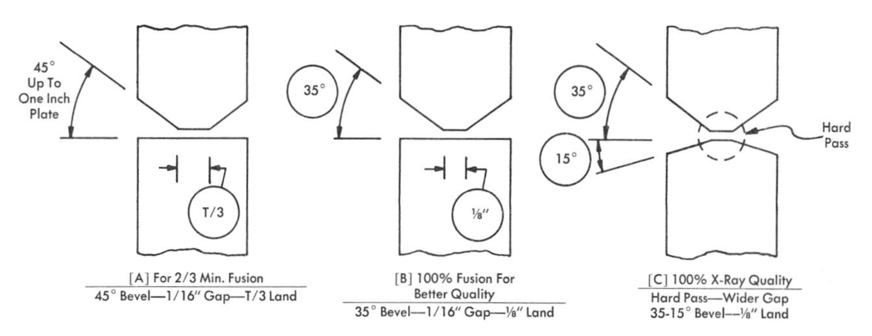

Fig. 84.13.–Horizontal field welding bevels for automatic welding of circumferential joints in vertical vessels.

between the bars are 100% fused to avoid built-in notches which sometimes cause failure.

Thorough inspection is particularly important in field-welded vessels, with emphasis on weld groove angles, root openings, back chipping or gouging and the use of qualified procedures. The degree of postvisual examination will depend upon the material, the customer requirements and Code specifications. These in turn are dictated by the intended service. Radiography and magnetic particle examination are widely used to varying degrees in field operations. Complete radiographic examination is becoming increasingly common for vessels subjected to severe temperature and pressure services, such as petroleum refinery vessels. In many cases, dye penetrant, magnetic particle and ultrasonic examination methods are employed to supplement standard methods.

Field testing large vessels often introduces unusual problems. Where the vessels and supports are adequate for liquid weight, hydrostatic testing is preferred. To permit an adequate test of the upper portion of such vessels, it is sometimes necessary to provide excess thickness in the design of the lower vessel parts to avoid a serious overstress. If a vessel is to be only partially full during normal operation it is sometimes impractical, for economic or other reasons, to design for a liquid load. In such cases pneumatic or a combination pneumatic-hydrostatic loading is employed. The testing of any vessel can be hazardous in the event of a sudden extensive failure. Because of the greater energy stored in a pneumatic test, a defect is proportionally more dangerous; precautions such as radiographic inspection of main joints, magnetic particle inspection of attachments and postweld heat treatment (stress relief) are advisable.

Postweld heat treatment of large vessels in the field presents unusual problems, but is being accomplished successfully and with good control as the occasions demand (Fig. 84.14).

The vessel is usually in its final position for field stress relief and enclosed in a

Fig. 84.14.–Field stress-relief operation on an oxygen converter and trunnion ring. The entire vessel was heated to 1150 F and held for the prescribed time in the postweld stress-relief temperature range.

temporary furnace or covered externally with an insulation. Heat may be internally or externally supplied, depending on the geometry of the vessel. For internal heating, gas or oil burners are fired through openings in the insulated vessel or hot air is blown into the vessel opening from an external furnace. Where the vessel diameter fluctuates or intricate internal fixtures are involved, hot air may be preferable or multiple burners employed for heat uniformity. The choice and location of burners and hot air flow details should assure even temperature distribution. Hazards in lighting burners and keeping combustion with dampers exist, and should always be considered in the field heat treatment of vessels. Electric heaters are also being successfully used to postweld heat-treat vessels.

At the present time, almost any size or shape vessel can be postweld stress-relieved in the field, as long as it is self-supporting at the selected temperature.

THIN-WALLED PRESSURE VESSEL MANUFACTURING

Although essentially similar with respect to fabrication of relatively heavy wall, the manufacture of welded vessels of wall thicknesses approximately 1/4 in. or less presents certain problems peculiar to this class of vessel.

The same welding processes used for thick walled vessels are also used for thin walled vessels. Automatic or semiautomatic equipment is usually employed and the weld is completed in a single pass. Manual arc welding when used will normally require joint preparation and welding from both sides in order to obtain full penetration of the welded joint.

Backing bars permanently affixed to the inside wall of the joint or removable chill bars are used to obtain complete penetration of the welded joint when the filler metal is added from one side. These should fit snugly against the inside surface across the joint and the abutting edges; the backing bar should be clean. The abutting edges of both longitudinal shell seams and circumferential groove-welded head seams should be in close alignment both from standpoint of appearance and sound practice. For this reason, the difference in circumferential measurement of the shell cylinder and the flanged head should be held within close tolerance.

Distortion of the cylinder by reason of the welding operation is readily corrected by rerolling to a true cylinder. It is important, however, that the plate adjacent to the welding seam be set to true circularity to help control the tendency to form a longitudinal peak at the weld. In automatic welding the welding speeds and parallel adjustment of the welding groove may be governed by experience on similar classes of work. In 1/4 in. and thinner plate there is usually no necessity for scarfing the abutting edges. These may be sheared square and the edges left open to permit thorough penetration at the joint.

Nozzles and bosses or pads are usually manually welded to the vessel wall. When welding pads on vessels of 14 gage or under, it may be necessary to provide chill pads on the opposite surface to prevent burning through.

Except when specified by the customer or by reason of the service conditions or the material used, radiography and postweld heat treatment are not required. Because of the relatively high cost of repair compared to making the single bead weld in thin vessels, the importance of careful planning, supervision and inspection cannot be overemphasized.

The quality control methods used by different tank manufacturers may seem on the surface to be very different; however, on close examination the underlying principles are similar. Continuous watching and checking of welding operators and machines is required, with close and continuous attention to small detail. Neat fits everywhere are imperative. It is difficult to bridge a gap, whether in a main seam, or at a boss or where a coupling or nipple is attached. Not only the surfaces to be welded but the metal itself must be clean. All necessary equipment for either machine or manual welding should be maintained in excellent condition. It is the experience of most manufacturers that for single pass welding it is more economical to do the job right the first time. Repair and retesting are more expensive than many realize, and it is more economical to spend a little more on the fabricating end if it will save time and money on the test floor.

QUENCHED AND TEMPERED STEEL VESSELS

Quenched and tempered steels are now commonly used in construction of pressure vessels. These steels are specified for vessels with critical applications and can take advantage of the improved notch toughness resulting from accelerated cooling produced by temperatures in the austenitizing range. The ASME Code also permits some fabrication of vessels from selected quenched and tempered low-alloy steels to take advantage of the higher tensile strength which permits reduced wall thickness and weight.

Mechanical Properties

The superior notch toughness of quenched and tempered steel is attributed to the favorable metallurgical structure developed in the base metal through accelerated cooling from a temperature range of about 1525 to 1775 F. This is accomplished by a uniform water-spray or dip-quench treatment in water or brine immediately after removal of the part from the furnace. Quench pits having a capacity of 152,000 gallons of water have been used for heat treatment of large pressure vessel subassemblies. It is understood that the proper tempering heat treatment must follow quenching to obtain the required combination of toughness and strength. Close control of the tempering temperature is sometimes required.

Fabrication

Successful fabrication of quenched and tempered steels does not normally require the use of unusual equipment and techniques. It is of primary importance, however, for the vessel fabricator to use methods which will avoid reduction of strength or notch toughness in previously quenched and tempered base metal. To accomplish this objective, special precautions must be observed in forming operations, selection of welding processes, weld preheat, interpass and postweld (stress-relief) heat-treatment.

The base metal may be either quenched and tempered at the mill, or the fabricator may desire to perform the heat treatment at his plant after the completion of preliminary operations. If the base metal is supplied in the quenched and tempered condition, care must be taken not to heat it above the tempering temperature except for localized operations such as welding and arc- or oxygen-cutting. Hot forming must not be performed because the properties resulting from the slow rate of cooling would not equal those obtained by quenching. Attempts to correct distortion by local heating must be carefully controlled. Attempts to restore original properties by local heating and quenching restrained areas must be avoided.

Reheat treatment of the entire subassembly following inadvertent reduction of mechanical properties through overheating may be performed subject to the following considerations:

1. Need for recertification of base metal mechanical properties. If this is necessary, sufficient excess metal must be provided for testing.
2. The high cooling rates associated with quenching welded sub-assemblies may distort them. If highly restrained components containing nozzles are quenched, the unequal strains produced during cooling could cause cracking.

It must be remembered that the quenching operation usually increases the strength of low-alloy steels, particularly when tempering temperatures at the lower end of the range are specified. The increased yield strength must be considered when selecting equipment for cold forming heads and shell sections.

If the fabricator chooses to purchase base metal plate in the hot-rolled condition, he is free to hot form it to the desired shape and then quench and temper it to provide the desired mechanical properties.

Welding

There is no difficulty in maintaining the required mechanical properties in the heat-affected zone of quenched and tempered metals providing care is exercised in selection of welding procedures that will keep the rate of heat input down to a reasonable level. It must be remembered that while it is desirable to keep preheat or heat input below some critical value to maintain notch toughness, these variables must be kept high enough to sufficiently restrict the cooling rate so that cracking does not occur.

Selection of Welding Process, Preheat and Interpass Temperature

Higher rates of heat input during welding reduce the thermal gradient between the weld and the base metal heat-affected zone. So does high preheat or interpass temperature, thin base metal sections, heavy weld beads or those deposited with considerable oscillation. On some steels, specific limits must be established for these variables in order to maintain impact values, and tables have been published itemizing the maximum permissible heat input permitted for various conditions of plate thickness and preheat.

Within limits that must be established for each type of steel, adjustment of the variables to produce faster cooling will enhance the notch toughness of the weld and heat-affected zone.

Selection of Filler Metal

When enhancement of toughness is the prime purpose of specifying quenched and tempered material, the filler metals used to join these base metals must have good impact properties. Low-hydrogen welding systems are universally specified. There are many suitable filler metals available for joining quenched and tempered steels by the manual shielded metal-arc, submerged arc, gas tungsten-arc and gas metal-arc (including flux cored) processes. Care must be taken to select filler metals that produce weld deposits where impact properties are not already affected by postweld (stress-relief) heat treatment. Some construction codes require testing of each lot or batch of filler metal under the conditions to be employed in construction.

Consideration must also be given to the effect of reheating within the austenitizing range of the mechanical properties of weld metals. During normal multiple pass welding, the weld beads are individually quenched, immediately after their deposit, by the comparatively cool surrounding base metal. For this reason, weld metal intended for use in the as-welded or stress-relieved condition usually has a relatively low-carbon content. The low carbon provides suitable hardenability for these conditions and minimizes weld metal cracking.

When weld metal is subjected to a quench and temper heat treatment after the entire joint is completed, the cooling rate from the austenitizing range will not be as great as it was when individual beads were deposited. It is therefore necessary to test the proposed weld deposit under these conditions to assure meeting the required mechanical properties.

MULTILAYER MILTIWALL AND BANDED VESSELS

Vessels constructed by these techniques are intended for use at high pressures and are generally built to users' or manufacturers' specifications. They are known as banded, Multilayer[1] and multiwall vessels depending on the method of fabrication employed.

In Multilayer design the shells are made up of a multiplicity of steel plate layers, with each layer about 1/4 in. thick except the inner shell which may be 1/2 in. to 1 1/4 in. thick (Fig. 84.15). Each successive layer is wrapped under tension around the preceding assembly, and the longitudinal seam is welded using the underlayer as backing (Fig. 84.16). Shrinkage in these welds further tightens the wrapping. Each shell course is completely assembled prior to making the girth welds, which extend continuously through the entire thickness. High residual welding stresses are not developed across the multiple thin layers of this Multilayer vessel. Longitudinal strains from girth weld shrinkage are isolated within the individual layers, since there is little shear transfer between layers.

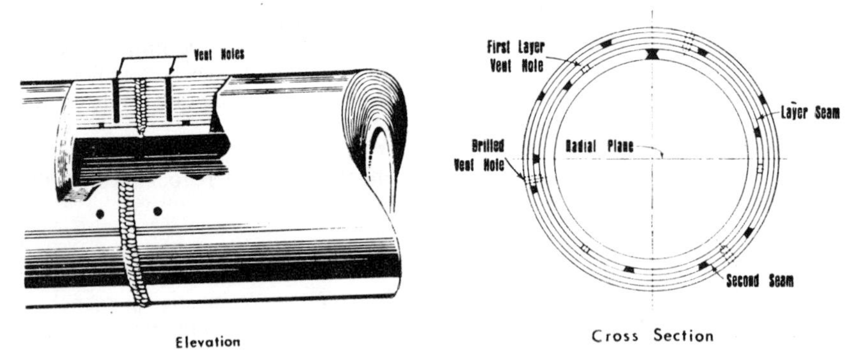

Fig. 84.15.–Cylindrical section depicting a design of multilayer construction: (A) Elevation (B) Cross section.

[1] Multilayer is a trademark of the Chicago Bridge & Iron Company.

Fig. 84.16– Wall thickness of vessel is built up by successive layers of thin metal plate wrapped tightly around inner cylinder and welded together.

Therefore, stress-relief of welds between adjoining shell sections, and between shell sections and nozzles or closures, is neither required or desired. Butt welds in the inner shell section and heads are fully radiographed. Progressive magnetic particle examination of the longitudinal welds in individual thin layers checks their soundness.

Banded construction is another design in common use. In the banded design the inner shell layer is fabricated from plate having a thickness adequate to carry the axial load of the vessel. Hemispherical heads are used to close the ends of this shell section. Conventional pressure vessel fabrication practices, including magnetic particle testing, radiography and stress relieving, are used in this assembly. Since the subsequent banding is used only to reinforce the vessel for hoop stress, the circumferential edges of these bands need not be welded together and usually are not. To facilitate installation without scoring, the bands are made slightly larger than the outside diameter of the inner vessel. Once the bands are positioned over the inner vessel, hydraulic pressure is applied and increased until the inner vessel yields in the hoop stress direction and plastically flows out into contact with the bands. Pressure is further increased far beyond the design pressure with a resultant elastic stretching of the bands. Upon release of pressure the springback of the bands exceeds that of the inner vessel. This condition prestresses the layers to produce a residual compressive stress at the inside surface and puts the layers in effective contact so that they behave as an elastic unit upon repressurization.

Multiwall[2] design uses a thermal shrink assembly method for fabricating high pressure vessel courses which are made of plate layers of from 1 1/4 to 2 in. thick. The inner shell is made by rolling, welding and rounding up and is subject to radiographic inspection. Successive layers are similarly fabricated with control of size by circumferential measurements and thermally assembled one over the other to obtain a metal-to-metal or slight interference fit between layers. Any number of layers may be assembled to obtain the total wall thickness. Figure 84.17A shows a shell course of five assembled layers and Fig. 84.17B shows the same course after machining the ends with welding bevels for joining courses together or to the vessel heads. All layer longitudinal seams and girth welds joining the courses are 100% radiographed. The inside layer can be of any material such as ferritic alloys containing chromium for resistance to hydrogen, stainless steel or other alloy for corrosion resistance. Outside layers are vented to the atmosphere by use of venting holes drilled through all layers except the inside one which holds the internal working pressure.

For a vessel design using inside layers of austenitic stainless steel, the layer

---

[2] Multiwall is a trademark of The Struthers Wells Corporation.

A *B*

Fig. 84.17–(A), A Shell course of five assembled layers. (B), The same course after machining the ends with welding bevels for joining courses together or to the vessel heads.

longitudinal welds are stress-relieved before or during heating or shrink assembly; girth seams may not be stress-relieved because of the differences in the coefficients of thermal expansion between steel and stainless steel. Vessels of steel, or those using stainless clad inner layers, are stress-relieved after vessel completion in accordance with Code requirements.

CRYOGENIC VESSELS

In general, cryogenic liquids are stored in double wall vessels. Inner and outer vessels are usually of the same basic configuration. Two basic pressure vessel shapes are shown in Fig. 84.18.

Spherical containers have been built with the inner vessel supported on a suspension system as shown and also with the inner sphere being column supported. Cylindrical vessels of double wall construction usually support the inner vessel on a suspension system.

Selection of the proper materials of construction for cryogenic service is one of the most important considerations facing the designer. Of primary concern is

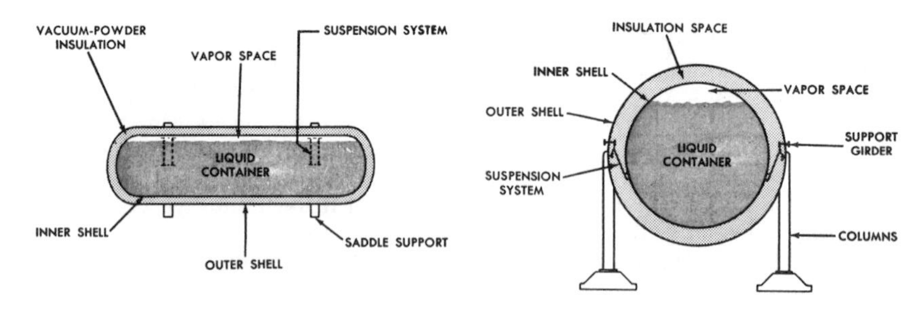

Fig. 84.18.–Cylindrical and Spherical Double Wall Vessel Designs for Cryogenic Liquid Storage.

how the material will perform at very low temperatures. Such properties as notch toughness, specific heat, coefficient of thermal expansion and coefficient of thermal conductivity at very low temperatures must be known, as well as the usual strength and ductility properties of the metal.

Most metals increase in strength with a decrease in temperature. Some, however, such as most grades of carbon steel, suffer an almost complete loss of toughness at low temperatures, making them useless for inner vessel construction. Copper, nickel, aluminum and most alloys of these metals exhibit no ductile-to-brittle transition and, therefore, are suitable for cryogenic service. Austenitic stainless steel of the 18% chromium-8% nickel classification also exhibits excellent low-temperature notch toughness.

INTERNALLY-INSULATED VESSELS

Vessels for certain applications involving high operating temperatures are now being built with special internal insulation to reduce the working temperature of the pressure shell. This allows the use of higher design stresses, particularly applicable to the high-strength quenched and tempered steels. Special attention must be given to details to prevent condensation or vapor flow through or behind the insulation, which would result in loss of insulation by erosion and increased conductivity. The amount of vapor flow through the insulation is a function of the flow velocity in the vessel or other factors which increase the pressure losses, such as tower packing and catalyst beds. Such flow may be reduced by metal vapor barriers incorporated into the insulation which, to be effective, must be welded to the shell. The conductivity is increased in proportion to the amount of metal used. Bonding cements and plastic insulations do not always provide continued adherence to the metal wall and, in general, the corrosion resistance and behavior of the materials at varying temperatures should be evaluated. Many gases such as hydrogen and hydrocarbons have higher thermal conductivity than air, thus increasing heat loss. This is further increased by the effect of pressure.

These types of internal insulation have been used:

- Unreinforced monolithic heat-resisting or insulating concrete
- Insulating block covered by concrete panels or supported refractory walls
- Insulating block or concrete faced with a vapor-tight metal shield sealed to the vessel wall at one end
- Insulating concrete covered by refractory concrete panels reinforced and attached to the vessel wall by clips or studs
- Insulating brick or insulating and refractory brick.

CORROSION-RESISTANT VESSELS

Various methods and materials are used in industry to provide a corrosion-resistant layer to internal surfaces of pressure vessels. The simpler methods make use of excess shell thicknesses and painting, spraying and high-temperature firing of both metallic and nonmetallic materials such as refractory, glass, rubber, lead or plastics. Wide use is also made of strip or sheet metal liners individually attached by welding to the internal surface of vessels used in the oil refining, chemical process and pulp and paper industries.

However, where a higher degree of attachment of the protective layer is required to meet the more severe requirements of vessels operating under cyclic conditions of temperatures and pressures, a continuously-bonded cladding is more desirable than applied liners. Continuously-bonded corrosion-resistant cladding is usually applied to pressure vessel steel plates by hot rolling (roll-clad), brazing, weld metal overlaying and explosion bonding.

APPLIED LINERS

Sheet and strip metal liners are most commonly used in thicknesses from 5/64 to 7/64 in. Sheet liners may vary in width from 24 to 48 in. and in length from 3 to 12 ft depending on the method of welding. These may be attached to the base metal by spot or seam resistance welding, by plug, slot or "plow-through" welds using submerged arc, gas shielded metal-arc or manual shielded metal-arc welding. Strip liners may average 3 to 5 ft in length and, because of their limited width (3 to 6 in.), are usually welded to the base plate at their edges only. The joint spacings between liners may vary to more than 1/2 in. Wider root openings are usually employed with strip liners.

In welding special alloy liners to a dissimilar base metal, careful selection of filler wire and welding process must be made to avoid the risk of weld cracking and possible leaks. The degree of leak tightness required of welded liners depends on the intended service; however, certain minimum requirements for pressure vessels are governed by the ASME Code. For more detailed information on joint types, method of welding and leak testing, see the chapter on applied liners.

ROLL-CLAD PLATE

Integrally-clad plate is manufactured in the steel mill by rolling a welded assembly of carbon or alloy steel slabs and stainless steel or other metal plates.

The hot rolling produces a continuous forge weld between the two metals and the resultant clad plate is heat treated, cut and formed as an integral plate. Welding and postweld heat-treatment are accomplished with due attention to the requirements of the individual composite as described in the chapter on clad metals.

BRAZE-BONDED CLADDING

Large, integrally-clad plates are made by a proprietary vacuum brazing process. Evacuation of a sealed assembly while being heated to a temperature high enough to relax the metal permits intimate contact and allows a 100% braze bond to take place. This process is used to manufacture a variety of ferrous and nonferrous clad plates for use in pressure vessels.

Tubesheets and small objects are clad commercially by a process in which platens of a press are contained within a furnace to provide both heat and special atmosphere.

EXPLOSION-BONDED CLADDING

Explosion bonding is a relatively new method of applying corrosion-resistant cladding which has a sound, continuous, metallurgical bond with the base metal.

The clad plate is properly positioned in relation to the base plate and an explosive charge, which is on top of the thinner plate, brings them together with intense impact.

Applications for explosion cladding in the pressure vessel industry which have been proven practical include integral attachment of cladding to sections of pipe from 2 to 9 in. in diameter and tubesheets for heat exchangers. Elbows for nuclear reactor piping up to three feet in diameter have been fabricated by joining elbow havles which have been explosively clad in the flat prior to oxygen cutting and hot forming to shape. Shell and head sections for pressure vessels may be hot-formed after explosion cladding.

Various metal combinations have been successfully bonded together with this process. Stainless steels and high nickel alloys have been clad to carbon and low-alloy steels. Other materials which have been employed in explosion cladding include copper alloys, titanium and aluminum. The metallurgical bond is highly reliable and does not separate when the clad plate is formed, welded, oxygen-cut or trimmed by any of the common methods.

Prior to the cladding operation the plates must be cleaned of scale and other contaminants to prevent their entrapment at the bond line. The plates must be carefully positioned at the specified distance from each other, which may vary between 3/32 and 1/4 in. depending upon other variables. The explosive charge, usually in sheet form, is placed on top of the cladding plate and is detonated. This brings the plates progressively together as the explosion radiates from the point of initiation. Surface oxides are squeezed out of the interface as the clad plate is progressively welded to the base plate under the advancing shock wave. The interface has a wave-shaped pattern that may include evidence of periodic local melting. This process is capable of producing clad plate of consistently uniform composition and thickness.

Essential variables which must be controlled to assure uniformity and reproducibility in explosion cladding include clad plate thickness, stand-off distance (between the cladding and the base plate), explosive charge density and type of explosive. Parts of different shapes may require different procedures and should be checked out in trial runs prior to the start of production.

WELD METAL OVERLAYS

Another method of obtaining a sound continuously-bonded cladding for pressure vessels is by weld metal overlays. This method provides the fabricator both with a means of cladding shapes and sizes not available from producers and with an alternate source for clad flat products which he may use depending upon economics.

A variety of submerged arc processes such as strip electrodes, series arc, single and twin wires with oscillation are being used successfully. The gas metal-arc process—both automatic and manual—is also used for certain applications and alloys. Weld metal deposition rates as high as 100 lb per hour are feasible. The choice of process is, as always, a function of accessibility, position of welding, size of components and other similar factors. Generally, for most applications, a 3/16 to 1/4 in. minimum thickness of overlay is required with both the submerged arc and gas metal-arc processes. Close control of filler metal composition and fluxes must be maintained as well as base metal dilution during welding to achieve the desired chemistry in the overlay.

Satisfactory overlays of austenitic stainless steel and certain high-nickel alloys have been obtained with these processes.

The chief problem in depositing overlays is the occurrence of incomplete fusion between beads. This can be eliminated by proper control of amperage, voltage, travel speed and bead spacing. For applications requiring complete integrity of bond, ultrasonic examination of the as-welded or machined overlay is desirable. Usually, a dye penetrant examination for surface defects is satisfactory for most commercial applications.

BIBLIOGRAPHY

"Review of Service Experience and Test Data on Openings in Pressure Vessels with Non-Integral Reinforcing," E. C. Rodabaugh; "Derivation of Code Formulas for Part B Flanges," E. O. Waters, *Welding Research Council Bulletin* 166 (October, 1971).

"Elastic-Plastic Deformations in Pressure Vessel Heads," F. A. Simonen and D. T. Hunter; "Summary Report on Plastic Limit Analysis of Hemispherical- and Toriconical-Head Pressure Vessels," J. C. Gerdeen and D. N. Hutula, *Welding Research Council Bulletin* 163 (July, 1971).

"PVRC Interpretive Report of Pressure Vessel Research, Section 3–Fabrication and Environmental Considerations," A. P. Bunk, *Welding Research Council Bulletin* 158 (January, 1971).

(1) "Interpretive Report on Oblique Nozzle Connections in Pressure Vessel Heads and Shells Under Internal Pressure Loading," J. R. Mershon; (2) "Elastic Stresses Near a Skewed Hole in a Flat Plate and Applications to Oblique Nozzle Attachments in Shells," F. Ellyin; (3) "Photoelastic Determination of the Stresses at Oblique Openings in Plates and Shells," M. M. Leven; (4) "A Photoelastic Analysis of Oblique Cylinder Intersections Subjected to Internal Pressure," R. Fidler, *Welding Research Council Bulletin* 153 (August, 1970).

"Elevated Temperature Fatigue Properties of Pressure Vessels Steels," R. A. DePaul, A. W. Pense and R. D. Stout, *Welding Journal*, 44 (9), 409s-416s (September, 1965).

"High Temperature Creep Tests on Model Pressure Vessels," PVRC Subcommittee on Effects of Fabrication Operations on Materials for High Temperature Service, *Welding Journal*, 44 (8), 347s–470s (August, 1965).

"Creep-Rupture Properties of Pressure Vessel Steels," V. S. Robinson, A. W. Pense and R. D. Stout, *Welding Journal*, 43 (12), 531s–540s (December, 1964).

"An Experimental Stress Analysis of a Thick Walled Cylindrical Pressure Vessel With a Hemispherical Head Closure," J. W. Dally and G. T. Schneider, *Welding Journal*, 43 (10), 461–464s (October, 1964).

"Correlation of Laboratory Determinations of Fracture Toughness with the Performance of Large Steel Pressure Vessels," E. T. Wessel, *Welding Journal*, 43 (9), 415s–424s (September, 1964).

"Internal Pressure Cyclic Fatigue Test of an HY-80 Marine Boiler Drum," I. Berman and D. H. Pai, *Welding Journal*, 43 (1), 24s–32s (January, 1964).

"Welding Thin Walled Titanium Pressure Vessels," G. Pagnotta and G. W. Hume, *Welding Journal*, 42 (9), 709–714 (September, 1963).

"Field Stress Relieving of Large Field Erected Vessels," J. B. Christofferson, *Welding Journal*, 42 (4), 294–301 (April, 1963).

"Aluminum Alloy (2219), Application to Missile Pressure Vessel Fabrication," C. H. Crane and W. G. Smith, *Welding Journal*, 40 (1), 335s–340s (January, 1961).

"A Story of Theories of Fracture Under Continued Stresses," I. Cornet and R. C. Grassi, *ASME Transactions*, Series D, pp. 39–45 (1961).

"Minimum Toughness Requirements For High Strength Sheet Steel," J. A. Kies, H. Romine, H. L. Smith and H. Bernstein, *ASME Transactions*, Series D, pp. 1–10 (1961).

"Stresses in a Spherical Vessel From External Moments Acting on a Pipe," P. P. Bijlaard, *Welding Research Council Bulletin*, 49 (April, 1959).

"The Making, Shaping, and Treating of Steel," United States Steel, 7th Edition.

CHAPTER **85**

INDUSTRIAL PIPING

PREPARED BY A COMMITTEE CONSISTING OF:

H. THIELSCH, *Chairman*
ITT Grinnell Corp.

R. BENNETT
Battelle Memorial Institute

F. C. BRAUN
Gulf Research and Development Co.

J. L. CAHILL
Retired

H. J. STOCK
General Electric Corp.

R. J. LANDRUM
E. I. DuPont de Nemours and Co.

W. R. SMITH
Bechtel Corp.

E. J. VANDERMAN
The Lummus Co.

CHAPTER 85

INDUSTRIAL PIPING

INTRODUCTION

In commercial piping systems, the designation piping includes pipe, tubing, fittings (elbows, tees, flanges, reducers), valves and headers. Other components include such items as expansion joints, flow nozzles, pumps, etc. Applicable welding procedures cover materials such as backup rings, consumable insert rings, covered electrodes, bare welding wire, welding fluxes and shielding and purging atmospheres. Bolts, gaskets, hangers, supports and insulation are not considered, as welding materials for these applications are not primarily involved in the integrity of the fabrication or erection of a welded piping system.

The number of ferrous and nonferrous alloys being used for piping is constantly increasing. This is due to industrial demands for better and more economic materials to meet growing operating requirements of power plants, nuclear plants, refineries, chemical and petrochemical plants, paper mills, textile plants, pharmaceutical plants and other industrial plants.

A discussion of the welding characteristics of all ferrous and nonferrous base metals now used commercially in piping systems is beyond the scope of this chapter. Only the more common commercial base metals are covered. These include (1) ferrous piping: carbon steels and wrought iron, chromium-molybdenum alloy steels, steels for low-temperature service, stainless steels, lined steels and (2) nonferrous piping: aluminum and aluminum alloys, nickel and nickel alloys, copper and copper alloys, titanium and titanium alloys, lead.

Nonmetallic piping involving polyvinyl chloride, fiberglass and asbestos materials are not treated in this chapter. Also excluded are metallic piping lined with nonmetallic materials such as glass, plastics, cement and wood.

CODES AND STANDARDS

Various codes and standards applicable to welded piping systems have been prepared by committees of leading engineering societies, trade associations and standardization groups. These are generally written to cover the requirements of quality and safety. Many applications, however, may require more conservative design and welding than prescribed in some or all of these regulatory documents. Recognition of service requirements in the evaluation and selection of appropriate standards is essential to a properly engineered and constructed piping system. Service failures frequently result when this is not done.

Many standards also specify inspection requirements in order to establish proper levels of soundness. Some code-writing committees have recognized that the level of quality attained in a welded piping system and the cost of producing this level, must be related to specific inspection requirements. Where full penetration weld deposits are required, 100% radiographic examination should be specified. Spot radiographic examination of, for example, 10% of the weld joints in a piping system should allow a lower level of weld quality and thus permit some incomplete joint penetration. This is recognized in ANSI B31.3 of the Code for Pressure Piping, which states for incomplete joint penetration:

The total joint penetration shall not be less than the thinner of the two components being joined, except to the extent that incomplete root penetration is permitted. The depth of incomplete root penetration or lack of fusion at the weld root shall not exceed 1/32 inch or half the thickness of the weld reinforcement, whichever is smaller. The total length of such incomplete root penetration or lack of fusion at the root shall not exceed 1-1/2 inches in any 6 inches of weld length."

"Unless otherwise specified by the engineering design, welds on which 100% radiography is specified, shall have complete joint penetration."

AMERICAN NATIONAL STANDARDS INSTITUTE

The American National Standards Institute (ANSI) is preparing or has issued the following sections of the ANSI B31 Code for Pressure Piping:

B31.1.0	Power Piping (1967)
B31.2	Industrial Gas and Air Piping (1968)
B31.3	Petroleum Refinery Piping (1966)
B31.4	Oil Transportation Piping (1966)
B31.5	Refrigeration Piping (1966)
B31.6	Chemical Industry Process Piping[1]
B31.7	Nuclear Power Piping (1969)
B31.8	Gas Transmission and Distribution Piping Systems (1968)

These sections deal primarily with adequate safety requirements for the selection and designation of dimensional standards, materials, design, fabrication, erection, testing and inspection of piping systems. The principal purpose of the Code is to serve as a guide to state and municipal authorities in drawing up their regulations and to serve architects, engineers, equipment manufacturers,

[1] Until publication of this new section, the 1955 edition of the Code for Pressure Piping (B31.1) should be consulted.

fabricators and erectors as a standard of reference, based on successful experience, for fabricated and erected piping systems.

AMERICAN SOCIETY OF MECHANICAL ENGINEERS

ASME Boiler and Pressure Vessel Code

This Code covers piping connected to boilers (Section I). It has been recognized by most states and many municipal authorities, who have made this Code a prerequisite to acceptance of the installation by the states and insurance companies.

The ASME Boiler and Pressure Vessel Code involves eleven sections of which the following are concerned with industrial piping:

Section I	Power Boilers. Formulates minimum specifications for the construction of piping directly connected to steam power boilers within prescribed limits. Where specific rules are not given in other sections, the rules in Section I may be used.
Section II	Materials.
Section III	Nuclear Power Plant Components.
Section V	Nondestructive Examination.
Section VIII	Unfired Pressure Vessels. Div. 1 and Div. 2 contain rules and fabrication and inspection requirements for the attachment of piping to pressure vessels.

Requirements for the qualification of welding procedures, welders and welding operators are set forth in:

Section IX	Qualification Standard for Welding and Brazing Procedures, Welders, Brazers and Welding and Brazing Operators.

The ASME Boiler and Pressure Vessel Codes have been adopted as regulations in many states and municipalities and in the provinces of Canada.

ASME Guide for Gas Transmission and Distribution Piping Systems

The purpose of this guide is summarized in the Code as follows:

"This Code states the requirements for materials, design, fabrication, installation, testing, and inspection of gas transmission and distribution pipe lines which are intended to conform to the Federal regulations governing the transportation of natural and other gas by pipeline."

AMERICAN PETROLEUM INSTITUTE

The American Petroleum Institute has issued a Standard for Field Welding of Pipe Lines (API Std. 1104) which includes weld quality acceptability limits, radiographic inspection requirements and welding and procedure test requirements.

AMERICAN WATER WORKS ASSOCIATION

Standards covering fabricated piping have been issued.

PIPE FABRICATION INSTITUTE

A number of welding standards have been prepared by the Pipe Fabrication Institute covering joint preparation, welding and preheating. A listing of the standards applicable to pipe welding is included in the bibliography.

AMERICAN WELDING SOCIETY

The American Welding Society has published monographs and several recommended welding practices. These are listed in the bibliography.

PROCEDURE AND WELDER QUALIFICATION

To obtain a satisfactory welded piping installation, it is considered necessary to establish and qualify a specific welding procedure. The procedure should cover base metal specifications; filler metals; joint preparation; pipe position; welding process or processes, techniques and characteristics (current setting, electrode manipulation); preheat, interpass and postheat time and temperature. Also, it is necessary to qualify the required welders and welding operators to demonstrate their ability to carry out that procedure, prior to welding of the piping components.

Even when a welder has been properly qualified, the maintenance of a specified quality level of welding requires proper supervision to ensure that all provisions of the applicable welding procedures are satisfied.

Several of the major code-writing organizations have established procedures for welder (manual and semiautomatic welding) and welding machine operator (automatic welding) qualifications. The requirements of the ASME Boiler and Pressure Vessel Code, Section IX, are most widely recognized by industry and insurance companies, and by state and municipal regulating bodies.

The ANSI Code for Pressure Piping B31 permits employers using identical welding procedures to interchange qualified welders under certain conditions. This practice is not generally considered advisable because of extensive abuses of the system which include the forging of certified qualification papers by unqualified welders. Insurance companies and inspection agencies usually require proof from the contractor that the pipe welds were made by experienced and properly qualified welders. The contractor can certify that all of his welders were qualified and tested under his supervision or by a reliable testing laboratory for the specific job involved.

The Heating, Piping and Air Conditioning Contractors National Association has formulated standard welding procedures for manual shielded metal-arc and oxyacetylene welding of pipe. The National Certified Pipe Welding Bureau is one of several proprietary organizations which supervises and certifies welder qualification tests in a uniform manner in accordance with the Association's standard procedures. The organization also supervises the interchange of qualified welders between contractors who have adopted the standard procedures, without the necessity of retesting them. A contractor, however, may also perform and maintain his own welder qualification certification, and would have to do so on work under the ASME Boiler and Pressure Vessel Code.

Included in most of the steel piping materials specifications of the American Society for Testing and Materials is a paragraph covering repair of defects by chipping or grinding and repair welding. Reference is also made to the ASME Boiler and Pressure Vessel Code by stating that the applicable welding procedures and welders shall have been qualified in accordance with Section IX.

The National Board of Boiler and Pressure Vessel Inspectors was organized to administer uniformly the rules of the ASME Boiler and Pressure Vessel Code. It is composed of chief inspectors of states and municipalities which have adopted the ASME Code. A pamphlet issued by the National Board, entitled "Recommended Rules for Repairs by Fusion Welding to Power Boilers and Unfired

Pressure Vessels (over 15 psi)," makes reference to seal welding and repairs to boiler tubes and pipe of low-carbon steels of known weldable quality with a carbon content of less than 0.35%. For welding alloy steel, reference is made to the ASME Boiler and Pressure Vessel Code. Appendices are included relating to welder qualification tests. No repairs by welding may be made without the approval of an inspector.

On military contract work, the welder qualification tests are normally made to the requirements of Military Specification MIL-STD-248. These are similar in scope and detail to the requirements of the ASME Boiler and Pressure Vessel Code. Piping on naval vessels is usually fabricated to Navy Department Specification NAVSHIPS 250-582, Fabrication and Control of Steel Pipe and Tubing for Welding Piping systems.

Very stringent qualification test requirements have been written into the Navy Department NAVSHIPS 250-1500-1 which includes pipe fabricated to nuclear standards under Navy nuclear contract work.

OTHER CODES

Specific requirements of various regulatory state and municipal codes have been summarized by the National Bureau of Casualty Underwriters' Boiler and Machinery Division, Synopsis of Boiler and Pressure Vessel Laws, Rules and Regulations.

BASE METALS

FERROUS

Material and Dimensional Standards

Dimensional standards, covering carbon steel pipe, cast and forged valves, fittings and flanges, were originally developed on the basis that they would be assembled into piping systems by means of mechanical joints. During the past thirty years, these standards have been rewritten and numerous new standards have been issued to take cognizance of the use of welding. Entirely new dimensional standards have been developed for butt- and socket-welded fittings. These standards provide fittings for high pressure service applications and involve alloyed carbon and stainless steel base metals equal in strength to corresponding seamless pipe base metals with considerable decrease in weight and space requirements over flanged fittings.

By far, the majority of piping systems in industry is now fully welded and contains relatively few, if any, threaded joints. Flanged joints are used in welded piping systems where some sections have to be opened for an occasional inspection of the pipe interior or for replacement.

The development of dimensional standards and tolerances by the American National Standards Institute and the Pipe Fabrication Insitute has standardized the supply of commercial fittings and pipe materials. The American Society for Testing and Materials, the American Petroleum Institute and other associations have prepared detailed standards covering the materials requirements for seamless and welded pipe, tubing, castings, forgings and flanges.

The concurrent standardization of electrodes and welding rods sponsored by AWS committees has been done without specific reference to the use of these electrodes and rods in welding of piping components. These standards are considered to be equally applicable to the welding of other wrought, forged and cast base metals. Where the applicable welding procedures for ferrous alloy base metals involve covered electrodes, those with low-hydrogen type coverings are generally preferred for welding pipe when optimum properties and quality are necessary.

Most ASME specifications for electrodes and welding rod are either identical with or similar to the corresponding AWS specification.

Weldability

The term weldability is used to indicate the relative ease or difficulty with which a material can be welded. Where specific welding conditions or other factors affect weldability adversely, greater precautions must be exercised.

In the welding of ferrous piping, a number of factors should be considered:

1. Tendency of the base metal, the weld metal or both to crack while cooling through the liquidus to solidus range (hot cracking).
2. Tendency of some metals to crack after solidification (cold cracking).
3. Mechanical properties of the weld metal and a change in mechanical properties of base metal caused by the effects of the welding operation.
4. A marked change in the metallurgical structure of the base metal as a result of the welding operation.
5. Changes in chemical composition by volatilization of alloying components, by reactions with air oxidizing out alloying elements, by reactions with shielding or purging gases or by diffusion.
6. Formation of refractory coatings which may require special fluxes or shielding gases.

Carbon steels in the carbon ranges of less than 0.30% C and wall thicknesses (less than 3/4 in.) encountered in most piping systems generally do not crack during or after welding, nor do any significant changes take place in their mechanical properties or metallurgical structure. When the component parts have a thickness of 3/4 in. or greater, however, there may be some difficulty in obtaining welds free from cracks because of welding thermal cycle effects on the mass of metal and the greater rigidity of the parts.

To reduce the adverse effects of the factors enhancing cracking, preheating may be considered advisable, particularly for the higher carbon (over 0.30% C) steel and thicker sections. Preheating is also considered advisable or essential on most alloy steels, when moisture may be present, when temperature of the material is below 60 F or when welding under restraint.

Postweld heat treatment is deemed necessary for carbon steel piping of heavy wall thickness and for most alloy steels. The principal beneficial effects of the postweld heat treatment are reduced stresses in the weld area and a more ductile metallurgical structure.

Recommendations for preheat and postweld heat treatment cycles are given in the subsequent paragraphs on the various major commercial piping base metals. These cycles represent commercial pipe welding practice. Occasionally, specific job conditions may require deviations from normal practice. Where such changes are deemed necessary, qualified welding engineers should be consulted.

The effects of the preheat and postweld heat treatment may also vary with the heating methods and equipment employed. The different heating methods used with pipe fabrication and erection are discussed in other sections of this chapter.

Each of the various codes for piping and pressure vessels has requirements for preheat and postweld heat treatment of welds for piping systems fabricated and erected in accordance with the rules of each code.

Carbon Steel Piping

Although a number of semiautomatic and automatic welding processes are being used successfully on carbon steel piping, the largest tonnage of commercial piping is still being welded manually by shielded metal-arc welding (SMAW). E6010 electrodes are widely used, although in recent years, increasing use has been made of the E7018 low-hydrogen iron-powder types of electrodes. The higher-strength carbon steel piping grades of the 70,000 psi mimimum tensile classifications are generally welded with E7010 or E7018 electrodes.

Full penetration welds required in critical piping are frequently made with split or solid backing rings or by welding the first pass by the gas tungsten-arc welding (GTAW) process. Since porosity and gas evolution tend to occur when welding carbon steel piping by the gas tungsten-arc welding process, consumable insert rings alloyed with deoxidizers are preferred to ensure soundness of the root welds. The consumable insert rings also improve the contour of the underside of the weld and minimize the tendency towards cracking.

The gas metal-arc welding (GMAW) processes are also being used increasingly on carbon steel piping, where full root penetration and complete fusion are not essential. Some problems in proper root penetration and fusion are frequently encountered in commercial shop and field welding, particularly at the tack welds. Shielding is normally done with gas mixtures of carbon dioxide and argon.

Oxacetylene welding of piping is gradually disappearing, except in some areas involving welding of gas distribution piping.

Submerged arc welding (Fig. 85.1) is used extensively in shop fabrication, whenever it is practical to rotate piping in the larger diameters and heavier wall thicknesses—usually beginning with the 8 in. od pipe sizes. Where backing rings are used and fit-up is good, the first pass may be made by the submerged arc welding (SAW) process. Otherwise, the first pass should be made by shielded metal-arc, gas tungsten-arc or gas metal-arc welding. Where carbon steel grades to the 70,000 psi minimum tensile requirements are employed, filler wires or fluxes should be used that produce weld deposits in the heat-treated condition matching the tensile properties of the pipe base metal.

Preheating carbon steel piping having a carbon content which is less than 0.30% C generally is not considered necessary. However, it is considered advisable by the ANSI Code for Power Piping that carbon steel base metals having a tensile strength of over 70,000 psi be preheated to 300 to 400 F prior to welding. Postweld heat treatment of pipe joints is generally considered necessary only when the wall thickness is 3/4 in. or over.

The ANSI Code for Power Piping, the ASME Boiler and Pressure Vessel Code in Section I (Power Boilers) and Section III (Nuclear Power Plant Components) and the PFI Standard ES-19 (September 1964)[2] have requirements for postweld heat treatment of piping. Section VIII (Unfired Pressure Vessels) of the ASME

[2] ES–19 (September 1964) "Heat Treatment of Ferrous Pipe Welds"

Fig. 85.1.–Submerged arc welding of piping.

Boiler and Pressure Vessel Code requires postweld heat treatment of joints that attach piping to a vessel under certain specified conditions. In the ANSI Code for Pressure Piping Section B31.8 (Gas transmission and Distribution Systems), postweld heat treatment is required when the wall thickness exceeds 1 1/4 in.

The postweld heat treatment cycle usually consists of heating to 1100 to 1250 F and holding for one hour per inch of wall thickness with a minimum holding period of one-half hour. Cooling in still air is normally considered adequate. However, the requirements of the applicable codes should be reviewed and followed for piping which is to meet the requirements of a particular code.

Wrought Iron

The carbon content of wrought iron is generally below 0.12%. Thus, its welding characteristics are essentially the same as mild steel. However, as a general rule, as little as possible of the wrought iron should be melted by the welding operation to minimize pickup of the silicate slag by the weld deposit. Lower welding speeds and currents are preferred. Preheating and postheating are not generally considered necessary.

All of the welding processes applicable to carbon steel are suitable for welding wrought iron piping. Most extensively used is the shielded metal-arc welding process. However, the oxyacetylene welding (OAW) process is still used in some field welding. Weld surfaces made by oxyacetylene welding may have a "greasy" appearance caused by melting of the silicate slag.

Those welding processes which tend to melt considerable base metal such as submerged arc welding and gas tungsten-arc welding should employ at least two weld layers to ensure sound pipe welds. The initial weld pass against the wrought iron may contain porosity produced by the silicate slag from the base metal. The soundness of the pipe weld can be improved upon remelting of part of the somewhat porous initial weld pass.

Carbon-Molybdenum Steels

Relatively little use is now being made of carbon-molybdenum steels in piping systems. Unfavorable experiences with these steels in applications involving service temperatures exceeding 800 F have been attributed to graphitization of the steel. The carbon, which tends to form graphite in the shape of nodules or flakes, substantially reduces the toughness of the steel.

The same welding processes discussed for carbon steels are applicable to the welding of carbon-molybdenum steels. Shielded metal-arc welding should be performed with E7010-Al, E7016-A1 or E7018-Al electrodes. Submerged arc welding is done with either carbon-molybdenum steel filler wires and neutral fluxes or with carbon steel filler wires and fluxes containing molybdenum. Either method produces satisfactory weld deposits.

Preheating of carbon-molybdenum steels is considered advisable when the wall thickness is greater than 1/2 in. Temperatures between 300 and 400 F are normally employed.

The ANSI Codes for Pressure Piping and the ASME Boiler and Pressure Vessel Code Sections I (Power Boilers), III (Nuclear Power Plant Components) and VIII (Unfired Pressure Vessels), Divisions 1 and 2, have requirements for postweld heat treatment of piping welded in accordance with the applicable code.

The postweld heat-treatment cycle usually consists of heating to 1100 to 1200 F and holding for one hour per inch of wall thickness with a minimum of one-half hour. Cooling in still air is normally adequate. In all cases, the requirements of the applicable codes should be reviewed and followed for piping which is to meet the requirements of a particular code.

When the service tremperature exceeds 800 F, the weld joints should be stress relieved for four hours at 1325 F. This so-called metallurgical stabilization heat treatment tends to suppress potential graphite formation.

Chromium-Molybdenum Alloy Steels

The chromium-molybdenum alloy steels represent the most important alloy steel group of piping base metals. These grades are used almost exclusively for service conditions in the 750 to 1100 F temperature range. For convenience, they are normally subdivided into the "low-alloy" group which covers the grades from 1/2 Cr-1/2 Mo to 3 Cr-1 Mo, and the medium alloy group including the 5 Cr-1/2 Mo to the 9 Cr-1 Mo grades. Welding recommendations for chromium-molybdenum pipe materials are given in Table 85.1.

Critical pipe butt welds requiring full penetration weld deposits are made by welding into backing rings or by welding the first pass with the gas tungsten-arc welding process. The weld is usually completed by shielded metal-arc welding or by submerged arc welding. Gas metal-arc welding is not widely used on the chromium-molybdenum piping steels.

Shielded metal-arc welding should be done with electrodes having low-hydrogen type coverings of the EXX15, EXX16 or EXX18 classification. Submerged arc welding should be done with chromium-molybdenum alloy steel filler wires of compositions corresponding to the base metals to be welded, using essentially neutral fluxes. Although acceptable weld deposits may be obtained by submerged arc welding, this practice is not considered sound with carbon steel filler wires and with fluxes containing chromium and molybdenum. Since a buildup in the interpass temperature tends to upset the alloy balance in the flux, the chromium and molybdenum content of the weld deposit may increase to more than twice that of the piping base metal.

Table 85.1–Welding recommendations for carbon-molybdenum and chromium-molybdenum high temperature alloy steels.

Steel	ASTM Specification (1)		Welding Recommendations		
	No.	Grade	Electrodes (7)	Preheat and Interpass Temperature (2) F	Postweld Heat Treatment (3) F
1/2 Mo	A155 A335 A369	several P1 FP1	E70XX-A1, E7016-A1, etc. (4)	Up to 300 (5)	1150–1350
1/2 Cr–1/2 Mo	A155 A213 A335 A369	1/2 Cr T2 P2 FP2	E70XX-B1, E80XX-B1, (4)	100–450 (5)	1150–1350
1 Cr–1/2 Mo	A155 A213 A335 A369	1 Cr T12 P12 FP12	E8015-B2, E9015-B2,	100–350	1150–1350
1 1/4 Cr–1/2 Mo	A155 A199 A200 A213 A335 A369	1 1/4 Cr T11 T11 T11 P11 FP11	E8015-B2, E9015-B2	100–350	1150–1350
2 Cr–1/2 Mo	A199 A200 A213 A369	T3b T3b T3b FP3b	E8015-B3, E9015-B3	400–450	1250–1350
2 1/4 Cr-1 Mo	A155 A199 A200 A213 A335 A369	2 1/4 Cr T22 T22 T22 P22 FP22	E8015-B3, E9015-B3, etc. (4)	400–450	1300–1400 (6)
2 1/2 Cr–1/2 Mo	A199 A200	T4 T4	E8015-B3, E9015-B3, etc. (4)	400–450	1300–1400 (6)
3 Cr-1 Mo	A199 A200 A213 A335 A369	T21 T21 T21 P21 FP21	5 Cr-1/2 Mo (E502)	500–600	1300–1400 (6)
5 Cr–1/2 Mo	A155 A199 A200 A213 A335 A357 A369	5 Cr T5 T5 T5 P5 – FP5	5 Cr–1/2 Mo (E502)	600–700	1350–1400 (6)
7 Cr–1/2 Mo	A199 A200 A213 A335 A369	T7 T7 T7 P7 FP7	7 Cr–1/2 Mo 9 Cr–1 Mo	600–800	1300–1400 (6)
9 Cr–1 Mo	A199 A200 A213 A335 A369	T9 T9 T9 P9 FP9	9 Cr–1 Mo 12 Cr (E410)	600–800	1300–1400 (6)

(1) This listing includes only the more important pipe and tubing specifications and alloys. Several additional specifications also cover castings, forgings and plate steels of these same compositions from which pipe may be produced by seam welding.
(2) Use lowest temperature for sections up to 1/2 in. thick and highest values for sections over 2 in. thick, with intermediate temperatures for intermediate thicknesses.
(3) 1 hr. per in. of wall thickness, 1/2 hr. minimum.
(4) Also the corresponding EXX16 and EXX18 low-hydrogen electrodes.
(5) May be necessary only to prevent cracks in root passes.
(6) After welding, the weld should be allowed to cool to below 600 F before the recommended postweld heat treatment is applied.
(7) Suffixes -B1, -B2 and -B3 refer to AWS subclassifications specifying electrodes containing 1/2 Cr–1/2 Mo; 1 1/4 Cr–1/2 Mo; and 2 1/4 Cr–1 Mo, respectively.

Preheating to the temperatures shown in Table 85.1 is considered essential. The higher temperatures in the preheating temperature range shown should be used for higher alloy compositions, heavier wall thicknesses and more complex structures.

Welding on the 1 Cr-1/2 Mo to 3 Cr-1 Mo piping steels may be interrupted at any time provided the following conditions are maintained:[3]

1. A minimum of at least 3/8 in. thickness of weld metal has been deposited or 25% of the welding groove is filled (whichever is the greater).
2. The weld is allowed to cool slowly from welding temperature to room temperature. A suggested manner of retarding cooling is to wrap the weld with asbestos and allow the joint to cool in still air.

Welding on the 5 Cr-1/2 Mo to 9 Cr-1 Mo piping material may be interrupted provided that:[4]

1. The recommended preheat temperature is maintained until welding is resumed, or
2. The partially completed weld is immediately postweld heat-treated at a minimum temperature of 1200 F (but below 1425 F) for at least 30 minutes.

In every instance, the weld area must be at preheat temperature before welding is resumed. Postweld heat treatment in accordance with Table 85.1 is generally considered essential.

On small diameter light-wall piping in the 1/2 Cr-1/2 Mo to 2 1/4 Cr-1 Mo grades, postweld heat treatment may be safely omitted in accordance with the ASME Boiler and Pressure Vessel Code, the ANSI Code for Power Piping and PFI Standard ES-8 (March 1964) provided that the outside pipe diameter is not over 4 in. and the pipe wall thickness is less than 1/2 in. This should not apply to cold-worked steels, since these may become very susceptible to stress-corrosion as a result of subsequent acid cleaning or service environments.

The medium chromium-molybdenum alloy steels should always be postweld heat-treated. For refinery applications, it is also desirable to specify a maximum Brinell hardness of 240 to ensure that the postweld heat treatment was properly made. If the steel is improperly heat-treated, metallurgical notches may form and contribute to failure as a result of thermal or mechanical cycling.

Low-Temperature Steels

For low-temperature service piping, nickel alloy steel, as well as several grades of fully-deoxidized carbon steels and austenitic stainless steels, are used. The lowest temperatures at which these steels are normally used are shown in Table 85.2. Welding recommendations for these steels are given in Table 85.3.

The 2 1/4% and 3 1/2% nickel steels are the most widely used grades for service temperatures down to -150 F. Although more economical, the Cr-Cu-Ni steels may become sufficiently notch-sensitive in the heat-affected zone so that the steel in this area would not meet the prescribed Charpy V-notch impact test requirements below -100 F.

Nickel increases the air-hardening characteristics of steel significantly so that preheating is necessary in the 2 1/2% and 3 1/2% nickel steels when the carbon content exceeds 0.15%. The 9% nickel steels should be preheated between 400

---

[3] PFI Standard ES–8 (March 1964)

[4] PFI Standard ES–12 (March 1964)

Table 85.2—Minimum temperatures at which a number of grades of deoxidized carbon steel and austenitic stainless steel are used

Minimum Temp. F	Type of Steel
- 50	Fine-grained, fully deoxidized steel
- 75	2 1/4 % Ni Steel
-150	3 1/2% Ni Steel, Cr-Cu-Ni Steel*
-320	9% Ni Steel, Stainless Steel

*Below -100 F Charpy impact properties in the heat-affected zone of the base metal may be impaired for some heats of these steels.

and 600 F. Postweld heat treatment between 1050 and 1150 F for one hour per inch of wall thickness is considered essential.

Stainless Steels

Austenitic chromium-nickel stainless steel pipe and tubing are far more extensively used than the martensitic and ferritic chromium stainless steel grades.

Austenitic Stainless Steel.—Almost all of the major commercial stainless steel alloys and grades and a number of special alloys are available as piping. These include Types 304, 304L, 316, 316L, 321, 347, 348 and 310. Welding recommendations for the common austenitic stainless steel grades are given in Table 85.4.

On stainless steel piping materials, the first (root) pass is generally made by the gas tungsten-arc welding process to ensure complete root penetration. For critical service applications, the inside of the pipe should be effectively purged with argon or helium to prevent oxidation of the underside of the weld during

Table 85.3—Welding recommendations for carbon steels, nickel steels and chromium-copper-nickel steels for low-temperature service

	ASTM Specification (1)			Preheat and Interpass Temperature (2) F	Postweld Heat Treatment (3) F
Steel	No.	Grade	Electrode (5)		
Deoxidized mild steel	A515 A516 A106		E6015 (4) E7015 (4)	60 Min	1175–1250 (over 3/4 in. wall)
2 1/2 Ni	A203	A, B	E8015-C1, (4) (5)	300–500	1175–1250 (over 1/2 in. wall)
3 1/2 Ni	A203 A333 A334	D, E 3 3	E8015-C2, (4) (5)	400–500	1175–1250 (over 1/2 in. wall)
9 Ni	A353		ENiCrFe-2	400–500	1175–1250
Cr-Cu-Ni	A410		E8015-C1, (4) (5)	300–500	1000–1100

(1) This listing includes only the more important plate and pipe specifications. Several additional specifications also cover steels of these same classifications.
(2) Use lowest temperature for sections up to 2 in. thick and highest values for sections over 2 in. thick, with intermediate temperatures for intermediate thicknesses.
(3) 1 hour per inch of wall thickness, 1/2 hour minimum.
(4) Also the corresponding EXX16 and EXX18 low-hydrogen electrodes.
(5) Suffixes -C1 and -C2 refer to AWS subclassifications specifying electrodes containing 2 1/2 Ni and 3 1/2 Ni, respectively.

Table 85.4–Welding recommendations for wrought austenitic stainless steel pipe grades

ASTM and AISI Desig-nation	Popular Designation	Chemical Composition (percent)				Recommended Electrode, Welding Rod[2]
		C	Cr	Ni	Others[1]	
304	19–9	0.08 max	18–20	8–12	–	E, ER308
304L	19–9 (extra low carbon)	0.03 max	18–20	8–12	–	E, ER308L
309	25–12	0.20 max	22–24	12–15	–	E, ER309
309Cb	25–12 Cb	0.20 max	22–24	12–15	Cb = 10 x C min	E309Cb
310	25–20	0.25 max	24–26	19–22	–	E, ER310
310Cb	25–20 Cb	0.25 max	24–26	19–22	Cb = 10 x C min	E,ER310Cb
310Mo	25–20 Mo	0.25 max	24–26	19–22	2.0–3.0 Mo	E, ER310 Mo
316	18–12 Mo	0.08 max	16–18	10–14	2.0–3.0 Mo	E, ER316
316L	18–12 Mo (extra low carbon)	0.03 max	16–18	10–14	2.0–3.0 Mo	E, ER316L or ER318
317	19–13 Mo	0.08 max	18–20	11–15	3.0–4.0 Mo	E, ER347
318	18–12 Mo Cb	0.10 max	18–20	10–14	2.0–3.0 Mo Cb = 10 x C min	ER318
321	18-8 Ti	0.08 max	17–19	9–12	Ti = 5 x C min	E, ER347 ER321
347	18–8 Cb	0.08 max	17–19	9–13	Cb-Ta = 10 x C min	E, ER347
348	18–8 Cb	0.08 max	17–19	9–13	Cb-Ta = 10 x C min	E, ER347, E, ER348

[1] Unless otherwise specified Mn is 2.00 max, Si is 1.00 max, P is 0.040 max and S is 0.030 max.

[2] AWS specifications. E means grade recognized by AWS as covered electrode; ER as bare electrode and welding rod.

root pass welding. Thus, stainless steel pipe for nuclear and power plant service and chemical plant applications is normally purged. However, in paper mills, where underweld cleanliness is frequently not considered a requirement purging is often eliminated, unless the piping system is considered critical and subject to severe corrosion or cleanliness requirements. Consumable insert rings are also considered essential where critical service applications are involved–particularly where excessive crevices "sink" along the underside of the weld.

On pipe having a wall thickness up to 3/16 in., the weld, started by gas tungsten-arc welding, is normally completed by this same process. On pipe with a greater wall thickness, the balance of the weld after the root pass is generally completed by the shielded metal-arc, the gas metal-arc or the submerged arc welding processes.

Martensitic and Ferritic Chromium Stainless Steels.–The chromium stainless steels (also called the straight chromium stainless steels) are generally separated into two further subclassifications: martensitic stainless steels and ferritic stainless steels. Only the martensitic steels can be hardened by heat treatment.

Martensitic Stainless Steels.–Martensitic or hardenable stainless steels usually contain up to about 15% chromium. Although these grades may have a ferritic microstructure, they are primarily martensitic when cooled in air or when quenched into a liquid medium from temperatures above 1500 F. Maximum hardening in most of these grades is achieved by cooling from temperatures above 1750 F to room temperature. These steels, therefore, might be compared in their response to hardening and tempering to the ordinary hardenable carbon and alloy steels with slight temperature modifications.

Welding recommendations for the commercial martensitic stainless pipe steels are given in Table 85.5.

Without preheat treatments, the highly hardenable, fully martensitic stainless steel grades generally are susceptible to cracking in the martensitic weld deposit and in the heat-affected zone, particularly when heavy sections are being welded. Light sections of the lower carbon grades may be an exception. For example, Type 410 in light wall thickness up to 1/8 in. ordinarily exhibits good welding characteristics, so that preheating provides little improvement and may be omitted.

Cracking may be minimized or avoided by preheating these steels to temperatures between 400 and 700 F, depending upon the hardening characteristics of the base metal and weld metal, and the intended service requirements. The preheat temperatures should be maintained during welding.

A postweld heat treatment between 1300 and 1450 F should follow immediately upon completion of welding for one hour per inch of weld thickness (minimum treatment of one hour). The postweld heat-treated sections should be cooled in air. Under these conditions, very ductile weldments may be obtained. In some cases, the postweld heat treatment may be done at any convenient time or may even be omitted and still result in completely satisfactory welds. Such procedures, however, should be specified only after consulting with properly qualified and experienced welding engineers.

The presence of some ferrite in the otherwise martensitic structure decreases the hardness developed in the steels as, for example, in the modified Type 410 grade (0.08% max C). This ferrite reduces cracking susceptibility. Nevertheless, cooling rates and interpass temperatures should be controlled. Preheat treatment between 300 and 500 F is generally advisable and should be followed by postweld heat treatment at 1300 to 1450 F. Only when welding thin pipe material below about 3/16 in. thickness may these preheat and postheat treatments be omitted.

When commercial Type 410 electrodes or welding rods are used to weld the Type 410 stainless steels, a preheat temperature between 500 and 700 F is advisable unless the joint thickness is less than 1/8 in. When low-carbon (modified) Type 410 grades (0.08 max C) are welded with Type 410 electrodes, this preheat temperature may be reduced.

Welding with either Type 310 or 309 austenitic stainless steel filler metals is preferred if a postweld heat treatment is not possible. However, a hardened heat-affected zone that is susceptible to cracking may still result.

Ferritic Stainless Steels.—Ferritic stainless steels usually contain between 10 and 30% chromium. Heating and cooling produce no significant structural changes in these steels and, as a result, they have a ferritic microstructure at all temperatures. These steels, therefore, can be hardened only by cold working, except for slight age-hardening characteristics shown by certain specially alloyed types.

Between 14 and 18% chromium, the separation between the martensitic and ferritic stainless steel classifications is approximate and depends primarily upon the carbon content of the particular stainless grade. Thus, a high carbon content results in a hard martensitic structure. With a lower carbon content, this steel would be primarily ferritic, as in Type 430. The addition of aluminum also tends to make the steel ferritic even at a chromium content of 11.5 to 14.5% as in

Table 85.5—Welding recommendations for wrought martensitic stainless pipe steels

ASTM and AISI Designations	Popular Designation	Chemical Composition (%)		Recommended Electrode or Welding Rod[1]	Preheat and Interpass Temperature	Postweld Heat Treatment[2]
		C	Cr			
410	12 Cr	0.15 max	11.5–13.5	E, ER410	600–700	Highly recommended
				E, ER310 or E, ER309	400–600	Recommended[3]
410 mod	12 Cr	0.08 max	11.5–13.5	E, ER410	300–500	Highly recommended
				E, ER310 or E, ER309	300–500	Recommended
420	13 Cr	over 0.15	12.0–14.0	E, ER410 or E, ER430	600–700	Highly recommended
				E, ER310 or E, ER309	400–600	Recommended

[1] AWS specifications. Characteristics of the commercial stainless steel electrodes and welding rods are summarized in Chapter 95.E means grade recognized by AWS as covered electrode; ER as bare electrode and welding rod.
[2] 1300–1400 F for 1 hour per inch of thickness.
[3] When preheat treatments are not employed, use small diameter electrodes.

Type 405, which contains 0.10 to 0.30% aluminum. On the other hand, 1 to 3% nickel makes these steels essentially martensitic as in Type 431, which contains 15 to 17% chromium and up to 2.5% nickel.

Welding recommendations for the common commercial ferritic stainless pipe steels are summarized in Table 85.6.

Since the completely ferritic stainless steels are not subject to air hardening, they are less susceptible to cracking in the welded section than the martensitic stainless steels. However, because these steels may become embrittled, their welding characteristics should be understood.

The chromium stainless steels which become fully ferritic at temperatures above 2100 F are generally susceptible to an embrittlement which is associated with solution of the carbide particles. This embrittlement is accompanied by severe grain growth. The embrittlement can be removed by annealing the steel for one hour between 1300 and 1450 F followed by quenching or air cooling, even though the grain size remains coarse. Such a postweld heat treatment is particularly important in single-pass welding where, without this postheat treatment, the ferritic weld metal, as well as part of the heat-affected zone, would be extremely brittle and would readily crack on subsequent deformation or bending operations at room temperature.

When postweld heat treatment is not possible, multipass welding with small-diameter electrodes, low current and stringer beads should be used to minimize this embrittlement. In these deposits, the subsequent weld beads produce annealing effects in the previous beads and reduce brittleness in the weld and heat-affected zones.

Since Types 405 and 430 tend to contain an average of about 50 to 70% ferrite, the balance being martensite, postweld heat treatments are usually required to reduce hardening.

In the welding of chromium stainless steels containing up to 23% chromium, satisfactory results are generally obtained with electrodes and welding rods having compositions similar to the base metal. However, Type 309 austenitic stainless steel electrodes and welding rods are also used extensively, although the trend is toward the use of the ferritic chromium stainless steel electrodes listed in Table 85.6. Preheat treatments are preferred. A postweld heat treatment at 1300 to 1450 F is essential if ductility is important, unless service at temperatures above 1000 F produces a similar effect. Chromium stainless steels containing more than 23% chromium are preferably welded with Type 310 or 309 electrodes.

Clad Steel Piping

In most applications, clad steel piping is selected for its corrosion resistance in environments which would attack mild or low-alloy steel. There are also some applications where the clad pipe steels are selected for their heat conductivity, oxidation resistance, wear resistance and other characteristics.

Although numerous liner combinations are possible, the majority of applications involve stainless-clad steels, where it has been determined that clad steels are more economical to use than solid stainless steel pipe.

On piping, where the weld must be made from one side only, special joint designs are required. Where backing strips can be used, suitable joint designs, which are illustrated in Fig. 85.2, may be employed. Welding is usually done by the shielded metal-arc process. Where a smooth surface contour on the blind side is desired, as in most piping systems, the joint should be prepared as shown in

Table 85.6—Welding recommendations for wrought ferritic stainless pipe steels

ASTM and AISI Designations	Popular Designation	Chemical Composition (%)			Recommended Electrode or Welding Rod[2]	Preheat and Interpass Temperature F	Postweld Heat-Temperature[1] F 1130–1400
		C	Cr	Other[1]			
405	12 Cr, A1	0.08 max	11.5–14.5	0.10–0.30 A1	E, ER430	Not necessary	Highly recommended
					E, ER310 or E, ER309	Not necessary	Recommended
					E, ER430	Not necessary	Highly recommended
430	16 Cr	0.12 max	14.0–18.0	....	E, ER310 or E, ER309	Not necessary	Recommended
446	27 Cr	0.20 max	23.0–27.0	0.25 max N	446	300–400	Essential
					E, ER310 or E, ER309	Not necessary	Recommended

[1] Unless otherwise specified, Mn and Si are 1.00 max., P is 0.040 max. and S is 0.030 max.
[2] AWS specifications. Characteristics of the commercial stainless steel electrodes and welding rods will be summarized in Chapter 94. E means grade recognized by AWS as covered electrode; ER as bare electrode and welding rod.
[3] 1130–1400 F for 1 hr. per in. of thickness.

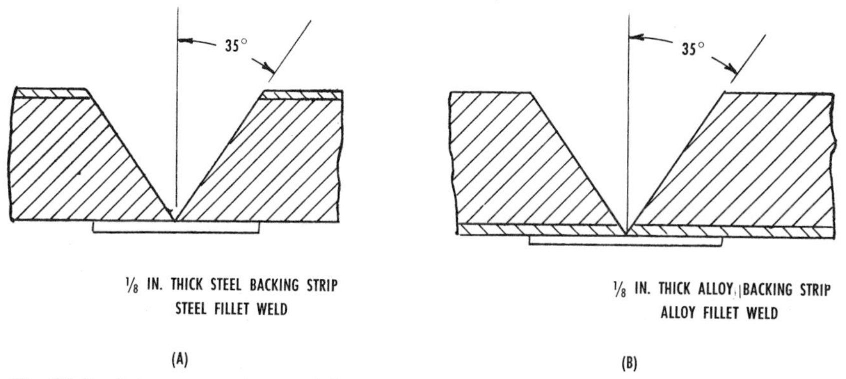

Fig. 85.2.–Joint preparation used for welding inaccessible joints: (a) steel side inaccessible (b) clad side inaccessible.

Fig. 85.3. The root is usually welded by the gas tungsten-arc welding process. On heavy pipe sizes, the balance of the weld is normally completed by shielded metal-arc welding. This practice is used in the piping industry, even on solid homogeneous materials and particularly in power piping installations.

Where the clad side is blind, the cladding should be welded with electrodes or welding rods of the same type of composition as the cladding metal, unless dilution must be considered. Special problems, however, may be involved in completing the weld.

On most metal pipe with wall thicknesses of 1/2 in. or less, the best practice is to complete the whole weld with the same electrodes or welding rods that were used to weld the cladding. This is not applicable to some nonferrous clad steels. On aluminum-clad steels, for example, the strength of the backup steel must be maintained. On such materials, special types of joints with overlap type root preparations may be desirable.

On wall thicknesses over 3/4 in., it is frequently desirable to complete the weld with the electrodes recommended in the previous tables for the corresponding base metals. This should occur after the clad root of the joint has been welded with the proper high-alloy steel or nonferrous electrodes. Care must be exercised in selecting the electrodes or welding rods and the welding procedure. They should not produce a brittle transition zone in the finished weld. It is frequently desirable to deposit a transition weld having properties somewhat between those of the cladding and backing steel. This is called buttering or surfacing. It is considered undesirable, for example, to deposit mild or low-alloy steel weld metal against the austenitic stainless steels. The transition zone would be rather hard, brittle and highly susceptible to cracking.

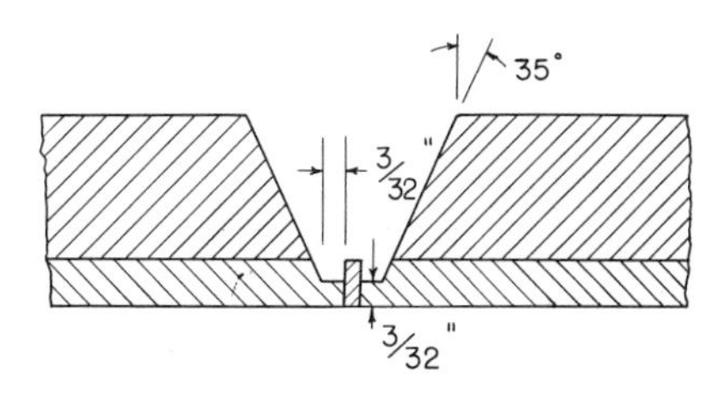

Fig. 85.3.–Recommended joint preparation for clad pipe joints requiring a smooth internal contour.

The proper heat-treating temperatures and procedures must be based on the characteristics of both the backing steel and the cladding alloy. Care should be taken to specify a heat treatment which produces beneficial effects in one material and yet will not harm the other. Cooling rates may have to be specified in cases where rapid cooling may adversely affect one of the base metals. It must be recognized that the coefficients of expansion of the respective metals may differ considerably and perhaps result in serious residual stresses.

Lined Piping

Applied liners consist of sheets, or relatively thin plates bonded intermittently by welding to a backing metal, usually carbon or low-alloy steel. Applied liner construction thus usually combines a low-cost carbon or low-alloy steel for strength of relatively heavy wall thickness with a high-cost stainless steel or nonferrous alloy of thin wall thickness.

In most applications, liners are selected for their corrosion resistance to environments which would attack mild or low-alloy steel. There are also many applications where applied liners are selected because of their wear resistance, oxidation resistance, heat conductivity or other characteristics.

Although numerous liner combinations are possible, the greatest tonnage produced involves stainless steel liner materials. In piping, applied liners may become more economical than solid materials or clad materials in wall thicknesses exceeding 3/4 in.

On pipe where lining methods are somewhat more costly than on large diameter tanks or pressure vessels (see Chapter 93), it is generally more economical to use solid stainless or other alloy base metals. However, there are applications where rather costly metals are required, such as the nickel-base alloys. When this is the case, the lining of carbon steel pipe with preformed liner plates may be more economical than the use of solid nickel-base alloy piping (Fig. 85.4). Availability of some of the special alloy metals is also a problem, since elbows or other fittings of nickel-base alloys, or other high-alloy base metals frequently are not readily available. Special fabrication of such shapes would involve considerable time and cost which might prohibit their use. For such conditions, lining may be accomplished without undue delay and often at a significantly lower cost.

In severely corrosive environments requiring refractory metals such as titanium, zirconium or tantalum, thin walled ferrous base metals may be lined

Fig. 85.4.–Elbow and reducer lined with nickel-base alloy strips.

Fig. 85.5.–Malleable iron 6 in. diaphragm valve body lined with zirconium by welding 0.060 in. thick strips of zirconium.

with the above metals to effect savings. Figure 85.5 illustrates a malleable iron diaphragm valve body lined by welding of 0.060 in. thick preformed zirconium strips.

Lining is generally done by the strip welding technique. The length of the strips usually varies from 2 to 10 ft, whereas the width may vary between 2 and 6 in., depending upon the service requirements. The values given in Table 85.7 are examples of strip widths used to fabricate stainless steel lined piping for different operating temperatures.

Table 85.7—Strip widths for strip-welded liners

Maximum Operating Temperature (F)	Maximum Strip Widths of Stainless Steels	
	Type 400 Series (in.)	Type 300 Series (in.)
Up to 450	Up to 6	Up to 6
450–750	4 1/2–5	3–4
750–850	4	2 1/2–3
850–950	3 1/2	2
Over 950	3	2

Narrower strips should be used when the service involves severe thermal cycling or steep temperature gradients. On the other hand, strips or widths of 24 in. have been used for essentially noncritical service applications at room temperature.

When lining pipe, the practical width of the liner strips will depend also on the pipe diameter or the inside curvature. For example, on 14 in. pipe, 4 in. wide liner strips would generally be considered practical.

Figure 85.6 illustrates the shingle strip lining technique recommended for 10 to 20 in. diameter pipe, employing 4 to 5 in. wide liner strips. Where liquids flow through the lined piping system, the ends of the shingles overlapping previously welded shingles should be on the downstream side.

At pipe inlets, manhole openings or wherever necessary, the exposed backing steel surfaces can be protected by weld metal overlays with filler metal alloys of proper composition. The effects of dilution must be considered in the selection of the proper filler metal.

At locations in lined pipe or fittings, because of inaccessiblity, it may not be practical to apply overlays. The ends of flanges may be overlayed or filled with weld metal when a gap exists between the liner strip and the flange face plate.

NONFERROUS MATERIALS

Welding of nonferrous piping and tubing most commonly involves the following materials:

1. Aluminum and aluminum alloys.
2. Nickel and nickel alloys.
3. Copper and copper alloys.
4. Titanium and titanium alloys.
5. Lead and lead alloys.

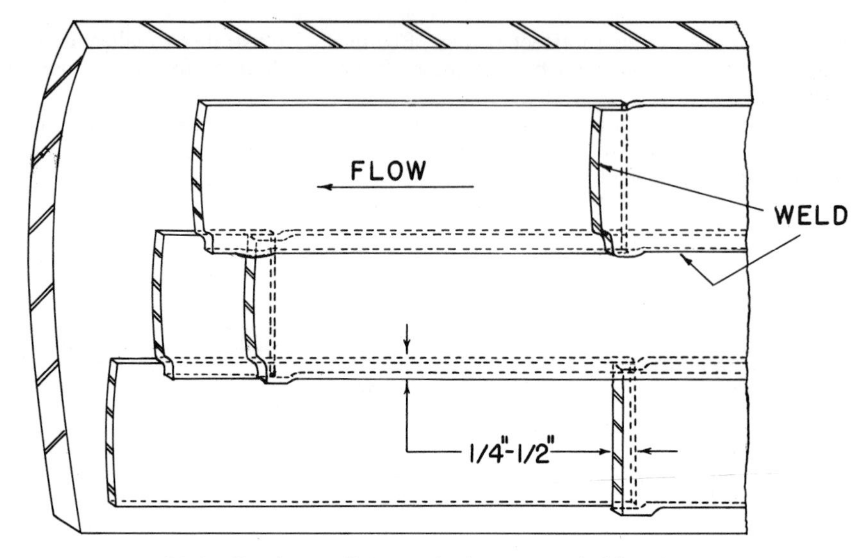

Fig. 85.6.—Shingle strip lining method recommended for piping.

Material And Dimensional Standards

The American Society for Testing and Materials has developed a number of standards governing the material requirements for aluminum, nickel and copper pipe and tubing materials in commercially "pure" and many alloy compositions. ASTM Specifications also cover plate, casting and forging materials from which pipe, tubing, valve bodies and fittings frequently are produced.

Dimensional tables for aluminum alloy pipe are given in ANSI H35.2 and include pipe sizes from 1/8 to 12 in. nominal diameter and Schedules 5, 10, 40 and 80. Permissible variations in diameter and wall thickness are included in several ASTM tube specifications applicable to aluminum and aluminum alloys.

Dimensional tables for nickel and nickel alloy materials are contained in a number of ASTM Specifications: B161, Nickel Seamless Pipe and Tube; B165, Nickel-Copper Alloy Seamless Pipe and Tube; B167, Nickel-Chromium-Iron Alloy Seamless Pipe and Tube. These specifications include Pipe Schedules 5, 10, 40 and 80 and dimensional tolerances.

ASTM Specification B251 provides General Requirements for Wrought Seamless Copper and Copper-Alloy Pipe and Tubes, and contains dimensional tables for regular and extra strong pipe which correspond to standard weight and extra strong iron pipe size and dimensions. Specific material requirements are given in ASTM Specifications B42, Seamless Copper Pipe, Standard Sizes; B43, Seamless Red Brass Pipe, Standard Sizes; B315, Copper-Silicon Alloy Seamless Pipe and Tube. ASTM Specification B302, Threadless Copper Pipe is written to cover deoxidized copper pipe for systems assembled with brazed-joint pipe fittings.

ASTM standards also exist for alloy castings and forgings. In the case of copper alloys, standards have been written specifically for use in manufacturing valves and fittings. Some of these standards cover the use of mechanical rather than welded joints.

ANSI standards have not yet been formulated to cover the dimensions of nonferrous pipe, valves, fittings and flanges other than screwed brass fittings, brass flanges and flanged fittings, and solder-joint copper and brass fittings. Since the nickel and aluminum alloys are used mainly for corrosion resistance or avoidance of fluid contamination for sanitary reasons, usually under mild conditions of pressure and temperature, the practice has been to construct valves, fittings and flanges to the dimensional standards used by manufacturers of comparable brass components.

Valves and fittings of the copper alloys are frequently soldered or brazed into piping systems particularly in applications involving the beverage industries. Increasing use of welding is made with aluminum- and nickel-alloy pipe materials and with copper. This includes substantial use of welding fittings. In the case of all but titanium and lead, such nonferrous welding fittings are often made to the same dimensions as wrought steel butt-welding fittings (ANSI B16.9). Because of the greater cost of these metals and the relatively mild pressure-temperature service conditions to which they are generally applied, such fittings and the pipe to which they are welded are used frequently in the thinner pipe schedules 5 and 10, involving thinner walls than their carbon steel counterparts. In other instances, and for the same reasons, matching fittings and tubing having different outside diameters than pipe, but with even thinner walls than lightweight pipe, are employed. These types of construction generally involve the use of transition pieces or adapters to permit connection to screwed and sometimes to brazed valves and other pieces of equipment.

Weldability

The weldability of ferrous piping materials is related primarily to their tendency to crack during the welding operations or during the subsequent cooling period, and to a lesser but highly important degree, to their state of internal residual stress as a result of welding. This last characteristic is usually of major importance with nonferrous materials.

Nonferrous metals, unlike some steels, generally do not tend to harden on cooling in air. They do not have suppressed transformation at temperatures below the plastic range and are not accompanied by appreciable shrinking. Thus, even though the ductility of the deposited metal may be lower than the base metal, the danger of cracking is remote in most alloys. The residual stresses are not usually of sufficient magnitude to warrant concern over possible stress-corrosion effects. Nevertheless, in some corrosive environments, nonferrous alloys may become susceptible to stress-corrosion cracking. The nonferrous piping materials do not experience the intergranular carbide precipitation which may be caused by the heat of welding in unstabilized austenitic stainless steels. There is usually no necessity for subsequent postweld stress relief or other heat treatment after welding, except as may be required by the ASME Boiler and Pressure Vessel Code or other applicable codes.

Aluminum and Aluminum Alloys

Relatively few of the many recognized aluminum alloys are commercially available as piping materials. Aluminum and aluminum alloys, from the standpoint of weldability, are characterized chiefly by their low melting points (1075 to 1210 F), high heat conductivity and high fluidity in the molten state. Training and experience are necessary before production welding should be attempted. Whereas aluminum and aluminum alloys can be joined by many of the commercial welding and brazing processes, those most commonly employed are the gas tungsten-arc and gas metal-arc processes. Some aluminum alloy piping and tubing are supplied in either a work-hardened or precipitation-hardened condition (obtained by cold-working or heat treatment, depending upon the alloy) which gives the material greater strength. In these alloys, the welding operation softens the metal for a distance of one to five times the pipe wall thickness on either side of the weld and materially reduces the tensile and yield strengths in this zone.

It is impractical to restore the original strength of fabricated assemblies by subsequent cold work or heat treatment. In spite of this, it is often desirable to use cold-worked or heat-treated pipe to otain increased resistance to denting or bending in handling and erecting. The bursting strength of pipe joined by circumferential butt welds is not reduced from that of unwelded pipe to the extent that might be expected, based on the actual local strength of the joint after welding.

Most widely used as filler metals are commercially pure aluminum (1100) and an aluminum alloy containing 5% silicon (4043). Care should be exercised in employing this latter filler metal because its corrosion resistance may be lower than the base metal. For instance, strong nitric acid will corrode 4043, but will not affect 3003. The 5% silicon filler metal is desirable for the 6061 and 6063 piping materials. More highly alloyed filler metals, however, are required for many aluminum alloys to provide a better match of mechanical or physical properties. The majority of commercial aluminum filler metals are covered by the AWS Specification on Aluminum and Aluminum Alloy Welding Rods and Bare Electrodes (AWS Specification A5.10).

The filler metals normally recommended for some of the more common commercial pipe materials are listed in Table 85.8.

Table 85.8—Welding filler metals recommended for the more common commercial aluminum piping materials

Base Metal	Bare Welding Rod	Covered Electrode
1100	1100	1100
3003	1100	1100
Alclad 3003		
3004	4043, 5154	4043
5050	4043	4043
5052	4043 5154	4043
5154	5154	4043
5254	5556	4043
5454	5556	4043
6061	4043, 5356	4043
6062	4043, 5356	4043
6063	4043, 5356	4043

The high fluidity of molten aluminum promotes sink, which tends to be most severe in horizontally positioned piping—particularly in the fixed postion. When not objectionable, aluminum backing rings are occasionally used. Collapsible stainless steel backup fixtures are also used, provided that they can be removed from the pipe.

Consumable insert rings are also used when welding with the gas tungsten-arc welding process, particularly when good root penetration is desired along the inside of the pipe. Preheating is desirable on aluminum pipe of diameters exceeding 2 1/2 in. for ease of welding. Typical welding recommendations for gas tungsten-arc welding of pipe are provided in Table 85.9.

Nickel and Nickel Alloys

This group of alloys exists primarily because of strength properties and good corrosion resistance to many acid environments. Nickel and high-nickel alloy pipe and tubing may be readily joined by welding the same materials to themselves or other dissimilar alloys such as ferritic and austenitic steels.

Nickel and high-nickel alloy castings of various compositions are obtainable. Regular foundry grade castings may be welded or repaired by the shielded metal-arc or gas tungsten-arc welding processes. When considerable welding is required, or when castings require welding to cast or wrought forms, such as fittings to pipe, casting chemistry must be modified to permit welding. In such cases, the order for castings should be clearly marked "Weldable Grade Castings Required" or by some similar statement.

Table 85.9—Welding recommendations for aluminum piping materials

Nominal Pipe Size, (in.)	Position	Preheat[1]	Current (A)	Electrode Size, (in.)	Argon Flow Cfh	Filler Rod Diameter (in.)	Number of Passes
1	H (rolled)	none	130	3/32	25	1/8	1
	H (fixed)	none	120	3/32	25	1/8	1
	V	none	125	3/32	25	1/8	1
1 1/2	H (rolled)	none	140	3/32	25	1/8	1
	H (fixed)	none	125	3/32	25	1/8	1
	V	none	130	3/32	25	1/8	1
2	H (rolled)	none	150	1/8	25	1/8	2
	H (fixed)	none	130	1/8	25	1/8	1
	V	none	140	1/8	25	1/8	2
2 1/2	H (rolled)	up to 400	170	1/8	25	1/8	2
	H (fixed)	350–400	140	1/8	25	1/8	1
	V	up to 400	150	1/8	25	1/8	2
3	H (rolled)	up to 400	190	5/32	25	5/32	2
	H (fixed)	350–450	150	5/32	30	5/32	1
	V	up to 400	170	5/32	25	5/32	2
4	H (rolled)	up to 400	225	5/32	30	3/16	2
	H (fixed)	400	175	5/32	35	5/32	1
	V	up to 400	200	5/32	30	5/32	3
6	H (rolled)	up to 500	250	3/16	30	3/16	2
	H (fixed)	400	190	3/16	40	3/16	1–2
	V	up to 500	220	3/16	35	3/16	3
8	H (rolled)	up to 500	275	3/16	35	3/16	3
	H (fixed)	400–600	225	3/16	45	3/16	2
	V	400–600	250	3/16	40	3/16	4
10	H (rolled)	up to 500	300	1/4	40	3/16	3
	H (fixed)	400–600	225	3/16	50	3/16	2–3
	V	400–600	275	3/16	45	3/16	5
12	H (rolled)	up to 500	325	1/4	40	3/16	3
	H (fixed)	400–600	250	3/16	50	3/16	3
	V	400–600	300	3/16	45	3/16	5–6

[1] Preheat temperatures above about 400 F and especially as high as 600 F will have deleterious effects on properties of alloys such as 6061, 6062 and 6063. Where preheat is needed for those alloys, the temperature should not exceed 400 F.

Welding Processes.—The welding processes used to join the nickel and high-nickel alloys are the same as used for joining steel. However, some processes are not applicable to certain nickel alloys. The welding processes applicable to each alloy are tabulated in the chapter on nickel in Section 4.

Since the nickel and high-nickel alloys are usually employed in severely corrosive or high-temperature service, the choice of welding process should be evaluated carefully. In some environments, it may be necessary to employ two different welding processes on the same joint. For example, when product purity is important and when elevated temperature service is expected, the gas tungsten-arc process must be used to make the root pass weld if the reverse side of the joint is inaccessible for slag removal.

Nickel and high-nickel alloys are susceptible to embrittlement by lead, sulfur, phosphorus and some low melting metals and alloys. These contaminators, singly or in combination, are liable to be found around piping installations in the form of grease, oil, machining lubricants, paint, marking crayons, marking inks, some temperature-indicating crayons and shop dirt. In welding the nickel and high-nickel alloys, all foreign materials are considered as contaminators unless proven ineffective. It is essential that joints to be welded, and adjacent surfaces, be cleaned mechanically and chemically to bright metal.

Before welding base metal which has been hot-worked in processing or fabrication, it is necessary to remove the thin, dark-colored oxide film from the immediate vicinity of the area to be welded, either mechanically by machining, sand-blasting, grinding or rubbing with an emery cloth, or chemically by pickling. Difficulty in the form of an unstable arc may be encountered if the oxide is not removed due to the repelling effect created by the high melting-point oxide.

Joint Design.—In general, the nickel and high-nickel alloys possess limited wetting properties. The molten weld metal is more viscous, requiring joints to be opened up to allow for filler metal manipulation. Except for a wider included joint angle (80 deg minimum), joint designs approximate those used for steel welding and are shown in other sections of this chapter.

Conventional backup rings are rarely used in nickel and high-nickel alloy piping. Crevices cannot be avoided and may promote corrosion in some environments. In some alloys, the notch effect can also cause root cracking. Figure 85.7 illustrates a typical root crack starting at a backing ring used in nickel-chromium-iron alloy. Consumable insert rings are generally preferred.

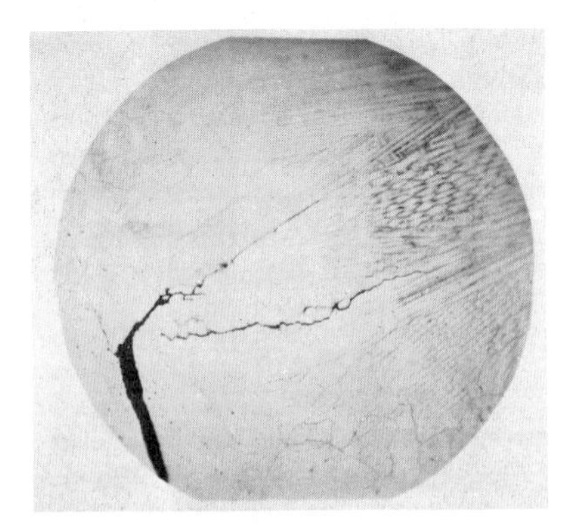

Fig. 85.7.—Root crack in nickel-chromium-iron alloy weld originating at backing ring.

The filler metals used to join the high-nickel alloys must exhibit equivalent, or nearly equivalent, properties if the weld is to perform satisfactorily in the same environment as the base metal. The nominal composition of the filler metal is dictated by the anticipated service performance, and should match as closely as possible that of the base metal. In most alloys, it is necessary to depart a little from the nominal composition in order to satisfy particular weldability requirements. Additions are frequently made, for instance, to control porosity or hot-cracking tendencies. To this extent, filler rod, wire and electrode deposits may not be identical in composition to the base metal on which it is intended they be used. The nickel and high-nickel alloy filler metals contain some alloying elements to control weld quality.

Nickel and high-nickel alloys may be welded in all positions with only slight modification of the welding procedures used in welding stainless steel. As with any welding job, it is best to position for downhand welding whenever possible. Vertical welding in the horizontal fixed pipe position should preferably progress in an upward direction.

There is a common tendency to use excessive current to promote fluidity in the nickel and high-nickel alloys. Operating outside the filler metal manufacturer's recommended amperage range may cause overheating of the last half of the electrode, resulting in loss of arc control, weld spatter and porosity.

Since nickel and high-nickel alloy weld metal is more viscous than stainless steel weld metal, careful electrode manipulation is necessary to obtain satisfactory weld bead contour. Generally, this favors a slight amount of weaving and accurate weld metal placement by the welder.

Where traces of residual welding slag might contaminate the contents within the welded pipe, the gas tungsten-arc welding process is used, at least, for the root pass. Consumable inserts are used extensively in making root pass welds in the nickel and high-nickel alloys.

Purging of the inside of piping underneath the weld is very important in root pass welding of the nickel and high-nickel alloys. Helium, argon, hydrogen and mixtures of these gases have been employed as purging gases. Helium as a purging gas has the ability to increase wetting. Nitrogen has also been used for purging.

The shielding gas is usually argon, helium or a mixture of the two. Some fabricators prefer adding a small amount of hydrogen to these gases. Hydrogen is thought to increase the heat of the weld pool so that any gas in the molten metal is more readily evolved. This may have some merit in welding root passes as the chilling effect from the mass of base metal on either side of the joint can otherwise cause fusion-line porosity. However, excessive hydrogen in the shielding gas may promote porosity in the nickel-base weld metal deposits.

The procedure used for manual gas tungsten-arc welding duplicates, in many respects, the procedures used for oxyacetylene welding in all positions. Filler metal is added in the same manner and the precautions on avoiding excessive agitation of the weld puddle, keeping the hot end of the filler metal in the protective gas atmosphere and relation of wire diameter to the material being welded are comparable.

The short-circuiting modification of the gas metal-arc process is being used extensively for the welding of nickel and high-nickel alloy piping. In common with other processes, procedure variables must be closely controlled if predictable results are to be obtained.

Minimum porosity is obtained in manually welded joints when the joint consists of at least 50% added filler metal.

Submerged arc welding may be used in shop piping fabrication on the nickel-chromium-iron, nickel-copper and copper-nickel alloys. In general, smaller diameter wires are used for these materials than normally used for submerged arc welding of steels. (1/16 in. diam wire is the most common size employed with 3/32 in. diam being used infrequently.) The same filler metals applicable to the gas metal-arc welding process are employed. However, suitable proprietary fluxes must be used. On backing bar joints, the first weld layer is usually made by shielded metal-arc welding, gas tungsten-arc welding or gas metal-arc welding. Single-Vee joints normally require the first two weld layers to be made using the shielded metal-arc, gas tungsten-arc or gas metal-arc process.

The oxyacetylene welding process is occasionally used for welding nickel and high-nickel alloy pipe and tube, particularly when the wall thickness is too thin for shielded metal-arc welding. Because of cleanliness requirements, the gas tungsten-arc welding process is used far more extensively than other processes for welding on thin piping materials.

The heat of welding does not produce deleterious effects on the base metal. The only known effect is annealing and a slight grain growth in the heat-affected zone. The nickel-moly alloys are an exception to this and for some service, must receive a solution heat treatment after welding.

Brazing.–Nickel and high-nickel alloys may be brazed with the same facility as other common base metals.

Before selecting a brazing alloy, it is important to ascertain whether the service requirements of the joint (strength or corrosion resistance) can be obtained with a given brazing alloy. Brazing alloys containing phosphorus should never be used on nickel and high-nickel alloys. Silver alloys are by far the most frequently used, and the lower melting point alloys are preferred. Copper alloys may also be used as brazing filler metals.

Fluxes satisfactory for other base metals are satisfactory for many of the nickel and high-nickel alloys. Special fluxes are required for the aluminum-containing nickel and high-nickel alloys. If a suitably dry, reducing, sulfur-free furnace atmosphere is available, flux may not be required.

The high-nickel alloys are subject to stress-cracking in the presence of molten low melting point brazing alloys such as those containing cadmium, zinc and silver. Severely cold worked parts should be annealed before brazing.

Copper and Copper Alloys

Deoxidized or oxygen-free copper pipe and tubing have been standardized. Oxygen-bearing electrolytic or tough-pitch copper seamless tubes are not usually available.

The chemical composition, high heat conductivity and high fluidity of copper and copper alloys when molten, have considerable bearing on their weldability. Oxygen-bearing electrolytic or tough-pitch copper is susceptible to the formation of gas bubbles when welded or heated above 1290 F in an atmosphere containing hydrogen. These coppers are not usually recommended for the arc and gas welding processes. They can, however, be readily soldered and also brazed, provided a silver brazing alloy having a flow point less than 1290 F is used, and the brazing temperature is carefully controlled. Piping of deoxidized copper or oxygen-free copper can be satisfactorily arc or gas welded, soldered or brazed in any position.

Because of the high fluidity of molten copper, shielded metal-arc and gas welds should generally be made with suitable backing rings. The gas tungsten-arc and gas metal-arc welding processes are also being used with good success.

All-position welding of piping by oxyacetylene welding is possible, but quite difficult because of the high fluidity of molten copper. Soldering and brazing are also commonly used, particularly in beverage industry applications. The high heat conductivity of copper usually requires oxyacetylene preheating when welding large diameter or unusually heavy copper piping.

Oxyacetylene welding of red brass and yellow brass can be satisfactorily accomplished in all positions with suitable backing rings. Shielded metal-arc, carbon-arc and gas metal-arc welding are not usually employed as the high heat intensity of the arc causes vaporization of the zinc content in the metal and sweating out and oxidation of any lead that may be present. Therefore, soldering and brazing are more generally employed. Large diameter and heavy wall pipe should generally be preheated from 300 to 500 F. Due to dezincification of yellow brass in some corrosive environments, red brass is generally used. Yellow brass pipe is not commercially available, and is made only on special order.

Copper-silicon alloys have a heat conductivity approaching that of steel. Their good welding characteristics are aided by a glass-like slag which forms on the surface of the molten weld pool. They generally do not contain chemical constituents which easily vaporize, hence they are more easily welded than any of the other copper alloys used in piping. The copper-silicon alloys are welded extensively by the gas tungsten-arc and the oxyacetylene welding processes. Shielded metal-arc welding is occasionally employed. Welds are readily made in the flat and vertical positions; the overhead position is more difficult. Backing rings are not required, although they are sometimes used.

The principal copper-nickel alloy used for pipe and tubing is that known as cupronickel 30, in which the number indicates the percent of nickel. Other cupronickel alloys with lower nickel proportions are available and are coming into increasingly popular use. The copper-nickel alloys offer superior resistance to the corrosive action of sea water, and are extensively used for water pipe and condenser tubing on ships. The most suitable processes for welding are shielded metal-arc welding and gas tungsten-arc welding. Oxyacetylene welding, with a special silicon-bearing welding rod, is also used successfully with the cupronickel 30, and to a somewhat lesser extent, with cupronickel 10. Brazing with other alloys is also utilized. However, care must be exercised to select a brazing alloy with a low-flow point to prevent intergranular penetration. This may occur at high temperatures if the base metal contains residual stresses set up during forming, shaping or cold bending.

Titanium and Titanium Alloys

Titanium can be welded by arc-welding techniques to produce joints with 100% efficiency. However, on titanium alloys (as on other refractory metals such as zirconium, tantalum and columbium), it is extremely important that the molten, hot weld and base metal area is effectively shielded from the atmosphere and from foreign materials. The molten weld metal and heat-affected zones must be shielded by a protective blanket of inert gases during welding. This can be done either by welding inside of special chambers or, partially, by means of the so-called trailing shield (Fig. 85.8). The gas tungsten-arc and gas metal-arc welding processes are most commonly used. To provide rapid cooling, it is also desirable to employ copper chill clamps wherever practical.

Light wall pipe and tubing are normally welded without filler metals in a single pass by the gas tungsten-arc welding process. As wall thickness increases

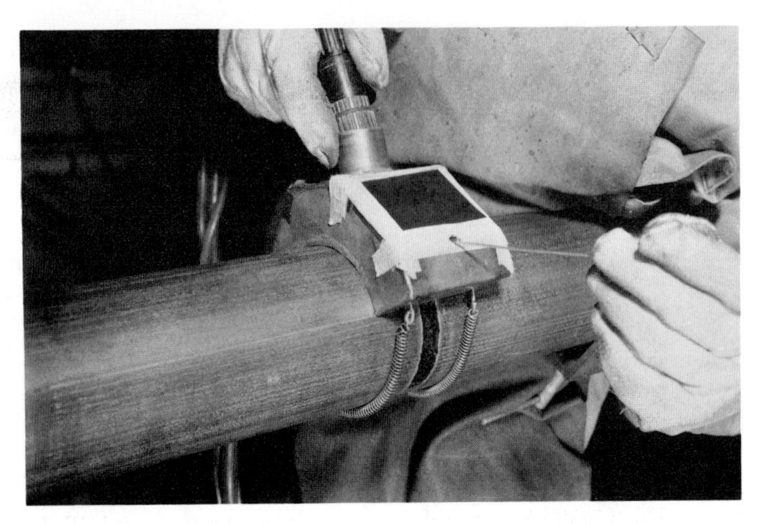

Fig. 85.8.–Trailing shield for welding of titanium tubing.

beyond 0.062 in., it will be desirable to add filler metal. The use of filler wire, however, may increase the possibility of contamination if the filler wire has not been thoroughly cleaned, or if the hot end of the filler wire is withdrawn from the protective gas and exposed to the atmosphere during intermittent disposition of the filler metal. Since there is a tendency in manual welding for some contamination of the weld to occur, resulting in embrittlement, it is particularly important that pipe and tube butt welds have complete penetration at the root of the weld. Otherwise, inadequate joint penetration in the weld root may propagate into a crack during service and result in a serious failure. (Fig. 85.9).

Fig. 85.9.–Cracking in titanium tube weld showing inadequate joint penetration and a brittle weld due to insufficient protection from the atmosphere.

Lead and Lead Alloys

The satisfactory welding of lead and its alloys requires a high degree of manual skill which is acquired only after substantial practice. This is due to low melting temperatures and high fluidity of the lead-base alloys.

The oxyhydrogen welding (OHW) process is most widely used. The oxyacetylene process is not generally used for wall thicknesses below 1/2 in., as this mixture of gases develops too high a temperature for welding thin sections. Air-hydrogen, air-acetylene or oxy-city gas may be used but should be avoided if oxyhydrogen is available.

Provisions for adequate ventilation should be made where lead is to be welded in closely confined quarters, to avoid the toxic effects of lead fumes.

Dissimilar Base Metals

The joining of dissimilar metals by any of the welding processes is dependent upon the relative melting point of the metals to be joined. If the melting points are close (some authorities state within 100 F), normal welding techniques are generally used.

The practice normally followed for welding dissimilar base metal combinations between carbon and low-alloy steel as recommended in PFI Standard ES–9 (May 1964) are summarized in Tables 85.10 and 85.11.

Table 85.10—Preheat recommendations for dissimilar steels welded to low-carbon steel

Types of Steel (%)	Preheats to be Applied: Before Tack Welding (F, Min)	Preheats to be Applied: Before & During Welding[1] (F)
C–Mo	70	70 to 400
1 Cr–1/2 Mo 1 1/4 Cr–1/2 Mo	150	300 to 600
2 1/4 Cr–1 Mo 3 Cr–1 Mo	250	400 to 700
5 Cr–1/2 Mo 7 Cr–1/2 Mo 9 Cr–1 Mo 12 Cr	350	400 to 700

[1] In general, the higher temperature within the specified ranges should be used for heavier thickness and more complex structures.

Electrodes

The electrode used for welding dissimilar metal joints may be chosen to produce deposited metal of the same nominal composition as either of the base metals or of an intermediate alloy content. Selection of weld metal analysis should receive careful consideration if the dissimilar metal joint is to be exposed to heat-treating or service temperatures where significant carbon diffusion occurs. In general, low-hydrogen lime-type electrodes of the AWS Classifications

Table 85.11–Postweld heat recommendations for dissimilar ferritic steel combinations

	C-Mo (F)	1 Cr–1/2 Mo 1 1/4 Cr–1/2 Mo (F)	2 1/4 Cr–1 Mo 3 Cr–1 Mo (F)	5 Cr–1/2 Mo 7 Cr–1/2 Mo 9 Cr–1 Mo 12 Cr (F)
Carbon Steel	1150 to 1300	1150 to 1300	1150 to 1400	1150 to 1400
C-Mo		1200 to 1350	1200 to 1400	1200 to 1400
1 Cr–1/2 Mo 1 1/4 Cr–1/2 Mo			1275 to 1400	1275 to 1400
2 1/4 Cr–1 Mo 3 Cr–1 Mo				1300 to 1425
5 Cr–1/2 Mo 7 Cr–1/2 Mo 9 Cr–1 Mo 12 Cr				

Note: 1. The holding time shall be a minimum of 1 hour per inch of wall thickness but in no case less than 1 hour.
2. In the case of joints between carbon and carbon-molybdenum steels, postweld heat treatment is not required when the carbon steel has a carbon content under 0.35% and a wall thickness under 3/4 in. and the carbon-molybdenum steel has a carbon content not exceeding 0.25% and a wall thickness of 1/2 in.

EXX15, EXX16 and EXX18 are preferred for welding dissimilar chromium-bearing ferritic alloy steels.

Austenitic stainless steel weld metal and other high-alloy compositions are sometimes employed in welding joints containing dissimilar ferritic steels. The possibility of stresses caused by the difference in the coefficient of expansion of ferritic and austenitic materials during thermal treatments or under service conditions should be given careful consideration. Carbon diffusion must be recognized where service temperatures exceed 800 F, particularly if heavy sections or cycling temperatures are involved.

For dissimilar welds involving nonferrous alloys, the selection of a welding process, filler metal and welding procedure must be made with care. The choice of filler metal should be based on an understanding of the metallurgical aspects of the proposed joint. The welding procedure and joint design should be based upon the known weldability characteristics of each of the dissimilar metals. The selection of welding process, filler metal, joint design and welding procedure should be evaluated by adequate procedure qualification tests.

Overlaying of the higher melting metal prior to final welding frequently is done. For example, when joining silicon-bronze pipe to carbon steel pipe, the end of the latter may be "buttered" with a silicon-bronze weld overlay.

Preparation of the welding procedure requires knowledge of the relative compatibility of dissimilar base metals. This includes consideration of the dilution which can be tolerated without resulting in a weld containing defects, brittleness or other undesirable characteristics. For example, most nickel and high-nickel alloys can be joined to other metals with tolerance for considerable dilution. The amount of dilution should be controlled for consistently satisfactory results.

Where a wide difference (usually over 100 F) in melting temperatures exists, it is often necessary to resort to brazing, braze welding or soldering techniques.

Each of the metals involved in the joint must individually be capable of being welded, brazed or soldered.

The design of joints for welding dissimilar metals is exactly the same as that for welding similar metals. Groove and fillet welds of the standard dimensions are normally used. The differences in coefficients of expansion, thermal conductivity, etc., influence the welding techniques and procedure more than the joint design. Actually, there are very few dissimilar basic types of metals which may be welded. Stainless steels are sometimes welded to mild or low-alloy steels, generally using stainless steel filler metal. Similarly, nickel and some nickel alloys may be welded to mild and stainless steels.

Brazing with silver-copper-cadmium-zinc alloys offers the best approach to joining dissimilar metals where the presence of this alloy is compatible with service requirements. Copper-phosphorus and copper-silver-phosphorus brazing filler metals are suitable only for copper and copper-base alloys. They should be used with care on dissimilar metals. They cannot be used on ferrous metals, nickel or nickel alloys. The standard joints employed in brazing similar metals apply equally well to dissimilar metals.

Soldering provides a means of connecting metals and alloys of widely different melting temperatures, particularly where one of the metals melts below the temperature necessary for brazing. Soldering is also helpful in those situations where it is necessary to avoid the high temperatures of welding and brazing. Joint designs for soldering dissimilar metals follow the principles and standards established for soldering similar metals.

JOINT DESIGN

GROOVE WELDS

The most common type of joint employed in the fabrication of welded pipe systems is the circumferential butt joint. It is the most satisfactory from the standpoint of stress distribution. Its general field of application is pipe to pipe, pipe to flanges, pipe to valve and pipe to fitting joints. Butt joints may be used for all sizes, but fillet-welded joints are often used to advantage for joining flanges, valves and fittings to pipe 2 in. and smaller in diameter.

Welding fittings and frequently pipes, valves and other standard components are usually furnished with end preparations which are made by machining in accordance with the ANSI Code for Power Piping as shown in Fig. 85.10. These bevels also appear in the American Standard for Steel Pipe Flanges and Flanged Fittings, ANSI B16.5, in the Standard for Butt Welding Fittings, ANSI B16.9 and the Standard for Butt Welding Ends, ANSI B16.25.

On piping, the end preparation is normally done by machining or grinding. On carbon and low-alloy steels, oxygen cutting (OC) is also used, particularly on pipe of wall thicknesses below 1/2 in.

BUTT JOINTS BETWEEN UNEQUAL WALL THICKNESS

When piping components of unequal wall thicknesses are to be welded, care should be taken to provide a smooth taper on the edge of the thicker member. Minimum lengths of taper are specified by the various sections of the ASME

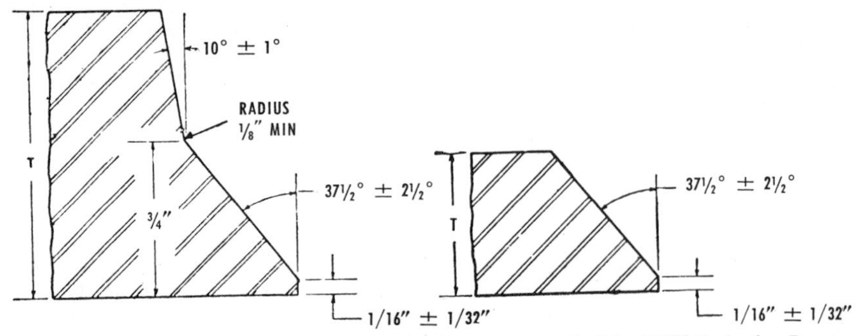

Fig. 85.10.–Standard pipe bevel preparation as recommended in ANSI Code for Pressure Piping. Butt welding ends having thickness (T) less than 3/16 in. for carbon steel, ferritic alloy steel or wrought iron, or 1/8 in. for austenitic alloy steel, may be square or have a slight chamfer; thicknesses greater than these, through 7/8 in. inclusive, should be beveled 37 1/2 deg, and the root face (land) should be 1/16 in. Thicknesses greater than 7/8 in. should be beveled 37 1/2 deg with a thickness beyond 3/4 in. beveled 10 deg.

Boiler and Pressure Vessel Code. The two methods of alignment which are recommended are shown in Fig. 85.11.

The wall thickness of cast steel fittings and valve bodies is normally greater than that of the pipe to which they are joined. In order to provide equal thicknesses for welding, the applicable sections of the ASME Boiler & Pressure Vessel Code and the ANSI Code for Pressure Piping permit the machining of the cylindrical ends of cast steel fittings and valve bodies to the nominal wall thickness of the adjoining pipe, provided these areas are finish-machined both inside and outside and are carefully inspected. The machined ends may be extended back in any manner provided the longitudinal section comes within the maximum slope line indicated in Fig. 85.11. The transition from the pipe to the fitting or valve end at the joint must be such as to avoid sharp reentrant angles and abrupt changes in slope.

END PREPARATION FOR GAS TUNGSTEN-ARC (ROOT PASS) WELDING

Whereas the pipe and bevel preparations shown in Fig. 85.10 are considered adequate for shielded metal-arc welding, they pose a problem when the gas tungsten-arc welding process is used. When this process is employed, extended "U" or "flat-land" bevel preparations are considered more suitable to minimize excessive sink. Although specific standards have not been developed, typical end preparations that are being used extensively in industry are illustrated in Fig. 85.12.

The end preparations apply to gas tungsten-arc welding of carbon and low-alloy steel piping, stainless steel piping and most nonferrous piping materials. On aluminum piping, the "flat-land" bevel preparations are preferred by some fabricators.

BACKING RINGS

Where pipe joints are welded primarily by the shielded metal-arc welding process, backing rings are employed in ferrous piping systems. In fact, a

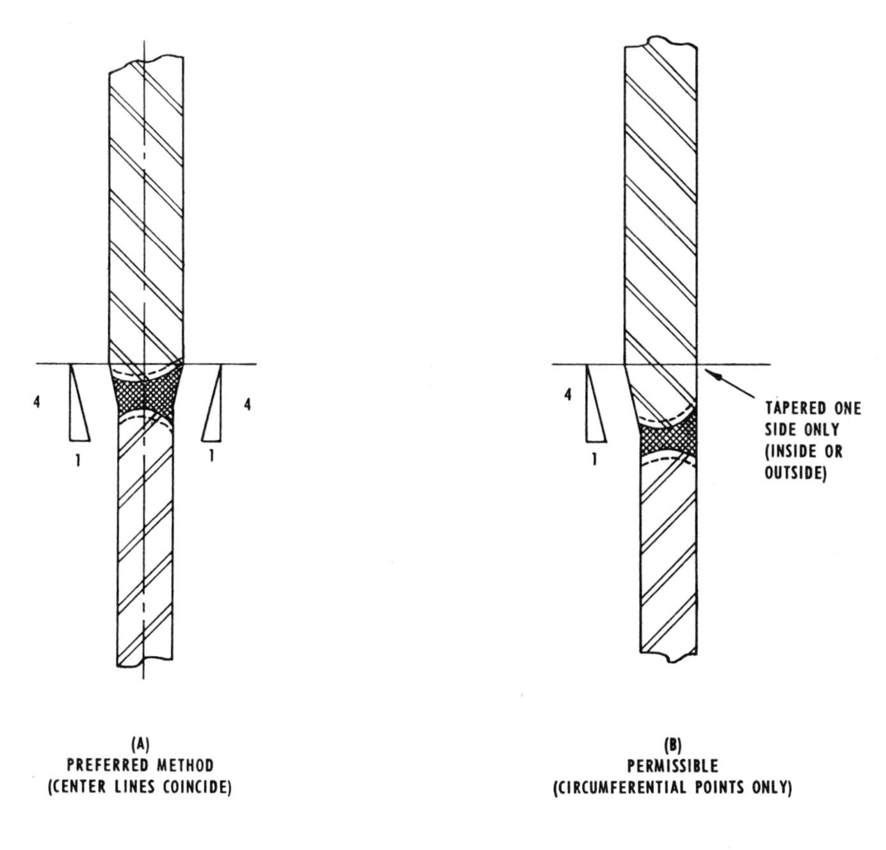

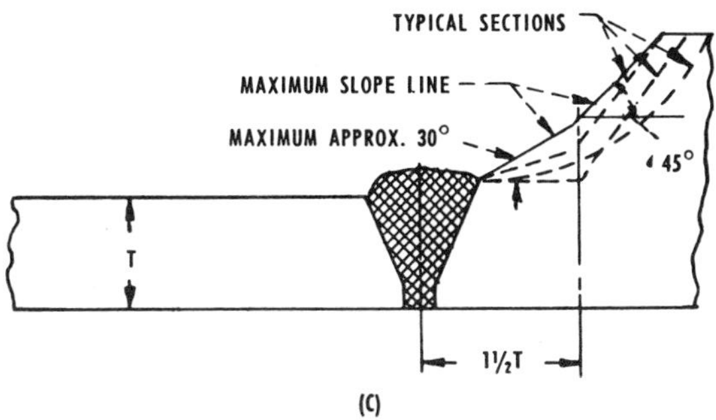

Fig. 85.11.–Methods of making butt joints for unequal wall thicknesses. (A) and (B) between pipe of unequal walls and (C) between pipe and valve or fitting.

significant number of pipe welds for steam power plants and several other applications are still made utilizing backing rings. Backing rings are rarely used in chemical and refinery piping.

The use of backing rings is primarily confined to carbon and low-alloy steel piping. They are generally made of a mild carbon steel having a maximum

Table 85.12—Dimensions for internal machining

Nominal Pipe Size (in.)	Schedule Number or Wall	Nominal od "A" (in.)	Nominal id "B" (in.)	Nominal Wall Thickness "t" (in.)	Machined id of Pipe "C" Tolerance +0.010 −0.00 (in.)	od of Backing Ring	
						Tapered Ring "DT" Tolerance +0.010 −0.000 (in.)	Straight Ring "DS" Tolerance +0.00 −0.010 (in.)
3	XXS	3.500	2.300	0.600	2.409	2.419	2.409
4	XXS	4.500	3.152	0.674	3.279	3.289	3.279
5	160	4.500	4.313	0.625	4.428	4.438	4.428
	XXS	5.563	4.063	0.750	4.209	4.219	4.209
6	120	6.625	5.501	0.562	5.600	5.610	5.600
	160	6.625	5.187	0.719	5.327	5.336	5.326
	XXS	6.625	4.897	0.864	5.072	5.082	5.072
8	100	8.625	7.437	0.594	7.546	7.554	7.544
	120	8.625	7.187	0.719	7.327	7.336	7.326
	140	8.625	7.001	0.812	7.163	7.173	7.163
	XXS	8.625	6.875	0,875	7.053	7.063	7.053
	160	8.625	6.813	0.906	6.998	7.008	6.998
10	80	10.750	9.562	0.594	9.671	9.679	9.669
	100	10.750	9.312	0.719	9.452	9.461	9.451
	120	10.750	9.062	0.844	9.234	9.242	9.232
	140	10.750	8.750	1.000	8.959	8.969	8.959
	160	10.750	8.500	1.125	8.740	8.750	8.740
12	60	12.750	11.626	0.562	11.725	11.735	11.725
	80	12.750	11.374	0.688	11.507	11.515	11.505
	100	12.750	11.062	0.844	11.234	11.242	11.232
	120	12.750	10.750	1.000	10.959	10.969	10.959
	140	12.750	10.500	1.125	10.740	10.750	10.740
	160	12.750	10.126	1.312	10.413	10.423	10.413
14 od	60	14.000	12.812	0.594	12.921	12.929	12.919
	80	14.000	12.500	0.750	12.646	12.656	12.646
	100	14.000	12.124	0.938	12.319	12.327	12.317
	120	14.000	11.812	1.094	12.046	12.054	12.044
	140	14.000	11.500	1.250	11.771	11.781	11.771
	160	14.000	11.188	1.406	11.498	11.508	11.498

Table 85.12 (continued)—Dimensions for internal machining

16 od	60	16.000	14.688	0.656	14.811	14.821	14.811
	80	16.000	14.312	0.844	14.484	14.492	14.482
	100	16.000	13.938	1.031	14.155	14.165	14.155
	120	16.000	13.562	1.129	13.827	13.836	13.826
	140	16.000	13.124	1.438	13.442	13.452	13.442
	160	16.000	12.812	1.594	13.171	13.179	13.169
18 od	40	18.000	16.876	0.562	16.975	16.985	16.975
	60	18.000	16.500	0.750	16.646	16.656	16.646
	80	18.000	16.124	0.938	16.319	16.312	16.317
	100	18.000	15.688	1.156	15.936	15.946	15.936
	120	18.000	15.250	1.375	15.553	15.563	15.553
	140	18.000	14.876	1.562	15.225	15.235	15.225
	160	18.000	14.438	1.781	14.842	14.852	14.842
20 od	40	20.000	18.812	0.594	18.921	18.929	18.919
	60	20.000	18.376	0.812	18.538	18.548	18.538
	80	20.000	17.938	1.031	18.155	18.165	18.155
	100	20.000	17.438	1.281	17.717	17.727	17.717
	120	20.000	17.000	1.500	17.334	17.344	17.334
	140	20.000	16.500	1.750	16.896	16.906	16.896
	160	20.000	16.062	1.969	16.515	16.523	16.513
22 od	40	22.000	20.750	0.625	20.865	20.875	20.865
	60	22.000	20.250	0.875	20.428	20.438	20.428
	80	22.000	19.750	1.125	19.990	20.000	19.990
	100	22.000	19.250	1.375	19.553	19.563	19.553
	120	22.000	18.750	1.625	19.115	19.125	19.115
	140	22.000	18.250	1.875	18.678	18.688	18.678
	160	22.000	17.750	2.125	18.240	18.250	18.240
24 od	30	24.000	22.876	0.562	22.975	22.985	22.975
	40	24.000	22.624	0.688	22.757	22.765	22.755
	60	24.000	22.062	0.969	22.265	22.273	22.263
	80	24.000	21.562	1.219	21.827	21.836	21.826
	100	24.000	20.938	1.531	21.280	21.290	21.280
	120	24.000	20.376	1.812	20.788	20.798	20.788
	140	24.000	19.876	2.062	20.350	20.360	20.350
	160	24.000	19.312	2.344	19.859	19.867	19.857

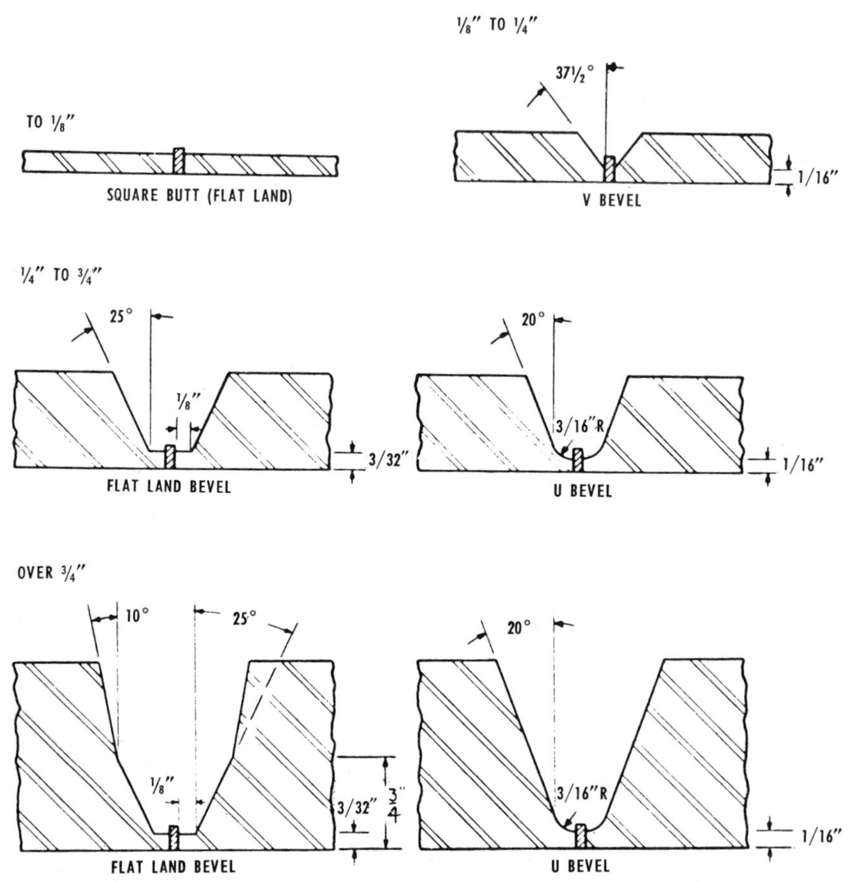

Fig. 85.12.–Typical end preparation for pipe which is to be welded with the gas tungsten-arc welding process.

carbon content of 0.20% and a maximum sulfur content of 0.03%. The latter requirement is especially important since high sulfur in deposited weld metal (which could be created by an excessive sulfur content in such rings) may cause weld cracks. Split backing rings are satisfactory for noncritical piping systems.

For the more critical service applications, solid flat or taper machined backing rings are preferred in accordance with the recommendation shown in Pipe Fabrication Institute Standard ES-1 (August 1965) utilizing pipe dimensions for internal machining, as shown in Table 85.12 and pipe end preparation as shown in Fig. 85.13.

When a machined backing ring is desired, it is a general recommendation that welding ends be machined on the id in accordance with the applicable PFI standard for the most critical services, and then only when regular pierced seamless pipe is used that complies with the applicable ASTM specifications. Such critical services include high-pressure steam lines between boilers and turbines, and high-pressure boiler feed discharge lines, as encountered in modern steam power plants. It is also recommended that the material of the backing ring

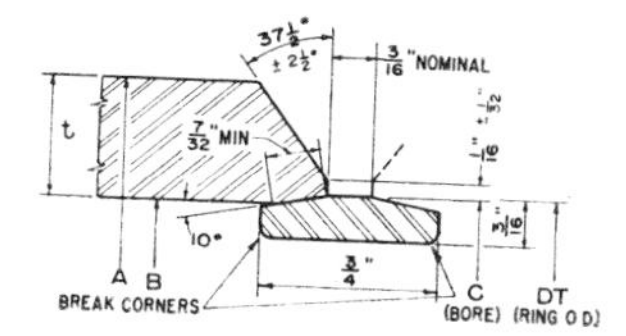

FOR WALL THICKNESS (T) 9/16" TO 1" INCLUSIVE AND TAPERED INTERNAL MACHINING

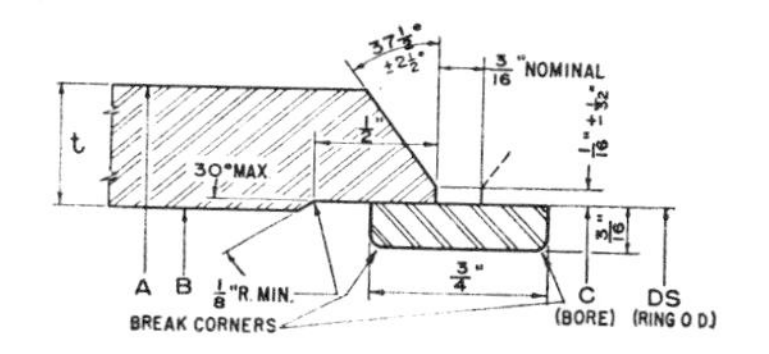

FOR WALL THICKNESS (T) 9/16" TO 1" INCLUSIVE AND STRAIGHT INTERNAL MACHINING

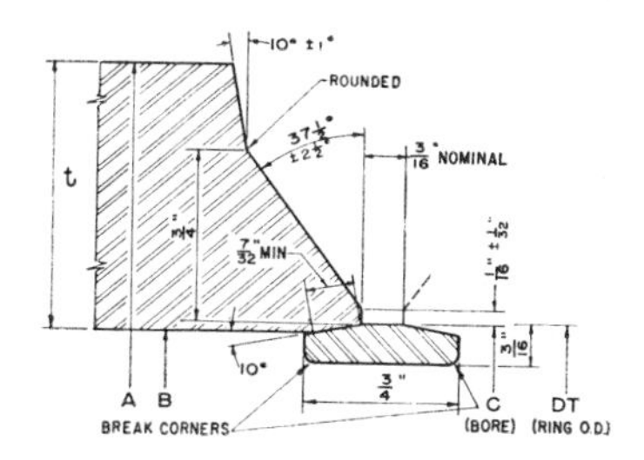

FOR WALL THICKNESS (T) GREATER THAN 1" AND TAPERED INTERNAL MACHINING

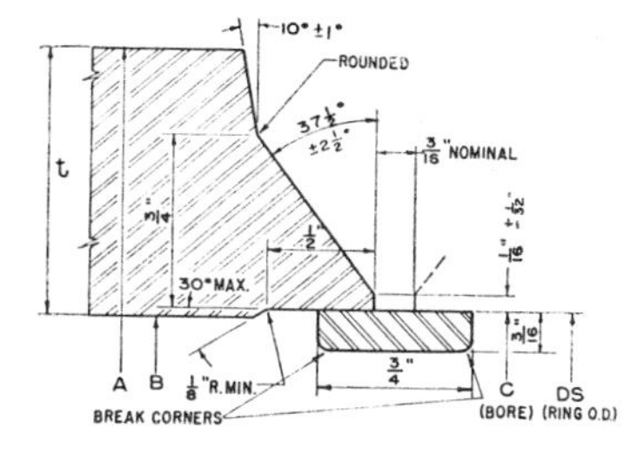

FOR WALL THICKNESS (T) GREATER THAN 1" AND STRAIGHT INTERNAL MACHINING

Fig. 85.13.–End preparation for critical service application employing flat or taper machined solid backing rings.

be compatible with the chemical composition of the pipe, valve, fittings or flange with which it is to be used.

On turned and bored tubing and fusion welded pipe, the design of backing ring and internal machining, if any, should be a matter of agreement between the customer and the fabricator. Regardless of the type of backing ring used, it is recommended that the general contour of the welding bevel shown in the PFI standard be maintained.

Wherever internal machining for machined backing rings is used on pipe and welding fittings in smaller sizes and lower schedule numbers than listed in Table 85.12, an additional operation may be required. This consists of depositing weld metal on the inside of the pipe in the area to be machined to assure satisfactory contact between the internal machined surface of the pipe and the machined backing ring. In such cases, the machining dimension should be a matter of agreement between the fabricator and the purchaser.

Whenever pipe and welding fittings in the sizes and schedule numbers listed in Table 85.12 have plus tolerance on the od, it may be necessary to perform an extra operation of depositing weld metal on the inside of the pipe or welding fitting in the area to be machined. In such cases, sufficient weld metal should be deposited to result in an id not greater than the nominal id given in Table 85.12 for the particular pipe size and schedule number involved.

Industry experience indicates that machining to dimension "C" for the pipe size and schedule number listed in Table 85.12 will, in most cases, result in satisfactory seat contact of 7/32 in. min (approximately 75% min length of contact) between pipe and 10 deg backing ring. In cases where this does not occur, it may be necessary to perform the additional operation of depositing weld metal on the inside diameter of the pipe or welding fitting to provide sufficient material for machining a satisfactory seat.

CONSUMABLE INSERT RINGS

The chemical composition of pipe is established primarily to provide it with certain mechanical, physical or corrosion-resisting properties. Weldability characteristics, if considered at all, are of secondary concern. On the other hand, the chemical composition of most welding filler metals is balanced with primary emphasis on producing a sound, high quality weld.

The steelmaking process employed in the manufacture of welding filler metals permits closer control of the composition range, which usually has a considerably narrower range than would be practical for pipe where much larger tonnages of steel are involved.

On many base metals, the fusing together by welding of the base metal compositions only may lead to such welding difficulties as cracking or porosity. The addition of filler metal tends to improve weld quality. However, in gas tungsten-arc welding, the addition of filler metal from a separate wire which the welder feeds with one hand while manipulating the gas tungsten-arc torch with the other may interfere with welding. The welder may leave areas having inadequate joint penetration, which generally are considered unacceptable as, for example, under the rules of the ASME Boiler and Pressure Vessel Code. Since some types of serious weld defects are detected only with difficulty during inspection, it is extremely important to provide the best welding conditions in order for the welding to produce quality welds.

One good technique to produce welds of the highest quality is to employ consumable insert rings of proper composition and dimensions. In piping for atomic reactors, where the weld joint quality is of extreme importance, welding authorities concerned with such fabrication generally agree that acceptable welds in stainless steel piping can best be made with consumable insert rings.

These are the three primary functions of the consumable insert ring: to provide the best welding conditions minimizing the effects of undesirable welding variables caused by the "human" element; to give the most favorable weld contour to resist cracking resulting from weld metal shrinkage and to eliminate notches at weld root; to produce the best possible weld metal composition for desired strength, ductility and toughness properties.

The best welding conditions are obtained where the flat land and extended U-bevel preparations are used. This joint preparation is particularly helpful where welding is done in the horizontal fixed position, since it ensures a flat or slightly convex root contour and provides the greatest resistance to weld cracking in those alloys particularly susceptible to microfissuring.

The weld root contour conditions to be expected from different bevel preparations when welding with or without consumable insert rings in different pipe positions are illustrated in Fig. 85.14. Where sink (excessive root reinforcement) is not acceptable, it is considered obligatory to use consumable insert rings with the special flat land or extended U-bevel preparation. In horizontal rolled and vertical pipe position welding, the consumable insert ring should be inserted into the beveled pipe in a concentric manner.

In horizontal fixed position welding, the consumable insert ring should be inserted eccentric to the centerline of the pipe as shown in Fig. 85.15. In this position, the consumable insert ring compensates for the downward sag of the molten weld metal and aids most in obtaining smooth uniform root contour along the id of the joint.

Welding Conditions	Consumable Insert Ring*	Position	Inside Pipe Contour			Permissible Concavity at inside of pipe
			Top	Side	Bottom	
"Flatland" Bevel	Yes	1G	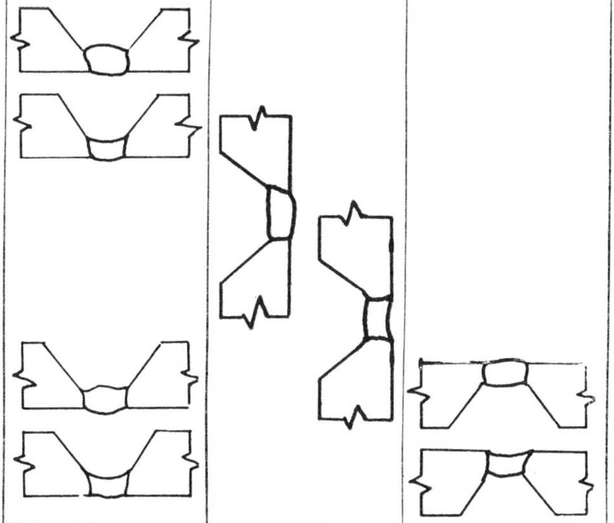			0
	No	1G				0
	Yes	2G				0
	No	2G				1/32
	Yes	5G				0
	No	5G				1/32
37½*	Yes	1G	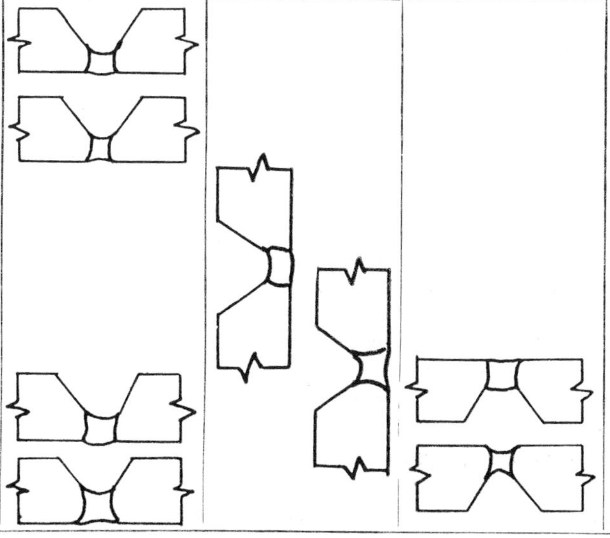			0
	No	1G				1/32
	Yes	2G				1/64
	No	2G				1/16
	Yes	5G				1/3
	No	5G				1/16

* In 5G (horizontal fixed) position welding the insert ring is positioned eccentric to the centerline of the pipe as illustrated in Fig. 85.15.

Fig. 85.14.–Root contour conditions which can be expected as a result of normal pipe welding with the gas tungsten-arc welding process.

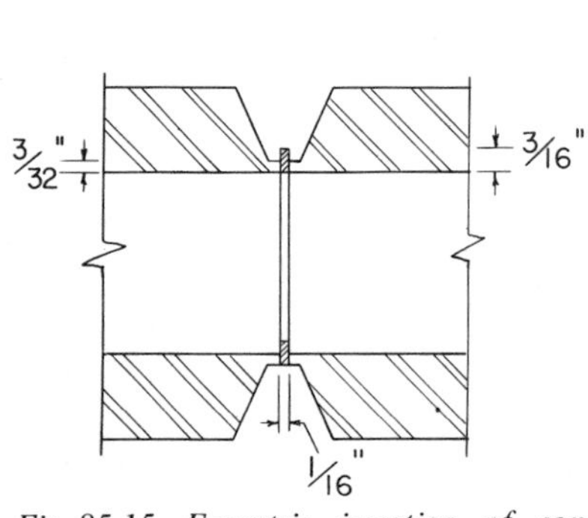

Fig. 85.15.–Eccentric insertion of consumable insert ring in pipe welded in the fixed horizontal pipe position.

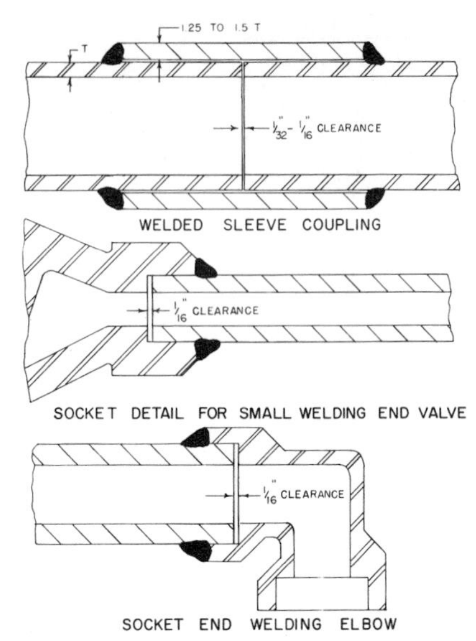

Fig. 85.16.–Examples of typical fillet-welded joints.

FILLET WELDS

For pipe sizes 2 in. in diam and smaller, circumferential fillet-welded joints are widely used for joining pipe to pipe, pipe to flanges, pipe to valves and pipe to socket joints. Backing rings would appreciably reduce the internal area of small diameter pipe which precludes their general use in small diameter piping. Figure 85.16 illustrates three typical fillet-welded joints. To minimize the potential for failure by thermal or mechanical fatigue, joint clearances of 1/16 in. are customarily allowed. Since this type of weld is subjected to shearing and bending stresses, adequate penetration of the pieces being joined is essential. This is particularly important with the socket joint, since the danger of washing down the end of the hub may obscure, by reason of fair appearance, the lack of proper penetration and soundness in the fillet weld. This condition is one which cannot be detected in the finished weld by visual inspection and may be very difficult to detect by radiographic inspection.

However, there are also service applications where socket welds are not acceptable. Piping systems involving radioactive solutions or gases or corrosive service with solution-enhancing crevice or stress corrosion generally require butt welds in all pipe sizes with complete weld penetration to the inside of the piping.

INTERSECTION TYPE JOINTS

Intersection joints such as medium and large size welded tees, laterals, wyes and openings in vessels are usually the most difficult to weld. Wrought carbon steel, carbon-molybdenum steel and chromium-molybdenum steel fittings conforming to ASTM Specification A234, Factory-made Wrought Carbon Steel and Ferritic Alloy Steel Welding Fittings, are preferred for accomplishing the layout of the piping system, since they provide bursting strengths equivalent to that of pipe of the same weight or schedule, and their installation involves butt welds only. Machines for oxygen cutting and beveling header openings and branch ends are available, but are not in general use. Cutting and beveling are usually done manually with an oxyacetylene cutting torch, though guided contour cutting machines are available.

Figure 85.17 illustrates examples of preparation of 90 deg intersection joints. In one, the header opening is sufficient to permit insertion of the branch, requiring beveling only of the header opening. In the other, the header opening is equal to the inside diameter of the branch, and only the branch is beveled. The latter form permits the use of a specially shaped backing ring if desired. Unreinforced branches, as illustrated in Fig. 85.17, are adequate only where the pipe is to be used at pressures considerably below its full capacity, usually between 50 and 75% of the pressure which the pipe can safely withstand. Reinforcing is required to make 90 deg intersections equivalent in strength to the pipe from which they are fabricated.

The intersection of two pipes at an angle of 45 deg or less produces a condition at the crotch of the intersection which makes it difficult, if not impossible, to secure the degree of penetration and soundness of weld deposit

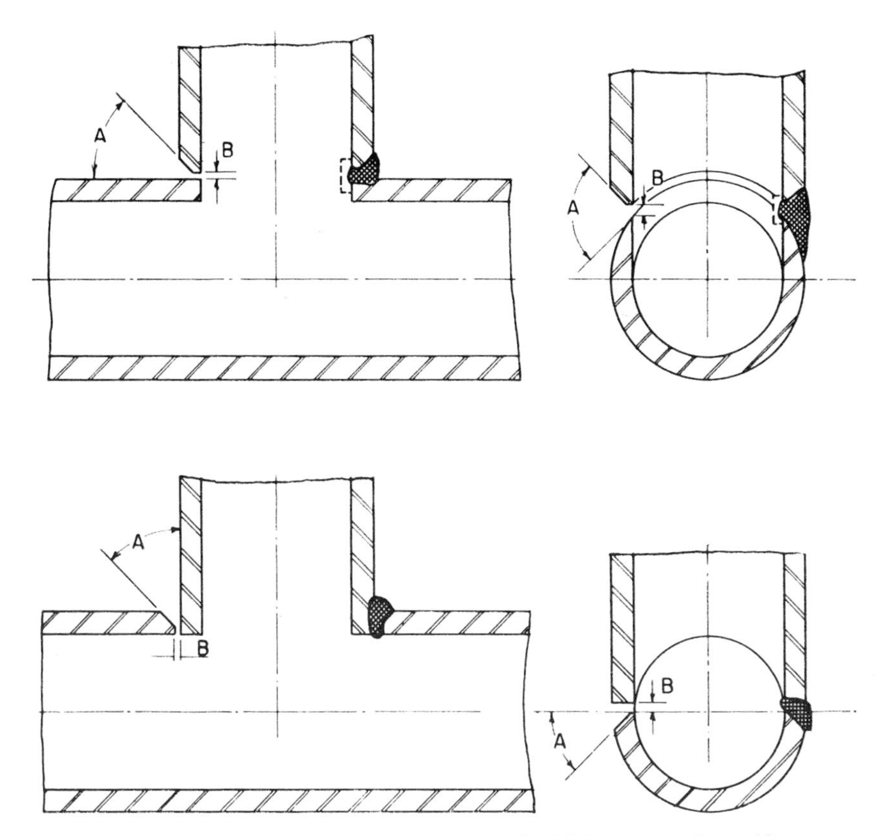

Fig. 85.17.–Recognized types of preparation for 90 deg intersection welds.

that is considered essential for severe conditions. Such intersection welds should be avoided. Piping should be designed, whenever possible, so that all intersection joints, regardless of angle, can be made under shop conditions where work can be positioned for welding in the flat position.

Manufacturers have simplified the problems of the fabricator by providing factory-made welding type nozzles, welding necks, welding outlets and welding tees, and other types of outlets especially for welding. These fittings eliminate a great deal of cutting, tacking and fitting previously required in this type of construction. However, care should be taken in selecting some fittings, particularly laterals, if full design pressure is to be maintained, since some are not provided with the required reinforcement. Typical nozzle, neck and welding outlets which can be used for almost all ranges of sizes are shown in Fig. 85.18.

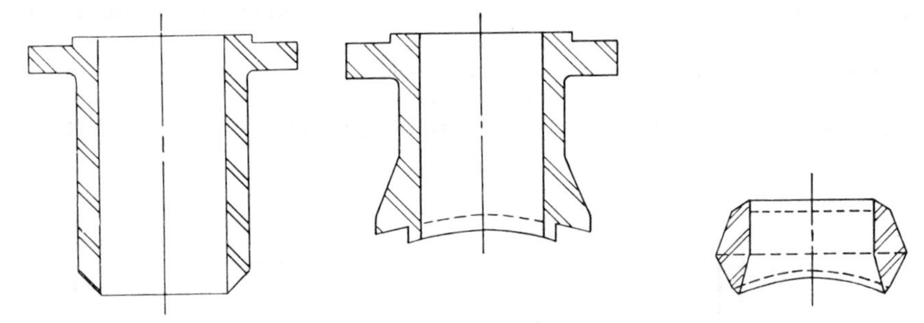

Fig. 85.18.–Examples of typical nozzle (center), neck (left) and welding outlets (right).

WELDING FITTINGS

Fittings are commercially available in most grades of ferrous material and in many nonferrous alloys. These fittings can be obtained in many combinations of size and thickness and they combine the best characteristics for unimpaired flow with ease of welding, and make the use of mitered construction generally obsolete. Some typical fittings for welding are shown in Fig. 85.19. Manufacturers' catalogs should be consulted for further details and dimensions.

Welding flanges are also available in the socket, slip-on and lap-joint types for use with welding stub ends or welding neck types. These are shown in Fig. 85.20. The use of slip-on flanges as covered by ANSI Standard (B16.5) includes all sizes in flange ratings to 900 lb; 1500 lb rated slip-on flanges are covered in sizes 1/2 to 2 1/2 in., socket-welding flanges are covered in sizes 1/2 to 3 in. in ratings to 600 lb, and sizes 1/2 to 2 1/2 in. in ratings up to 1500 lb. Butt and lap-joint types for use with welding stub ends are not subject to such restrictions.

BRAZED JOINTS

Some forms of the lap or shear type joint generally are necessary to provide capillary action for brazing of connecting pipe. Square-groove butt joints may be brazed but the results will be unreliable unless the ends of the pipe or tube are prepared accurately, i. e., plane and square, and the joint aligned carefully as in a

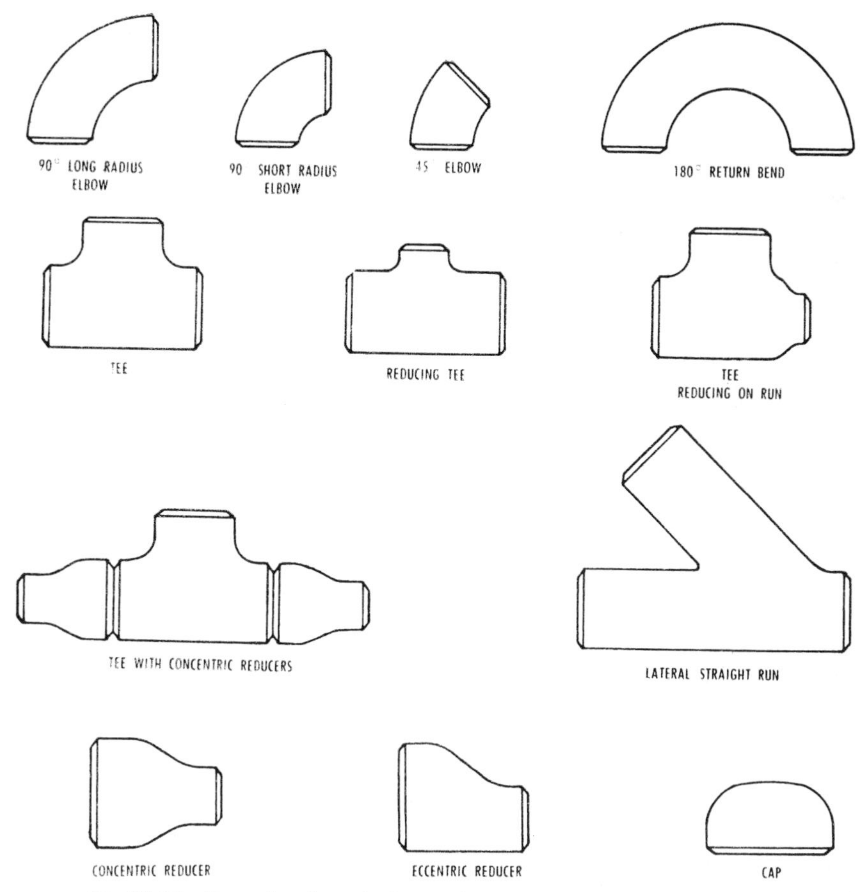

Fig. 85.19.–Examples of standard manufactured commercial welding fittings.

jig. High strengths may be obtained with butt joints if properly prepared and brazed.

The alloys generally used in brazing exhibit their greatest strength when the thickness of the alloy in the lap area is below approximately 0.005 in. Thin brazed alloy sections also develop the highest ductility. For brazing ferrous and nonferrous piping with silver- and copper-base brazing alloys, the thickness of the brazing alloy in the joint should be 0.003 to 0.006 in. Thicknesses less than 0.003 in. will make assembly difficult, while those greater than 0.006 in. will produce joints having lowered strength. The brazing of certain aluminum alloys is similar in most respects to the brazing of other materials. For aluminum, a clearance of 0.005 to 0.010 in. is satisfactory.

The length of lap in a joint and the shear strength of the brazing alloy are the principal factors determining the strength of a brazed joint. The shear strength of the joint may be calculated by multiplying the width by the length of lap and by the shear strength of the alloy used. An empirical method of determining the lap distance is to make it three times the thickness of the thinner or weaker

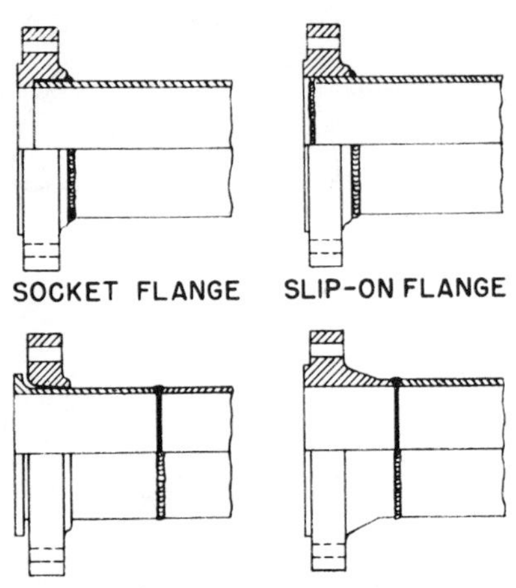

Fig. 85.20.–Examples of commercial welding flanges.

member being joined. Normally, this will provide adequate strength, but in cases of doubt, the fundamental calculations should be employed.

Generally, such detailed determinations are unnecessary for brazed piping since commercial fittings are available in which the length of lap is predetermined at a safe value. For brass and copper pipe, cast or wrought brass and wrought copper fittings are available. A bore of the correct depth to accept the pipe is provided. Midway down this bore may be a groove into which, at the time of manufacture, a ring of copper-silver-phosphorus brazing alloy is inserted. Since the alloy is preplaced, separate feeding of brazing alloy by hand is generally unnecessary.

Another type of brazing fitting available for pipe is provided with a simple bore of correct diameter and depth. Brazing with this fitting requires feeding of the alloy from a hand-held wire or strip. This fitting is available in brass and malleable iron, the former for use with copper and brass pipe, the latter with steel and wrought-iron pipe. In the case of the brass fitting, the copper-silver-phosphorus or the copper-phosphorus type of brazing alloy is employed with copper and brass pipe. For malleable fittings with ferrous piping, the brazing alloy should be of the silver-copper-cadmium-zinc type. Brass filler metal (60Cu-40Zn) of the type commonly employed in braze welding may also be used with these ferrous fittings and pipe.

Cleanliness of the brazing area is an important consideration in brazing pipe joints. As a result, the pipe and fitting must be thoroughly cleaned with steelwool, sandpaper, emery cloth or with a power-wire brush. Following the cleaning, brazing flux suitable for the brazing alloy to be employed must be applied to both the fitting and the pipe in the braze area. After assembling the cleaned and fluxed pipe and fitting, heating may be accomplished with an oxyacetylene flame or other methods that will provide the necessary heat and temperature for effecting the melting and flowing of the brazing alloy.

SOLDERED JOINTS

Solder is generally an alloy of lead and tin in various proportions, depending upon the type of soldering to be performed. Other metals (principally silver or cadmium) may be added for specific properties. Generally speaking, the standard 50-50 solder (50Sn-50Pb) is used in making soldered joints, while the higher lead alloys are used for wiped joints.

Solder is characterized by intrinsically low strength as compared with welding and brazing filler metals. As a result, soldered butt joints are rarely used and instead, the lap or shear type joint is employed. As in the case of brazing, the thinner the cross section of solder (within practical limits) the stronger the joint. Best current practice requires that joint clearances be on the order of 0.003 to 0.006 in. The strength of joints, in terms of base metal strength, may be calculated using the shear strength of the particular solder employed.

Soldered connections in piping are used principally on brass and copper tubing. For this service, a number of fittings such as elbows and tees are available. Some of these fittings may be wrought copper, others cast brass, but all are specifically designed for soldering. The bores of these fittings provide the proper clearance for the tubing. If annealed tubing is used and it is out of shape or size, tools are available for reshaping.

Another application of soldering to piping is for sealing threaded joints. In this case, the solder is not expected to provide any strength but merely seals the threads to ensure against leakage. The threaded joint is made up in the usual manner and solder is applied at the juncture of the pipe and fitting in the form of a fillet.

LEAD WELDED JOINTS

The design of welded joints for lead pipe joints is essentially the same as for other forms of welding. For thicknesses less than 3/16 in., the square-groove butt joint is employed. Thicknesses greater than 3/16 in. require a 10 to 15 deg bevel, resulting in a groove angle of 20 to 30 deg. The pipe ends are first cut square and then beveled. A typical single-Vee groove butt joint for lead pipe is shown in Fig. 85.21.

Other types of joints are sometimes employed under special conditions. One of these is the cup joint (Fig. 85.22) which is similar to the bell and spigot joint used for cast iron. Still another joint is the welded flange joint pictured in Fig. 85.23. In this joint, a fillet weld is employed. Since lead is a soft metal, these joints may be prepared readily with simple hand tools.

Occasionally, situations are encountered where the underside of the pipe joint is inaccessible for welding. In metals other than lead, such conditions should be avoided by careful field planning. In welding lead pipe, however, the solution is relatively simple by use of a split joint as shown in Fig. 85.24. Such manipulation is possible with lead because of its extreme ductility.

Cleaning the joint before welding lead is very important. The edges and about

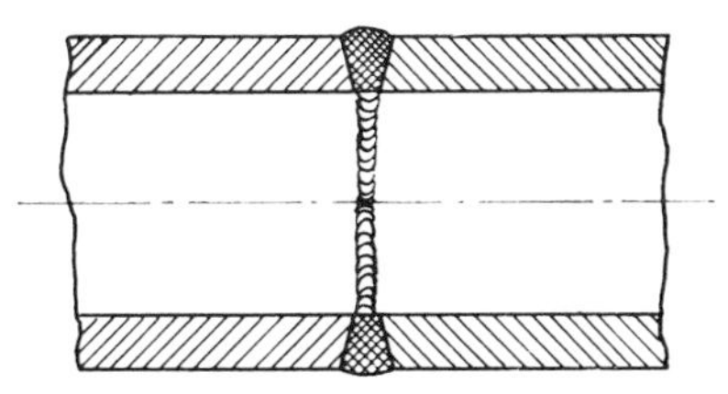

Fig. 85.21.–Butt joint for lead pipe.

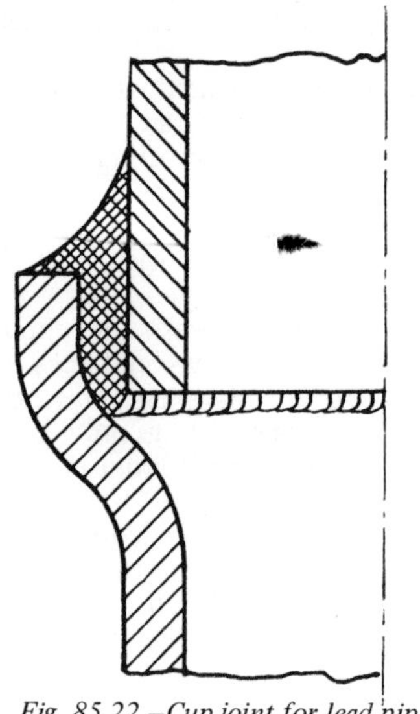

Fig. 85.23.–Flange joint for lead pipe.

Fig. 85.22.–Cup joint for lead pipe.

1/2 in. on each side should be scraped and welding should proceed within a short time after the cleaning.

WORKMANSHIP

A properly equipped pipe fabricating plant is the most efficient means for securing workmanship of the quality essential in industrial piping. These facilities should include power-driven machines for cutting and beveling

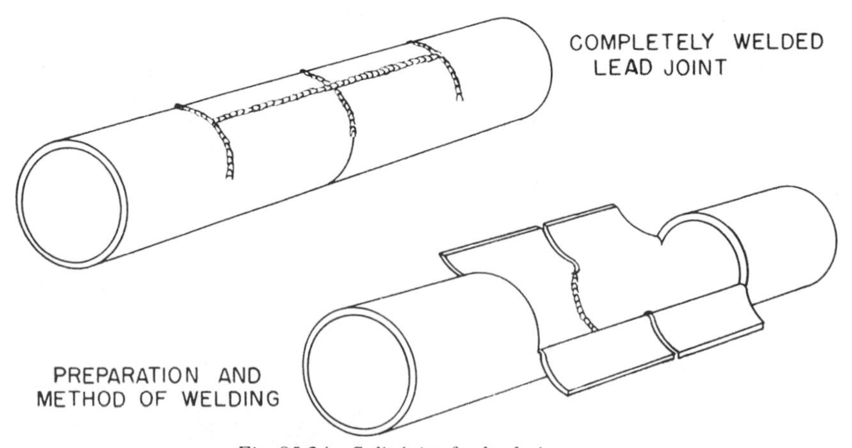

Fig. 85.24.–Split joint for lead pipe.

operations, accurate layout equipment, power-driven welding positioners, modern welding equipment for the manual, semiautomatic and automatic welding processes, heat-treating equipment and efficient material handling devices. Just as important as good equipment is the employment of properly trained and experienced personnel.

Inspection and testing operations must be done under close supervision, and control of weld quality should be maintained through laboratory research, testing and inspection. Inspection equipment includes X-ray facilities and radioactive isotopes, ultrasonic inspection, magnetic particle equipment, dye penetrant inspection, etc. Bend and tensile equipment should be available for the evaluation of welding procedures and welder test pipe coupons. Notch impact testing is required occasionally for the qualification of welding procedures applicable to piping for low-temperature service. Trepanning, saw cutting and other "destructive" sectioning are also required occasionally to visually check pipe weld quality.

ASSEMBLY AND PREPARATION FOR WELDING

The following procedures are recommended to achieve pipe welding of good quality.

Joint Preparation

The end preparation of parts to be welded should be done so as to conform as nearly as possible to the joint design set up in the procedure specification. Cutting should be done with some form of machine tool or mechanically-guided oxyacetylene or oxy-fuel gas torch. The latter methods are limited to those metals which can be so cut and which are not adversely affected by the heat involved. On thin walled pipe, cutting is also done with high-speed air or electric motor-driven saws with cut-off disks mounted in place of the saw blades.

All welding faces and adjoining surfaces for a distance of at least 1/4 in. from the edge of the welding groove, or from the toe of the fillet in the case of socket-welded or fillet-welded joints, should be thoroughly cleaned of rust, scale, paint oil or grease. Oxygen-cut surfaces should be ground reasonably smooth to remove all traces of scale and any cutting irregularities. Depending on the metal being welded, grit or shotblasting or chemical treatment may be necessary for thorough cleaning of parts and to avoid contamination during the welding.

Layout and Assembly

Layout and assembly involve the fitting together of the various component parts comprising a subassembly, or a complete piping system preparatory to welding. Dimensional allowances may have to be made for shrinkage during welding, particularly on metals of high coefficients of expansion. The parts to be joined should be carefully spaced, aligned and tack-welded together so that after welding, the entire assembly will conform to the required dimensions within reasonable tolerances. Normally, in ship fabrication, an end-to-end tolerance of ± 1/8 in. is considered the maximum acceptable.

In making up subassemblies, the usual procedure is to set up the largest component, either on adjustable support "horses" or on a level top "layout" table, with its longitudinal axis in a horizontal plane. The longitudinal axis and one end of the member are then used as base lines to which the locating dimensions and setting of the smaller parts can be referred, using a rule, steel tape, hand level, squares, straight edge or bevel protractor as required.

In the layout and assembly of complete piping systems in place, the same general procedure applies. The largest horizontal components are erected, and then these members are used as reference points for fitting and assembling of the smaller components. In such systems, the components, whether they be pipe, valves, fittings or prefabricated assemblies, must be carefully aligned with respect to other structures, equipment and each other. In cases where thermal expansion, contraction or other movement is not involved, the aligning operations present no particular difficulty. Where such movements are a consideration, care must be exercised in both the placement and alignment to make sure that the piping system, after it is welded and subjected to service conditions, is located consistent with the design and, where applicable, the isometric erection drawings.

In assembling butt-type joints, the use of accurately fitted backing rings facilitates spacing and alignment. Where backing rings are not used, the ends of the components to be welded should be carefully aligned and spaced preparatory to tack welding which, in open joints, may be facilitated with a spacing gage.

Intersecting joints should be carefully laid out with templates, standard layout curves or jigs. After the parts of the joint have been accurately cut and before they are beveled, it is advisable to make a trial fitting of the joint. This will permit a visual gaging to be made of the corrections needed for root spacing, alignment, etc., and will reveal any irregularities in the cutting of the parts that may require attention.

If the parts to be welded can be handled manually, the alignment spacing and in many cases, assembly, can be done without the use of external holding devices. In setting up heavy sections, the use of external clamps or other means of holding the parts in correct relationship to each other is essential, not only for maintaining proper alignment but also to facilitate handling.

In production welding, where a number of identical assemblies are involved, as when welding flanges to pipe, the use of set-up fixtures is of considerable advantage (Fig. 85.25). The design of such fixtures will vary, but the essential requirement is to provide a means for quickly setting component parts in proper relation to each other, ready for tack welding, without requiring individual clamps for each joint.

The dimensional allowances in the layout and assembly of piping systems to compensate for shrinkage during welding are established principally by experi-

Fig. 85.25.–Setup of slip-on flange with pipe flange aligner and dial-angle flange level.

ence. While it is possible, theoretically, to determine shrinkage for specific cases, the application of such calculations to all cases enountered in pipe welding is impractical and perhaps of questionable accuracy. Many variables affect shrinkage and cannot be determined by calculation. Size of electrode, type and size of bevel opening, welding process and procedure, interpass temperature and between-pass grinding affect shrinkage. Repair welding in local areas may intensify shrinkage and distortion.

In the usual types of groove welded joints in carbon steels, or low-alloy steel piping of Sch. 40 and Sch. 80 thicknesses, experience indicates that the longitudinal shrinkage in each joint will equal about one-half the root spacing. Radial shrinkage is negligible and may be disregarded for all practical purposes. Austenitic stainless steels experience about twice the shrinkage and also may suffer substantial radial shrinkage.

In socket-welded or fillet-welded joints, both transverse and longitudinal shrinkage are negligible and generally are not considered in the layout and assembly operations.

Shrinkage in the welding of intersecting joints results principally in angular distortion of the branch pipe and bowing of the main pipe. In these cases, it is virtually impossible to establish even approximate shrinkage values that would hold for all cases. The steps taken to control shrinkage are just as important as making dimensional allowance for such shrinkage in the layout and assembly operations.

Tack Welding

After the joint is properly lined up, short tack welds are made in the joint prior to actual welding. Tack welds should be sufficient in number and of suitable proportion to hold the part in place during ordinary handling. Usually, on piping up to and including 4 in. nominal pipe size, two 1/4 to 1/2 in. long tack welds are made on opposite sides of the pipe. On larger pipe sizes, tack welds are made at 4 to 6 in. intervals. They should be made by a properly qualified welder of a quality equal to that specified by the welding procedure, so that they need not be removed during subsequent welding of the joint.

For shielded metal-arc and gas tungsten-arc welding, the tack welds should penetrate and fuse through the full thickness of the land. Where tack welding is done by the gas tungsten-arc process, and consumable inserts are utilized, "skin" tacking is usually preferred without fusing through the land to the pipe inside.

Prior to welding, the tack welds should be ground to remove slag and possible crater cracks. Particularly with gas metal-arc welding, it is important to provide a reasonable taper towards both ends ("feather edge"), facilitating a sound tie-in with the subsequent weld.

Where consumable insert rings are used, spot tack welding of the insert rings to one side of the joint in two or more equidistant places is sometimes desirable before assembly of the joint. One method is to strike the arc on a small copper plate placed inside the pipe near the bevel and to touch the arc to the edge formed between the ring and the inside of the pipe. The two pipe ends are then brought together and tack-welded on the outside of the joint. The tack welds should not be fused throught the full thickness of the insert ring and land. Before breaking the arc, the tungsten electrode should be manipulated slowly toward and up the side of the bevel. This will help to avoid weld craters and cracks.

Frequently, in setting up joints in steel pipe of heavy wall thickness, the need for tack welds in the joints is eliminated by the use of heavy steel bridge "C" clamps.

WELDING PROCESSES

Although increasing use is made of semiautomatic and automatic welding processes, the process still extensively used in pipe shop welding and particularly in field welding is the manual shielded metal-arc process. Automatic and semiautomatic submerged arc, gas tungsten-arc, gas metal-arc and oxyacetylene welding are used to varying extents. The fundamentals of all of these processes, together with the rules governing their qualification and the qualification of welders, are treated in detail in other sections of the Handbook. The following discussion deals with the application of the various welding processes to pipe fabrication and pipe erection, and bears directly on workmanship.

Shielded Metal-Arc Welding

This process may be used for nearly all ferrous and nonferrous metals involved in piping systems. The equipment required for its application is comparatively simple and compact, readily portable, safe to use and, with ordinary care, requires little maintenance. Electric power supply for the operation of either dc or ac is readily provided in the permanent shop and is generally available for field erection. If electric power is not available, the use of gasoline or diesel-engine driven welding machines gives satisfactory service.

The number of passes required for welding joints in ferrous piping varies with the wall thickness of the pipe, the welding position of the pipe, the size of the electrode used and the welding currents employed. In welding low-carbon and low-alloy steel pipe in the rolled or horizontal position, under generally accepted procedures, the number of weld layers is approximately one per 1/8 in. of pipe thickness. The electrodes used in these procedures vary from 1/8 to 5/32 in. diam for the first pass, and 5/32 in. diam for intermediate and final passes. Occasionally, 3/16 or even 1/4 in. diam electrodes are used for pipe welding, particularly where heavy groove or fillet welds are required.

In welding medium-carbon and higher alloy steels in the rolled or horizontal position, the number of weld layers may be increased 25 to 30% by the use of smaller electrodes to lessen the heat concentration and to ensure complete grain refinement of the weld metal.

Electrode manipulation also varies with the type of covering on the electrode. For example, electrodes with EXX18 coverings generally require a wider weave than the EXX10 or EXX16 types.

With the pipe in a vertical fixed position, deposition, as described here, is not possible. Instead, the metal is deposited in the form of a series of overlapping stringer beads, the number of which, when using 5/32 in. maximum diam electrodes, may be approximated by assuming 25 to 30 beads per square inch of weld area.

The number of passes for welding joints in nonferrous piping will vary to a considerable extent because of the wide differences in fusion temperatures and heat conductivities of the metals, factors which have considerable bearing on the welding technique employed.

It is a common practice to do as much welding as possible in the flat or downhand position, using suitable power-driven rotators for continuously rotating the work at a speed consistent with the rate at which the filler metal can be deposited properly. This assures uniformity, both in depositing the filler metal and in distribution of heat, thus lessening distortion. The extent to which flat position welding can be applied is limited by the dimensions, shape and weight of the component parts to be welded, the facilities for rotating the work, the amount of handling involved and sometimes, the heat treatment required.

Fixed position welding requires special care in depositing successive beads or layers (as called for by the welding procedure) uniformly around the joint in order to avoid excessive stress concentrations or distortion due to uneven heat distribution. The assembly or piping system should be positioned so that the joints are readily accessible to the welder from all points. When preheating or postweld stress-relieving is to be done, ample space should be provided for applying and removing the heating equipment.

In shielded metal-arc welding with covered electrodes, the preferred direction of welding is generally to start at the bottom and progress upward. Some welding, especially of thin or medium thickness pipe, is done in the opposite direction. The danger in this procedure is that slag is more likely to be entrapped in the downward progression of welding. Ordinarily, more metal per layer is deposited when welding upward, and the requirement that the layers be thin enough to undergo complete grain refinement must be observed. On the other hand, welding downward is considered to require a higher degree of manual skill to secure adequate fusion with the side walls and to avoid trapping of slag.

Each layer of deposited weld metal must be thoroughly cleaned prior to the deposition of the following weld metal layer. Wire brushing, especially when done with a power-driven wire brush, is effective in removing the slag deposited by covered electrodes used in arc welding. On stainless steel pipe, stainless steel wire brushes should be used. Surface defects which would otherwise affect the soundness of the weld should be chipped out. The surface of the deposited metal and the faces of the base metal should be prepared for the following layer by removing bumps and sharp corners or grooves which might otherwise be difficult to fill without risk of slag inclusions. The grinding required, if any, will vary with the type of electrode used. Since the slag comes off readily on weld deposits properly made with EXX18 electrode types, grinding may be eliminated with these electrode types. Such cleaning may also be done with power-operated chipping tools.

Gas Tungsten-Arc Welding

This process is being applied increasingly to pipe welding and is extensively used on almost all of the ferrous and nonferrous piping materials. Substantial quantities of carbon steel piping are welded in the root with the gas tungsten-arc welding process. For practical reasons, the thinnest section that can be handwelded is approximately 1/32 in. The maximum thickness that may be welded is limited by available equipment and will vary for different metals. For pipe wall thicknesses over 1/4 to 3/8 in. (depending on the pipe metal), it is generally more economical to complete the pipe weld, with other welding processes, such as gas metal-arc or submerged arc welding, after the gas tungsten-arc root pass has been made.

Accurate preparation and good fit-up are particularly important considerations. Depending on the metal being welded, the cleaning operations preparatory to welding should be carefully performed and, in some cases, may require the use of chemical treatments, especially on aluminum and titanium because of their tendency to oxidize readily. Some type of gas or metal backing is beneficial. Depending on the metal weld, care should be exercised in the selection of suitable materials or shielding gases to avoid contamination or other undesirable effects on the welding.

Gas Metal-Arc Welding

In recent years, variations of the gas metal-arc welding process have become popular for pipe welding, particularly those using "fine-wire" or small diameter wires. Gas shielding for carbon and low-alloy steels is primarily done with carbon dioxide, or with argon-carbon dioxide mixtures. Due to the oxidizing nature of carbon dioxide shielding gas in the arc zone, it is important that the filler metal contain deoxidizing elements. Such wire is readily available. For other metals and alloys, pure argon, argon-oxygen, argon-helium and argon-helium-carbon dioxide mixtures may be employed. In welding the low-carbon stainless steels, carbon dioxide in the gas mixture must be held to a low level to avoid carbon pickup in the weld deposit. Procedure qualifications for such applications should include a chemical analysis of the weld deposit to assure that the carbon content does not exceed 0.03%.

Welding is done on open pipe joints with a root spacing of approximately 3/32 in. ± 1/32 in. When welding in the horizontal fixed pipe position, the first pass is normally started at the top and continued downward to the bottom. Subsequent weld passes can either start at the top and proceed downward or at the bottom and proceed upward.

Tack welds need to be carefully prepared. Even when the edges of the tack welds are ground, it may be difficult to obtain an X-ray quality weld, since at the edge of the tack welds inadequate joint penetration, weld craters or burn-thru may result. At present, the application of commercial gas metal-arc welding to pipe welding is limited by many fabricators to essentially noncritical piping where weld quality requirements permit weld defects such as incomplete fusion. This would be unacceptable under the inspection requirements of the ASME Boiler and Pressure Vessel Code and other applicable codes involving critical applications, such as nuclear and critical steam power plant piping and some chemical process piping.

Submerged Arc Welding

Submerged arc welding is extensively used in shop welding of carbon steel, chromium-molybdenum alloy steel and stainless steel piping. This process can be automatically or semiautomatically employed and is generally applicable to longitudinal seam welding of rolled plate piping, circumferential welds on piping 8 in. in diam and larger, and straight fillet welds of fabricated plate attachments. In addition, semiautomatic submerged arc welding is also being used to some extent in welding intersection joints. Shop facilities include 600 to 1200 A dc generators or ac transformers, work positioners and, for the fully automatic method, equipment for supporting and positioning the welding head. For some applications, backing rings are used for groove welds. Where fit-up is poor, the first weld pass on the backing ring should be made by manual shielded metal-arc welding. In the place of backing rings, the root pass of the weld may be made by the gas tungsten-arc welding process. Accurate preparation and good fit-up of the joints are essential since the process allows little, if any, flexibility to compensate for these irregularities.

Since the interpass temperature of butt-welded joints tends to increase, it is particularly desirable that on alloy piping, the alloying elements are contained in the filler wire and not in the flux.

Oxyacetylene Welding

Pipe welding applications involving manual oxyacetylene welding are decreasing, although up to 20 years ago, this process was used for welding most of the ferrous and nonferrous metals involved in piping systems. Its use is now generally confined to small-size piping. The equipment required for oxyacetylene pipe welding is less expensive and more portable than that required for the shielded metal-arc welding process. Consequently, it provides an inexpensive means for welding of piping systems in instances where the lack of power facilities or the cost of electric welding equipment precludes the use of arc welding. However, it is frequently difficult to obtain qualified pipe welders, particularly for field erection.

As in the case of manual shielded metal-arc welding of piping, oxyacetylene welding techniques may differ in some details, but the fundamentals relating to joint design, number of passes, direction of welding and cleaning are the same. The number of passes required for welding joints in ferrous piping varies with the wall thickness of the pipe, the position of the pipe when welded and the size of the welding rod used.

While the methods of depositing weld metal for the various pipe positions are similar to those employed in manual shielded metal-arc welding, the thickness of the deposited weld layers is usually greater. For piping that can be rolled or when the work is in a horizontal position, one pass is usually deposited for wall thicknesses up to 3/8 in., using the backhand technique. Two passes are preferred for wall thicknesses 3/8 to 5/8 in., three for wall thicknesses 5/8 to 7/8 in., and four for wall thicknesses 7/8 to 1 1/8 in. With the forehand technique, the weld is usually deposited in a single pass. With the pipe in the vertical fixed position, the number of passes varies widely. However, for most applications, determination of the number of passes by the rule of one pass for each 1/8 in. thickness of pipe wall is considered good practice.

With regard to direction of welding, the backhand technique in oxyacetylene welding usually involves starting at the top of the pipe and finishing at the bottom, while the forehand technique starts at the bottom of the pipe and works upward to the top.

Brazing, Braze Welding and Soldering

The application of the processes involving brazing, braze welding and soldering techniques has been briefly discussed elsewhere in this chapter. More detailed information on the processes is provided in Sections 3A and 3B in the chapters on brazing and soldering. The principal factors involving joint preparation, filler metals, fluxes and heating are applicable to piping.

Repair of Welds

When a pipe weld is to be repaired, removal of the defect is normally done by grinding, chipping, machining or by oxyacetylene cutting or air carbon-arc gouging provided that these methods are suitable for the respective piping and welding metals involved. In heavy wall steel pipe, weld defects near the weld root are normally removed by oxyacetylene or air carbon-arc gouging which generally permits closer control.

After removal of the defect, the surface upon which the repair welds are to be deposited should be ground and cleaned to ensure that a sound weld repair may be obtained. Where the quality of the completed pipe weld is determined by radiographic examination, it may be desirable to perform radiographic, magnetic particle or dye penetrant examination of the grooved area to ensure that all of the defective metal has been removed.

Shrinkage and Distortion

Precise calculations of shrinkage in welded pipe joints are difficult and are not usually made. Shrinkage will depend on the metals' coefficients of expansion, the thickness of the heaviest section being welded, the heat input and the degree of restraint imposed by fixtures, if used. Another problem of shrinkage of pipe welds, particularly in metals having high coefficients of expansion, involves the upsetting of the pipe in the weld area and gradual decrease in id with an increasing number of weld passes. The theoretical aspects of the subject of shrinkage and consequent distortion, shrinkage control and of the stresses set up in welded joints due to the thermal action or mechanical restraint are treated in the chapter on the physics of welding in Section 1.

The angular distortion in circumferential and intersecting pipe welds can be controlled and reduced to a minimum by (1) proper application of the welding procedure, particularly with respect to uniformity of heating (2) uniform preheating of the parts prior to and during the welding operation (3) skip welding (4) the use of fixtures for holding the parts in fixed relation to each other during welding. The first two methods permit free movement of the parts and generally result in lower residual stresses, while with the third method, the opposite is true. Residual stresses can be relieved to some extent by peening or, more completely, by thermal treatment.

In fabricating headers or pipe assemblies containing various types of welded joints in close proximity to one another, the control of distortion due to shrinkage is quite complicated. It involves consideration of the layout and fitting of the assemblies, the sequence of welding, positioning of the welds and the avoidance of heat concentrations, from the standpoint of not only their individual effect, but also their relative effects on the entire assembly. The means employed to control such distortion have already been mentioned, but their effectiveness is difficult to estimate and depends largely on the experience of the pipe welding fabricator.

HEAT TREATMENT

A very important phase of the successful welding of piping involves heat treatment. Heat-treating considerations include (1) preheating prior to welding (2) interpass temperature-controlled heating during welding (3) postweld heat treatment upon completion of the weld (4) straightening by heating of distorted piping sections. Satisfactory service performance of welded piping frequently depends upon the performance of proper heat treatments. Because of the importance of heat treatments, the specifications of the heat treatments should be included in the respective pipe welding procedures. Many Codes such as the ASME Boiler and Pressure Vessel Code (Section IX), the ANSI Code for Power Piping and others include in their procedure qualification and welder qualification requirements consideration of the preheat and postweld temperatures. Changes in the preheat and postheat temperatures by 100 F may require requalification (welding and testing) of the applicable welding procedure.

Proper heat treatment should also include consideration of the heating and, even more critical, the cooling rates. Specification of the heating and cooling rates should include a consideration of the weld joint configuration, the wall thicknesses involved, and the heating methods employed. More rapid heating and cooling rates may be permissible on circumferential pipe butt welds than on angular nozzle welds and other attachment welds between sections of dissimilar wall thicknesses. More rapid heating and cooling rates may also be permissible with induction heating where heating is produced within the pipe wall, than with the surface heating methods depending on thermal conduction through the outer skin and then through the pipe wall. Occasionally, special consideration in specifying heat treatments of piping must be applied to such items as valves and flow nozzle sections, where the heat treatment may damage sensitive materials such as valve seats and flow nozzle chambers.

Fast rates of heating and cooling, uneven heating and cooling or uneven heating within the wall of heavy wall pipe assemblies, particularly reinforced branch welds, may result in weld cracking. Cooling of complicated pipe assemblies from high temperatures, particularly at rapid rates, may also result in severe distortion. This is frequently a problem with austenitic stainless steels where water quenching from temperatures over 1900 F may be required.

HEATING METHODS

Various methods are used in the heat treatment of piping. These include (1) torch heating involving various gases and gas mixtures such as oxyacetylene, oxypropane or butane (2) electric resistance heating (3) induction heating (4) furnace heat treatment. Each method has its own advantages and limitations.

Torch Heating

Torch heating is done either with a single burner or with ring burners (Fig. 85.26). Heating with single burners is normally limited to piping of smaller diameters and lighter wall thicknesses. For pipe heat treatment, the methods of

Fig. 85.26.–Postheat treatment of pipe weld with ring burner.

employing "softer" flames such as propane or butane torches are normally preferred over heating with oxyacetylene torches. Ring burners are also available for various gas mixtures including propane, oxypropane and oxyacetylene.

Whereas sufficiently uniform preheating to temperatures up to 500 to 600 F is readily done with single burner torches, postweld heat treatment at stress-relieving temperatures over 1000 F on circumferential joints should preferably be done with ring burners. Uniform heating with single burner torches is difficult and may result in metal upsetting and even cracking.

Resistance Heating

Electric resistance heating now represents the most widely used method applied to pipe welds, particularly at field-erection sites. It is usually done with special finger elements or blankets using nichrome wires heated by passing a welding current through them. Shorting-out of the electric current with the base metal is avoided by means of insulating beads placed around the wires. The preheat and interpass temperatures can be controlled by means of magnetically attached surface thermometers containing electric "on-off" contactors (Fig. 85.27), or by the use of thermocouple-operated controllers attached to the pipe.

Induction Heating

Induction heat-treating methods are extensively used for the heat treatment of pipe welds, particularly at field-erection sites. This method is applied particularly on the heavier pipe wall thicknesses from about 2 to 4 in. and over. The relatively low frequencies generally used are 25, 60 and 400 Hz. The 60 Hz equipment is most widely available. In torch and electric resistance heating, the heat to the inside wall of the pipe is slowly conducted from the point of application on the outside surface. Induction heating generates heat essentially within the pipe wall. On pipe of the heavier wall thickness, this has the advantage of providing more uniform heat throughout the wall and a smaller temperature difference between the outside and inside surfaces in the weld area. The lower the frequency, the deeper the penetration of the induction heat. The differences between 60 and 400 Hz heating are insignificant for wall thicknesses

Fig. 85.27.–Automatic control of preheat and weld interpass temperature with magnetically attached "on-off" dial thermometer.

up to about 1 1/2 to 2 in. On wall thicknesses over 2 in., more uniform and rapid heating is obtained with the 60 Hz or 25 Hz equipment.

The electric field is normally obtained by wrapping copper conductors around the weld to be heated. Special fixtures (Fig. 85.28) are also available to permit more rapid attachment and removal of the induction heating coils.

Furnace Heat Treatment

The most satisfactory method of postweld heat treatment of welded pipe sections involves heating in a furnace. Large furnaces for this purpose are available in commercial pipe fabrication plants (Fig. 85.29). Heating and cooling rates and holding temperatures are maintained uniformly by means of automatic furnace controls. Heating of such furnaces is usually done either by natural gas or with oil of low vanadium or sulfur content.

TEMPERATURE CONTROL

The control of temperature in both preheating and postweld heating operations is especially important and frequently influences equipment selection.

For torch heating, temperature indicating crayons are widely used. They should be sulfur- and lead-free. Preheating temperatures up to 700 F are also checked with direct-reading, magnetically-attached surface thermometers.

In electric resistance heating, surface thermometers or electrically operated pyrometers are used to control automatically the current flow to the heating units. Thermocouples are usually attached to the metal to be heated by induction heating. The thermocouple wires are then connected to control equipment which may automatically control the time-temperature cycle and even program the heating and cooling rates of the metal.

In preheating or postheating welds in any of the alloys that have air-hardening tendencies, as the chromium-molybdenum and nickel-alloy steels, special precautions should be taken to exclude, as much as possible, the circulation of

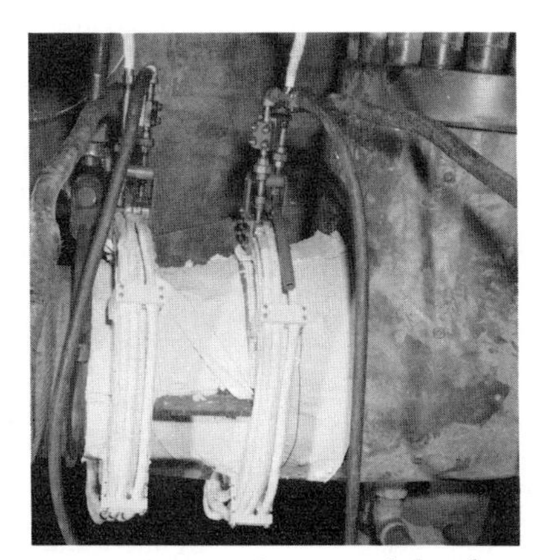

Fig. 85.28.–Use of water-cooled coils to stress relieve heavy wall piping by induction heating.

Fig. 85.29.–Stress-relieving furnace in a large pipe fabricating plant.

air around the joint during both heating and cooling periods. This can be done by avoiding drafts and by closing open ends of the parts being welded to prevent air circulation through the interior of the piping during the heating period. During the cooling cycle, the joint should be covered with a suitable insulating jacket, usually involving layers of asbestos sheeting, immediately after the welding and heat treatment are completed.

METALLURGICAL HEAT TREATMENTS

It is sometimes necessary to heat treat complete welded piping assemblies not only for stress relief, but also to restore the original structure of the metal or to attain other desirable properties. This may involve annealing, normalizing or quenching treatment, depending on the alloy welded. For proper control, extensive furnace equipment is required. For example, the medium and high-chromium alloy steels are often normalized at 1650 F and tempered at about 1300 F, or annealed at 1700 F followed by furnace cooling. Specific austenitic stainless steels may require annealing at 1950 F followed by rapid cooling to obtain maximum corrosion resistance. The furnaces used are usually gas- or oil-fired, pyrometrically controlled, and of such construction and proportion as to allow the treatment of a number of assemblies in one charge.

The various metals used in piping require the supervision of a competent metallurgist and a rather extensive laboratory for the evaluation of procedures and the analysis of results.

Pipe Support During Heat Treatment

The temperatures involved in postweld heat treatment and other heat-treatment operations are often in a range where the ability of the metals to resist deformation and distortion are lowered to an appreciable extent. It is, therefore, necessary to support the welded pipe sections during the heat-treating operation in a manner that will reduce to a minimum the possibility of deflection in any direction. This is accomplished in the shop by placing adjustable roller-type supports under the parts being welded as near to the joint as possible, allowing sufficient space for the placement of the heating apparatus over the joint. In field work, where the welds are made in position, chain falls or other suitable rigging secured to the building or other supporting structures are used to accomplish the same purposes.

Where postweld heat treatment or other heat treatments of a number of assemblies are done in a heat-treating furnace, special care should be exercised in placing the assemblies to make sure that they are uniformly supported in all directions. Temporary structural bracing, tack-welded to the assemblies, if necessary, may have to be used to accomplish this purpose.

STRAIGHTENING

The welding of piping assemblies, due to the expansion and contraction of the metal, tends to distort the original setup and alignment of parts. The amount and degree of distortion depend upon the size, shape and thickness of the metal welded, the welding process, procedure and sequence and the care by the welder. Rewelding of localized weld areas where defects have been removed by grinding or gouging can also contribute to distortion. Much of the distortion can be controlled during welding; some of it may be corrected during a postweld heat treatment, although it may be necessary to correct the effect independently.

One method employed to correct misalignment in welded pipe assemblies involves the controlled alternate heating and cooling of areas adjacent to the weld so that the consequent expansion and contraction of the metal will tend to draw the parts into proper alignment. The application of this method is limited to those metals and service applications upon which the effect of the alternate heating and cooling is not particularly detrimental to the properties or structure of the metal. For example, suppose a length of reasonably straight pipe is clamped firmly at one end to a rigid table, and an area 6 in. or more in length around the pipe circumference adjacent to the clamping device is heated to a red heat. If a stream of cold water is then suddenly applied to one side of the pipe, the sudden contraction will cause the free end of the pipe to deflect in the direction of the side of the pipe thus cooled.

The application of heat and of the cooling medium requires considerable skill and judgment, if satisfactory results are to be obtained. Such skill is only acquired through extensive experience since it is practically impossible to calculate the deflection resulting from such heating and cooling operations.

Occasionally, it is possible to correct piping assemblies that have been furnace-annealed or normalized by setting up the assembly in a facing head with the normal axis of the piece in line with the center lines of the machine. By then machining the flanges or pipe ends so that their faces are square with the normal axis of the pipe and with each other, they will meet the overall dimensional requirements of the piece section.

The foregoing methods of squaring and aligning are, of course, applicable in cases where the assemblies, after welding, are distorted to a considerable extent. Another method of correcting misalignment is by the gradual application of force, commonly known as cold pulling. This method has limited application, however, due to the possibility of setting up excessive stress concentration in an assembly, and should be employed only in cases where the effect of such stress concentrations would be negligible.

The final step in the straightening operation is the checking of all parts and dimensions of the assembly with the requirements of the drawing from which the assembly was constructed. While this procedure is essentially a function of inspection, much additional handling can be avoided if the work is done before the assembly is removed from the straightening table.

CLEANING AND FINISHING

Cleaning of welded piping is of considerable importance. The extent of cleaning and methods used is, of course, dependent upon the type of work involved, and may range from merely blowing-out by compressed air or steam to shotblasting or pickling and passivating.

Unless piping is heavily scaled as a result of heat treating or other causes, it can be adequately cleaned on the inside with turbine cleaners which function in much the same manner as boiler tube cleaners. If such equipment is not available, tapping the outside of the piping and blowing it out with air or steam will suffice, although this method takes considerable time and additional handling and is not always reliable.

If the piping has been heavily scaled, the inside should be sand or shotblasted.

(Shotblasting of stainless steel piping is not recommended). Sometimes sandblasting of the outside may also be desirable. If sandblasting of the inside is not feasible, a turbine technique may be employed in some cases, provided that special care is taken to make sure that the turbine heads are not cutting into the base metal.

The cleaning methods usually employed on alloy steel piping are similar to those used for carbon steel piping.

Pickling of the completed pipe assemblies may be desired, particularly on stainless steel or some nonferrous piping. This can be accomplished either by immersion in pickling vats or by adequate swabbing of the piping with pickling solutions. In either case, however, the assemblies must be passivated and then washed with water. Since pickling does not always effectively remove all scale from the inside of carbon and low-alloy steel piping and over-pickling may be undesirable, sand or shotblasting is generally preferred for these metals.

The solutions used for pickling or passivation vary with the metals treated. The methods employed are dependent on available facilities and the practices of the manufacturer. Reference should be made to standard texts on the subject.

INSPECTION AND TESTING

The purpose of inspecting and testing welded pipe joints is to provide assurance that the work meets the purchaser's specifications and that work which is within the scope of a specific mandatory code complies with the requirements of that code. It is generally accepted that such assurance can be obtained by (1) the use of a welding procedure that has been proven satisfactory (2) control of all operations to make certain that the established procedure will be followed exactly throughout the job (3) frequent inspections to establish that operations are being performed correctly and that materials meet specifications (4) occasional confirming tests of actual joints.

PRELIMINARY INSPECTION

Before any actual welding is done, the inspector should make sure that the welding procedure is one that has been properly qualified and that its use will produce welds of the proper mechanical, physical and metallurgical characteristics. The inspector must satisfy himself that the welders have, by actual tests, demonstrated their ability to make sound welds when following the prescribed procedures. The joint design should be checked to ensure that it conforms to the qualified welding procedure. The inspector should assure himself that the materials used are suitable for the purpose and conform to the requirements of any contract, code or specification governing the work. The electrodes, filler wires and fluxes should be identified properly and stored under conditions which will not be adverse to the materials. For example, the low-hydrogen electrode types now used extensively in pipe welding should be stored in heated or humidity-controlled rooms or cabinets. Drying ovens are also commercially available for this purpose.

VISUAL INSPECTION OF WORK IN PROGRESS

An important duty of an inspector is the inspection of individual joints prior to actual welding to ascertain that each joint has been properly prepared, set up and fitted. The alignment of parts, fit of the backing or consumable insert ring, if used, and spacing of the pipe ends are especially difficult and may prevent a welder from producing a thoroughly sound joint.

The inspector must make sufficiently frequent observations to determine whether the prescribed procedure is being properly followed by each welder, particularly as to welding current, number of passes being deposited, interpass temperature control and cleaning between passes.

Completed welds should be inspected visually for general appearance, noting the amount of weld reinforcement and the presence or absence of undercutting. When the underside of a weld is accessible, it should be inspected to determine if complete penetration has been obtained and if excessive burn-thru has occurred or "icicles" have been formed.

The inspector should make certain that, where required by code or specifications, the welds are die-stamped with the welder's symbol and any other identifying symbol stamp that may be required, provided that die-stamping is not considered to affect the metal adversely. Die-stamping is considered acceptable on carbon steel piping and, with low-stress dies, on the chromium-molybdenum piping grades containing up to 10% chromium.[5] Die-stamping is not advisable and should be substituted by marking with electric pencils on the martensitic, ferritic and austenitic stainless steels, the low-temperature steels and the nonferrous piping materials. In instances where extensive permanent markings are required, as in the case of ASME Boiler and Pressure Vessel Code piping, it may be more economical, as well as metallurgically satisfactory, to stamp the required data on a light gage plate which should be permanently affixed to the fabrication in an approved manner. The method is acceptable for all of the above materials.

NONDESTRUCTIVE EXAMINATION

Nondestructive test methods are used extensively to establish compliance with the requirements of applicable codes and specifications and conformance to good workmanship considerations.

Nondestructive examination, in many respects, is not understood adequately by a great many of those who specify inspection techniques or equipment. In many instances, costly repairs have been made in pipe welds where defects were of such minor magnitude that failure would have been extremely unlikely. However, there are also a number of instances where detectable, serious defects were not observed simply because the inspector did not use the proper equipment or operated the equipment incorrectly.

Good inspection equipment is available. However, it is essential that the inspector, who must be qualified, fully understands the characteristics, advantages and limitations of the equipment and applicable inspection methods. It is also essential that the inspector be familiar with the various types of defects which may be present in the weldment which he is examining. Finally, he should understand the materials involved, their relative tendency toward embrittlement and crack propagation at room and service temperatures, and must be able to predict whether or not a defect is critical. Proper interpretation must consider also the effects of service conditions, fatigue, creep and corrosion.

---

[5] PFI Standard ES–11 (August 1968), "Permanent Identification of Piping Materials"

Without this training, a minor noncritical defect may be removed at substantial cost, only to be replaced with a far more critical condition which, in time, may result in a costly or disastrous service failure. For example, where repairs are made by welding, the repair weld or subsequent heat treatment may institute a severe notch or damage the adjacent base metal.

It should be recognized that nondestructive testing techniques may frequently only indicate an area of suspicion or may miss a hazardous condition altogether. Destructive techniques may then be necessary to supplement the nondestructive examinations performed.

Commerical examination of pipe welds usually involves examination by (1) radiography with X-ray and radioactive isotopes (2) ultrasonic method (3) dye penetrant methods (4) magnetic particle method. Reference should be made to Chapter 6, Section 1, for details on these nondestructive testing methods.

DESTRUCTIVE TESTING

Several destructive test methods are also employed to maintain control over weld quality. Destructive tests may be performed to evaluate, in detail, questionable defects observed by radiographic or ultrasonic inspection. The removed coupons permit a more accurate evaluation to determine if the nonhomogeneities are actually defects and a weld repair is necessary, or if the questionable area can be considered acceptable. Reference should be made to Chapter 6, Section 1, for details on destructive testing methods.

LEAK TESTS

Hydrostatic testing of welded piping components and pipe lines is universally used. Extensive systems are often tested by this method in sections, the separation being made by valves. On flanged joints, closure may be made with blind flanges.

Hydrostatic tests are also applied to shop-fabricated pipe assemblies. These may have flanged or screwed ends permitting closure of the ends with gasketed blind flange plates or screwed caps.

Where the service pressure is severe and full hydrostatic testing on assemblies fitted with welding ends is mandatory, temporary heads or caps on the connecting ends of the assemblies may be welded so that the required test pressure can be safely applied. In such cases, the welding ends of the assemblies are not finished to dimensions until after the testing has been completed. Increasing application is also being made of cone lock pipe stoppers (Fig. 85.30) which, when tightly closed into the pipe ends, permit test pressures to 2000 psi and over.

For most work, clean water is used. The temperature of the water should be no lower than that of the atmosphere. Otherwise, sweating will result and proper examination will be difficult. Bleeder valves or petcocks should be provided at the highest point in the line to permit venting of all air in the piping during the filling operation.

The hydrostatic pressure should be applied gradually and should be left on for a sufficient length of time to permit making a complete examination of all welds. The pressure gage should be checked to be sure that it is functioning properly.

The test pressure should be that called for by the applicable code,

Fig. 85.30.–Test plugs for hydrostatic testing of fabricated piping assemblies.

specification or contract. Lacking such instructions, it should be at least the maximum working pressure to which the piping will be subjected. For power piping, the test pressure should never be more than two times the primary service pressure rating of any valves included in the test. Globe valves should not be subjected to a test pressure under the seat greater than one and one-half times the primary service rating. Full pressure may be applied above the seat or with the valve open.

It may be necessary to protect expansion joints and relief valves during testing, particularly when testing of the sections of the system is done prior to the completion of the entire system. Expansion joints of the slip type should be secured against possible separation. Other types should be studied to determine whether they need extra protection or support during testing.

Any evidence of leaking or other weakness under test should be marked. Following release of the pressure, defective areas should be repaired and another hydrostatic test made.

As the expansive force of compressed air or other gas is relatively great, and since there is always a possibility of a failure under test, the use of air or other gas for testing is discouraged. It should be employed only under extraordinary circumstances and under competent supervision.

Air testing, therefore, is employed to a limited extent in general shop fabricating practice and usually only to satisfy the requirements of a particular specification. The larger fabricating shops necessarily have facilities for supplying compressed air for other plant services and usually can arrange to test with compressed air up to 100 psi.

Freon tests or soap-bubble tests are less frequently used. Also, mass spectrometer testing is occasionally employed. The high cost of this method does not make it a very practical tool for the examination of pipe welds.

SERVICE TESTS

The methods of testing which have been described are designed to assure the quality of the welding before the system is put into service. In addition to such testing and inspecting, many engineers place particular reliance on a service test, essentially a trial operation of a system under normal service conditions.

While the procedure for service testing varies with the installation, there are certain precautionary measures, practically always applicable, that should be taken. These include the removal of all temporary test blanks and test connections, the checking and setting of all valves and regulating devices for proper operation and the checking of all hangers and anchor structures to be sure that they are functioning properly.

The results required, as far as the welding of the system is concerned, are a demonstration that the welds are free from leaks and are capable of withstanding the stresses imposed on them by the service to which they are subjected. The fact that it is usually the contractor's responsibility to repair any leaks that develop during a service test and to assume the liability for any damage to other work or equipment resulting from such leaks is an incentive for him to make certain as the work progresses, that the welds are satisfactory for the service conditions.

BIBLIOGRAPHY

"Weldability of Steels," R. D. Stout and W. D. Doty, Welding Research Council, 1971 Edition.

"Recommended Practice for Welding of Transition Joints Between Ferritic and Austenitic Stainless Steels," ES23, Pipe Fabrication Institute (April, 1969).

Piping Handbook, S. Crocker and R. King, McGraw-Hill Publishing Co., New York (1967).

"End Preparation and Machined Backing Rings for Butt Welds," ESI, Pipe Fabrication Institute (November, 1966).

"Welding of Austenitic Chromium-Nickel Steel Piping and Tubing," D10.4, American Welding Society (1966).

Defects and Failures in Pressure Vessels and Piping, H. Thielsch, Reinhold Publishing Corp., New York, N. Y. (1965).

"Welding of Chromium-Molybdenum Steel Piping," D10.8, American Welding Society (1961).

"Recommended Practices for Gas Shielded-Arc Welding of Aluminum and Aluminum Alloy Pipe," D10.7, American Welding Society (1960).

Nondestructive Testing Handbook, R. C. McMaster, The Ronald Press, New York (1959).

"Welding Ferrous Materials for Nuclear Power Piping" D10.5, American Welding Society (1959).

"Gas Tungsten-Arc Welding of Titanium Piping and Tubing," D10.6, American Welding Society (1959).

CHAPTER 86

TRANSMISSION PIPELINES

PREPARED BY A COMMITTEE CONSISTING OF:

R. B. GWIN, *Chairman*
Columbia Gas System Service Corp.

E. A. JONES
Bethlehem Steel Corp.

R. W. MIKITKA
Chicago Bridge & Iron

J. HUNTER
Transcontinental Gas Pipeline

R. R. WRIGHT
Associated Pipeline Services

CHAPTER 86

TRANSMISSION PIPELINES

INTRODUCTION

Welding assumes considerable importance in the growth of the pipeline industry. The underground network for transporting natural gas, crude oil and refined petroleum products extends to all sections of the United States and totals over one million miles of pipeline. Most of these lines are welded, and all new pipelines are completely welded.

Welded steel pipe is used extensively in domestic water supply and distribution lines by municipalities, in transporting irrigation water, and in penstocks in hydroelectric power development. These water pipelines are not as long as the gas and products lines, but many of them are much larger. The Hoover Dam penstocks, for example, are 30 ft in diameter and have a wall thickness up to 2 3/4 in.

This chapter covers the fabrication and field welding of pipe for penstocks and transmission pipelines of all sizes for natural gas, crude oil, refined petroleum products and water. The first part of the chapter covers pipelines for gas and liquid products. The last part of the chapter covers pipelines for water and penstocks for power development.

OIL, GASOLINE AND GAS PIPELINES

Steel pipe suitable for welding, in accordance with API or ASTM Specifications, is ordinarily used for long transmission pipelines that conduct either liquid products or natural gas. The carbon and manganese contents of the pipe are limited to the amounts required for weldability. Seamless steel pipe is used extensively for all pressures and services, but is limited in size to a 26 in. outside diameter (od). Electric resistance and automatic submerged arc-welded pipe are also used extensively for all services and operating pressures. Continuous welded pipe is used for gas distribution lines. Oil pipelines as large as 40 in. od have been constructed, but smaller pipelines are generally employed. For gas transmission over long distances, lines of 42 in. od have been built, but smaller pipelines of 24, 30 and 36 in. od are common.

The field welding and joining of these pipelines are done by the manual shielded metal-arc and gas metal-arc welding processes. The manual shielded metal-arc welding process is generally employed, but the semiautomatic gas metal-arc welding process is also used. Several fully automatic welding machines are being developed. Oxyacetylene welding is used to some extent on short lines and smaller diameters. On large diameter pipe, the present trend is to double joint, even triple joint (see page 86.11), by automatic welding at the pipe mill or at predetermined unloading and distribution points; this reduces the number of field welds by one-half or two-thirds, respectively. Special aligning and automatic welding equipment is employed. The submerged arc welding process is commonly used. These longer sections are then trucked along the right-of-way.

GOVERNING CODES AND SPECIFICATIONS

The Hazardous Materials Regulations Board of the Department of Transportation prescribes minimum safety standards for the design, construction, operation and maintenance of pipelines carrying liquid petroleum products, natural and other gases. The standards for liquid pipelines are contained in Part 195 of Title 49, Code of Federal Regulations. Standards for natural gas are contained in Part 192 of Title 49. Both parts contain references to numerous other standards and material specifications including API, ASTM, ANSI, MSS and NFPA. The important standard for welding is API 1104, Standard for Welding Pipelines and Related Facilities. Standards are periodically reviewed and updated to reflect the latest welding technology.

The vast majority of steel pipe for transmission pipelines has been manufactured to the API Specifications for Line Pipe. API 5L Specification covers pipe grades with minimum specified yield strengths to 35,000 psi, and API 5LX Specification covers pipe grades with minimum specified yield strengths to 65,000 psi. Other conventional grades require 42,000, 46,000, 52,000 and 60,000 psi yield strengths.

Prior to 1927, a maximum design working stress of about 12,000 psi was considered good engineering practice. Higher design working stresses have become common practice with improved field welding, higher quality welded pipe, higher strength metals and improved nondestructive testing techniques.

The above mentioned federal standards presently permit a working stress in rural areas equal to 72% of the specified minimum yield strength.

The working stress levels of seamless and welded line pipe and longitudinal joint efficiency of welded pipe have greatly increased in the past ten years. It is expected that the strength properties of line pipe will continue to be increased and, with improvement in nondestructive testing, higher working stresses will be permitted in the future. These developments will improve the economic factors that must be considered in the construction of transmission lines. There is some research being conducted on pipe steels with a minimum yield strength of 100,000 psi. One pipeline has been built from this high-strength steel. The increased demand for energy in the United States makes this research effort extremely important.

MATERIALS

Pipe for Liquid Products

Pipe for oil and gasoline or petroleum products is usually manufactured to API Specifications However, ASTM Specifications such as A53, A106 and A381 are also employed. When high-strength line pipe is desired, it is available under API Specification 5LX.

Electrodes for the shielded metal-arc welding of low-carbon steel pipelines should conform to Specifications for Mild Steel Arc-Welding Electrodes (AWS A5.1-69). Any of the E60XX classifications of electrodes are satisfactory, although there is a strong preference for the E6010 classification. High-strength piping requires weld metal of equivalent strength; electrodes for these materials are covered by Specifications for Low-Alloy Steel Arc-Welding Electrodes (AWS A5.5-69). The E7010, E7011, E8010 and E8011 classifications of electrodes have been used for both groove and fillet welds. On steels which exhibit hardenability, the E7015, E7016 or E7018 classifications (or higher strength classifications of these types) may be used. Electrodes of these types, known as low-hydrogen electrodes, minimize the tendency to underbead cracking and reduce the need for high preheating. The low-hydrogen electrodes are used almost exclusively for fillet welding.

For oxyacetylene welding, the welding rods should conform to the Standard Specifications for Iron and Steel Gas Welding Rods (AWS A5.2-69). Classification RG60 is recommended for all standard, low-carbon steel pipe; where higher strength welds are needed, RG65 welding rods may be used.

Pipe for Natural Gas

Gas transmission service normally requires the use of the larger sizes of pipe in the standard range through 42 in. od with growing interest in even greater diameters of 56 or 60 in. Fabricated pipe with a welded longitudinal seam is available in all of these sizes through 56 in.; fabricated pipe with a spiral seam is available in diameters in excess of 56 in. Since standard-weight pipe satisfies part of the demand, the same specifications apply in this field as those for gasoline and oil pipe. For large-size pipe, it is important to consider any means for saving steel, and the use of high-strength, light-wall pipe is quite common for the larger transmission pipelines. The increase in strength is achieved by cold working or through the addition of alloying elements.

The principal difference between pipe for liquid products and pipe for natural gas is in the size. Large diameter pipes are fabricated by welding cylindrical sections formed from flat plate. Thirty and thirty-six in. od are typical pipe sizes used for gas service, and a description of their manufacture will provide an understanding of this method.

The operations start with steel plates 40 ft long with widths approximating the pipe circumference and with thicknesses appropriate for the operating pressure. After flattening and trimming to size, the edges are prepared for welding. The plates are press-formed into first a "U" shape and subsequently, into an open "O".

The next step is the welding of the longitudinal seam by one of several welding processes. In submerged arc, for example, the cylinders are continuously fed, end-to-end, through an automatic welding machine. The first weld pass is backed up inside by water-cooled backing bars. Generally, the weld bead is deposited on the inside diameter (id) and then on the od of the pipe. However, some manufacturers deposit the first weld bead on the od. This is double (two-pass) submerged arc-welded pipe. After welding, the pipe, which is less than nominal od, is placed in dies and subjected to internal hydrostatic or mechanical pressure which expands it to an exact od. This operation cold works the pipe to increase its yield strength and, at the same time, ensures uniform roundness and straightness; it also provides a proof test of the pipe. Pipe manufactured by this method, after thorough inspection, has been used in pipelines throughout the world.

In the spiral welding process, the plate or coil edges are first treated by trimming to width and preparing with the appropriate bevel or chamfer. A helix is then formed continuously by a modified three-roll bending arrangement supported by internal or external cage rolls. Submerged arc welding, using single electrodes or two electrodes in tandem, is applied to the id pass first and then to the od, 180 deg away. Stock is fed continuously to the unit by submerged arc welding plates or coils together on the id surface prior to forming. After seam welding, the appropriate length of pipe is cut off and the od cross weld made.

In the manufacture of resistance-welded pipe, coils of steel may be used for the smaller size pipe and welded in a continuous length. After welding, nondestructive testing devices are used to evaluate the quality of the weld. The pipe is cut to the desired length after welding.

In the manufacture of butt-welded pipe for distribution systems, coils of steel skelp are also used. Unlike resistance-welded pipe, the steel skelp is heated to a welding temperature in a furnace, and the edges are welded with pressure in suitable rolls. In this manner, large quantities of pipe, not exceeding 4 1/2 in. od, are produced in a continuous manner at a speed exceeding 1000 ft per min.

In the manufacture of seamless pipe, solid round billets of the proper diameter and length are first examined for surface defects which are removed by scarfing. These billets are then heated, pierced and formed into pipe. The piercing operation merely makes a hole in the solid billet. Successive passes through the plug-rolling mill elongates the pipe, reducing the wall thickness to its approximate finished dimension.

A reeling machine rounds the pipe and smooths the inside and outside surfaces. The pipe then passes through a series of sizing rolls for a final sizing which is followed by a straightening operation. The pipe is then expanded, hydrostatically tested and beveled before presenting it to final inspection.

DESIGN

The principal difference between natural gas, crude oil and refined petroleum products pipe is the range of sizes. In general, pipe for the transmission of liquid products will be smaller in size than pipe for natural gas. However, there is now a 40 in. od products pipeline, and plans are underway for a 48-in. crude oil line. The range of pipe wall thicknesses is substantially the same for both classes of service. Therefore, in considering the design of joints for welded transmission pipelines, it is logical to treat these two groups as one.

The physical dimensions of pipelines are governed by the operating conditions, and, as a result, the required flow through the line will dictate the size of pipe needed. The thickness of the pipe wall is largely governed by operating pressure, but may be further influenced by such factors as external loading, bridges and road crossings. In all cases, it is necessary to use a protective coating and cathodic protection.

Since virtually all transmission pipelines are buried underground, it is not necessary to consider stresses which are essentially of a structural nature and which occur when a pipeline is carried above ground. Similarly, expansion and contraction do not present the problems encountered in overhead steam piping, for example. With the pipe buried at appreciable depths, generally below the frost line, and with the continuous passage of fluid through the pipe, it may be considered to be operating at an essentially constant temperature. If temperature changes are involved, they are relatively small in magnitude and of such a protracted nature as not to cause any serious stresses. In addition, the methods of installing pipe in the trench, under compression, tend to reduce the opportunities for imposing direct longitudinal tension stresses on the line. In fact, the stresses induced by the usual operations involved in pipeline construction will be of greater magnitude than those encountered in service.

Since the fundamental design considerations are the same, whether the pipeline is installed with mechanical or welded joints, it follows that the welding design problems are concerned principally with joint designs. As compared with other welding operations, welded pipe joints create another problem in that they are generally accessible from the ouside only. Diameters are normally too small to permit welding from the inside. Reference to the chapter on welded joints in Section 1 will disclose that joints welded in this manner are difficult to make with complete penetration to secure optimum strengths. Pipeline welders, however, have developed a technique to overcome these problems by depositing welds that are equal and, in many cases, superior to shop welding.

Through the years, many attempts have been made to develop joint designs which would provide the assurance of strength and soundness. Among the early designs were adaptations of the bell and spigot joint long familiar in cast-iron pipe practice. However, the weld used in this joint is a fillet weld, and the joint strength is inferior to that obtained with a butt joint. Another early type was the double-bell joint with an internal liner or backing ring. This backing ring was of the same dimensions as the pipe, and the ends of the pipe were belled sufficiently to accept this ring. The bell type has not survived because it is costly to control the tolerances of the pipe ends and provide the ring. There is also the difficulty of inserting off-standard lengths in a line in the field, since no means are available for making the belled ends in the field.

The joint design for pipeline welding, as shown in Fig. 86.1, consists of a single-Vee groove with a 30 deg bevel, a 1/16 in. root opening and a 1/16-in. land. There have been a number of discussions concerning the angle of the bevel for pipeline welding, mostly in reference to the 30 deg versus the 37 1/2 deg bevel. The 30 deg bevel is by far the most common in the pipeline industry, and consistently produces a satisfactory welded joint which requires less weld metal. The 30 deg single-Vee groove is acceptable for pipe thicknesses up to 3/4 in. The single-U or modified groove weld should be used for heavier pipe (see the chapter on welded joints, Section 1). The beveling of the pipe ends is readily performed in the pipe mill, shop or in the field. In the mill or shop a pipe lathe is commonly used, while in the field an automatic oxyacetylene cutting torch (commonly called a pipe-beveling machine) is often used. Manual oxyacetylene

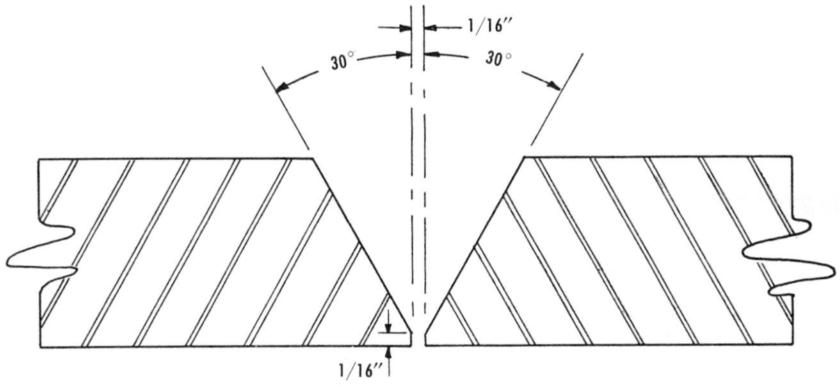

Fig. 86.1.–Typical joint design for welded pipeline.

cutting, while frequently employed, does not produce as clean or as accurate a cut as automatic cutting.

The use of backing rings is largely controlled by the service requirements and by maintenance procedures. If it is essential that complete fusion to the root of the joint be assured, without protrusions (icicles) within the pipe bore, then the use of backing rings is indicated. However, if internal pipe cleaning devices are to be employed, as in oil pipelines and many gas pipelines, then backing rings may introduce difficulties. These rings constitute an obstruction in the pipe and reduce the cross-section area. They also add to the cost of making the joint. The current pipeline practice of depositing a root bead using 1/8 in. or 5/32 in. diam electrodes of classification E6010 will produce thoroughly reliable joints through which cleaning devices may pass freely.

CONSTRUCTION PRACTICES

In welded construction of pipelines, certain collateral operations are commonly involved whether arc or gas welding is used. The quality of the finished line and the speed with which it is installed are directly influenced by these operations; hence, proper control and coordination of them are of vital importance in expediting the completion of a welded pipeline.

The preliminary preparations for pipeline construction follow the general pattern for all engineering work. Thus, surveys of the proposed units of the line are made. These surveys may be made by the conventional transit line method or the newer aerial techniques. Following this, the necessary plans are prepared together with specifications. These specifications will encompass all phases of the work, equipment, materials and testing, and will cover all the detail requirements for welding, including welder qualifications. The type of steel pipe to be employed is usually very carefully specified to assure steel of weldable quality. The amounts of preheating, if any, may also be specified.

With the location of the line established by the surveys, and the right-of-way secured, the next step is to clear the right-of-way of trees and obstructions and to construct a pre-designed grading of hill or mountain passes. Following this, the ditch for the pipe is dug. Ditch digging machines are almost universally used for this operation. Usually, the next operation is that of stringing the pipe along the right-of-way. Sometimes, stringing may precede the ditching operations. Trailer trucks are employed for moving the pipe from the railroad cars to conveniently located distribution points along the line. At these points, two pipe lengths may be welded together before distribution along the right-of-way. Side-boom tractors in conjunction with trailer trucks distribute the individual or double lengths of pipe along the right-of-way. Using specialized bending equipment, a crew measures and bends the pipe to meet the profile of the ditch. It should be noted that all of these operations are preliminary to actual beside-the-ditch welding.

Manual Shielded Metal-Arc Welding

Welding sets the pace in the construction of a pipeline. The speed of the subsequent operations is geared to the speed of welding. Therefore, the "stovepipe" method of pipeline construction was developed in order to move the welding operations along the right-of-way as rapidly as possible. In this method, the pipe joints are lined up and the stringer bead (first pass) is deposited, as shown in Figs. 86.2 and 86.3. The line-up men and welders then move on to the next joint while a second group of welders deposits the second (hot) pass. They then move on, and a third group of welders completes the weld. Since there is much more welding involved in completing the weld, the third group (firing line) will consist of a number of men, each welding on a separate joint.

One of the problems in producing a good welded joint originates with the difficulty in getting a good line-up. The line-up crew is charged with the responsibility of lining up the pipe with the assistance of an internal line-up clamp. The line-up crew must position the pipe in order for the first crew of welders to be able to make the stringer bead. The line-up crew is shown in Fig. 86.2. The number of stringer bead welders varies between two and four. The number of welders in this crew is dependent upon the pipe size and the daily progress to be made. Obviously, the more stringer bead welders, the faster pipe may be laid.

Fig. 86.2.–Line-up operation with an internal line-up clamp.

Fig. 86.3.–Photograph showing the line-up crew depositing the stringer bead.

As can be seen from the joint design in Fig. 86.1, all the welding must be done from the outside. A good stringer bead welder will make a weld with a smooth, flush internal bead about 1/16 in. high and 3/16 in. wide. The bead will be so smooth that it is difficult to tell that the bead was deposited from the outside. It is necessary to produce a high quality stringer bead in order to provide complete joint penetration.

Immediately following the stringer bead crew, another welding crew follows to put on the first filler pass, commonly known as the "hot pass". The hot pass welders also work in groups of two to four. The hot pass is put on immediately following the stringer bead, to take advantage of the residual heat from the stringer bead. The immediate hot pass is necessary to eliminate underbead cracking in high-strength pipe alloys, such as the API 5LX Grades. The second pass also provides additional structural strength for the welded joint. For example, the pipe with only the stringer bead and hot pass in the welded joint must withstand the expansion and contraction due to temperature changes, because it may be several hours or days before the remaining filler passes are completed. "Firing line" welders deposit the remaining filler passes and cover pass.

A typical welding procedure for the horizontal fixed position is shown in Table 86.1. This typical procedure outlines the number of passes and the electrode size for each pass. A common procedure specifies the E6010 electrode for the stringer bead and cover pass, and the E7010 electrode for all other passes. Another common combination used by many companies is the E6010 electrode for the stringer bead and the E7010 electrode for all subsequent

Table 86.1—Welding procedure—horizontal fixed position

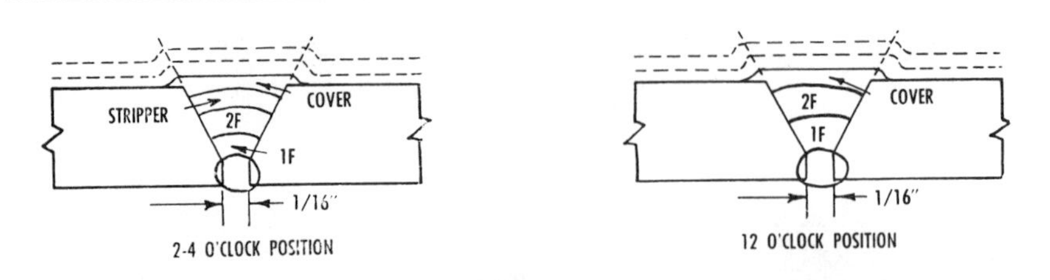

CROSS SECTION OF WELD
EACH PASS TO BE DEPOSITED BY THE DOWNHILL METHOD

ELECTRODE SIZE FOR VARIOUS PASSES

Wall Thick-ness	B Stringer	1F 1st Filler	2F 2nd Filler	Stripper	3F 3rd Filler	Stripper	Cover
0.154	3/32	1/8		1/8			1/8
0.187	1/8	1/8		1/8			5/32
0.219	1/8	1/8 or 5/32	5/32	5/32			5/32 or 3/16
0.250	1/8 or 5/32	1/8 or 5/32	5/32	5/32 or 3/16			3/16
0.281	5/32	5/32	5/32	5/32 or 3/16			3/16
0.312	5/32	5/32	5/32	5/32 or 3/16			3/16
0.375	5/32	5/32	5/32	5/32 or 3/16	3/16	3/16	3/16
0.438	5/32	5/32	5/32	5/32 or 3/16	3/16	3/16	3/16
0.500	5/32	5/32	5/32	5/32 or 3/16	3/16	3/16	3/16
0.563	5/32	5/32	5/32	5/32 or 3/16	3/16	3/16	3/16

passes. Then, there are a few companies that use the E7010 electrode for all passes.

An important part of the welding procedure is the weld cleaning. Much of the weld quality depends on the thoroughness of cleaning between passes. Many times inferior welds have been caused by improper cleaning. While the strength of the welded joint does not require that the slag be removed from the cover pass, it is still removed for weld inspection.

Gas Metal-Arc Welding

The gas metal-arc welding process was first introduced to pipeline construction in the late 1950's. It received widespread acceptance and was used on a number of major pipelines. Unfortunately, many people, including some highly skilled manual shielded metal-arc welders, were using this process without understanding it and without receiving proper training. The equipment was more complicated than the manual equipment and because it was new, it was subject to many field breakdowns. Properly trained mechanics were not available. As a consequence of the problems encountered, the use of the equipment has declined from its peak in the early 1960's.

Basically the same methods and sequence of operation as described for manual shielded metal-arc welding are used. The method of pipeline construction described in the previous section is used as is the joint design shown in Fig. 86.1. Some welding procedures do not require the immediate application of the hot pass because of the heavy stringer bead. However there are others that do require an immediate hot pass as in manual welding. Interpass cleaning is not necessary with the gas metal-arc welding process because there is no slag.

Generally, a 300 A dc constant potential welding machine is used. These are powered by gasoline- or diesel-driven generators. The diameter of the welding wire commonly used is 0.035 in. The shielding gas varies from plain carbon dioxide to combinations of gases such as carbon dioxide and argon. The strength of the completed weld depends upon the type of welding wire and shielding gas selected.

The defects associated with the gas metal-arc welding process are similar to those found in the manual shielded metal-arc welding process with the exception of slag. There are no slag inclusions because there is no flux or electrode covering. Incomplete fusion, however, is a common defect because of the shallow penetrating characteristics of the process (short-circuiting operation).

This process aided in the development of automatic welding which is discussed in the next section.

Automatic Arc Welding Equipment

The development of automatic welding equipment for use in the construction of pipelines has been slow for a number of reasons. The present manual shielded metal-arc welding procedure for cross-coutnry welding of large diameter pipelines is quite flexible and welding production is phenomenal. On 36 in. diam pipe, it is not unusual to weld one mile of pipe per day (approximately 125 welds). In any automatic welding process, it is difficult to compete with the speed and flexibility of the manual procedure. In addition to this, the manual skill of a welder is required to make adjustments for the numerous conditions encountered in getting complete penetration around the entire circumference of the welded joint. The problem with many automatic welding processed tried so far on pipeline welding has been the inability of the automatic equipment to travel around a stationary pipe.

Automatic Submerged Arc Welding Equipment

In commonly used automatic submerged arc equipment, it is necessary to rotate the pipe in order to make the weld. This equipment is normally set up near the pipeline, and used to weld two lengths of pipe together which are then hauled out on the right-of-way. The two lengths of pipe are positioned on pipe racks and then onto rollers. The pipe ends are butted together and the welding head, as shown in Fig. 86.4, is moved into position. The turning rolls rotate the pipe while the automatic welding head makes two complete passes on the outside of the pipe.

When these two external passes are completed, the line-up clamp is removed. An automatic welding head has been developed to go on the inside of the pipe. This equipment is then positioned in the inside of the pipe and one complete pass is made on the inside. The cross section of a typical weld made by this process is shown in Fig. 86.5. One of the important advantages in these submerged arc welding automatic processes is the ease and extent to which the equipment can be controlled. In spite of the inherent lack of process flexibility, as compared to manual shielded metal-arc welding, the automatic submerged arc

Fig. 86.4.–Automatic submerged arc welding equipment employed for welding pipe.

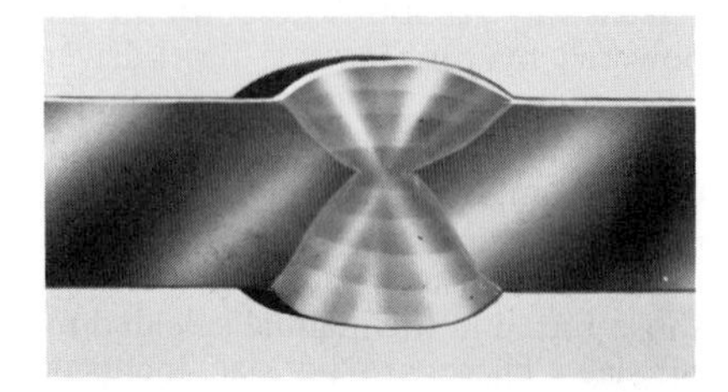

Fig. 86.5.–Cross section of weld made by the automatic submerged arc welding process.

welding process can continually produce circumferential welds of consistently good quality. The significance of this feature is that these high quality welds can be made on weld joints with wide variations in dimensions.

All of the automatic equipment described previously requires that the pipe rotates along an axis. Coming into use are several automatic units where the welding head travels around the pipe while the pipe remains stationary.

GAS METAL-ARC WELDING EQUIPMENT

One piece of equipment utilizes the gas metal-arc welding process. This welding system is comprised of three pieces of equipment: a pipe end preparation machine, a welding unit that operates on the inside of the pipe to deposit one welding pass and a small portable machine that operates on the outside of the pipe to complete the weld.

Since the first weld pass is made from the inside of the pipe, a special joint preparation is required to provide a small Vee-groove on the inside of the pipe. The groove on the outside of the pipe is prepared with a 20 deg included angle rather than the standard 60 deg to reduce the amount of weld metal to be deposited. An end preparation machine is used ahead of the welding operation.

The internal welding unit is a combination line-up clamp and welding machine. Four welding heads are mounted at about 90 deg spacing, and two of the heads operate simultaneously to weld one-half of the pipe. Welding proceeds from the top of the pipe downward. The opposite two heads then counter rotate to complete the opposite half of the root pass. The welding wire is 0.035 in. diam and the shielding gas is 75% argon-25% CO_2.

The external welding units are small portable machines that are used in conjunction with a tracking band that is pre-positioned on the pipe at a fixed distance from the weld. Each unit is self-driven around the pipe on the tracking band. Usually two units are used at each weld joint, one operating on each side of the pipe. Welding proceeds from the top of the pipe downward. Also, these units may be operated simultaneously with the internal unit. This system has been used on various construction projects both onshore and offshore in the U. S., Canada and in England.

Another unit under development utilizes the flux cored arc welding process. This system also utilizes an end preparation machine. The root pass is deposited from the outside of the pipe over a copper backup attached to a specially designed internal line-up clamp. All welding proceeds from the top of the pipe downward. The welding wire is 5/64 in. diam; no external gas shielding is used. The joint design consists of a 29 deg bevel, 3/32 in. land and 0.60 in. root opening.

The machine used for root passes contains two welding heads mounted approximately 180 deg apart on the same travel gear. Welding speed on the root pass is typically 70 ipm. The machine used for completing the weld contains four welding heads, spaced at approximately 90 deg, with typical welding speeds of 30 ipm. This system is currently being used in pipeline construction.

RIVER CROSSINGS AND SPECIAL CONSTRUCTIONS

The finished pipeline, buried in the ground, must be connected to pump or compressor stations located at convenient intervals. The installation of a large amount of piping at these stations is handled by special crews. Pipeline crossings under roads and railroads involve special construction and are generally built by crews especially equipped for this work.

River crossings are always solid-welded throughout. The pipe is usually roll-welded in long sections of more than 200 ft. If the river is comparatively narrow, the long sections are welded together on the banks and the entire line pulled across the river in one section, after which it is welded into the main line on the banks. On very wide rivers, the pipeline is lowered from a barge as shown in Fig. 86.6. The pipe is usually welded into 80-ft sections before it is placed on

Fig. 86.6.–Typical river-crossing operation. Pipeline is welded on a barge and lowered into the water. Concrete coatings or anchors are usually used to weight the pipe.

the barge. It is then lowered into the river, one section at a time, the section on the barge being welded to the section that is practically in the water. Extra heavy pipe, often 1/2 in. thick or more, is used. Concrete coatings or anchors are usually used to add weight to the pipe.

Another type of construction used extensively for large pipelines involves crossing a river on a cable suspension-type bridge. Spans between towers may range up to 2000 ft. Suspension-type bridges are also used across deep canyons and dry washes.

Pipeline Construction For Underwater Service

Pipelines constructed for underwater service may be divided into two very broad construction types. These are marsh or shallow bay crossings and offshore construction. Both types are constructed from specially designed barges. Because of the tremendous expense of offshore barge equipment and the distance from land, offshore construction is usually performed around-the-clock in two twelve-hour shifts with the construction crews living on board the barge. In marsh construction, the crews are usually transported to the construction barge by boat and may only work one shift.

The greater percentage of underwater pipelines is welded by the manual shielded metal-arc process; however, recently, an automatic gas metal-arc process has been used. In both types of underwater construction, the welders work at "stations" and the pipe is moved to a station for a specific welding pass to be applied.

Pipe for underwater construction is coated for corrosion protection. At an approximate 12 in. nominal od or larger, it becomes more economical to concrete coat the pipe to provide a negative buoyancy than to use an excessive wall thickness to achieve the same purpose.

Marsh Construction

In marsh construction, a "spud" barge is usually employed. This consists of one or more barges joined together to form a floating work platform. The name spud barge refers to the anchor system. A spud is a piling that may be lowered into the bottom to anchor the barge and raised when the barge is to be moved. Two or more spuds may be used on a barge of this type.

Pipe is sent onto the spud barge from a supply barge by a crawler-type crane. A side boom is used to stab each new pipe joint into place over the internal alignment clamp. As a new joint is stabbed, a winch pulls the completed joint off the back of the barge. Barrels or other flotation devices are attached to the pipe and it is floated down a predug ditch.

A typical spud barge operation to lay 24 in. to 36 in. od pipe would use 11 or 12 welders at five welding stations. A typical operation would require three stringer bead welders, two hot pass welders, four filler welders at two stations and two cap welders at the final welding station. There would be roller cradles between each welding station to support the pipe to the end of the barge.

When a weld is completed, the joint is radiographed and coated. When these operations are completed, a horn blows, the pipe is pulled and a new joint is added.

Offshore Construction

Offshore construction is generally considered to start in water depths of 15 to 20 ft. At this point, large pipeline construction barges, 300 to 400 ft in length, begin construction where the marsh construction type spud barges stop. Offshore pipeline "lay" barges are held in position and moved forward during construction through the use of 20,000-lb anchors. The barges usually have heavy cranes, up to 500 ton-capacity, for handling the pipeline and other offshore pipeline facilities. Room and board are supplied on the lay barge for all personnel working on the pipeline job.

In offshore construction, the concrete-coated pipe is welded through the use of up to five welding stations. The stringer bead and hot pass are supplied at welding station one while welding stations two through four apply filler passes. At station five, the cover pass is applied by the fifth group of welders. As each welding pass is completed, the barge is moved forward and a new joint of pipe is added. Behind welding station five, there are radiographic stations (usually two) and a welded joint coating station. After welding, radiography and coating, the pipe passes down a semi-buoyant "stringer" which supports the pipe until it is within a few feet of the sea floor. After laying, the pipe is buried in the sea bottom.

To date, large diameter (20 in.–24 in.) pipelines have been laid by this method in waters up to 350 ft deep. Larger diameter (36 in.) pipe has been laid to a water depth of 150 ft.

INSPECTION AND TESTING

Inspection, as a general phase of pipeline construction, includes many operations, some of which are not exclusive to the welding operations. An extensive discussion of the details involved in the inspection of weldments, including typical weld defects, will be found in the chapter on inspection of welding in Section 1.

Qualification of Welders and Procedures

It is essential that all welds in a pipeline be of good quality. Therefore, a great effort has been made to develop proper welding procedures and to assess the capabilities of the welders. All welders must be qualified in accordance with a qualified welding procedure before they start actual welding operations on the pipeline.

In the welding of pipe, the welding position refers to the position of the axis of the pipe and not to the orientation of the weld, as is the case with plate. Therefore, a horizontal pipe weld is one in which the axis of the pipe is in a horizontal plane. This definition applies even though the weld is in a vertical plane. The standard test positions for pipe are the horizontal rolled position, the horizontal fixed position, the vertical (fixed) position, and the inclined position. Welding in the horizontal fixed position, for example, requires (in order, starting at the top) flat position welding, vertical position welding and overhead position welding; if the direction is from the bottom upward, then this sequence will be reversed.

The procedure and qualification tests must be accomplished in accordance with one of the following standards: (1) API Standard 1104, Standard for Welding Pipelines and Related Facilities; (2) ASME Boiler and Pressure Vessel Code, Section IX; (3) ANSI B31.8 Code for Gas Transmission and Distribution Piping.

Many companies have adopted one of these specifications as written, while others have used them as guides in the establishment of their own specifications.

When this is done, the minimum requirements of the above standards must be met. After a procedure is once qualified, it can be used thereafter without further qualification. A welder must requalify if there is reason to question his ability.

There are a number of variables that affect the soundness and strength of the completed weld. These include such things as pipe material, diameter, wall thickness and welding position. The above standards list several of these variables and indicate that a welder or welding procedure must be requalified if there is a change in any of the variables.

On pipeline construction where the work can be segregated, it is sometimes possible to arrange the work so that a given crew will weld in only one position, e.g., the horizontal fixed position, in which case qualification need only be for that position. If, however, the welder is to work in all positions, he must then be qualified for all positions.

In a qualification test, a sample pipe weld is made, and coupons with the weld in the middle are removed by oxyacetylene cutting. These coupons are then subjected to various tests, such as the tensile, nick-break, root- and face-bend tests. The welder or procedure is qualified if these coupons meet the requirements established in the above mentioned codes.

Inspection During Welding

The welding inspector usually conducts the qualification tests for all welders whose work he will be inspecting on the construction job. Therefore, he becomes familiar with their capabilities and skills and also learns of their weaknesses. During actual production, one of the main functions of the welding inspector is the observation of the men at work, noting their compliance with the specified welding procedure. Generally, he will observe all men several times a day. Of particular interest will be the welding of the stringer bead and hot pass—the most critical to quality and soundness. Good weld quality depends, to a large extent, on joint line-up; this operation must therefore be checked at frequent intervals. A further area of importance is the cleaning and preparation of the bevel edges before alignment. On large pipeline construction jobs, a number of welding inspectors should be present, one on each important phase.

A source of welding difficulty may be the physical condition of the electrodes. Here the inspector must see that the correct type and sizes are used, that the electrodes are clean and undamaged and that they have been properly stored so that no deterioration has occurred. A further source of difficulty is the condition and cleanliness of the deposited bead; care must be exercised that interbead preparation is properly controlled to eliminate inclusions and discontinuities.

Welding machines and other equipment must receive daily inspection to assure proper functioning. Improperly operating equipment should be taken out of service until repaired. Careful inspection and conscientious assistance to the welders in the performance of their work and improvement of weld quality are the basic functions of the welding inspector.

Inspection After Welding

The two major nondestructive testing methods used on modern pipelines are visual and radiographic inspection. The significant features of visual inspection were described in the previous section.

Fig. 86.7.–Radiographic equipment which is propelled through the pipeline.

Radiographic inspection employs either X-rays or gamma rays as a source. For larger diameter pipe, the source can be propelled through the pipeline on battery-powered or gasoline-driven crawler units. Such a unit is shown in Fig. 86.7. Film belts are wrapped around the entire circumference of the welds to be radiographed. The crawler units have mechanical and radiological methods of locating and thus stopping at a circumferential weld. The units are programmed before placement into the pipe as to speed, exposure time, etc. Such units can travel for miles through a pipeline, thus reducing the number of open ends that the contractor must leave.

The radiographic inspection of small diameter pipe must be from the outside. Here the source is placed on one side of the pipe and the film, 180 deg opposite. A minimum of three such exposures must be made to adequately cover the entire circumference. External radiography takes more time than internal because of the increased number of exposures per weld and the increased time per exposure. This results from both the expanded source-to-film distance and material thickness (double wall vs. single wall).

Darkrooms mounted on trucks follow the radiographic crew down the right-of-way thus providing the developing facilities needed to supply the inspector with developed radiographic film as soon as possible after the welds are completed.

Radiography is used almost exclusively for the nondestructive testing of pipelines. Magnetic particle, dye penetrant and other forms of nondestructive testing find little use on the pipeline because of the need to inspect the weld through its entire thickness. Ultrasonics has this capability but has been handicapped by the irregularities of the manual shielded metal-arc welded stringer bead and cover pass. Shear waves are used with the transducer moving around the pipe, parallel to the weld. The stringer and cover irregularities do reflect sound and cause interpretation difficulties. Also the acceptable limits for slag and porosity cannot be determined without a follow-up radiograph. However, ultrasonics will find increasing use as fully automatic welding equipment is developed. The automatic welding has smooth stringer and cover passes, and there is no slag. Also, incomplete fusion, which is more common in gas-shielded arc welding than in manual welding, is difficult to detect with radiography. Therefore, ultrasonics is a likely way to inspect the automatic welds of the future.

Occasionally, destructive tests are used. For these tests, a weld is cut from a line and subjected to mechanical tests similar to those required for the qualification test. Such tests are generally made as a check of the welder's quality when nondestructive testing is not used such as on low pressure distribution lines. It should also be noted that all welds are visually inspected after completion, and, based on the judgment of a qualified inspector, such inspection can give a great deal of information on the quality of the weld.

Regulatory bodies and codes now require that pipelines be pressure-tested prior to being placed in service. These stringent, proof testing requirements are to assure that, on the basis of adequate testing, the safest possible pipeline is placed into service. Hydrostatic testing to stress levels equal to the actual yield of the base metal is becoming common.

Testing with gas was used extensively in the past. However, the experience gained indicates that testing with a noncompressible fluid, such as water, is inherently more desirable than testing with a compressible fluid, such as natural gas. As a result, hydrostatic testing is the predominant method used. Being capable of testing safely to higher stress levels is one of the more important reasons for use of water as a test medium.

WATER PIPELINES

Welded steel pipe is used extensively for pipelines conveying water for domestic, industrial and irrigation purposes. Municipalities use it for their larger supply lines to reservoirs and in distribution systems.

For flow lines to small municipalities, for distribution systems in municipalities and for industrial purposes, pipe size diameters range from six inches to a few feet. For large supply lines, the diameter in which the pipe may be fabricated is almost limitless.

GOVERNING CODES AND SPECIFICATIONS

Pipe for conveying water is usually purchased to conform to AWWA C-202 Standard for Mill-Type Steel Water Pipe of any size produced to meet finished pipe specifications or AWWA C-201 Standard for Fabricating Electrically Welded Steel Water Pipe fabricated from plates or sheets. Pipe purchased under ASTM Specification A 134 Electric-Fusion (Arc)-Welded Steel Plate Pipe (16 in. and over) or ASTM A 139 Electric-Fusion (Arc)-Welded Steel Pipe (4 in. and over) is satisfactory for water service, as is pipe conforming to API Specification 5L for Line Pipe.

Welded steel water pipe is manufactured in either a pipe mill or a pipe fabricating plant. The processes used for welding are resistance-seam welding, flash welding and automatic electric arc welding. In addition to pipe welded by these processes, seamless steel pipe is commonly used.

For large diameter pipe having wall thicknesses greater than 1 1/4 in. or for water pipe placed above the ground and supported by stiffener rings, it is advisable to use the specifications and design practice as discussed in the section of this chapter on penstocks.

Water pipes usually have some kind of coating inside and outside to protect them against corrosion. For mildly corrosive conditions, a good quality of coal-tar paint may be used. For severe corrosive conditions, either inside or outside, the pipe is usually protected with a coal-tar enamel lining or coating applied in accordance with AWWA C-203 Standard for Coal-Tar Enamel Protective Coatings for Steel Water Pipe or AWWA C-205 Standard for Cement-Mortar Protective Lining and Coating for Steel Water Pipe or AWWA C-602 Standard for Cement-Mortar Lining of Water Pipelines in Place.

Other standards commonly specified are AWWA C-206 Field Welding of Steel Water Pipe Joints, AWWA C-207 Steel Pipe Flanges and AWWA C-208 Dimensions for Steel Water Pipe Fittings.

BASE METALS

Pipe for water supply service is usually made of low-carbon steel. Water pipelines normally operate under lower pressures than pipelines for gas and petroleum liquid products, so it is not necessary to use the high-tensile strength steels.

DESIGN

In designing steel water pipelines, it is necessary to have a topographical profile made from a survey of the line. From the information contained on the profile and other information, such as quantity of water to be carried, surge conditions, nature of the soil from a corrosion standpoint and any unusual operating conditions that may exist, it is possible to determine (1) diameter and wall thickness of the pipe (2) location and size of air valves and blow-off valves (3) type of protective coating to be used on the inside and outside of the pipe.

After determining the diameter of the pipeline by the conventional design formulas, the next procedure is to calculate the internal working pressure by adding to the pressure shown on the profile such additional pressure as might be caused by water hammer. External loading from earth pressures and live loads or pressure from a negative pressure gradient must be considered with the pipe empty. Wall thickness is selected from the largest thickness determined to resist internal pressure or to resist external pressures.

Slight bends may be made in field joints of water pipelines to fit the profile of the ground. For making bends in large diameter pipe or for making bends of several degrees in small pipe, special elbow sections of the correct degree are usually fabricated at the pipe mill or in the pipe fabrication plant.

WORKMANSHIP

If the pipeline is to be buried, it is assumed that the general procedure will be carried out as with any other pipeline construction–the right-of-way will be cleared, access roads provided and a suitable trench dug out in which to lay the

Fig. 86.8.–A section of 96 in. X 0.875 in. water pipe being welded in the ditch. Internal line-up clamp can be seen inside of pipe.

Fig. 86.9.–Steel water pipe 96 in. outside diameter (od) X 0.875 in. wall thickness being welded with the shielded metal-arc welding process.

pipe. The pipe is distributed along the right-of-way, and any special fittings are distributed at their proper position (Fig. 86.8). If the pipe is enamel-lined or externally-coated, considerable care must be employed in the hauling and distributing operations so that there is no damage to the inside lining or the outside coating. Trucks for hauling this type of pipe are usually provided with padded bunks cut out to fit part of the circumference of the pipe. When coated pipe is distributed along the right-of-way in rocky ground, the end of the pipe should be placed on wood blocks so that none of the coating on the outside of the pipe rests on rocks or gravel.

Whether field-welded joints or sleeve-type compression couplings are used, the pipe is usually picked up, one length at a time, and joined together in the bottom of the trench (Fig. 86.9A). Bell holes or scaffolds are provided for welding and coating field joints.

Pipelines crossing rivers require special considerations as to their location. If the elevation permits, such pipelines are frequently carried across the stream on piers. The procedure for welding field joints on steel water pipelines is covered by the AWWA C-206 (AWS D7.0) Standard Specification for Field Welding of Steel Water Pipe Joints.

For general purposes, water pipe is welded in the same way as other large size transmission or distribution pipe. There are some differences, but they are largely in fabrication. Since most water pipe is in built-up or populated areas, provision must be made for more changes of direction and elevation than in cross-country pipelines. These provisions are usually designed in advance, and the pipe is fabricated to plan and the fabricated pieces delivered to the proper location. These pieces are seldom more than 40 ft long because of the difficulty of handling longer pieces under city conditions. Also, there are usually more outlets than in other large size piping, and these are often fabricated at the pipe mill.

There are many economic advantages in welded steel pipelines for water service since longer lengths, therefore fewer joints, are used than in many other installations. The greatest advantage of welded pipelines for water service is the resistance to breaks, cracks and leaks. These can cause major damage when they occur under streets or roads and can cause considerable inconvenience when service is interrupted.

FIELD INSPECTION AND TESTING

Inspection and qualification requirements for water pipelines are prescribed in the AWS–AWWA Specification for Field Welding of Steel Water Pipe Joints. Recommendations given under Inspection and Testing in the Oil, Gasoline and Gas Pipelines section of this chapter are applicable to the inspection phase of field welding of water pipelines.

PENSTOCKS AND LARGE DIAMETER WATER PIPES

Penstocks are used for the conveyance of water under pressure to hydraulic turbines for the generation of electric power. Smooth, streamlined interior surfaces are necessary in reducing hydraulic losses to conserve the available pressure head. The design and construction require a high degree of reliability. The strength and flexibility of steel make steel penstocks especially suitable for the variety of pressure fluctuations met in turbine operation. Large diameter penstocks are usually made from steel plates rolled into cylindrical sections with the ends of adjacent sections welded together. This welding is done in the shop for sections whose diameters are within shipping limits and in the field for large diameters.

The same principles of design and construction which govern penstocks apply equally well to very large high-pressure water pipelines.

GOVERNING CODES AND SPECIFICATIONS

At present, there are no nationally recognized specifications for the design and fabrication of penstocks. Because of this, it is the usual practice to prepare detailed specifications for each job. Design requirements for such specifications may be based on the standards for design of pressure vessels, described in the ASME Boiler and Pressure Vessel Code, Section VIII, Division 1 or 2. More liberal design, as permitted by rules in Division 2 of the ASME Code Section VIII, can be utilized when all stress-producing factors are considered. In all cases, fabrication, inspection and welding requirements should conform to the standards set out in the ASME Boiler and Pressure Vessel Code, Section VIII, Division 1 or 2. Fabrication requirements should cover in detail the type of steel to be used, the specific workmanship required and the inspection and testing to be employed.

Welding procedures should be qualified before being used, and all welders should be qualified before being permitted to weld penstocks. Specification requirements should also include detailed instructions regarding assembling and laying of pipe sections, as well as painting requirements. Since actual soil conditions at the site of installation and the adaptability of metals and equipment during construction may materially affect the economy of the installation, contract specifications should contain provisions adapting the construction to the particular conditions of the job.

BASE METALS

Welding processes and procedures have been developed to the extent that it is possible to weld a large variety of steels in all of the thicknesses used in penstocks and pipeline fabrication. This includes the steels which operate under high internal working pressures and diameters that vary up to 45 ft. Choice of base metal should be determined by service requirements.

The steels generally considered suitable for pipeline and penstock construction can be grouped into structural quality and pressure vessel quality steels as outlined below:

ASTM STRUCTURAL QUALITY STEELS

A 283 Low and Intermediate Tensile Strength Carbon Steel Plates of Structural Quality.
A 36 Structural Steel
A 131 Structural Steel for Ships
A 572 High-Strength Low-Alloy Columbium-Vanadium Steels of Structural Quality

ASTM PRESSURE VESSEL QUALITY STEELS

A 285 Low and Intermediate Tensile Strength Carbon Steel Plates for Pressure Vessels
A 442 Carbon Steel Plates with Improved Transition Properties for Pressure Vessels
A 516 Carbon Steel Plates for Pressure Vessels for Moderate and Lower Temperature Service
A 537 Carbon-Manganese-Silicon Steel Plates, Heat Treated, for Pressure Vessels.
A 517 High-Strength Alloy Steel Plates, Quenched and Tempered, for Pressure Vessels

The mechanical properties are given in Table 86.2. For chemical composition and other detailed requirements, reference should be made to the particular ASTM Standard.

Pressure vessel quality steels are suggested for use in penstocks because they are subjected to frequent pressure surges and, in many cases, to freezing temperatures.

High quality, fine-grain steels are particularly desirable in complicated weldments, such as laterals and tees having heavy reinforcing elements. In these cases, particularly where plate thicknesses exceed 1 1/2 in., as well as in locations which experience freezing temperatures, notch toughness characteristics of the base metal must be considered and impact tests may be required.

Table 86.2—Mechanical properties requirements—ASTM steels for pipelines and penstocks

ASTM Designation	Tensile Strength (ksi)	Yield Strength (ksi, min)	Elongation in 8 in. (%, min)	Elongation in 2 in. (%, min)	Maximum Plate Thickness Available (in.)
Structural Quality					
A 283					
Gr A	45.0/55.0	24.0	27	30	4
Gr B	50.0/60.0	27.0	25	28	4
Gr C	55.0/65.0	30.0	22	25	4
Gr D	60.0/72.0	33.0	20	23	4
A 36	58.0/80.0	36.0	20	23	4
A 131					
Gr A	58.0/71.0	32.0	21	24	1/2
Gr B	58.0/71.0	32.0	21	24	1
Gr C, E, CS	58.0/71.0	32.0	21	24	2
A 572					
Gr 42	60.0 (min)	42.0	20	24	4
Gr 45	60.0 (min)	45.0	19	22	1 1/2
Gr 50	65.0 (min)	50.0	18	21	1 1/2
Pressure Vessel Quality					
A 285					
Gr A	45.0/55.0	24.0	27	30	2
Gr B	50.0/60.0	27.0	25	28	2
Gr C	55.0/65.0	30.0	23	27	2
A 442					
Gr 55	55.0/65.0	30.0	21	26	1 1/2
Gr 60	60.0/72.0	32.0	20	23	1 1/2
A 516					
Gr 55	55.0/65.0	30.0	23	27	12
Gr 60	60.0/72.0	32.0	21	25	8
Gr 65	65.0/77.0	35.0	19	23	8
Gr 70	70.0/85.0	38.0	17	21	8
A 537					
Gr A	70.0/90.0	50.0	18	22	2 1/2
Gr B	80.0/100.0	60.0	–	22	2 1/2
A 517					
Gr A, B, C, D, J	115.0/135.0	100.0	–	16*	1 1/4
Gr G, H, K, L, M	115.0/135.0	100.0	–	16*	2
Gr E, F, P	115.0/135.0	100.0	–	16*	2 1/2

**Reduction of area: 35% min.*

In choosing base metals for penstocks, it should be remembered that the quality of the completed installation will be influenced by the type of steel and filler metals used, the welding current, welding speed, joint design, the skill of the welders and the weather conditions. The combination of these variables will determine the quality of the finished product.

The structural quality steels ASTM A 283 and A 36 are suitable for the smaller diameters and lower pressure pipelines where conservative stress levels are used and plate thicknesses are 5/8 in. or less. The A 283 steels are rolled in Grades A, B, C, and D of which Grades B and C are considered most suitable for penstocks. Although the maximum carbon content is not specified for A 283, it is advisable to limit the carbon content to below 0.30%.

The ASTM A131 steels include several grades with increasing improvement in notch toughness. The Grade A, available up to 1/2 in., is comparable to A 283; the Grade B, available up to 1 in., has limits on carbon and manganese content. Grades C, CS and E are made fine-grain; normalizing can be specified for Grade C over 1 3/8 in. thick and is a requirement for all thicknesses for Grades CS and E. Grade E requires the performance of an impact test.

The ASTM A 572 grades of steel are higher strength steels where the increase in strength is obtained by the addition of alloying elements columbium, vanadium and nitrogen. The major application for these steels has been in pipelines; however, they have been used for supports and other structural attachments on penstocks. It is advisable to limit maximum thickness of these steels to 1 1/2 in.

The pressure vessel quality steels listed in Table 86.2 give a family of steels with increasing strength and quality as reflected in notch toughness and uniformity. The pressure vessel steels are purchased to the requirements of ASTM A 20 which requires testing of individual plates.

The ASTM A 285 steels have had wide acceptance in pressure vessel work and to a lesser degree in penstocks. The steel is not particularly noted for good notch-toughness at low temperatures, and its use should be restricted to thicknesses 3/4 in. or less.

For large diameter high pressure penstocks, ASTM steels A 442, A 516, A 537 and A 517 should be considered. ASTM A 442 is made with definite limits on manganese and silicon contents so as to hold carbon to the lowest practical amount. Plates over 1 in. thick are made to fine-grain practice and normalizing can be specified. Supplementary requirements that can be specified include vacuum treatment, product analysis, simulated postweld heat treatment of test coupons, additional tension test, Charpy V-notch impact tests, drop-weight tests, ultrasonic examination and magnetic particle examination of plate edges.

The ASTM A 516 Specification gives four grades of steel. The Grades 55 and 60 overlap the Grades 55 and 60 in the A 442 Specification, but A 516 is made to fine-grain practice in all thicknesses. Normalizing is a requirement for thicknesses greater than 1 1/2 in. and may be specified for thicknesses less than 1 1/2 in. The Grade 70 requires a 38.0 ksi yield strength (min) and a 70.0 ksi tensile strength (min) with thicknesses up to 8 in. which cover the needs of a wide range of applications. The same supplementary requirements are available with the A 516 Specification as with the A 442 Specification. The Grade 60 is essentially identical to ASTM Specification A 201 Grade B which is no longer produced.

The ASTM A 537 Grade A and B steels afford two readily weldable steels with all the desirable characteristics of a steel for penstock application, particularly where thick sections are required. Strength is one step above the A 516 steels. Both Grades A and B are made to fine-grain practice. Grade A is normalized in all thicknesses and Grade B is heat-treated by quenching and tempering. The superior quality of Grade A has been reflected in selection of this steel for many of the high head and large diameter penstocks built in the last decade in this country, as well as internationally.

The ASTM A 517 high-strength alloy quenched and tempered steels are appropriate selections for the very high heads and high head-large diameter combinations. The use of A 517 will be determined by economics, and by thicknesses of low-carbon steel plates requiring costly postweld heat treatment, generally in the 2 in. range. The use of this steel is generally avoided where an extensive amount of fillet welding is required due to the dangers of underbead cracking. To ensure the integrity of the finished penstock, skilled welders and competent supervision coupled with full inspection of butt welds are necessary when A 517 steel is used.

DESIGN

The diameter of a penstock is determined from an economic analysis in which the cost of the penstock is balanced against the value of power lost due to flow in the pipe. Economical velocities have been found to vary from 10 to 20 ft/s, depending on the head and other variables. The diameter resulting in the lowest annual cost, including depreciation, value of power lost and fixed charges, is the most economical to use.

The efficiency of arc-welded joints varies according to the type of joint and the weld examination used. Longitudinal joints of penstocks subjected to the full operating head, complete fusion and complete penetration double-welded butt joints, either single- or double-beveled, provide the most efficient connection, not only in strength and impact resistance, but also hydraulically. The ASME Code, Section VIII, Division 1 specifies permissible joint efficiencies depending on the type of joint and degree of radiographic examination employed. For the higher strength and quality metals, full radiographic examination is warranted for designs based on the higher stress basis permitted by the Section VIII, Division 2 Code, ASME. Details for fittings and other penetrations should be of the same quality level used for the shell plates.

On high pressure lines employing thick plates over 1 1/2 in. or in other complicated weldments, postweld heat treatment becomes a consideration. Often, a higher strength steel permitting thinner plates can be selected in those cases where postweld heat treatment is not feasible or, in some borderline cases, special and more stringent requirements can be imposed on the welding procedures.

Preheat is required for all quenched and tempered steels in all thicknesses. The preheat will increase the cost of construction; the cost is further increased by the higher price of these steels as compared to the carbon steels. Unless such increased costs are balanced by a reduction in plate thickness and weight made possible by the use of higher design stresses, the use of these steels will not be economical.

In addition to the steel plates, various other ferrous metals are used in the construction of penstocks. The structural steel supports for the penstocks are usually fabricated from steel conforming to ASTM A 36 Structural Steel for rolled shapes and A 283 Grade C for plate. Stiffener rings may also be made from this type of steel. All permanent attachments welded to shell plates regardless of size should be of metals having an allowable design stress not less than the calculated shell stress and suitable for the same service as the shell.

Flanges for connection to gates, valves or turbines are usually made from forged or rolled steel, conforming to ASTM A 105 Specification for Forgings, Carbon Steel, for Piping Components. This steel has a tensile strength of 70,000 psi (min), a yield strength of 36,000 psi (min), an elongation in 2 in. of 22% (min) and a reduction of area of 30% (min).

The welding electrodes used for manual shielded metal-arc welding should be of the low-hydrogen type. Electrodes should be selected for the position in which welding is to be performed, and of the type required for the metal to be welded.

Among the nonferrous metals used in penstock construction, there are bronzes and brasses for rocker supports, nickel cladding and flax packing for expansion joints and other materials which are described in other sections.

A hydrostatic pressure test is often required on the completed penstock. Test pressure is generally established between 1.25 and 1.50 times design pressure.

Pressure surges due to water hammer are produced by the rapid closure of the turbine gates, and can be computed for simple conduits in accordance with standard formulas or charts developed for that purpose. Under certain operating conditions, as with oncoming loads, the turbine gates will open rapidly, and subnormal pressures may develop in the penstock. These may also result from pressure oscillation due to surges or when the water column abruptly parts and rejoins in the penstock. To prevent a possible collapse of the penstock under such conditions, it may be necessary to adjust its profile or to reinforce the shell with stiffeners at critical points along the profile. The impact loads due to surge conditions make it desirable to use ductile metals which will absorb the abnormal stresses by yielding, thus avoiding rupture in the plates and welds. The scope of this chapter does not permit a full discussion of the water hammer phenomenon which is treated in special publications on hydraulics.

The design of support rings for large self-supporting steei pipe is based on the elastic theory of the thin cylindrical shells. The pipe shell is subjected principally to beam and hoop stresses, and the loads are transmitted to the support rings by shear. Pipes adequately reinforced for the bending moments at the points of support may be carried on long spans without intermediate stiffeners. If penstocks are continuously supported at a number of points, the bending moment at any point along the penstock may be computed, assuming that it is a continuous beam, using applicable beam formulas.

The length of the span to be used is a matter of economy. Long spans, over 75 ft, are usually uneconomical because the plate thicknesses must be increased and heavy rings used at the supports to carry the large reactions. See the bibliography for the listing of a special publication on design of support rings. In areas subjected to seismic disturbances, the supports should be designed for a horizontal force of 0.1 to 0.2 of the gravity load, depending on the severity of earthquake shocks reported for the area.

A typical support ring consisting of two 6 by 3/4 in. thick steel bars is shown in Fig. 86.10. The rings are welded to the pipe shell with two full-length fillet welds and are tied together with diaphragm plates. Two support columns made from 8 in. wide flange beams are welded between the rings with fillet welds and plug welds in order to carry the pipe and water load to the rocker supports and concrete piers.

The pipe may also be supported with saddles made to fit the contour of the pipe and generally covering 120 deg of the invert. The saddles can be built up from steel plate and shapes or formed with reinforced concrete. Span length between saddle supports will generally be less than with the ring girder support because of stresses that develop in the shell over the horns of the saddle. Stiffener rings placed adjacent to the saddle can be added to strengthen the shell and thus permit longer spans. With saddle supports, some means must be provided to permit thermal expansion of the pipelines to take place and to reduce friction forces. Sliding low-friction bearing plates can be used at the base

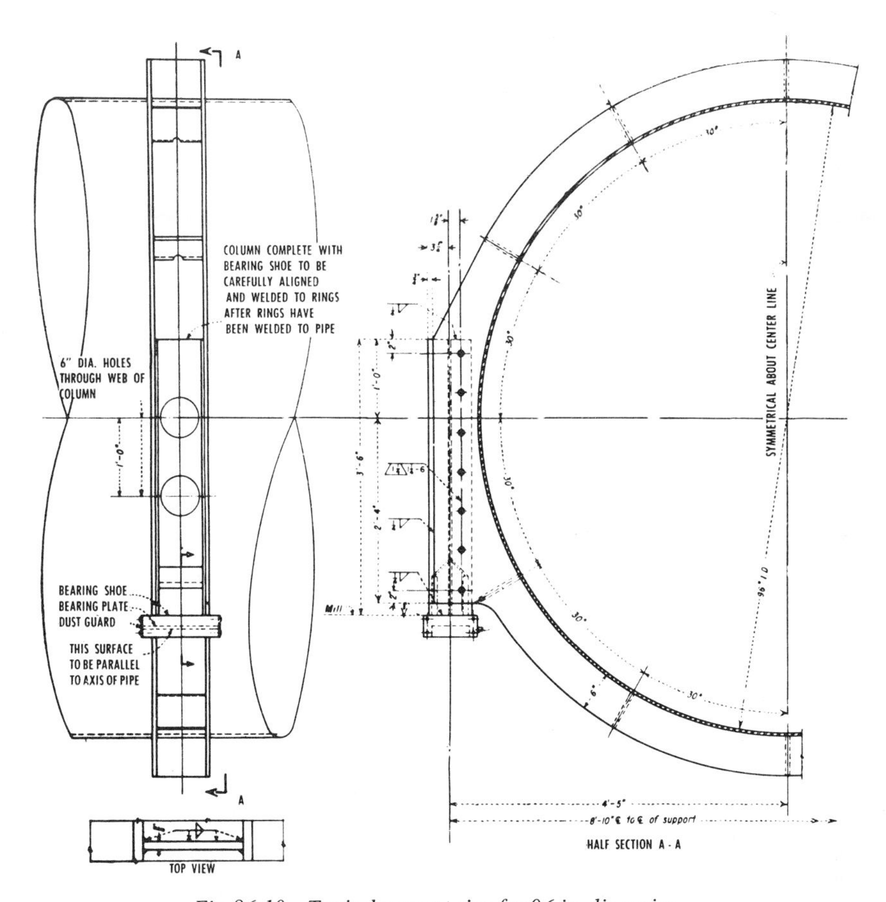

Fig. 86.10.–Typical support ring for 96 in. diam pipe.

of the saddle, or graphited asbestos sheet packing can be placed between saddle and pipe to reduce friction forces.

Changes in profile or alignment of a penstock are effected with fabricated segmental bends, using small deflection angles to reduce hydraulic losses due to flow. Bends may be of constant diameter or they may be formed as reducing bends in which the diameter is reduced in the downstream direction.

When a penstock header supplies two or more turbines or is diverting a portion of the flow to an irrigation line, branch outlets or lateral connections are used between the main line or header and the branch pipe. Such branch outlets or lateral connections require special reinforcement to compensate for the metal removed for hydraulic reasons. Figure 86.11 shows one such branch outlet on a 102 in. diam penstock. The branch outlet is of conical shape and diverts a portion of the flow to a 50 in. diam branch pipe leading to an outlet valve which controls irrigation release. The intersection of the branch pipe with the penstock is provided with a curved Tee-beam, reinforced with two 6 in. diam tie bolts welded to the beam at both ends.

Exposed penstocks are subjected to atmospheric temperature changes when

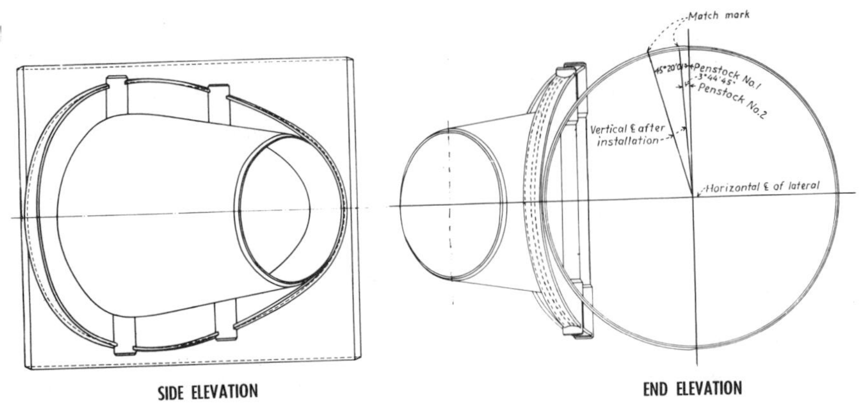

Fig. 86.11.–Branch outlet for 102 in. diam penstock.

empty, and to the temperature of the water when in operation. If fully restrained against movement, the temperature variations will produce a unit stress equal to the coefficient of expansion times the modulus of elasticity of the steel, or 195 psi per degree F temperature change. In order to permit free movement and prevent excessive temperature stresses in exposed penstocks, expansion joints are provided.

There are several types of expansion joints, among which the sleeve-type is the most generally used. Figure 86.12 shows a typical all-welded sleeve-type expansion joint for a 96 in. diam steel penstock. The packing consists of several rings of square, braided flax, compressed with a packing gland and take-up bolts to make the joint watertight. Where penstocks pass through construction joints separating two concrete masses, as between the dam and the powerhouse, provision should be made for movement in both longitudinal and transverse directions. In such cases, a so-called double-acting expansion joint is used. This joint permits longitudinal movement due to temperature changes and vertical displacement due to foundation deflection.

Penstocks are usually provided with filling lines, manholes, vents, drains, piezometer connections, service outlets and flanged connections which must be given special attention in the design. Finished openings in the pipe shell must be reinforced as required by the ASME Code to prevent concentrations of stress.

WORKMANSHIP

Fabrication and Erection

Transportation problems usually determine whether welded plate steel penstocks should be fabricated in a contractor's shop or at the site of installation. Since rail transportation of pipe exceeding 12 ft in diam is not feasible, such sizes are either completely fabricated in the field, or the plates are cut to size and the edges prepared for welding and rolled in the shop. They are then shipped for assembly and welding in the field. The fabrication of penstocks includes a number of special operations performed on such equipment as plate rolls, presses, oxygen-cutting equipment, welding machines, stress-relieving equipment, testing machines and radiographic and handling equipment. In

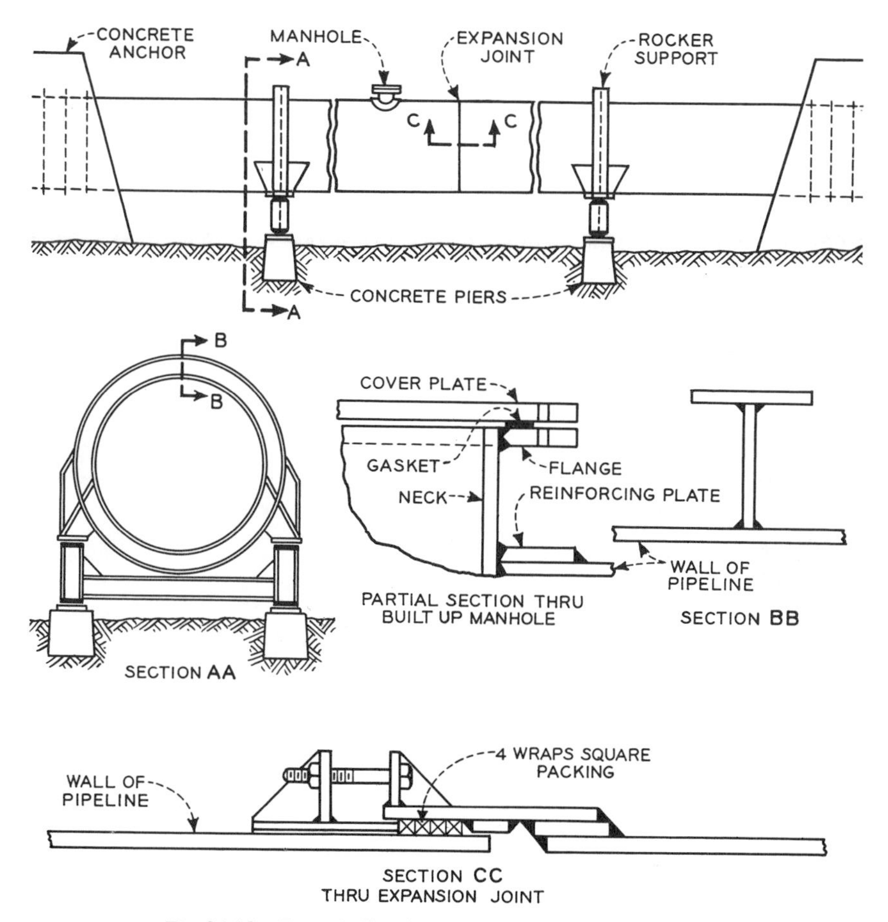

Fig. 86.12.–Typical all-welded sleeve-type expansion joining.

addition, sandblasting and coating equipment are used when coating the completed pipe.

Plates of the desired size and thickness for large penstocks are usually ordered from the steel mill. Upon arrival in the shop, the plates are oxygen-cut or planed to exact width and length, and the edges are prepared for welding.

The preparation of the plate edges depends on the welding process to be used. Welds of more uniform quality can be produced more easily with automatic submerged arc welding than with manual welding because full control of the welding speed and current can be maintained. Therefore, this type of welding is used on penstocks where possible. Since automatic welding is not well-suited for irregular circumferential joints and for joints which are to be welded in-place during erection, manual welding is used. Heavy plates for manual field-welded joints are usually provided with double-Vee grooves. For automatic welding, the plates should be provided with single-Vee or double-Vee grooves, depending on the weld backing or plate thickness which also determines the details of the welding procedure to be employed.

After the plates are cut to size and prepared for welding, the edges are pressed to shape on a plate press. The plates are then rolled into pipe courses the lengths of which are equal to the plate widths. Depending on their diameter, the pipe courses are made from one or more plates with one or more longitudinal joints. After rolling, the longitudinal seams are welded on the automatic welding machine, completing a pipe course. Two or more pipe courses are then welded into pipe sections or laying lengths of 20 to 40 ft.

For long runs of pipe, the accumulated weld shrinkage in girth joints should be taken into consideration. Often shrinkage is approximated as equal to the root opening in the joint so that plate widths are prepared assuming zero root opening in the joint. Actual variations in accumulated shrinkage that occur in the field can be corrected for by varying the root opening slightly or by providing extra length on one pipe that is then trimmed in the field to suit the closing space.

Fig. 86.13.–Pipe courses for a 15 ft diam penstock.

Figure 86.13 shows some pipe courses and sections for a 15 ft diam penstock during fabrication. The length of pipe sections is dependent on plate width, loading economy, the capacity of testing machines or handling equipment. Sheared carbon-steel plates are usually rolled at the mill in thicknesses from 3/16 to 2 in. and in widths from 24 to 148 in. For some types, as for example carbon-silicon steel, plates up to 12 in. thick are rolled. Welding economy indicates the use of steel plates of the greatest width practicable; however, the cost of plates increases rapidly with an increase in width. Therefore, both welding and plate costs must be considered in determining the economical width to be used in penstock fabrication.

Arc welding on penstocks requires a continuous supply of electric current of proper characteristics. Either dc or ac may be used, the former being the more common. DC may be furnished from large motor generator sets.

For the manual welding of penstocks and large diameter water pipe of plain carbon steels with tensile strength less than 65,000 psi and thicknesses less than 3/8 in., and where 100% radiography is not required, E6010 electrodes are used. For the higher strength steels and thicker plates of carbon steel, low-hydrogen electrodes of the EXX18 classification are used in a strength grade that matches the base metal.

Baking of electrodes for one hour at temperatures on the order of 800 F or as recommended by the supplier of the electrodes is a requirement. After bake-out, the electrodes are transferred to a holding oven maintained at a minimum temperature of 200 F until ready to use.

A minimum preheat of 200 F is necessary when welding steels with a tensile strength of 70 ksi or less and with a thickness exceeding 1 1/4 in. Preheat, warm to the hand, is necessary on thinner metal when ambient temperatures fall below about 50 F. A preheat of up to 350 F maximum may be required in all thicknesses for the high-strength steels whose tensile strength exceeds 70 ksi.

Postweld heat treatment is a requirement on butt welds in the thicker plates and in other highly restrained assemblies such as bifurcations. Peening is seldom used. However, in certain highly restrained thick sections, it may be used on intermediate weld passes to control distortion and help relieve welding stresses.

The quality of field work is dependent on the accuracy of shop fabrication. Proper edge preparation for welding and accurately formed plates and pipe sections of proper length, if protected against deformation during handling and shipment, will facilitate field installation work. The welding is generally limited to girth joints and must be performed in position. The fitting of field joints prior to welding is important. On the top half, the pipe ends may be fitted from the outside, and on the bottom half, from the inside, using temporary fitting lugs welded at intervals across the joint as shown in Fig. 86.14. All fitting lugs are on the top side of the joints, leaving the overhead side fully accessible for the first weld deposits.

The actual welding of a girth joint may be accomplished with two welders, one stationed on the inside and one on the outside of the pipe, as outlined in Fig. 86.15. Making the first pass using a backstep sequence and welding uphill in opposite quadrants simultaneously will help to maintain the root opening and ensure uniform shrinkage in the joint. When two complete layers of weld metal have been deposited in the overhead position, the fitting lugs on the top side can be removed and the joints backchipped. The balance of the welding is then completed. Chipping should be done with round-nosed tools or with the oxygen or air carbon-arc gouging processes (see chapters on cutting in Section 3A) until all visible defects are removed.

The sections to be welded should be accurately matched and retained in position during welding by the use of expandable spiders on the inside of the pipe, or by exterior butt straps permitting all welding to be performed from the inside of the pipe. This latter practice may be necessary in tunnels having limited

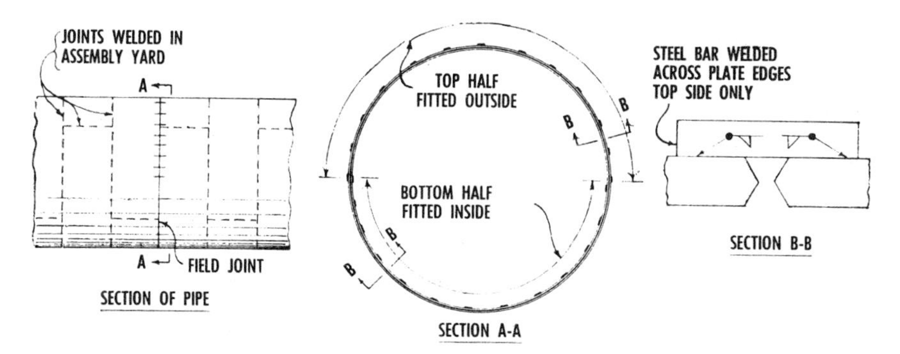

Fig. 86.14.–Method of fitting field joints.

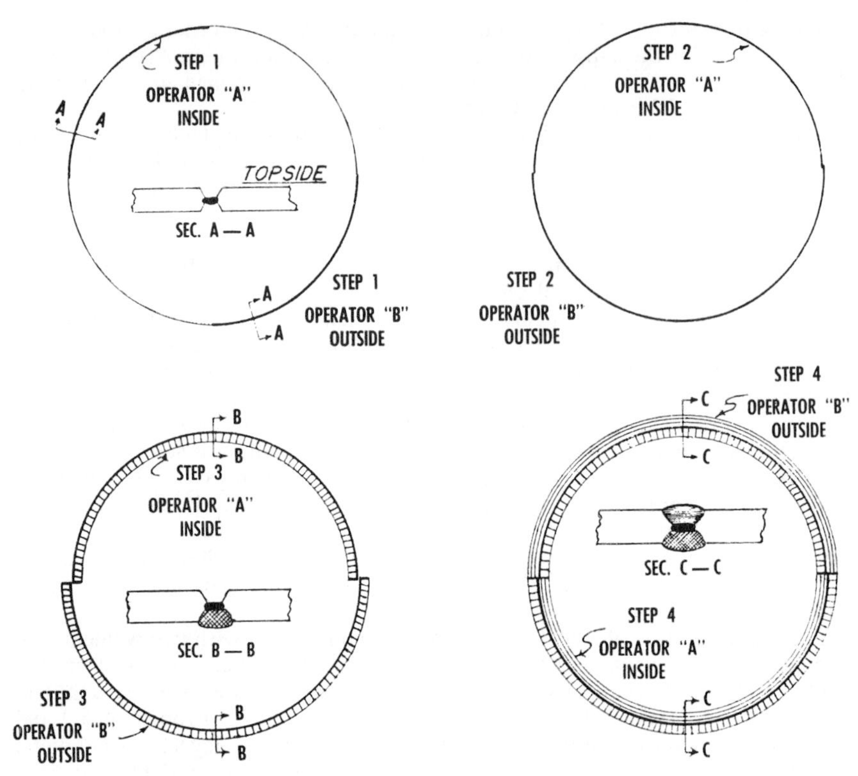

Fig. 86.15.–Welding sequence for field joints.

clearance around the pipe. Whenever possible, however, all joints should be welded from both sides. Weld metal originally deposited at the bottom of the weld groove should be chipped or gouged to obtain a clean surface for the deposit of the first bead from the opposite side of the joint. Tack welds, unless made in a manner which facilitates their inclusion in the final weld, should be removed so that they will not become a part of the final joint.

The weld metal should be deposited in successive layers; each layer should be thoroughly cleaned to sound metal before applying the next layer. A weld reinforcement not exceeding 1/8 in. should be used. Peening of weld deposits is not required except in special cases when necessary to compensate for weld shrinkage. The completed weld must show complete joint penetration without undercutting or other imperfections. When welding in cold weather, it is advisable to preheat (to about 100 F) the plates before welding.

INSPECTION AND TESTING

If required by the purchaser, the welding procedure to be used by the penstock fabricator should be qualified in accordance with the Standard Qualification Procedure, AWS B3.0. This includes reduced-section tension and free-, root-, face- and side-bend tests for groove welds and fillet-weld soundness, and longitudinal and transverse shear tests for fillet welds.

All welding on penstocks must be performed by welders who have passed qualification tests in accordance with the AWS Standard Qualification Procedure. The tests for groove welds include root-, face- and side-bend tests. Tests for fillet welds include fillet weld soundness tests in each of the different positions of welding (flat, horizontal, vertical and overhead) for which the welder is to be qualified.

A check on the quality of production welding is maintained by the requirement that one test plate be prepared for a specific completed length of seam (Table 86.3). Test plates should be of the same base metal and of approximately the same thickness as used for the penstock. The edge preparation, electrodes and welding procedure used for the test plates should also be the same as used for the penstock.

All fabrication, testing and installation work on penstocks should be subject to rigid inspection by qualified welding inspectors to ensure construction in accordance with the specification requirements. The inspector should check the tolerances in pipe joints, expansion joints or other features which will affect field installation and fitting.

When a pipe section is completely welded, the joints are inspected by radiographic, ultrasonic, magnetic particle or liquid penetrant methods. Radiographic inspection is the preferred method, particularly on butt joints, because of its wide acceptance, ease of intepretation and primarily because it affords a permanent record of the weld. Cracks, incomplete fusion and undercuts are not acceptable. The acceptability of porosity, slag inclusions and cavities can be determined from the size and distribution of the defects compared with standard charts. Unacceptable defects must be removed and the welds repaired and inspected again to ensure the quality of the repair weld. Figure 86.16 shows the

Table 86.3—Length of penstock seam weld per test plate

Plate Thickness (in.)	Linear Feet
Under 1/2	300
1/2 to 1 1/4	200
Over 1 1/4	100

Fig. 86.16.—Circumferential pipe joint being inspected by radiography.

radiographing of a circumferential seam in the shop. (See the chapter on inspection in Section 1 for further information on radiography and other inspection methods). Radiographic inspection of welds in the high-strength steels is usually delayed a minimum of 24 hours following completion of welding to be sure of detecting any delayed cracking that might develop during this period.

Hydrostatic proof testing generally follows completion of all work on a section. Figure 86.17 shows the set-up for testing a pipe section. The test pressure is normally established at 1.25 to 1.5 times maximum design pressure including water hammer. The pressure test of individual penstock sections may be replaced by a proof test on the entire penstock which is of more value since all components are tested. Heating of test water is sometimes necessary when poor or unpredictable notch-tough metals are used to ensure the metal is in the ductile range. All unacceptable defects found by visual inspection, radiography or pressure-testing should be removed by chipping, gouging or grinding. The resulting cavities should be refilled with weld metal in the same manner as the original grooves. Weld repairs should be re-radiographed or tested again as required to prove the quality of repairs. For penstocks requiring radiographic inspection of circumferential joints welded in place during installation, portable radiographic equipment is used.

All radiographs should be examined by the inspector to determine the acceptability of the welds and the extent of repairs. In addition to witnessing all procedures and welder qualifications and production tests, the inspector should see the preparation of test plates and, when possible, the testing of specimens. The hydrostatic pressure test of each penstock section, or the proof hydrostatic test of the complete installation should also be witnessed by the inspector,

Fig. 86.17.–Hydrostatic testing of a pipe section.

together with all cleaning and coating operations on the pipe. A complete record of all inspections and of all weld repairs should be maintained, the films marked with identification numbers and maintained in the project office.

BIBLIOGRAPHY

"Better Inspection Procedures Will Yield Better Line Pipe," Robert R. Wright, *Oil and Gas Journal* (November 29, 1971).

"Welded Steel Water Pipe Manual," Steel Plate Fabricators Association, Inc. (1970).

"Stress Analysis of Wye Branches," United States Department of the Interior, Bureau of Reclamation, Engineering Monograph No. 32.

"New Design Criteria for USBR Penstocks," Harold G. Arthur and John J. Walker, Journal of the Power Division, Proceedings of the American Society of Civil Engineers, Vol. 96, No. PO1, Proc. Paper 7034, 129–143 (January, 1970).

"Welded Steel Penstocks," United States Department of the Interior, Bureau of Reclamation, Engineering Monograph No. 3.

"Penstock Codes–U. S. and Foreign Practice," Andrew Eberhardt, *Journal of the Power Division*, Proceedings of the American Society of Civil Engineers, Vol. 92, No. PO2, Proc. Paper 4752, 137–155 (April, 1966).

"Welding High Strength Pipeline Steels," A. G. Barkow, *Metal Progress* (April, 1966).

"Welding of Quenched and Tempered Steels," W. D. Doty, *Welding Journal*, 44 (July, 1965).

"Why Hydrostatically Test to Yield," L. E. Brooks, *Pipeline Industry* (Nov., 1965 and Jan., 1966).

"Steel Pipe Design and Installation," AWWA Manual M11, American Water Works Association, Inc. (1964).

"Semiautomatic Welding of Transmission Pipelines," G. J. Williams, V. H. Demont and J. Snider, *Welding Journal*, 43, 111–117 (February, 1964).

"CO_2 Welding 'Equals or Exceeds' Earlier Work," L. J. Cunningham and J. Baker, *The Oil and Gas Journal* (Sept. 21, 1964).

"Penstocks and Scroll Cases for Niagara Power Plants," John N. Pirok and J. Edgar Revelle, Transactions ASCE, Vol. 128, Part III, Paper 3453 (1963).

"Higher Strength Steels for Pipe," Robert S. Ryan, *The Oil and Gas Journal* (September 23, 1963).

Gas Transmission and Distribution Piping (ASA B31.8), American Society of Mechanical Engineers (1963).

"Automatic Pipeline Welding," Robert S. Ryan, *The Oil and Gas Journal* (April 10, 1961).

"Water Hammer Design Criteria," John Parmakian, Symposium on Penstocks, ASCE Proceedings Symposium Series, #4 (March 1961).

"Some Problems of a Penstock Builder," John N. Pirok, Symposium on Penstocks, ASCE Proceedings Symposium Series, #4, (March 1961).

"Penstock Design and Construction," G. R. Lathan, Symposium on Penstocks, ASCE Proceedings Symposium Series, #4, (March 1961).

"Multi-Layer Penstock and High Pressure Wyes," Ewald F. Schmitz, Symposium on Penstocks, ASCE Proceedings Symposium Series, #4, (March 1961).

"Kemano Pressure Conduit Engineering Investigations," F. L. Lawton and J. C. Sutherland, Symposium on Penstocks, ASCE Proceedings Symposium Series, #4, (March 1961).

"Penstock Experience and Design Practice," Gordon V. Richards, Symposium on Penstocks, ASCE Proceedings Symposium Series, #4, (March 1961).

"Large Spiral Casings of T-1 Steel," E. L. Seeland, Symposium on Penstocks, ASCE Proceedings Symposium Series, #4, (March 1961).

"Design of Large Pressure Conduits in Rock," F. W. Patterson, R. L. Clinch and I. W. McCaig, Symposium on Penstocks, ASCE Proceedings Symposium Series, #4 (March, 1961).

"Determination of Stresses on Anchor Blocks," M. R. Bouchayer, Symposium on Penstocks, ASCE Proceedings Symposium Series, #4, (March, 1961).

"Submerged Arc Welding on the Pipeline," William B. Handwerk, *Welding Journal*, 38, 672–765 (July, 1959).

Oil Transportation Piping (ASA B31.4), American Society of Mechanical Engineers (1959).

"Qualification Tests for Welders," Robert R. Wright and Robert S. Ryan, *Pipeline Construction* (October, 1959).

"Aluminum Pipe Line," Elton Sterrett, *Ibid.* (October, 1959).

"A Welding and Pipelining History," *Welding Engineer* (May, 1959).

"Factors in Fabricating Branch Connections," Robert S. Ryan, *The Oil and Gas Journal* (March 9, 1959).

"Technical Advances in High-Strength Pipe," W. A. Saylor and A. B. Wilder, *Pipe Line Industry* (January, 1959).

"Hydraulic and Weight Features of Welded Spiral Cases," R. A. Sutherland, Symposium on Spiral Cases, *Transactions of ASME*, Vol. 80, No. 5 (July, 1958).

"Welded Spiral Case Construction Procedures," G. R. Latham, *Ibid.* (July, 1958).

"T. V. A. Experiences–Welded Versus Riveted Spiral Cases," R. M. Gardner, *Ibid.* (July, 1958).

"Pressure Embedment of Spiral Cases," Bradley G. Seitz, *Ibid.* (July, 1958).

"Pipeline Welding and Quality Control Methods," A. G. Barkow, *Proceedings of the American Society of Civil Engineers* (June, 1958).

"Stresses in Large Horizontal Cylindrical Pressure Vessels on Two Saddle Supports," L. P. Zick, *Welding Journal*, 30 (September, 1951).

CHAPTER 87

NUCLEAR POWER

PREPARED BY A COMMITTEE CONSISTING OF:

T. B. CORREY, *Chairman*
Battelle-Northwest Laboratories

N. C. BINKLEY
The Babcock & Wilcox Co.

W. F. BROWN
Hanford Engineering Dev. Laboratory

H. R. CONAWAY
Huntington Alloy Products Div.

J. P. CORLEY
Battelle-Northwest Laboratories

W. J. FARRELL
Sciaky Bros., Inc.

J. M.FOX,JR.
Douglas United Nuclear, Inc.

R. B. HALL
WADCO

F. HETZLER
Akiebolaget ASEA-ATOM

T. B. JEFFERSON
Publisher, Welding Engineer

V. D. LINSE
Battelle-Columbus Laboratories

C. B. MCKEE
Hanford Engineering Dev. Laboratory

D. M. ROMRELL
Hanford Engineering Dev. Laboratory

C. B. SHAW
Hanford Engineering Dev. Laboratory

G. M. SLAUGHTER
Oak Ridge National Laboratory

R. W. STRAITON
Bechtel, Inc.

C. C. TENNENHOUSE
Wall Colmonoy Corp.

E. P. VILKAS
Astro-Arc Co.

W. V. WATERBURY
Revere Copper and Brass, Inc.

R. N. WILLIAMS
Welding Design and Fabrication

CHAPTER 87

NUCLEAR POWER

GENERAL

The current generation of power reactors is of the boiling water (BWR) and pressurized water (PWR) type (Fig. 87.1). However, there are a few liquid metal and gas-cooled types in use. The power reactor of the next generation is a liquid sodium metal-cooled reactor which produces more nuclear fuel than it uses.

While individual systems may vary from the general pattern, most of the present reactor systems use a container–a reactor pressure vessel or an equivalent assembly of pressure tubes in a calandria–to contain the nuclear reaction. This reactor vessel contains a reactor core which consists of the fissionable fuel enclosed in fuel elements and the necessary structural support suitably disposed with respect to the moderator, instrumentation control rods and reactor coolant. The fuel elements generally contain uranium enriched in the fissionable isotope uranium 235 or combinations of uranium and plutonium 239 in the form of oxides and in a quantity to form a critical mass under appropriate conditions.

The fuel material is generally sealed in a cladding of zirconium, aluminum, stainless steel or possibly some other material to contain the fission material. The reactor vessel will usually contain a material such as hydrogen, deuterium, carbon or compounds of these, called a moderator. The reactor has an array of neutron-absorbing material contained in automatically operated, movable control rods which are inserted in whole or in part into the core and through the pressure vessel wall by sealed positioning rods. Circulating through the reactor core and the external system is a reactor coolant, which may be water, heavy water, steam or a liquid metal or a gas such as air, helium, nitrogen or carbon dioxide.

Many modifications or combinations of these materials and components can

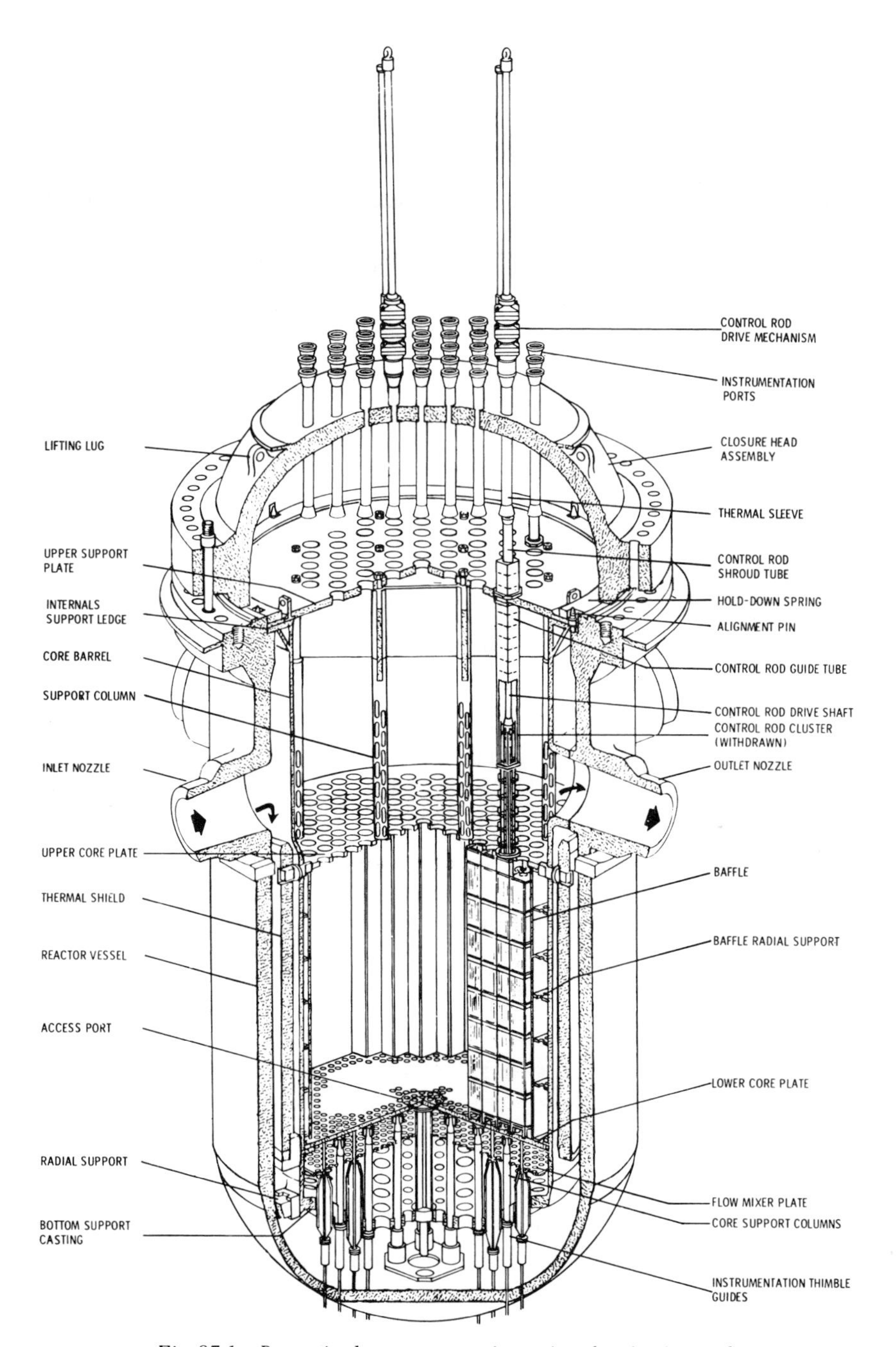

Fig. 87.1.–Pressurized water reactor in section showing internals.

exist in reactor systems. For example, it may not be necessary to provide a protective cladding for reactor fuels if noncorrosive gas coolants are used. The coolant generally passes from the reactor to the heat exchanger where steam is generated for the development of electrical power by relatively conventional methods, or it may be passed directly into the turbine as steam from the reactor. The chemical and nuclear characteristics of the reactor system, the power generation requirements and the environmental conditions of temperature, pressure and radiation will influence the size, configuration and structural materials employed in the reactor vessel and in the associated equipment which the primary coolant may contact. In addition to the reactor vessel and heat exchanger or steam generator, the associated equipment will include piping, valves and pumps, as well as purification, spent fuel replacement and waste disposal equipment. Welding is involved in the fabrication of practically all of these components and in the production of clad fuels.

High radiation levels exist in the operating reactor vessel as well as in its immediate vicinity. Radiation fields of varying intensity may also exist in other parts of the system due to the transport, by the coolant, of fission products which may be introduced by leakage or rupture of fuel elements. Corrosion products may also contain a substantial amount of induced radioactivity and transport it to remote parts of the reactor system. Failure of a welded reactor system component thus may result in the release of quantities of radioactive materials to the external environment; it may, in some cases, result in a loss of coolant and consequent melt-down of the reactor core which could lead to a still greater release of radioactivity outside the system. Power reactor systems are provided with suitable containment vessels, with size, volume and construction sufficient to contain the energy and the radioactivity resulting from the maximum predictable accident to the system.

Nevertheless, the failure of individual components and the release of radioactivity can provide imposing and expensive problems of decontamination, remote maintenance and repair. For these reasons, it is generally accepted that a high and consistent level of fabrication quality is required in any welded or other pressure-containing reactor system component whose failure might result in the release of radiation. At the same time, it should be recognized that no single set of rules is applicable to all nuclear systems or components. Some components place very demanding requirements on materials and fabrication, but every component of every nuclear reactor system is not necessarily as critical. A considerable amount of sound engineering judgment must be exercised to provide against overly restrictive requirements.

FUEL RODS

Rod-type cluster fuel elements are used virtually exclusively in power reactors (Fig. 87.2). Typical fuel rod design consists of pelletized ceramic fuel encapsulated in a metal cladding tube and closed with plugs welded at both ends (Fig. 87.3). Some experimental fuel incorporates vibratory compacted particulate fuel. However, the problems of reducing fuel rod moisture content to low ppm levels prior to seal welding presently dictate the use of pelletized fuel.

Fuel rod outer diameters are typically about 1/4 in. for liquid metal breeder reactors (LMBR), 3/8 to 5/8 in. for light water reactors (LWR's) and about 5/8 in. for an advanced gas-cooled reactor (AGR). Thickness of the cladding varies

Fig. 87.2.–Boiling water reactor nuclear fuel bundle being lowered into the guide tube on its way into the core.

from 0.020 to 0.040 in. depending on the diameter of the tube, material and use.

Typical lengths for BWR and PWR fuel rods are about 12 ft, although shorter lengths have been used in previous generation reactors. It is too early in the development period of the LMBR's to specify a typical length, although 8 ft are being used on the FFTF. Modern AGR fuel rods are about 3 ft in length with eight joined end-to-end to form a rod about 24 ft long.

Fuel rods for modern reactors consist of accurately dimensioned cladding

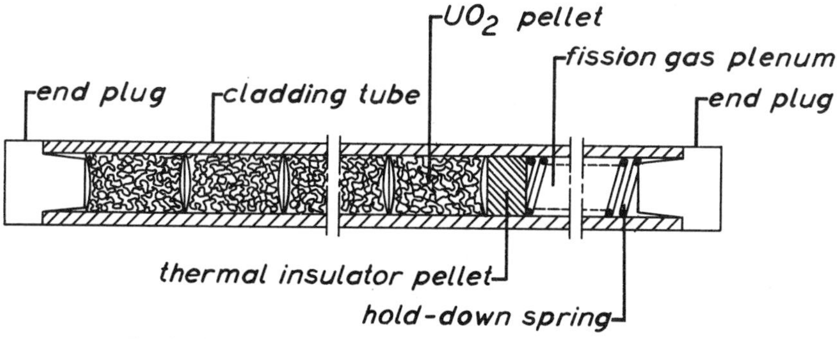

Fig. 87.3.–Sketch of section of a uranium oxide pellet fuel tube.

charged with precision ground ceramic pellets (Fig. 87.3). A lower end plug serves as a supporting base for the stack of ceramic pellets. The pellets are held in place by a hold-down spring between the uppermost pellet and an upper end plug. The hold-down spring serves the purpose of keeping pellets in place during handling and transport. The spring is positioned in an otherwise empty volume provided for gaseous fission products. Insulating plugs may be used on one or both ends to minimize axial heat dissipation. Helium is introduced during the manufacturing process for improving the heat transfer between the fuel pellets and cladding.

Virtually all of the power reactors in general use and all of those currently planned use rod-type fuel elements arranged in a tight cluster or bundle with spacers to provide a finite channel for the liquid coolant. The geometry of the bundles is such that they make a tight-fitting core for the reactors. The bundles are either rectangular or hexagonal. Spacing is designed into the bundles or into the spacing between the bundles to permit the insertion of the control rods. The completed core assembly is a tightly packed mass that has uniform coolant channels throughout. Packing of the core assembly is such that flow of the liquid coolant and its gas, in the case of boiling water reactors, does not produce movement of any of the components against another causing fretting corrosion. Fuel bundles and control rods are arranged in the core of the reactor in such a manner that any single bundle may be removed and replaced with a like unit.

CLADDING METAL

Fuel cladding for current commercial power reactors is a zirconium alloy such as Zr-2, Zr-4 and Zr-Cb (Ozhennite) used principally for neutron economy. Liquid metal-cooled reactors such as the FFTF use Type 304 stainless steel.

END CLOSURE DESIGN

Many types of end closure caps and plugs have been developed since the introduction of the rod- or pin-type fuel. Each one was developed to overcome

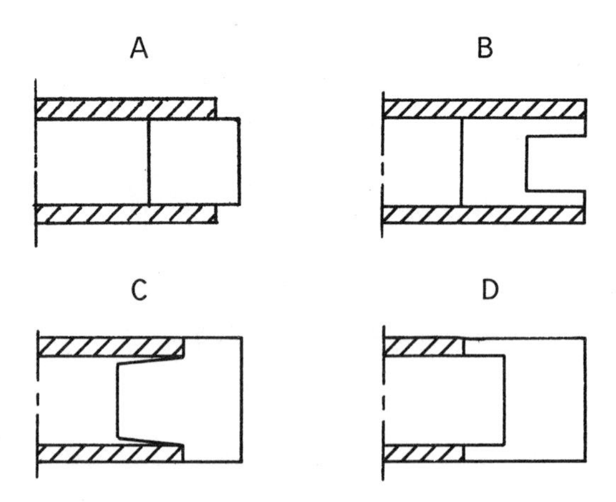

Fig. 87.4.–Sketches of fuel tube terminating end closures.

an inherent weakness in a design applicable to a particular welding process and to satisfy a positioning requirement in a fuel bundle.

End closure design for fuel rods includes at least four primary considerations:

1. Assembly and fixture of end plugs.
2. Provision of an adequate heat sink.
3. Control of fuel rod atmosphere.
4. Quality control.

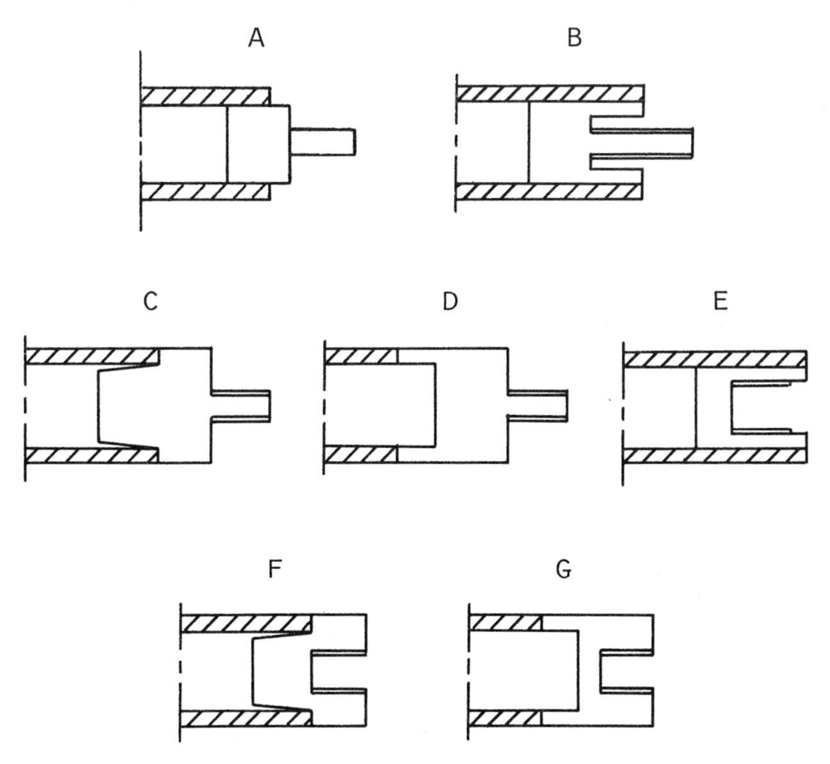

Fig. 87.5.–Sketches of connecting or positioning end closures.

The type of end plug shown in Figs. 87.4C, 87.5C and 87.5F is the earliest design that was used in rod- and pin-type fuel tubes. With the conventional constant current gas tungsten-arc welding process, it was virtually impossible to obtain full root coalescence on all closures without overheating the fuel rod end. With the advent of the precision pulsed current gas tungsten-arc welding process, it appears that this problem has been overcome. Modified designs of the cap at the tube interface to provide filler metal have overcome weld surface depression. This type of joint is readily examined by automatic ultrasonic testing techniques.

The type of end closure shown in Figs. 87.4C, 87.5C and 87.5F is ideally suited for electron beam welding for the following reasons:

1. A concentric tight-fitting joint is readily obtained.
2. The beam passes into the solid cap.
3. Filler metal, if required, can be designed into the cap or readily added.

In addition, the weld joint is readily examined by automatic ultrasonic testing techniques. Shown in Fig. 87.6 is a representative electron beam welded closure and in Fig. 87.7 typical sections. The advantage of an interference fit for the end plugs shown in Figs. 87.4C, 87.5C and 87.5F is fuel rod handling simplicity since end plugs will not fall out or be forced out of position by a hold-down spring. The disadvantages are that cladding tube and plug tolerances must be tightly controlled which increase farication costs. Run-out (axial parallelism) could be a problem, but it can be regulated easily by good dimensional control.

In the case of a small number of end closure welds, it may be most economical to provide a final machining and/or threading operation. Interference tolerances above a specific limit, which must be experimentally determined, will cause outward bulging of the tube during welding when the metal becomes plastic. Unless the end cap is inserted after the fuel atmosphere is set, there is a problem of uniform fuel atmosphere. End plugs shown in Figs. 87.4C, 87.5C and 87.5F, with a clearance fit, require fixturing to hold the plug tightly in alignment during

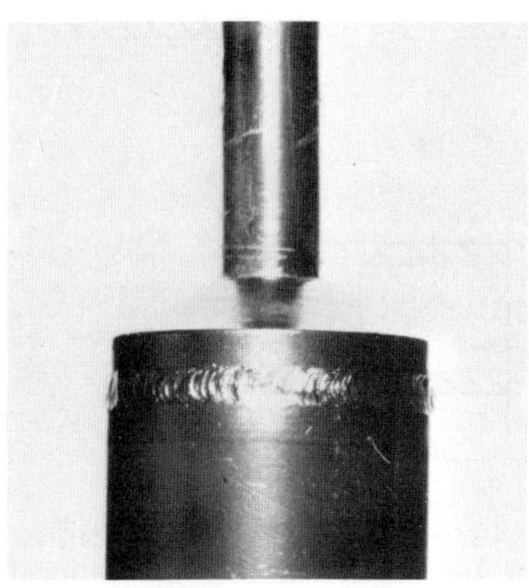

Fig. 87.6.–A Fig. 87.5C electron beam welded end closure. 1X. Material Ozhennite. (Enlarged to 235% in reproduction.)

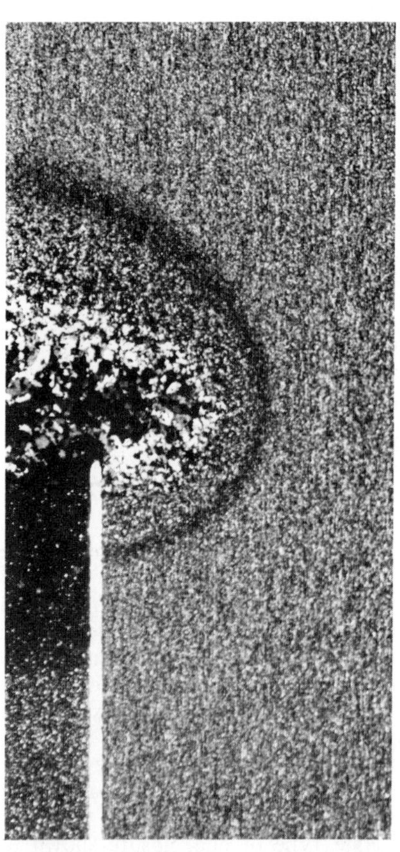

Fig. 87.7.–Macrophotographs of longitudinal sections of Fig. 87.5C electron beam welded closures. 10X. Material 0.020 and 0.030 Ozhennite. (Enlarged to 165% in reproduction.)

welding. Advantages include optimum control of dimensional run-out when well-designed fixturing is employed, a straight forward approach to the design of automatic plugging equipment and improved degassing of the weld joint for hard-vacuum electron beam welding. A disadvantage is that fuel rod handling difficulties are created by a loose end plug.

End plugs of the type shown in Figs. 87.4D, 87.5D and 87.5G are suitable for gas tungsten-arc, electron beam and upset (resistance) welding. Upset welding of this type of end plug, by either a conventional or magnetic force resistance welder, extrudes a collar of metal around the joint which must be removed by a machining operation.

PROVISION OF AN ADEQUATE HEAT SINK

Tubing for fuel rod manufacture is of the highest grade with reference to dimensions and strictly specified metallurgical properties. A good end closure weld design will minimize the heat-affected zone (HAZ) disturbance and accompanying metallurgical restructuring.

In the case where a fuel pellet column hold-down spring is in contact with the end plug, care must be taken that melting and inter-alloying of the spring and plug are avoided. This is particularly critical when the spring and end plug are of dissimilar alloys. It is possible, for example, to create a severe corrosion problem by contaminating a Zircaloy weld fusion zone with nickel-alloy material. The problem can be eliminated by minimizing surface area contact between spring and plug or by protective plating of the spring with, for example, chromium or by providing an adequate heat sink. Copper chill blocks are an excellent tool in many applications.

Gas tungsten-arc welding delivers heat energy to the surface which is distributed more or less equally in all directions. For plugs shown in Figs. 87.4B, 87.5B and 87.5C, the heat transmitting path provided by both tube and plug is equal thereby assuring fusion with minimum energy input. The plugs shown in Figs. 87.4C, 87.5C and 87.5F offer an uneven heat flow path. Heat flow via the cladding tube is restricted, but an open path is provided by the plug. This would cause preferential melting on the tube side of the joint. A peripheral notch machined in the end plug equal or more in depth to the thickness of the tube and located from the joint by about the thickness of the tube (Fig. 87.8) is effective in controlling axial heat flow. Heat input is thereby concentrated and HAZ reduced.

CONTROL OF FUEL ROD ATMOSPHERE

It is important that a hermetically-sealed fuel rod contain a carefully controlled atmosphere. Fission gases released during fuel service life cause increasing internal gas pressure as a function of time. A fission gas plenum, (generally the volume shared with the hold-down spring) is provided for high performance fuel.

Fuel rod internal gas composition is equivalent to welding grade helium. Helium has excellent heat transfer characteristics for an inert gas and provides optimum heat transfer in the gap between fuel pellets and the cladding tube. In addition, the helium provides an effective quality control medium for the detection of welding or material flaws by mass spectographic helium-leak detection.

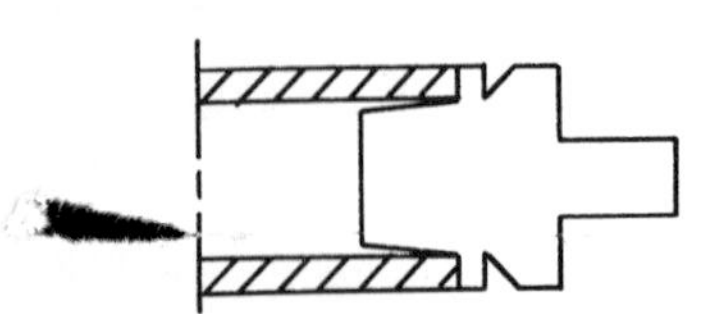

Fig. 87.8.–Heat flow barrier notch to provide heat balance for gas tungsten-arc welding.

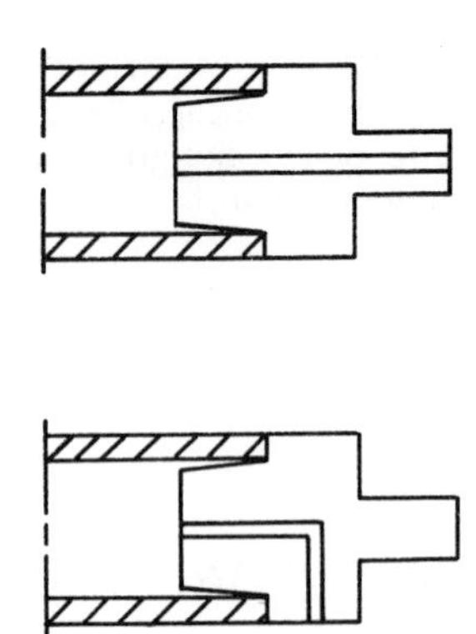

Fig. 87.9.–Two types of vented closures to permit evacuating and backfilling of fuel tube with helium.

Impurity gases such as water vapor must be limited to low amounts to prevent unwanted internal gas pressure buildup and a corrosive attack of the fuel cladding tube (particularly for the zirconium alloys). Values well below 50 ppm are normally specified. This requires special attention prior to the final hermetic sealing step. Two general routes are available: (1) drying of individual components immediately proceeding final assembly and helium gas filling (2) closure welding and vacuum drying of the assembled, but vented fuel rod.

The process whereby assembled fuel rods are vacuum dried requires apparatus where vacuum and temperature can simultaneously be obtained. Process parameters are typically $10^{-1} - 10^{-3}$ torr and above 200 F for periods up to two hours. This is followed by cooling to room temperature and filling the rod with helium to atmospheric pressure (an exception is that some PWR fuel is internally pressurized to values exceeding 400 psig). The final operation is hermetic sealing.

Processes to perform the foregoing operation are highly varied and usually consist of closely guarded proprietary information. The simplest method is to perform all operations in a welding chamber equipped with vacuum, helium, internal heaters, plugging and welding apparatus. The problem becomes increasingly complex in relation to production rationalization and requires close teamwork by fuel design, welding and vacuum equipment specialists.

End closures shown in Fig. 87.9 allow a tube-to-plug weld to be made while providing a passage for vacuum drying and helium backfilling. The passage is later seal-weld fused to provide hermetic sealing. This is the process commonly used for electron beam welding since the tube-to-plug welds are made under vacuum. Operations following the second tube-to-plug weld are performed in separate equipment and hermetic sealing is accomplished by gas tungsten-arc welding. A representative electron beam welded closure in a section of this type is shown in Fig. 87.10.

Production rationalization where gas tungsten-arc welding is applied allows simplification of the gas evacuation and backfill operation. The end closure shown in Fig. 87.11 has a clearance passage machined on one side of the plug (shown by added line). The final peripheral weld seal and hermetic sealing can be accomplished simultaneously.

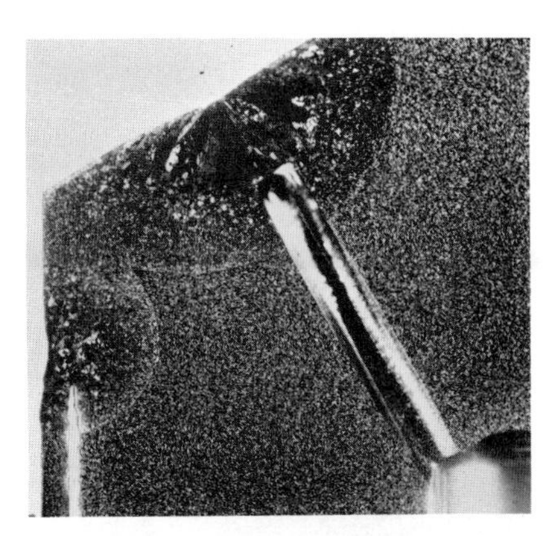

Fig. 87.10.–Longitudinal section of an electron beam welded closure with backfilling hole closed with gas tungsten-arc spot weld. 10X. Material Ozhennite. (Reduced to 66% in reproduction.)

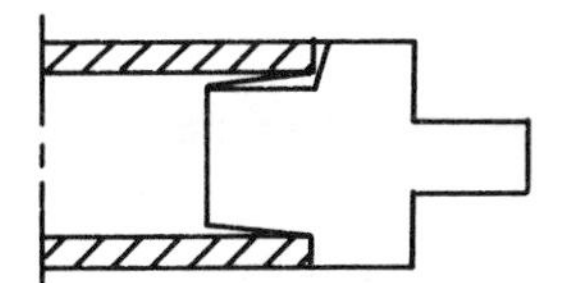

Fig. 87.11.–End closure with gas passage machined on one side permitting peripheral and seal weld to be made simultaneously.

END CLOSURE WELDING

All of the known rod-type fuel elements are closed by welding. Welding techniques have been very highly refined by the world's nuclear fuel manufacturers. Each section of the world has selected essentially one primary welding method to satisfy its own general requirements:

1. The United States uses the gas tungsten-arc process.
2. Canada uses the upset (resistance) butt welding process.
3. Europe uses the electron beam welding process.

Fuel rod weld joint design should be considered as a pressure vessel weld. It has to contain the helium gas, which is introduced at the time of manufacture at pressures to 400 psig, plus the fission gases at the pressures produced by the operating temperature of the fuel in the pin or rod. The maximum allowable cladding temperature from loss of coolant is 2200 F.

In order to disseminate the improvements in the welding of fuel rod and pin end closures, representative samples of the three common closure welding processes are described in detail later.

TYPES OF CLOSURE WELDS

The types of fuel rod end closure welds can be divided into two well-defined types: single closures and double closures. In the single closure type, the end plug or cap is joined to the end of the cladding tube by a welding or brazing process (Fig. 87.12).

The advantages of the single-welded closure are:

1. Low cost compared to other methods.
2. Demonstrated dependability.

The disadvantages of the single welded closure are:

1. There is no nondestructive test that can be automated or is quick and sure for some end closure designs.

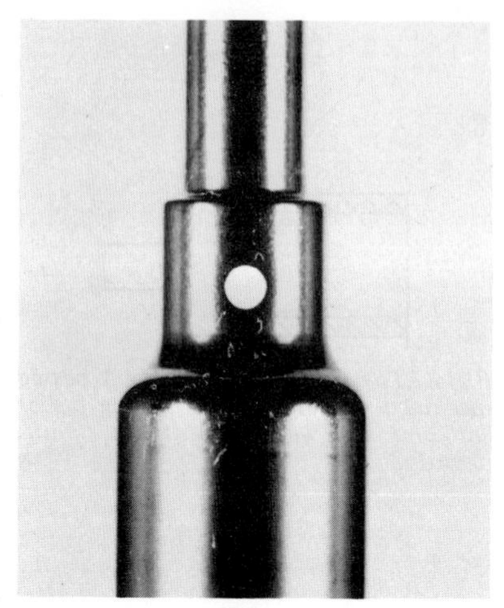

Fig. 87.12.–A Fig. 87.5A gas tungsten-arc welded end closure. 1X. Material Zircaloy-2. (Enlarged to 205% in reproduction.)

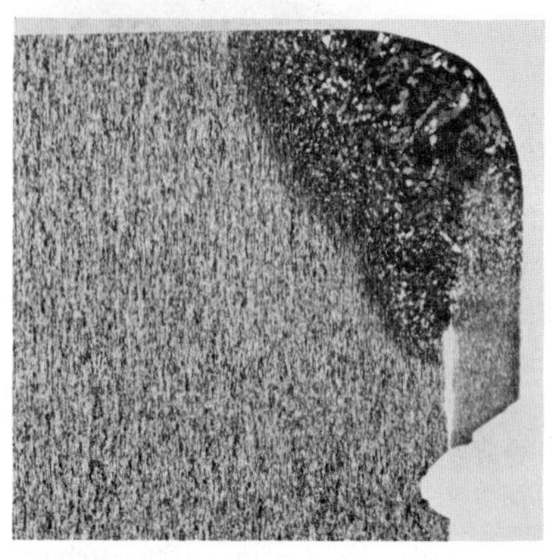

Fig. 87.13.–Electron beam welded end closure with one deep penetrating weld followed by one shallow weld. 10X. Material Zircaloy-2. (Reduced to 84% in reproduction.)

2. Some types are very difficult to make without weld thinning or porosity using the gas tungsten-arc welding process.

The double-closure type end closure uses one and in some cases two welding processes in which the second weld is fused circumferentially to the first weld or braze without melting through the first weld or braze (Figs. 87.13, 87.14 and 87.15). Examples of double closures are:

1. Deep penetrated electron beam-welded closure followed by a shallow broad focused electron beam-welded closure (Fig. 87.13).
2. Magnetic force or resistance-welded closure followed by a gas tungsten-arc or electron beam-welded closure (Fig. 87.14).
3. Brazed closure followed by a gas tungsten-arc welded closure (Fig. 87.15).

The advantages of the double-welded closure are:

1. Virtual assurance that there will be no coolant penetration through the end closure welds.
2. Nondestructive testing of the end closure welds is minimized or eliminated.
3. The double-welded closure is readily inspected with ultrasonics.

The disadvantage of the double-welded closure is its additional cost in time and equipment. With the exception of the brazed closure, all of these closures can be made with minimal heat-affected zones. No double-welded closures are known to have been used on commercial power reactor fuel elements.

Gas Tungsten-Arc Welding

The pyrophoric nature of zirconium alloys, the high oxidation rate of stainless steel at welding temperature and the radioactive fuel dictate that the final and sometimes the first end closure welds on fuel rods and pins be performed in an inert atmosphere chamber.

Helium improves the heat transfer from the fuel core through the cladding and is therefore generally the shielding gas used when welding the second end closure

A B

Fig. 87.14.–End closure made with magnetic force weld followed by gas tungsten-arc weld. 50X. Material Zircaloy-2. (Reduced to 90% in reproduction.)

on ceramic fuel rods. Argon or argon-helium mixtures may be used when welding the first end closure, and the welding may be performed in a localized chamber depending on the cladding material and weld quality requirements.

The electrode is 2% thoriated tungsten rod without primary shielding or cooling through the electrode holder. Welding currents from 11 to 28 A are used, the finite value being dependent on cladding material, thickness, diameter

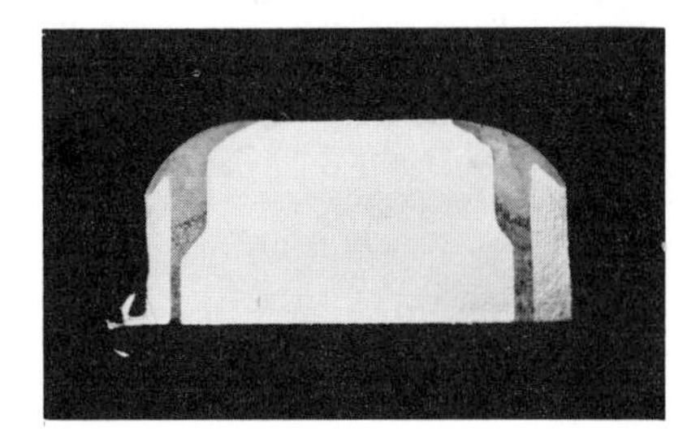

Fig. 87.15.–End closure of fuel brazed with 5 Be-Zr followed by gas tungsten-arc weld. The wide brazes are to prevent cracking produced during welding of a narrow braze. 3X. Material Zircaloy-2. (Enlarged to 110% in reproduction.)

and joint configuration. For good arc control, the electrode-to-work gap is set at 0.011 to 0.013 in. before arc initiation. The short weld time and arc current-voltage relationship preclude the use of an automatic arc voltage control unit.

A precision type of single range d-c precision power supply with approximately 150 V open-circuit and a high frequency type arc starter is most desirable. The trend is toward the use of pulsed current welding power supplies for making the end closure welds. The advantages of pulsed current welding can be summarized as follows:

1. Increases depth-to-width ratio.
2. Virtually eliminates lack of coalescence of the weld root.
3. Minimizes the width of the heat-affected zone.
4. Surface finish is virtually as good as constant current.

Arc starting is very difficult when using pure helium as a shielding gas and a 2% thoriated tungsten electrode. The helium has a high thermal conductivity and thus cools the electrode, tending to prevent the high frequency from heating a spot to the electron-emission temperature. Also, the kinetic energy of the helium atom is too small to dislodge the metal vapors from the previous weld which are condensed on the electrode tip and cover up the thoria that is the primary source of electrons to start and support the arc. The mass of argon atoms is great enough to dislodge the film in a manner similar to sandblasting (Fig. 87.16). By careful shaping, an electrode tip design has been developed that overcomes the problem of arc starting for short runs (Tables 87.1 and 87.2). By making the included angle of the tip very small and the point very sharp, the R. F. or impulse used to start the arc is sufficient to heat the point to an adequate electron emission temperature.

A representative end closure welding schedule is:

Material–Zircaloy-2.
Diameter–5/8 in. od.
Wall thickness–0.032 in.
Shielding gas–helium.
Arc length–0.011-0.013 in. preset.
Welding current–28 A.
Rotational speed–7 rpm.

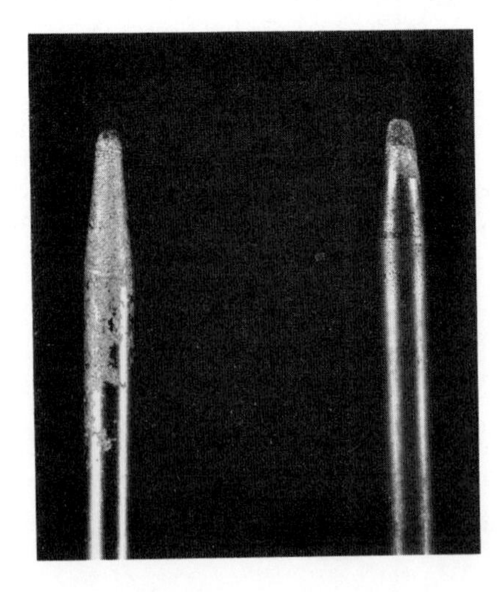

Fig. 87.16.–Difference in electrode tip surfaces when welding with identical conditions on Zircaloy-2. Left, argon shielding showing surface cleaned by sputtering. Right, helium shielding showing metal vapors condensed on tip. 4X. Material 2% thoriated tungsten 1/16 in. diam.

Table 87.1—First end closure welding procedure

Process	GTAW Vac Melt		
Alloy	316 Stainless Steel		
Power Supply		Weld Settings	
Mfg.		Prepurge Time, s	1-2
Model	50 A	Start Current, A	18-19
Serial No.		Upslope Time, s	0-2
		Fixture Delay, s	0-1
Torch		Weld Current, A	18-19
Mfg.		Voltage, V	10-20
Model	HW #20	Type/Polarity	DCSP
Serial No.	#1-2 & 3	Weld Time, s	15-16
Cup Size	Gas Screen #4 Cup	Downslope Time, s	6-9
		Finish Current, A	2-5
Shielding Gas		1 Rev Time, s	11-12
Type	75% He 25% Ar	Weld Speed, ipm	3.6 Ref.
Volume	8 cfh Helium Gage	Fixture Speed Dial	∿55-65 Ref.
		Postpurge Time, s	9-12
Backup Gas		Weld Position	Horizontal
Type	He Grade (A) Comm	Arc Start Method	H.F.
Volume	20 cfh	Arc Start Intensity	N/A
Electrode			
Type	2% Th		
Diam, in.	1/16		
Ext., in.	.250 ± .010		
Shape, deg	20 to 40		

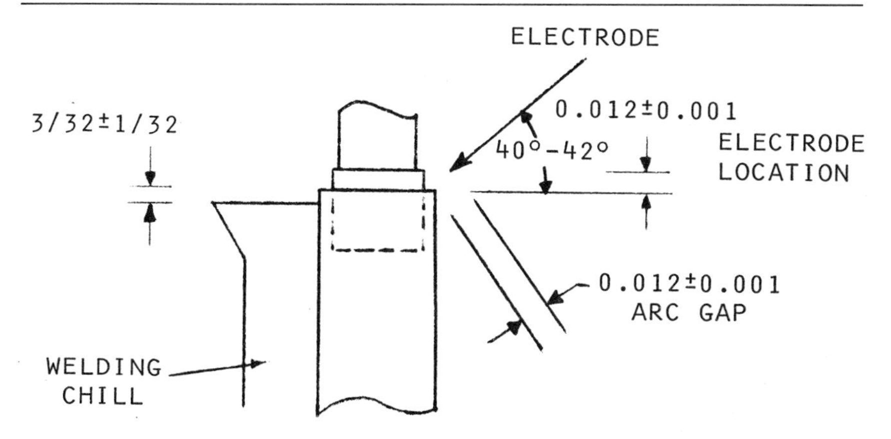

SKETCH

Table 87.2—Second end closure welding procedure

Process	GTAW Vac Melt	Drawing No. FTR Reference	
Alloy	316 Stainless Steel		
Power Supply		Weld Settings	
Mfg.		Prepurge Time, s	None
Model	50 A	Start Current, A	11-12
Serial No.		Upslope Time, s	0-2
		Fixture Delay, s	0-1
Torch		Weld Current, A	11-12
Mfg.		Voltage, V	15-25
Model	HW #9	Type/Polarity	DCSP
Serial No.	N/A	Weld Time, s	15-16
Cup Size	None	Downslope Time, s	6-9
		Finish Current, A	2-5
Shielding Gas*		1 Rev Time, s	11-12
Type	He (A) Comm	Weld Speed, ipm	3.6 Ref.
Volume	Welding Chamber	Fixture Speed Dial	Graham Drive ∿21.8 Ref.
Backup Gas		Postpurge Time, s	None
Type	None	Weld Position	Vertical
Volume		Arc Start Method	H.F.
		Arc Start Intensity	N/A
Electrode			
Type	2% Th		
Diam, in.	0.040		
Ext., in.	1/16 - 3/16		
Shape, deg	20 - 40		

*Note: Welding chamber shall be evacuated to at least 1 x 10^{-4} torr prior to backfilling for welding. A gas sample shall be taken for each lot of pins welded. The gas sample shall be analyzed for purity using a mass spectrometer.

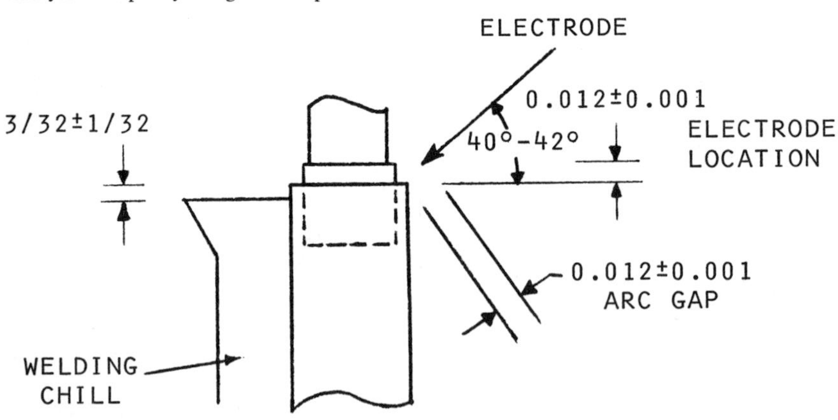

,SKETCH

In the welding of zirconium alloys, there is enough variation in the composition of the alloys that it is necessary to give the welding current as a range and adjust the current to provide a satisfactory weld with each heat of alloy.

In the welding of end closures in fuel rods and capsules, an area has been found in which it is virtually impossible to make a gas tungsten-arc welded closure without a blowout at the closure point. It is determined primarily by the ratio of the diameter of the cap to the volume of contained gas in the capsule. If the ratio is very large, and thus the confined volume of gas is very small, there is virtually no problem. As the ratio decreases slightly, it is very difficult to make a weld without a blowout at the closure point. In this region, it is usually best to use a vent hole and close it with a gas tungsten-arc spot weld. Further decreases in the ratio produce a point after which the problem of blowout at the closure weld ceases. So far, no known numerical ratios have been developed to define these areas, leaving the specific application to trial and error.

Electron Beam Welding (EBW)

Electron beam welding is advantageous for large production fuel rod fabrication where automation, dependability and quality acceptance are prime requirements. A typical rejection rate for EB welding is 0.01%. High power density electron beam welding is directly applicable to the production of end closures on rod-type fuels. Its primary advantage is that the large force of the electron beam assures a fully penetrated joint. The electron beam diameter is small compared to the width of the fusion zone, therefore a tight fit-up is required as the beam will go through a crack several thousandths of an inch wide without producing fusion. The problem of spiking occurs with welds that are blind or not fully penetrated. Spiking is the non-uniform root penetration profile that is produced in EB welds that are not fully penetrated. As the depth of penetration increases, the spiking problem increases. As long as the welding procedure is set up to assure full penetration of the cladding tube wall, there is no problem with spiking in the EB welding of nuclear fuel end closures. Soft vacuum, 1-500 micron pressure EB welding is not applicable to the welding of zirconium alloy clad fuel rod end closures as it destroys the corrosion-resisting properties of the welds.

For a finite penetration and welding speed, the diameter of the electron beam is a function of the accelerating voltage and the beam current. As the accelerating voltage goes up, the beam current and diameter decrease. The converse is also true. These relationships are not linear. As the thickness to be welded increases, the power required increases and the width of the fusion zone increases. Considered separately in relation to penetration with one parameter being held constant, there is approximately a linear relationship between beam current and beam voltage and the penetration produced. The quality of an EB weld is at least as good as a gas tungsten-arc weld while the surface is somewhat rougher. Application of a cosmetic weld pass with a defocussed broad beam will produce an equal surface finish.

The low-power input and high-welding speed avoid a major problem encountered in welding under vacuum, i.e., boiling-off of alloying elements. Zircaloy-2 and Zircaloy-4 contain 1.2 to 1.7 weight percent tin. Tin has a higher vapor pressure than zirconium and consequently boils off and deposits in a cold area of the welding system. One function of tin is to provide corrosion resistance. It is therefore critical to limit the time that the alloy is in a molten condition. For normal high-speed welding of fuel rod end closures, a normal welding pass of 1 1/4 to 2 1/2 revolutions does not effect corrosion-resistance performance. Not until a weld has been rewelded approximately six times does the corrosion behavior appreciably deteriorate.

End closure plugs, shown in Figs. 87.4C, 87.5C and 87.5F, are directly applicable to EB welding:

1. The weld is readily examined by in-line high-speed mechanized ultrasonic testing.
2. Tight fit-up of the faying surfaces is readily obtained.
3. Filler metal, if required, can be designed into the end plugs.
4. There is massive solid material to terminate the electron beam penetration.

Producing the first end closure by EB welding poses no problem as the material is clean and empty. The weld may be made immediately after a fast pumpdown to at least as low as 1 x 10^{-4} torr. The second end closure weld requires that the end plug contain a gas passage (Fig. 87.9). Welding is performed in a vacuum thereby preventing helium backfilling. The necessity to guarantee an extremely low moisture content in the fuel rod atmosphere makes the gas passage a production asset. Pumping time to 1 x 10^{-4} torr will be longer due to outgassing of the fuel pellets through a small diameter hole.

After the second end closure is welded, vacuum drying time for backfilling with helium and sealing the vent hole can be minimized by connecting the outgassing chamber to the EB welder with a transfer valve or by transferring the fuel rods from the EB welder into an evacuated transfer cylinder for storage or transfer to the outgassing chamber. The inherently small fusion zone and HAZ produced by EB welding allow greater control over run-out between cladding tubes and end closure plug axes (parallelism) with a minimum of fixturing. This permits high design tolerance demands to be fulfilled without production penalties.

A major advantage of the small fusion zone and HAZ is that end closure welds may be monitored automatically by ultrasonic testing. Well-designed EB equipment, complemented by automatic ultrasonic testing, can produce fully controlled end closures at rates greater than one end closure per minute. The advantages are two-fold since the traditional quality control incorporating X-ray inspection is replaced with a more objective testing method. Time-consuming film exposure, development and subjective test interpretation are avoided. In addition, X-ray control for end closure welds provides a two-dimensional inspection whereas ultrasonic provides a three-dimensional inspection.

The electron beam is insensitive to minor changes in alloying constituents which may occur in different heats of clad tubing and end plugs. This is a distinct advantage in comparison to gas tungsten-arc welding where welding machine performance and quality may be affected.

Since EB welding requires only an occasional adjustment, it is practical to have highly qualified personnel set up a welding program and then turn the equipment over to a production operator. Weld quality can be continuously monitored by nondestructive testing techniques while weld penetration can be followed by statistically-based metallography.

The present state of ultrasonic inspection indicates flaws in a weld to the depth of the cladding tube wall thickness. This is the case when the ultrasonic transducer is positioned to feed signals along the cladding tube to the weld zone. This is not a disadvantage since small defects at depths greater than 100% of the cladding tube wall thickness are not considered detrimental to fuel rod dependability. The ability to gage ultrasonic signals and thereby discriminate flaw size is a great advantage for automation of equipment. Root flaws can generally be eliminated by defocusing the beam slightly.

Typical electron beam welding procedures are:

Low Voltage Equipment

Material	Zircaloy-2.
Wall thickness	0.030 in.
Penetration	0.060 in.
Voltage	48 kV
Beam current	22 mA
Surface speed	100 in./min
Focus setting	5.35
Gun-to-work distance	4 in.

High Voltage Equipment

Material	Ozhennite*	
Wall thickness	0.020	0.035 in.
Penetration	0.037	0.075
Voltage	97 kV	120 kV
Beam current	3 mA	3mA
Surface speed	24	24 in./min
Focal point	Surface	Surface
Spot size	Smallest	Smallest

*Ozhennite is a zirconium-base alloy containing 0.2% tin, 0.1% columbium, 0.1% nickel, 0.1% iron, balance zirconium.

Upset Welding

Upset welding of the end closures for rod-type nuclear fuels was successfully developed on the earliest experimental fuel rods with ceramic fuels. The principal work was performed with magnetic force resistance welding on fuel cladding ranging from sintered aluminum powder through stainless steel, zirconium alloys, molybdenum, tungsten, beryllium and graphite. The advantages of upset welding are:

1. A very low reject rate.
2. A very fine-grained weld structure with no evidence of melting.
3. A good production rate.
4. A low capital equipment cost.
5. A direct application for pressurizing fuel rods and pins to any required pressure.
6. The process permits making double-welded closures by an alternate process.

The disadvantages of upset welding include the following:

1. The process leaves external flash that must be removed by machining (Fig. 87.17).
2. It requires a second operation for low moisture and gas contaminations.
3. It is difficult to examine by nondestructive testing methods.

Straight resistance welding of the fuel rod end closure is essentially a projection welding technique. A ring projection is formed on each tube end. The thin walled tubing is contained in the machine's bottom arm in a split clamp type electrode. The end plug is retained under vacuum in the upper electrode die. The proper combination of heat and forging force causes the surface oxides

Fig. 87.17.–Section of straight resistance-welded fuel tube end closure. 60X. Material Zircaloy-2. (Enlarged to 129% in reproduction.)

to break up, and recrystallization of the Zircaloy-2 occurs at temperatures below the melting point, thereby producing a strong metallic bond, as shown in Fig. 87.18. Proper joint design with exacting weld procedure repeatedly produces joints of high integrity. Utilizing inert gas, a simple purging arrangement within the upper electrode, protects the Zircaloy-2 from internal atmospheric contamination. Initially, cleanliness and care in handling components were rigidly

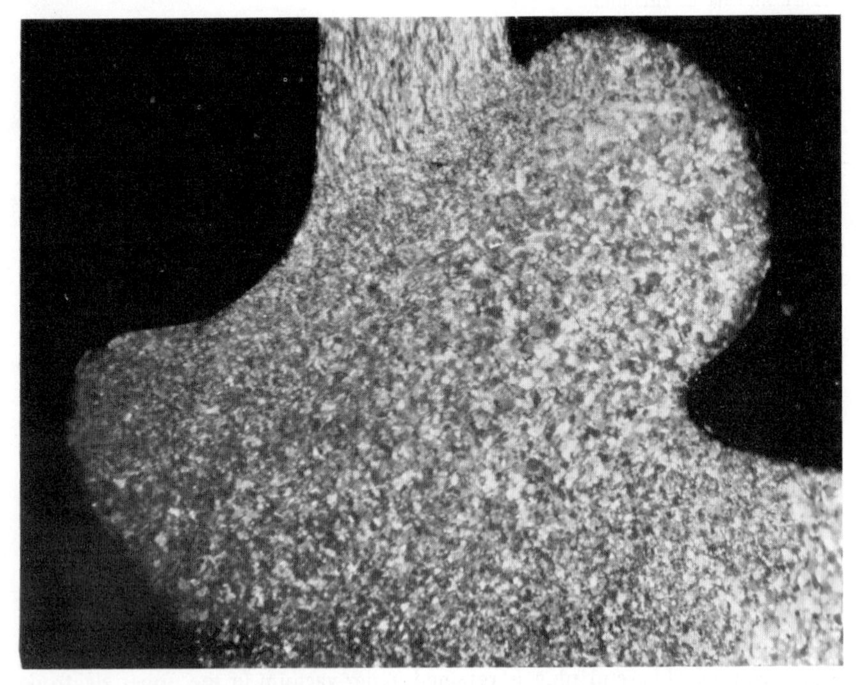

Fig. 87.18.–As-welded and longitudinal section of as-welded straight resistance welded fuel tube end closures. 2X. Material Zircaloy-2.

enforced, but with improved techniques, these requirements have been relaxed to a level suitable for specialized industrial production.

Metallographic examinations show that the weld zone is made up of a fine-grain equiaxed structure with no indication that melting has occurred. In most cases, the original interface of components cannot be detected, but occasionally it can be seen as a very fine line across a small portion of the joint. For purposes of evaluation, this "weld line" is regarded as a zone of non-bonded material, and the joint quality is defined by expressing the length of the fully bonded joint at the plane of the interface as a percentage of the tube wall thickness. The value is maintained by careful control at better than 175% at any joint, although it is by no means certain that a joint with less than 175% fully bonded width would result in a fuel defect.

All elements are nondestructively tested with a helium mass spectrometer. In assessing mechanical strength, however, the most convenient way is to treat the empty fuel element as a pressure vessel and apply an internal hydraulic pressure. Under these conditions, the fracture will always occur in the tube wall in a typical hoop stress failure pattern at a burst pressure dependent upon the strength of the tubing. Because the welded joints never failed in these circumstances, the test was modified by fitting a thick walled sleeve over the tube so that the structure was forced to fail in longitudinal stress. Now fracture occurred at the edge of the heat-affected zone, not in the weld zone, as seen in Fig. 87.19.

Fig. 87.19.–Resistance welded end cap fractured adjacent to the weld with hydraulic pressure. The tube was restrained with a thick sleeve. 3X. Material Zircaloy-2.

Fig. 87.20.–Magnetic force resistance welded ring-type joint. 5X. Material Zircaloy-2.

Magnetic force resistance welding of the fuel rod end closures is either a ring-type projection (Fig. 87.20) or a tapered plug (Fig. 87.14). In the ring-type projection, a ring is formed on the end closure cap and the end of the fuel tube acts as the other projection. The end closure cap is held in the moveable electrode holder with a copper resistance welding alloy collet while the tube is held in a similar collet, with the end of the tube projecting a short distance through the collet. A proper combination of current and forging force disrupts and displaces the interface producing a fine-grained joint with no evidence of melting (Fig. 87.21). The plug-type weld is formed with the same tooling except that the fuel tube is flush with the end of the collet. This will prevent the forcing of the end cap into the tube from expanding the end of the tube. Testing methods and quality control are virtually the same as for the straight resistance welded closure. As with the straight resistance welded closure, the excess metal is removed by a machining operation (Figs. 87.14 and 87.22).

Laser Beam Welding (LBW)

The development of Nd-glass and Nd-YAG continuous and pulsed lasers has provided economical equipment for the welding of fuel rod end closures in either an inert atmosphere or vacuum. Lucite, by a factor of ten, is the best material for the windows to transmit the laser beam as it tends to repel metal vapors. For continuous laser beam welding of 0.805 in. diam, 0.027 in. wall stainless steel tubing, an end closure weld can be made in one revolution at 180 to 340 watts and 4 rpm. It can also be used for end cap welding of pressurized fuel rods. A continuous laser can be programmed like gas tungsten-arc welding to accommodate heat buildup and crater filling.

Furnace Brazing(FB)

Furnace brazing is often used as a method of fabricating reactor parts, particularly when dealing with assemblies having multiple joints. These mainly include fuel plate to side plate joints, fuel rod support grids, control rod guides, heat exchangers and back brazing of welded heat exchangers.

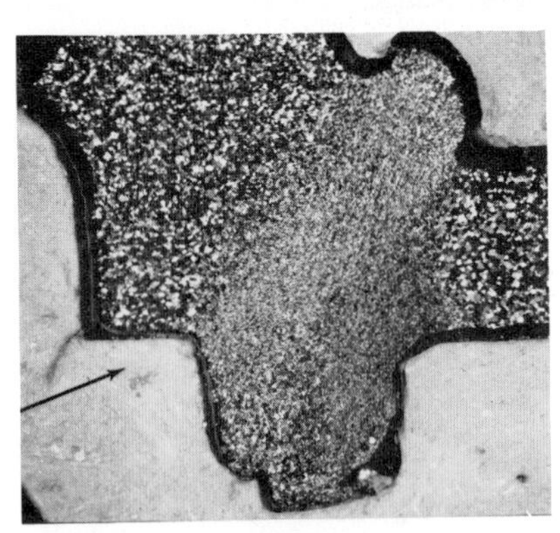

Fig. 87.21.–Microphotograph of magnetic force resistance welded ring-type end closure. 40X. Material Zircaloy-2. (Reduced to 51% in reproduction.)

Fig. 87.22.–Magnetic force resistance welded end closure in as-welded condition. IX. Material Zircaloy-2. (Enlarged to 156% in reproduction.)

Furnace brazing is done by assembling the parts with brazing filler metal in the form of powder, wire or sheet located in or adjacent to the joints. The entire assembly is then placed in a brazing furnace having a controlled atmosphere of hydrogen, inert gas or vacuum. When the assembly reaches the flow temperature, the brazing filler metal melts and flows by capillary action into the joint. Fluxes or subsequent cleaning operations are not required due to the action of the controlled atmosphere. Nearly all materials used in nuclear reactors have been successfully joined by this basic method. Table 87.3 gives recommended brazing filler metals for furnace brazed joints of AISI type 304, 310 and nickel-base alloy for use in various reactor environments.

Table 87.3—Recommended brazing filler metals for furnace brazing AISI types 304, 310 and nickel-base alloy for use in various reactor environments

Brazing Filler Metal AWS 5.8-69	Pressurized Water to 600 F	NaK to 1500 F	Still Air to 1500 F
Gold-base filler metals			
BAu 1	Not Recommended	Not Recommended	Not Recommended
BAu 2	Not Recommended	Not Recommended	Not Recommended
BAu 3	Not Recommended	Not Recommended	Not Recommended
BAu 4	Recommended	Recommended	Recommended
Nickel-base filler metals			
BNi 1	Recommended	Recommended	Recommended
BNi 2	Recommended	Recommended	Recommended
BNi 3	Not Recommended	Recommended	Not Recommended
BNi 4	Not Recommended	Recommended	Not Recommended
BNi 5	Recommended	Recommended	Recommended
BNi 6	Recommended	Not Recommended	Not Recommended
BNi 7	Recommended	Recommended	Recommended

Fires may occur when BeZr brazing filler metals on zirconium alloys are heavily etched in HNO_3 or a solution of HNO_3, H_2SO_4 and $CuSO_4$. The preferential dissolution of beryllium leaves a spongy mass of zirconium that may be ignited when shocked electrically or mechanically. It detonates in air with a loud report and a brilliant white flash of light. Weld bead fires do not propagate to the Zircaloy-2 cladding and leave no ash or other residue when they burn out.

Brazing is directly applicable to zirconium alloy rod-type fuel elements, but the techniques for gas tungsten-arc, electron beam and both types of resistance welding have been so highly developed that there is little incentive to develop brazing as a method of closing rod-type fuels. Table 87.4 shows the brazing filler metals that have been developed for this use. Only the first filler metal listed has had extended in-reactor exposure. The latter four have satisfactory corrosion resistance, mechanical strength and brazing characteristics for nuclear fuel applications.

The following are the principal disadvantages of brazing end closures:

1. The known filler metals with a minimum brazing temperature of 1050 C is a beryllium-containing metal.
2. Beryllium control zones are required for storing, handling, cleaning, brazing, machining, welding and maintenance.
3. Zirconium alloys are left in the modified alpha phase at the brazed end.

4. Care must be exercised on the time-temperature relationship or grains will extend through the tube wall.
5. Maintaining oxygen and carbon levels low enough in the beryllium-containing brazing filler metal to provide good flowability is difficult and costly.
6. Brazing has to be performed in a high-purity atmosphere chamber, either vacuum or inert gas.
7. If a fusion weld is to be made over a BeZr brazed joint, a wide braze joint 0.020-0.040 in. wide must be used to prevent cracking in the remelted braze that is diluted with some cladding metal.

Table 87.4—Brazing filler metals and brazing temperatures for zirconium and its alloys

Brazing Filler Metal Composition In Percent	Brazing Temperature C
Zircaloy-2 - 5 Be	1050
Zr - 5 Be	1050
Zr - 50 Ag	1520
Zr - 29 Mn	1380
Zr - 24 Sn	1730

CHAMBERS FOR WELDING AND BRAZING NUCLEAR FUELS, FITTINGS AND FUEL BUNDLE ASSEMBLY

The costs of an atmosphere chamber (with its auxiliary equipment) necessary in producing a weld with a finite atmospheric contamination are virtually the same using any joining process.

The problem of keeping the chamber and piping immaculately clean, gasketed and threaded, the joints helium-tight and the vacuum pumping system in the best operating condition cannot be stressed too much. The degradation of any or all of these items degrades the ultimate vacuum, the inert atmosphere and thus the quality of the weld or braze.

Atmospheric chambers for gas tungsten-arc welding rod-type fuels vary greatly. They range from a very simple chamber with glove ports, a viewing window and low vacuum pumping equipment to exotic chambers with multiple glove ports and viewing windows, high vacuum pumping equipment, circulating gas purifying systems, controls for maintaining constant inert gas pressure in the chamber during welding and very precise instrumentation for monitoring and recording gas contaminants (Fig. 87.23).

Circulating gas purification systems that will maintain the oxygen, moisture and nitrogen levels at not more than 1 ppm with butyl gloves are available as catalog items. Lower levels are limited strictly by economics and requirements. They may have dual absorption chambers to permit continuous operation of the chamber (one absorber is being regenerated while the other is in use).

Provisions for handling the fuel rods for welding and tungsten electrode positioning and changing may be gloves in glove ports, mechanical manipulators or a completely mechanized system. The choice is limited by throughput rate, product quality requirements and economics.

Fig. 87.23.–Chamber with high vacuum pumping equipment and without gas purification or monitoring equipment for welding stainless steel clad fuel rod end closures.

PRESSURIZING FUEL RODS AND PINS

The technique for pressurizing fuel rods and pins was evolved to a high level early in the development of rod-type ceramic fuel. A vent hole was used to prevent a blowout of the weld when making a second closure. It is now being used to pressurize fuel rods after EB welding the end closures. Since the pressure in these early fuels was virtually atmospheric, the gas tungsten-arc spot welding process was used. A bare 1/16 in. 2% thoriated tungsten electrode, with an 18 deg included angle point in a fixed insulated holder, was used for the torch. A representative backfilling hole closed by this method is shown in Fig. 87.10. The welding conditions were:

Hole diameter	0.030 in.
Initial slope delay	11 cycles
Initial current	40 A
Initial slope time	6 cycles
Spot welding current	110 A
Spot time	51 cycles
Final slope delay	125 cycles
Final current	10 A
Final slope time	120 cycles
Weld depth	0.040 to 0.056 in.

The pressure within a fuel rod or pin is controlled for the following reasons:

1. To eliminate a corrosive atmosphere.
2. To introduce helium to improve heat transfer from the fuel to the cladding.
3. To prevent the coolant pressure from collapsing the cladding against the ceramic fuel pellets.

Internal pressures up to 400 psig helium at the time of manufacture are being used to prevent cladding collapse during operation. The most apparently desirable process for sealing the vent or backfilling hole in the end closure would be the gas tungsten-arc process. However, there is virtually nothing known about welding in inert atmospheres above approximately 50 psig. The performance and arc instability problems in argon with tungsten electrodes are well-known at pressures to approximately 50 atmospheres and they are under study to 1000 atmospheres by physicists. At 400 psig, an argon arc between tungsten electrodes is very wild. No known work has been done with helium by physicists or others using welding currents at much above atmospheric pressure. Physicists prophesy that the tungsten arc in helium at 400 psig will be more stable than in argon. All available information indicates that it is too unstable for welding. At these pressures, arc starting is very difficult requiring separation of the electrodes or a capacitor charged to a high voltage and discharged between the electrodes. Resistance welding is not known to be affected by elevated pressures so it appears to be a good prospect for use in sealing the backfilling hole in the fuel rod end closure. The problems of entering the pressurized chamber with the welding electrodes and the handling of groups of fuel rods can be solved by known technology and judicious design.

A pulsed Nd-glass laser is being used for drilling and closing the backfilling holes for pressurizing fuel rods. The fuel rod end closure caps are attached by either laser, gas tungsten-arc, electron beam or resistance welding. The end of the fuel rod containing the compression spring is inserted into a small high-pressure chamber having an evacuation system, helium supply at the required pressure and a window for the entry of the laser beam. An "O" ring seals the rod at the entry point into the chamber.

Design of the end closure cap is such that the drilled hole will pass through at least 0.050 in. of material. Experience has shown that a hole with a 0.008 in. nominal diameter can be drilled 0.090 in. deep with a single pulse and welded closed at least 0.040 in. with a single pulse in 400 psig helium. The total time to drill, pressurize the tube and weld the hole closed is approximately fifteen seconds. Evacuation time is dependent on chamber size, pumping rate and final gas purity level required. A crater approximately 0.003 in. deep and 3/16 in. in diam is produced on the external surface.

Magnetic force welding is being used to make end closure welds with fuel tubes pressurized up to 500 psig. Welding is accomplished in a chamber that accommodates a single tube using one cycle of current. Floor-to-floor time per tube is dependent on pumping rate, chamber size and final gas purity level required. For versatility, a 100 kVA magnetic force welder with a 75 kVA synchronous independent magnet and control is recommended. Other types of magnet controls produce equal quality of welds but have limited flexibility for development work.

A technique that appears to have been overlooked is the pinch seal in which a small diameter tube section is brazed or welded into the backfilling hole and after pressurizing, the tube is pinched off either hot or at room temperature producing a pressure weld. This technique is applicable inside high-pressure chambers with an external operation for production groups of fuel rods.

There is no known way of getting an electron beam into a high-pressure chamber and preventing its scattering by the high density gas.

FUEL ROD BUNDLE SPACERS

Fuel rod bundle spacers perform these functions:

1. Maintain precise spacing between the fuel rods.
2. Maintain the fuel rod bundle configuration and spacing in the reactor.
3. Provide a guide for the control rods in some types of reactors.

Spacers are fabricated of very close tolerance sheet stock of stampings whose dimensions are carefully controlled (Fig. 87.24). Spacers are assembled by welding to very close tolerances after which those for the BWR's and PWR's are annealed in vacuum on a sizing fixture. The spacers may require from approximately fifty to several hundred precision placed and penetrated welds to provide the dimensional tolerances and strength required.

Corrosion problems caused by etchants, used after welding, which produce materials between overlapping surfaces that are not fused and are corrosive in reactor atmospheres have initiated joining studies by other methods.

With the development of the pulsed Nd-Glass and Nd-YAG lasers, equipment

Fig. 87.24.–Zircaloy-2 fuel bundle spacer.

is available for laser welding of grid spacers. An inert atmosphere or vacuum chamber is required.

Gas Tungsten-Arc Welding

Joining of fuel bundle spacers by the precision autogenous gas tungsten-arc spot welding process has been so simple in its development and dependable in its application that virtually no other joining process has been explored. Precision welding power supplies used for end closure welding are the same as those required for fuel end bundle spacer welding. They may require an additional control or timing device and a gas tungsten-arc spot welding torch. Judicious joint design precludes requiring filler metal.

Inert gas atmosphere chambers for gas tungsten-arc welding of fuel bundle spacers are required to produce a weld that shall be free of corrosion in its exposure period in a nuclear reactor. Just which chamber is required is determined by whether the spacer to be welded is stainless steel, a nickel or a zirconium alloy and how much oxidation if any can be tolerated on the weld metal and surrounding area. Stainless steel spacers have been successfully welded in air by using a gas tungsten-arc spot welding torch nozzle that encases the weld area.

Electron Beam Welding

Electron beam welding has not been used for spacer joining due to high welding equipment costs. With the small precision low power units now available, an incentive should develop with the stricter corrosion requirements. This approach is being explored abroad to eliminate an etching operation after welding of zirconium alloy spacers.

Resistance Welding

Resistance welding is just starting to be used in the joining of fuel bundle spacers. It is currently being used for the joining of cross members. All of the cross members on both sides of a spacer are welded simultaneously producing uniform, high quality joining in a single operation. A continuing study is being made of the design of spacers and tooling to permit resistance welding of the side members.

Furnace Brazing

Furnace brazing is now in use and also under intense study for the fabrication of reactor core associated hardware of stainless steel and nickel alloys. These include principally fuel plate to side plate joints, fuel rod support grids and control rod guides. In Fig. 87.25 is shown a fuel bundle that has been furnace brazed. Table 87.3 gives recommended brazing filler metals for furnace brazed joints of Types 304 and 310 stainless steels and NiCrFe alloy for use in various nuclear reactor environments.

Furnace brazing has not been used for zirconium alloy spacer joining for principally the same reasons that it has not been used for fuel rod end closures. The brazing filler metals available now have high in-water corrosion resistance at elevated temperatures, flow well and form good fillets. One brazing filler metal, 5-BeZr, has had a very long and successful in-reactor exposure. For details of this and additional brazing filler metals for zirconium alloys, see Table 87.4.

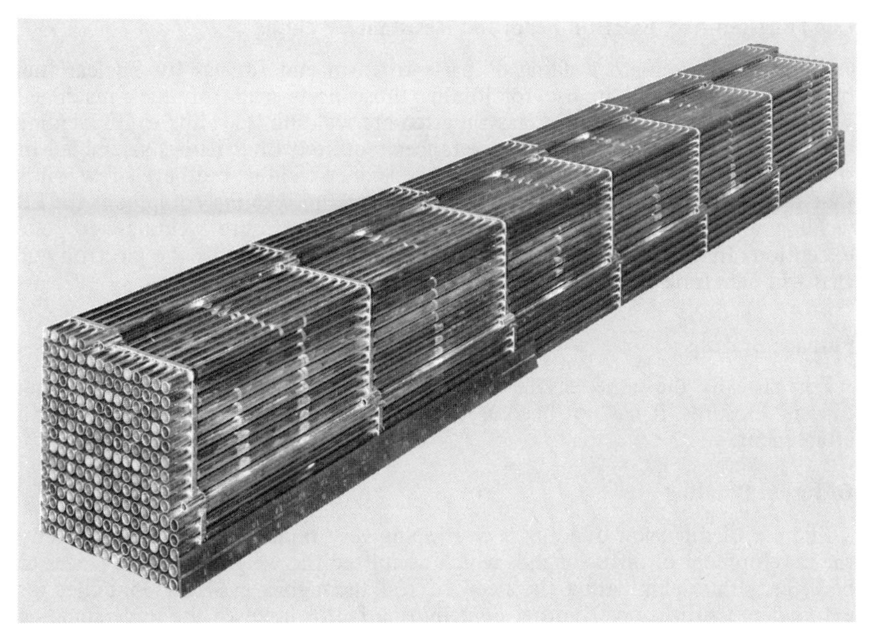

Fig. 87.25.–Full-scale mock-up of type 18-8 stainless steel fuel bundle furnace brazed with AWS BNi-7 brazing filler metal.

SPACER SPRINGS

Spacer springs are assembled into the fuel rod bundle spacers to hold the fuel rod in its exact position in the spacer and prevent fretting corrosion between the fuel rod and fuel rod bundle spacer. Originally the springs were joined by resistance welding but corrosion problems between unfused surfaces presented problems. Gas tungsten-arc welding is used in a fixture on a bench, depending on the torch shielding gas for the required shielding. EB and laser welding would be applicable except for the high equipment costs.

FUEL BUNDLE END FITTINGS

The end fittings on a fuel bundle provide the following functions:

1. The eye(s) in the upper fitting provides a point of attachment for the lifting device which loads into and unloads from the reactor.
2. The upper end fitting provides a strong massive fitting to hold the upper end of the fuel bundle in place.
3. The upper end fitting may provide an exit path for the coolant to the upper portion of the reactor vessel.
4. The end of the lower fitting is designed to guide the lower end of the fuel bundle through the core and into a structure that positions and supports the bottom end of the core.
5. The lower end fitting provides a strong massive fitting to hold the lower end of the fuel bundle in place.
6. The lower end fitting provides an entrance path for the coolant into the reactor core.

Gas Tungsten-Arc, Electron Beam and Resistance Welding

The gas tungsten-arc welding of parts to form end fittings for nuclear fuel bundles is decreasing in use for joining those parts made of thick machined sections. Distortion caused by gas tungsten-arc welding is leading to EB welding to provide the close dimensional tolerances required. End fittings assembled of sheet metal parts are generally joined by resistance welding but have a few joints that require gas tungsten-arc welding. Chambers and welding equipment for EB welding are the same as those required for end closure welding with one exception. In EB welding, it is essential to select a rating for the electron gun that will penetrate the thickest section anticipated.

Furnace Brazing

For basically the same reasons that brazing is not used on zirconium alloy fuel rod end closures, it has not been used on fuel bundle end spacer fabrication and attachment.

Diffusion Bonding

The use of diffusion bonding is developing very rapidly. This was initiated by the development of diffusion aids which permitted the weighting of the pieces to be joined, thus eliminating the need to seal them in a gas-tight container for external pressurization. Another contributing factor involves the understanding of surface preparation and cleaning methods. Essentially, if the surface films between the two pieces to be joined will not dissolve into the substrate during the bonding operation, they must be removed. Depending on the material, cleaning may be performed with etchants, solvents, abrasives, wire brushes of the same or similar material or cathodic etching in an inert atmosphere.

Diffusion bonding may be applied to many joining operations of nuclear components that are now joined by fusion welding. The addition of a diffusion aid usually permits the joining operation to be performed at a lower temperature. By holding the parts at temperature, the diffusion aid will diffuse away leaving an integral joint that is not detectable with an optical microscope. For this type of bonding the diffusion aid should be less than 0.0005 in. thick. Again, depending on the material to be bonded, the diffusion aid may be placed by electro-plating, vapor deposition or a thin foil. Properly prepared diffusion bonded joints will have properties equivalent to the base metals.

CONTROL RODS

Control rods for power reactors fall in two groups: (1) cruciforms (Fig. 87.26) or T types for boiling water reactors (2) rod types for the pressurized water reactors. Both types generally use boron as the poison or neutron absorber in the form of a carbide in the boiling water reactors and in the form of a silicate in the pressurized water reactors.

Fabrication and welding techniques are very similar to plate and rod-type fuel. Actual fabricating techniques are very difficult to obtain as the core fabricators consider this information highly proprietary. There is considerable brazing activity in the assembly of control rods of stainless steel and nickel alloys. Brazing filler metals for brazing Types 304 and 310 stainless steels and NiCrFe alloy for use in various nuclear environments are shown in Table 87.3.

Fig. 87.26.–Cruciform-type control rod.

For basically the same reasons that brazing is not used on zirconium alloy fuel rod end closures, it has not been used in assembling zirconium alloy control rods.

QUALITY CONTROL

The implications of fuel failure and subsequent electrical generating station shutdown point to strict quality control requirements. The traditional quality control methods available for final inspection of end closures are visual inspection, X-ray, ultrasonic and mass spectographic helium-leak detection.

It is difficult to obtain X-ray sensitivity greater than 2% of the thickness of the material being examined. An end closure design optimized for X-ray inspection can increase test sensitivity and provide flexibility for economic rationalization of testing procedures. If internal obstructions are not present (pellet-spring), the closures shown in Figs. 87.4D, 87.5D and 87.5G would be the most desirable, and the end closures shown in Figs. 87.4C, 87.5C and 87.5F, the least desirable for X-ray inspection.

Ultrasonic inspection of end closure welds has proven to be a fast, efficient method adaptable to a wide range of end plug geometries. The method, however, has been found to be compatible only with EB welding which produces a minimum of metallurgical disturbance.

PRESSURE-CONTAINING COMPONENTS

All reactors require the use of coolants to remove the heat produced by fission. For power-producing reactors, the heat transferred by the coolant is converted to output power. The coolant, usually water, a gas, molten sodium or an organic liquid at high temperature, is pumped through the piping system, absorbing heat from the reactor and transferring it to the coolant in a heat exchanger, thus generating steam which drives a turbine. In boiling water reactors, steam generated in the reactor goes directly to the turbine. Primary coolant cycle operating pressures require that the reactor fuel elements and auxiliary components be enclosed in a heavy-walled pressure vessel. In addition, external shielding in the form of high-density concrete may be necessary for protection from neutrons, gamma rays and heat. Water is often used as a shielding material, as are lead and neutron absorbers such as cadmium and boron. For thermal insulation, reflecting metallic insulation, glass or mineral fiber insulations are used. Many reactor systems are totally enclosed in a containment vessel, a heavy metal or concrete shell designed to contain the materials and energy which might be released by a failure of the primary system. The welding operations employed to fabricate or join the individual components of a nuclear power plant are discussed in the section devoted to that component.

CODES AND SPECIFICATIONS

Reactor vessels and components in this country have generally been fabricated to the requirements of the ASME Boiler and Pressure Vessel Code, including Section III, Rules for Construction of Nuclear Vessels and the ANSI B31 Code for Pressure Vessel Piping, together with its Code Case Interpretations.

FABRICATION OF HEAVY WALL PRESSURE VESSELS

Design and Service Requirements

Nuclear reactor vessels and auxiliary vessels for water-cooled reactors and auxiliary equipment currently in commercial service generally operate at temperatures under 650 F. However, some experimental and prototype reactors, using nuclear or fossil-fueled superheat, and reactors using molten fluorides as coolants operate at 1000 F and higher. Liquid metals are also used for coolants. The designer of reactor vessels must take into account the stresses imposed by the operating pressures, temperature gradients and thermal and mechanical cycling. He must consider the effect of shock loading for mobile power plants. These design criteria affect the choice of materials and overall design. The operating pressures have been up to 2500 psi for pressurized water reactor vessels. The design must minimize stress concentrations, and through engineering specifications, require fabrication procedures and nondestructive tests to minimize the possibility of fabrication and metallurgical defects.

Metals

Metals available to the designer include plates, extruded shapes, forgings, bars and castings and in some instances less conventional types of fabrication such as powder metallurgy products or weld-deposited shapes. These metals are joined by welding and the properties of the weld metal, heat-affected zones and base metals must all be considered.

The designer's aim should be to choose a metal and design the overall part so that the most economical metal and fabrication procedure will be utilized. In well-established services, design criteria are sufficiently well-known to meet this goal. In newer services, overdesign either leads to a successful performance at excessive cost or to the neglect of some important design factor. It is the purpose of the designer, metallurgist and welding engineer to eliminate this latter possibility. Among the structural metals used in pressure vessels are plain carbon and low-alloy steels, austenitic stainless steels, clad steels, nickel alloys, zirconium, aluminum and columbium.

A nickel-chromium-iron alloy is used for some components of nuclear power plants because of its resistance to chloride ion stress-corrosion cracking. This alloy has good weldability properties; however, in nearly all weldments involving some restraint or sections over gage thicknesses, additions of filler metals must be made to ensure sound weld deposits. The properties and welding procedures for nickel and high-nickel alloys are discussed in Chapter 67, Section 4.

The NiCrFe alloy can be welded by any of the usual automatic and manual welding processes, including shielded metal-arc, gas tungsten-arc, automatic submerged arc, plasma-arc, electron beam and gas metal-arc welding.

Specific discussions of the various welding processes will be found in Sections 2, 3A and 3B of this Handbook.

Plates–. For large capacity vessels designed for up to 2500 psi pressure, the wall thickness required for an 80,000 psi tensile strength, vacuum-degassed metal may range from 6 to 11 in., approaching the capacity of the plate mills to produce economically sized plates meeting the specification requirements. Metal of this thickness and strength also approaches the generally available limitations of commercial fabrication practice. Generally speaking, the shells of the vessels are made from plates or forgings shaped into cylinders and the heads may be made from plate and forgings, dependent upon size and thickness.

Forgings–. Most designs for reactor vessels require large flanges to bolt the portions of the vessels together. These flanges must be rigid enough to withstand the bending stresses around the bolt periphery and must be welded to the shell section and to the head section or to a forging or plate comprising other portions of the head. These flanges are presently made from forgings having a cross section of approximately 2 ft^2 with a bore diameter of 10 to 20 ft or more. The present-day forging technology permits the mills to make such forgings to readily meet the 80,000 psi tensile strength requirements. A vacuum pouring technique is used to degas the metal.

Impact Properties

Investigating laboratories have determined, through their work on brittle ship plates and bridge and pressure vessel failures, that the low-alloy ferritic metals exhibit a reduction in capability to deform under certain applied stresses as the temperature is reduced. The maximum temperature at which a metal exhibits essentially no ductility under certain arbitrary specified test conditions has been given the name of "nil-ductility transition temperature" (NDT temperature). These metals, when highly stressed below this temperature, may fail in a brittle manner starting at a crack of sufficient size. Such a critical crack can be an actual metal flaw or a fatigue crack resulting from the extension of a preexisting defect or geometric stress concentration by cyclic stressing. If at no time the vessel is stressed below some temperature frequently specified as the NDT plus 60 F (or higher), a fairly large crack can be present without initiating brittle fracture; but under sufficiently severe conditions–excessive pressure combined with an extremely large crack–it may still fail in a brittle manner. The size of the crack is dependent on several factors described more completely in the extensive literature on fracture behavior.

Application of the Charpy V-notch test to the various low-alloy steel plates and forgings originally chosen for this work indicated that very close metallurgical control was necessary to obtain a finished product which would meet the requirements. This consisted of controlling the cooling rate from the austenitizing temperature which considerably improved impact values. The faster the cooling rate, which in all cases was followed by a tempering operation, the better the impact properties, within limits. With thin sections, air-cooling will sometimes suffice, while heavier thicknesses require accelerated cooling, accomplished by fan cooling, spray quenching or dip quenching.

For the very thick cross sections required in large forgings and in the increasingly thick plate sections to be required by future larger vessels, it is becoming increasingly difficult to develop the desired tensile and impact properties; research work is being conducted to establish metals that will develop the desired properties with slower cooling rates or to select higher strength steels that could be used in thinner sections. To obtain optimum properties at the most highly stressed finished machined surfaces, the forgings are contour-machined close to these surfaces before quenching; this allows the later removal of as little metal as possible which has been cooled at the faster rate. The notch-ductility considerations, stated in NDT terms, are used in setting hydrostatic test temperatures and start-up heating and pressurizing rates for maximum safety considerations.

The same notch-ductility considerations apply to weld metal and the heat-affected zones of welds. Contributions to the technology have been necessary to develop weld metal which would have a satisfactory balance between tensile properties and impact strength. The specified physical properties of the heat-affected zone of presently used metals have been met because the HAZ cools rapidly and, following postweld beat treatment produces a metal which will have high-impact strength.

Effect of Neutron Irradiation on Material Properties

The properties of the various materials have also been investigated for the effect of neutron irradiation. It has been found that the tensile strength and hardness are increased and duçtility decreased when the metals are irradiated at relatively low temperatures. As the temperature of the metal increases, however, and it is subjected to neutron irradiation, the change in properties is lessened and tests above approximately 600 F have shown less cumulative damage, presumably due to a partial annealing-out of the effects of radiation. Prolonged irradiation at low temperatures raises the NDT temperature.

Corrosion Resistant Cladding or Linings

A combination of metals may be used in reactor pressure vessels in order to utilize the high strength of the low-alloy metal. For example, a cladding or lining can be used on the internal portions of the vessel which will be more corrosion-resistant to the application than is the backing metal. Vessels for pressurized water service are usually clad with an austenitic stainless steel or with a NiCrFe alloy by weld over-lays or roll bonding. A variety of welding processes are used to produce arc weld-deposited cladding. Explosion-bonded cladding has been used for some applications.

Welding Groove Designs

Some conventional types of welding grooves are shown in Chapter 84, Fig. 84.6. The single-U has a 1/8 in. root face at the bottom, curving with a 1/4 to 5/16 in. radius to a straight wall, and sloping approximately 1/8 in. per in. to the outside surface of the plate. Steps in welding with the single-U groove are illustrated in Fig. 84.6B. The single-Vee-groove has straight sloping sides, a root opening and is generally backed with a bar for welding. It has the advantage of being easily made by thermal-cutting techniques when machining equipment is not available. This is also true of the double-Vee groove which finds extensive use in field-welding applications.

The single-U and the single-Vee grooves lend themselves to automatic submerged arc welding from one side as well as manual welding. For relatively small diameter vessels, where preheat for welding is required, the best practice is to design a joint that can be welded from the outside.

Selection of Filler Metal

When enhancement of toughness is the prime purpose of specifying quenched and tempered material, the filler metals used to join these base metals must have at least matching impact properties. Low-hydrogen welding systems are universally specified. There are many suitable filler metals available for joining quenched and tempered steels by the manual shielded metal-arc, submerged arc, gas tungsten-arc, gas metal-arc and flux cored arc welding processes. Care must be taken to select filler metals that produce weld deposits which are not adversely affected by postweld (stress-relief) heat treatment. Some construction codes require testing of each lot or batch of filler metal under the conditions to be employed in construction.

Consideration must also be given to the effect of reheating within the austenitizing range of the mechanical properties of weld metals. During normal multiple-pass welding, the weld beads are individually quenched, immediately after their deposit, by the comparatively cool surrounding base metal. For this reason, weld metal intended for use in the as-welded or stress-relieved condition usually has a relatively low-carbon content. The low carbon provides suitable hardenability for these conditions and minimizes weld metal cracking.

When weld metal is subjected to a quench and temper heat treatment after the entire joint is completed, it is necessary to more closely match the chemistry of the base metal and to test the proposed weld deposit under these conditions to assure meeting the required mechanical properties.

Welding

The permissible carbon and low-alloy steels shown in Table 87.5 are weldable in plate thicknesses up to and beyond 12 in. and forgings to 24 in. Electroslag welding is used for main longitudinal seam welds, where it is possible to quench or normalize and temper afterwards; the submerged arc and manual shielded metal-arc processes are used in longitudinal and circumferential weld joints.

Figure 87.27 shows an etched cross section of a submerged arc butt weld. The weld metal may be a different analysis if it meets the required mechanical properties. For welding attachments and nozzles to such vessels, manual shielded metal-arc welding has been used to a considerable extent, as have submerged arc, gas metal-arc and gas shielded flux cored arc welding processes.

Table 87.5—Chemical compositions of steels commonly used for heavy wall nuclear pressure vessels

ASTM Designation	Maximum or Range, %							
	C	Mn	S	P	Si	Mo	Ni	Cr
Plate								
A-516, Grade 70	0.27 (a) 0.28 (b) 0.30 (c) 0.31 (d)	0.85-1.20	0.04	0.035	0.15-0.30			
A-533, Grade B	0.25	1.15-1.50	0.040	0.035	0.15-0.30	0.45-0.60	0.40-1.00	
Forgings								
A-266, Class I	0.35	0.40-0.90	0.04	0.04	0.15-0.35			
A-182, Grade F1	0.30	0.50-0.85	0.045	0.045	0.15-0.35	0.44-0.65		
A-508, Class I (and Case 1332 - 4 Par 1) (similar to A-541, Class 1)	0.35 0.30	0.40-0.90 1.35	0.025	0.025	0.15-0.35			
A-508, Class 2	0.27	0.50-0.80	0.025	0.025	0.15-0.35	0.55-0.70	0.50-0.90	0.25-0.45
A-508, Class 3	0.15-0.25	1.20-1.50	0.025	0.025	0.15-0.35	0.45-0.60	0.40-0.80	

(a) 1/2 in. and under
(b) Over 1/2 to 2 in., inclusive
(c) Over 2 to 4 in., inclusive
(d) Over 4 to 8 in., inclusive

Note: Percentage is maximum unless range is given

Roll-bonded explosion clad plate and specialized forgings are commonly used. They compete with weld deposited overlay clad metals. Most weld cladding is deposited with the submerged arc, plasma-arc and gas metal-arc welding processes. The choice of process is dependent upon welding position, size of part, accessibility and similar considerations. For economy, cladding should be kept as thin as practicable. The minimum required thickness is approximately 1/8 in. This cladding operation is controlled to avoid excessive dilution with the base metal. Special analyses of electrodes and fluxes are generally considered necessary to meet chemistry requirements.

Techniques and parameters for welds, including amperage, voltage, speed of travel, type of flux, dryness of flux, preheat temperature and postheat temperature, are controlled. This control will ensure a deposited weld metal which will meet simultaneously the impact properties, tensile requirements, soundness requirements of strength welds and, in addition, the corrosion-resisting properties and analyses of the weld overlays.

Within limits that must be established for each type of steel, adjustment of the variables in butt welds to produce faster cooling will enhance the notch toughness of the weld and heat-affected zone. Cladding techniques are described in Chapter 93.

Relatively thin layers or beads of weld metal permit a partial progressive grain refinement of preceding layers of ferritic weld metal. This minimizes the columnar structure characteristics of single beads of weld metal as deposited, and improves ductility and impact resistance. Satisfactory welding results may be obtained with either dc or ac. However, welding of magnetic materials with ac has a distinct operating advantage over d-c welding in that occasional, troublesome arc blow is minimized.

In rigid type structures—joints in heavy wall vessels or nozzles attached to thick shells—it is advisable to weld continuously until the joint is finished to avoid the tendency for cracks to occur in the partially welded joints. When this is not possible, it may be necessary to eliminate stress raisers from backing rings or side wall undercut or to utilize an intermediate stress-relief heat treatment to prevent cracking during or after cooling. It is usually better to maintain a constant and uniform preheat until welding is completed than to perform heat treatment on a partially completed nozzle weld unless the rate of heating is limited.

The main seam welds and all the full joint penetration welds, such as through nozzles (Fig. 87.28) and inserted type nozzles, are fully radiographed. The weld flaws which are presently allowable consist of porosity and slag—in no case greater than those permitted by the ASME Boiler and Pressure Vessel Code. No detectable cracks or incomplete fusion are permitted. The very heavy welds are examined by isotope or high-voltage betatron or linear accelerator radiography. Accelerators, capable of detecting some defects of about one-quarter of 1% of the thickness of the heavy welds, give an extremely sensitive examination. The present standards generally require that a penetrameter of 1% of the weld thickness be visible on the film when radiographing welds over 6 in. thick; this graduates to 2% as the thickness decreases. Any defects exceeding the Code limitations are required to be removed and repaired, usually by rewelding. The repaired areas are then reheat-treated and reinspected.

Postweld Heat Treatment

After nondestructive examination and repair of defects, the weldment is then ready for its postweld (stress-relief) heat treatment. During the course of the

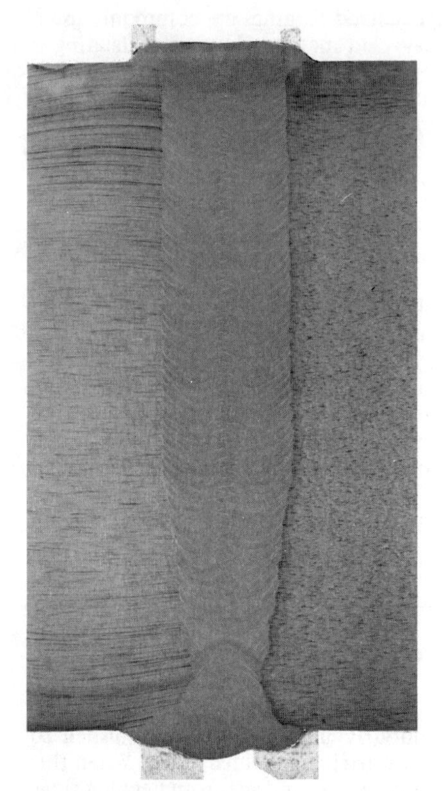

Fig. 87.27.–Cross section of groove weld joining forging and plate materials.

Fig. 87.28.–Typical weld between shell and nozzle wall.

fabrication of the thick, highly restrained low-alloy steel weldments, certain interstage heat treatments may be required. These heat treatments frequently consist of preheating for welding, maintenance of a specified minimum interpass temperature during welding and possibly an interstage stress relief following the completion of a subassembly. The purpose of this interstage heat treatment is to lower restraint stress in heavy sections. It is frequently not necessary, however, to maintain the welding preheat temperature until insertion into a furnace for stress-relief heat treatment. For weld overlay cladding, most weld repairs and many strength welds, an acceptable and proven alternative is to hold the subassembly at a temperature of 300 to 400 F for approximately four hours after the completion of welding. This procedure is sufficient to prevent cracking by allowing diffusion of most of the hydrogen present in the weld and heat-affected zone. This procedure is helpful since each shell course will have several other courses added to it in the fabrication sequence and may be subject to several interstage heat treatments prior to the final heat treatment.

Where weld overlay cladding is a part of the vessel, the heat treatment must be compatible with this construction to avoid excessive diffusion of carbon from the low-alloy base metal into the alloy cladding interface which may result in poor bend ductility. In some cases where machining is not performed after final heat treatment and no scale or oxidation of the part is permissible because of cleanliness requirements, it may be necessary for heat treatment to be performed in a controlled atmosphere.

Austenitic stainless steel weldments may or may not require heat treatment, depending largely on dimensional stability requirements. In some cases, close tolerance machining is required after welding, and heat treatment is required to attain suitable dimensional stability. Where the removal of peak stresses for dimensional stability combined with a minimum of distortion is the principal requirement, a heat treatment of approximately 900 F may be used. However, a solution anneal at 1850 to 1900 F followed by rapid cooling may be used to substantially relieve all the residual stresses and to improve resistance to intergranular corrosion in some environments. The heat treatment used is dependent upon the required tolerances and metallurgical and corrosion considerations. Heat treatment of austenitic stainless steel weldments between 1100 and 1200F may cause carbide precipitation. In most cases, no heat treatment is used.

Hydrostatic Test

After fabrication is completed, a hydrostatic test is required, usually at 30 to 60 F above the maximum nil-ductility transition temperature determined for the base metal in the vessel to avoid the possibility of brittle failure during the hydrostatic test. This 150% over-pressure test is also credited at times with providing a "mechanical stress relief." Similarly, this relationship between nil-ductility transition temperature and test temperature applies to control of start-up of the vessel in actual service.

Additional Requirements

Many design, metallurgical and fabrication controls are required to produce work of sufficient quality and integrity. Designs are rigidly analyzed for stress and may be checked by strain gage tests of the finished product. Parts going into reactor equipment are analyzed for mechanical properties and chemical composition; records are kept of the test results.

Special fabrication techniques are required in some cases to prevent cracking of a part in fabrication. Stainless steels, for instance, after welding under high restraint followed by a high-temperature heat treatment, may sometimes crack during heat treatment due to low-stress rupture ductility. In such cases, it may be helpful to use special filler metals which have superior high-temperature strength and ductility.

Parts of auxiliary components, particularly rotation of moving parts subject to wear and corrosion, may require the welding of dissimilar metal combinations. Very hard materials, such as nickel-base alloy X-750 and precipitation hardening stainless steels, are used to avoid galling conditions, and nitrided stainless steels have been used for this purpose. Corrosion-resistant stainless steel or nickel alloy piping is required for such purposes. For resistance to stress corrosion, NiCrFe alloy base metal and similar alloy weld overlay cladding are sometimes required.

In austenitic stainless steels and nickel alloy welding, the finished weld metal deposit analysis is controlled to minimize microfissuring. Special welding techniques are often used to lessen distortion. This requirement is no different for nuclear application than for any other application, except that complicated designs may require more close tolerance welding.

HEAT EXCHANGERS

Design of the components for heat transfer is an important consideration in power reactor systems. Heat exchangers differ in design, performance and the conditions under which they must operate. Water, heavy water, molten sodium, sodium-potassium alloy, various gases and fluid fuels are employed as primary coolants. Coolant systems may include single-, double- and triple-heat loops. An example of a single-heat loop is the direct BWR system delivering steam to a turbine. Two loops are usually employed in water-cooled PWR systems, with the primary water transferring heat in a boiler to produce steam for the turbine loop. Sodium systems may also include a third or intermediate loop containing sodium or an alloy to separate primary sodium from water.

Operating temperatures and pressures in heat exchanger components vary from system to system. In the PWR system, the coolant is maintained at about 2000 psi to prevent bulk boiling in the reactor. Thus, heat can be carried in liquid water at about 530 F to a boiler in which steam is generated for the turbine loop. In the sodium-cooled, graphite-moderated reactor system, for example, sodium emerges from the reactor at 925 F to generate steam in a heat exchanger at 825 F and 815 psi.

Heat exchangers for nuclear service are frequently more compact and complicated than those for conventional use. The relatively high heat transfer coefficients, obtained with such coolants as liquid metals and boiling water, make it desirable to minimize tube wall thickness and employ highly conductive materials. Heat exchangers for nuclear power application have many uses and vary in size from quart bottles to very large types that are either too long or too large in diameter to be transported by railway cars. Their range of use varies from such mundane tasks as cooling drinking water, cooling lubrication oil and disposal of waste heat from the steam exhaust of turbines to the highly critical task of transferring the heat from the reactor core coolant to the steam generator. In the BWR system, the reactor core coolant supplies the steam directly to drive the turbine. Since the shells enclosing the heat exchanger tube sheet and tubing are unfired pressure vessels, they are covered in the section under Fabrication of Heavy Wall Pressure Vessels in this chapter and in Chapter 84, Pressure Vessels and Boilers. The operations in fabricating a heat exchanger that will be covered in this section are (1) weld overlay cladding (2) weld joint designs (3) methods of welding.

WELD OVERLAY CLADDING

The nature, temperature and pressure of the two heat-carrying mediums in a heat exchanger determine the thickness and material of which the tube sheet is made. In many cases, the tube sheet is made of carbon steel and is overlaid on one side with a layer of highly corrosion-resistant alloy that may be as thin as 1/4 in. or as thick as 4 in. The thickness of the overlay is not only determined by the corrosive fluid and its temperatures and pressures, but by the joint design for attaching the heat exchanger tubes. The joint designs vary all the way from the tube extending just through the front face of the tube sheet to a nozzle being trepanned from a very thick overlay on the back side of the tube sheet. Suffice it to say that depending on the thickness of the cladding and thickness of the base metal, the cladding is attached by either roll bonding, arc welding or explosion welding. For details of these processes, see Chapter 93, Clad Steel and Applied Liners.

BASE METALS

The base metals used in the heat exchanger tubes and tubesheets for a nuclear power operation will vary from carbon steel and nickel-chromium alloys in the primary heat exchanger to copper for oil coolers, and copper and copper-nickel alloys for the steam condenser for the turbine. The BWR system utilizes stainless steel in many heat exchangers.

JOINT DESIGN

The design of the weld joint for a tube-to-tubesheet weld is determined by the heat exchange medium, pressures, temperatures and exposure to thermal shock. For applications handling non-corrosive media at relatively low pressures and temperatures, very economical weld joint designs are available. However, as the heat exchange media become corrosive or the pressures and temperatures become elevated, stress- and crevice-corrosion becomes a problem and the weld joint design becomes more difficult.

One of the problems faced in heat exchanger designs is the evaluation of potential crevice corrosion in the unbonded area between the tube and tubesheet. Thus, the trend in the design of weld joints has been to design away from the problem of crevice corrosion.

The nature of heat exchangers is such that the tubes are relatively small in diameter and the wall thickness relatively thin. Thus the combination of dimensions and materials permits the use of the gas tungsten-arc welding process for tube-to-tubesheet welding.

WELDING METHODS

Gas Tungsten-Arc Welding

Virtually all of the tube-to-tubesheet welding is with the automated gas tungsten-arc welding process, with or without the addition of filler metal.

Electron Beam Welding

There is at least one fabricator of heat exchangers using electron beam welding for tube-to-tubesheet welding. A typical EB welded joint employs a tube flush with the tubesheet where the electron beam is directed down the interface between the tube and tubesheet. It is focused so that the toe or bottom of the weld has a radius that is adequate for preventing cracking with the particular metal.

Two types of tube-to-tubesheet welders have been developed for the gas tungsten-arc welding process. The first type, which makes only face welds at the face of the tubesheet, is mounted on the tubesheet with the front face of the tubesheet in the horizontal position. The welding torch is mounted on a pantograph fixture that orbits the welding torch around the weld joint. This unit can weld tubes from 1/8 in to 3 in. od. The diameter of the electrode path is continuously adjustable by an adjusting screw. The precision programming control and the fixture making a tube-to-tubesheet weld are shown in Fig. 87.29.

Fig. 87.29–Orbiting front face welding head showing programming controller and the fixture making a tube-to-tubesheet weld.

The second type of tube-to-tubesheet welder is shown in Fig. 87.30. This type has an extendable rotary quill that permits making welds from the front face of the tubesheet through to the back face of the tubesheet. The use of quill-type heads available from stock is essentially limited to tubes having an id from 3/4 to 2 in. However, manufacturers will design units for smaller or larger sizes on special order.

Most of the gas tungsten-arc tube-to-tubesheet welding appears to have been done with constant current. The use of pulsed current should further improve the high yields of this process.

The gas tungsten-arc welding power supplies and programming units for tube-to-tubesheet welding require the same precision equipment as is used for nuclear fuel tubes, tubing and pipe welding. These are described in detail in Section 2 of this Handbook.

Explosion Welding

Explosion welding is one of the newest methods of making tube-to-tubesheet welds. In Fig. 87.31 is shown the relation of the explosive charge to the surfaces to be welded, and in Fig. 87.32, the steps in making an explosion weld. A tube-to-tubesheet section fabricated by explosion welding is shown in Fig. 87.33.

The most critical part of the setup is the gap between the two pieces to be joined. This gap is essential in providing the proper jetting action which produces the weld. Different types of bond lines can be formed depending on the flow

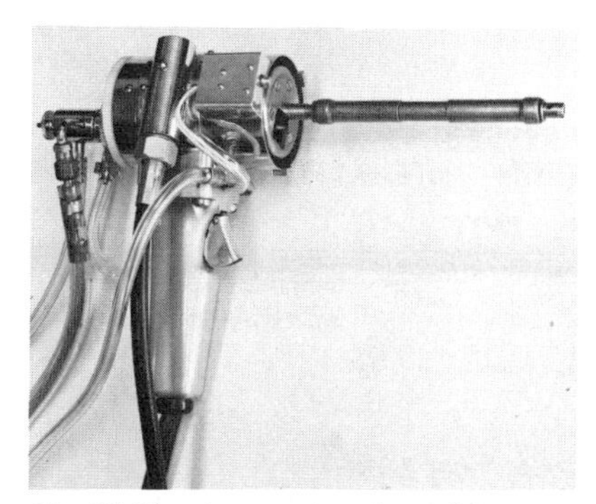

Fig. 87.30.–Automatic tube welding gun.

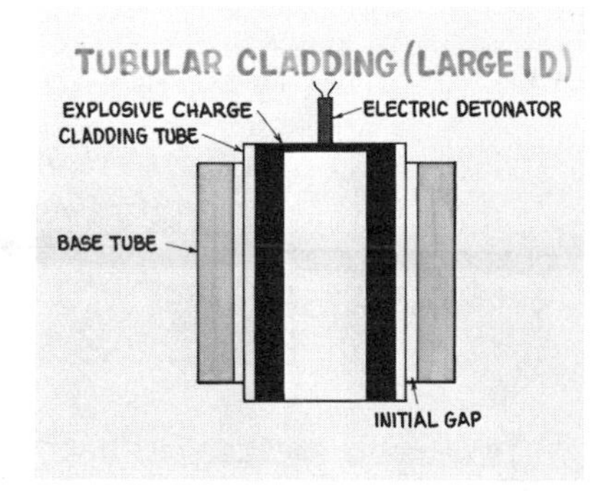

Fig. 87.31.–Relation of the explosive charge to the surfaces to be welded.

behavior of the jet. If the flow is unstable–this is the case normally encountered–the jet oscillates above and below the interface plane, producing a characteristic wavy or rippled interface (Fig. 87.34). If the flow is stable, the material flow is laminar and a straight interface results.

These two examples are idealized cases, as in practice considerable differences are noted in the wave pattern (that is, wave shape, wave amplitude and wave length) and in the volume and distribution of jet material which is retained or trapped in the interface. Most strong, explosion welds have interfaces which are wavy, and for this reason, the rippled interface has become the hallmark or signature of explosion welding. A good explosion weld will contain a minimum of jet material, and will have a strength which equals or exceeds the strength of the weaker base metal component.

CORE INTERNALS

The technical term for the internal structure of a nuclear reactor is "core internals." Core internals are the structures that position and support nuclear fuel and control rods inside a reactor vessel. They also channel the coolant that carries off the heat of fission to help generate steam, then electricity.

The core internals are contained in the core barrel of the reactor. The largest barrel is 14 ft in diam, 30 ft long and weighs 80 tons. Other core parts add about 100 tons. The barrel is usually constructed of Type 304 stainless steel, as are most other parts of the reactor internals which include the core plates, support columns, deep beams, guide tubes and rods, formers, baffles and thermal shield. An additional part is the loop which is the industry jargon for a steam generating system. A reactor with four loops has four steam generating systems.

Most core parts are welded. All of them are built to extremely close tolerances–a necessity if the core is to function properly. Formers–labyrinths of plates–must fit barrels within 0.005 in.; welds between upper and lower barrel segments must have controlled shrinkage within 0.030 in. and the barrel must not be more than 0.020 in. out of parallel.

The need for accuracy and dependability in core internals is a matter of necessity. Combined fabricating errors must be small enough to permit an exact and smooth-working relationship between guide tubes, control rods and instrumentation.

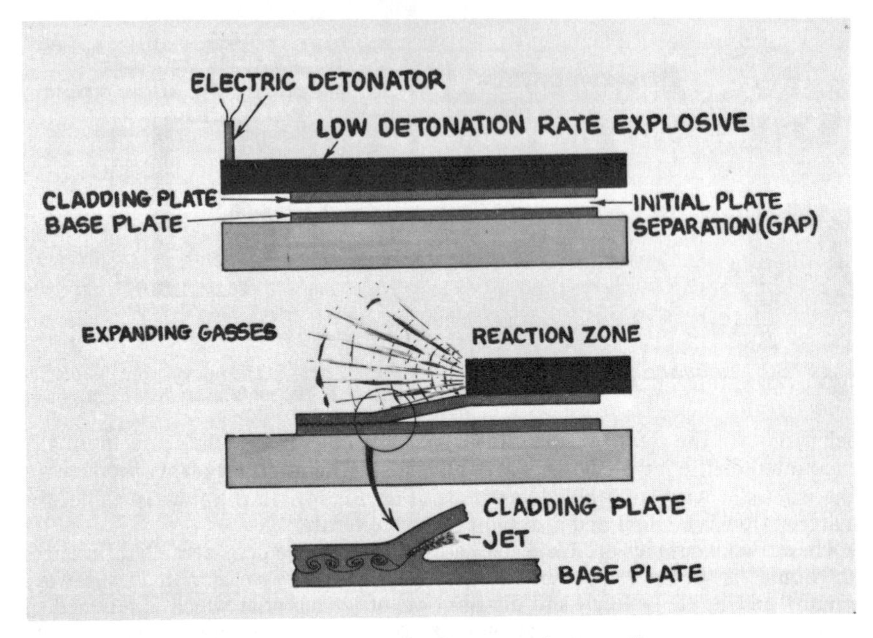

Fig. 87.32.–Steps in making an explosion weld.

The most critical welds in the core internals are the final assembly welds on the barrels. Previously, barrels were welded with shielded metal-arc and automatic submerged arc welding processes. Today, barrels are being welded with the hot wire gas tungsten-arc welding process.

Figure 87.35 shows that a barrel has five parts: top flange, upper barrel cylinder, two lower barrel cylinders and a lower support casting. In the turning roll setup, barrel parts turn under hot wire gas tungsten-arc welding heads held by manipulators. Welding operators work either outside on elevators (so they can work safely without disturbing the arc), or they tread the barrel inside as it turns under the arc.

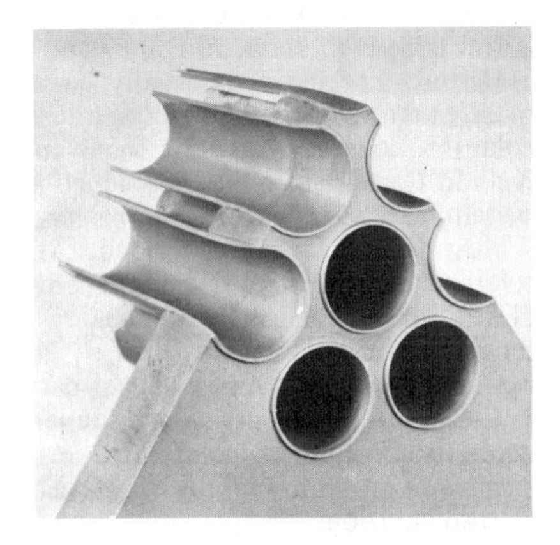

Fig. 87.33.–Tube-to-tubesheet section fabricated by explosion welding.

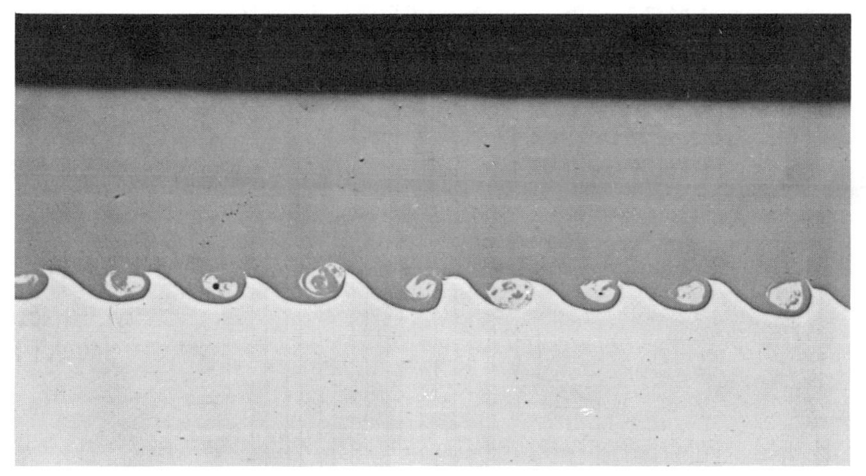

Fig. 87.34.–Characteristic wavy or rippled interface.

The reason for the change from shielded metal-arc and submerged arc to gas tungsten-arc welding was to improve welding accuracy and speed and to do it in final assembly with all the core parts already trial-fitted to the barrel being joined. Welding barrels with one arc tended to twist and warp the barrel. Some twisting and warping can be accepted in subassembly welding, but not in final assembly welding when all core parts have been trial-fitted.

The new three o'clock welder eliminates twist by welding on two sides of a barrel at once. It is so fast that it has cut barrel welding by 80%. It welds all sizes up to 18 ft in diam and 40 ft long. It is called a three o'clock welder because it welds horizontally while the barrel (and internals) turn on end. Elevators lift welding operators to welding heights 10 to 40 ft above the turntable on which the barrels turn. After welding, the welds are stress-relieved with electrical resistance finger-type heating units.

OTHER WELDMENTS

Altogether, cores take 5,000 pounds of weld–3,800 pounds of it deposited with hot wire gas tungsten-arc, the rest with gas tungsten-arc and shielded metal-arc welding processes.

Deep beam is another example of hot wire gas tungsten-arc welding. In Fig. 87.36, a 10 ft square beam is made of intersecting stainless plates fillet-welded together. It must finish square within 1/16 in.

Upper support columns (Fig. 87.37) which hang from the core's upper support plate provide another example. They are tubular weldments with three hot wire gas tungsten-arc girth welds. During welding, the columns turn on rolls under a fixed welding head. Soft vacuum electron beam welding of support columns has been shown to weld better and faster than hot wire gas tungsten-arc welding and eliminates much of the edge preparation.

Barrel nozzles are another example wherein nozzles are shrunk-fit into the barrel openings using liquid nitrogen. Welding operators then deposit multipass welds in a partial groove, and crown groove welds with heavy fillets. Pressure in the reactor makes nozzles self-sealing; full-depth welds are unnecessary.

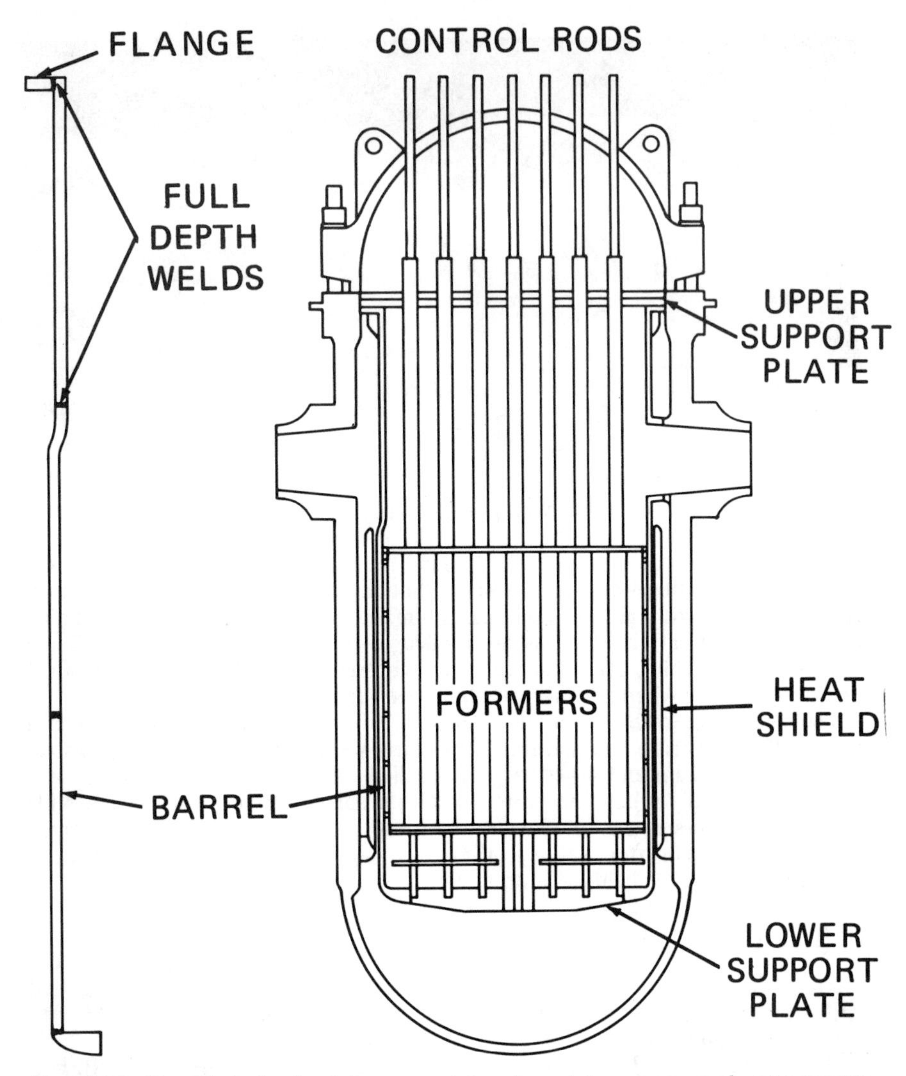

Fig. 87.35.–Simplified sketch of how core internals go into a pressurized water reactor vessel. The barrel weighs 250 tons and can be as wide as 18 ft and as long as 40 ft. The bottom support plate is a heavy welded casting that is often called "Swiss cheese" because it is full of holes.

Another method involves alignment pins for core vessels which are surfaced using the oxyacetylene process and Co-Cr alloys for resistance to wear.

A final example is thermal shields, which like barrels, are welded on rolls. A shield for a two-loop core is 3 1/2 in. thick and weighs 30 tons.

Many core parts, like formers, deep beams and barrel cylinders are cut from sheet and plate stock and require edge preparation before welding. Most of the parts are plasma-arc cut from light gage sheet to 4 1/2 in. thick plate. Other edge preparations are performed with machine tools such as milling machines, lathes and drill presses. Application of tape controls to all of these operations is advantageous.

TUBE AND PIPE WELDING

Tube and pipe in a nuclear power plant provide two rather well-defined functions: the operation of instrumentation and controls and the transportation of fluids. In the first category, the dependable and safe operation of the complete nuclear reactor electrical generating system is dependent on the integrity of perhaps tens of thousands of welded joints. The second category covers the piping which transports bulk quantities of fluids. These fluids may be as innocuous as chilled drinking water or as corrosive as sea water or sodium metal at a high temperature. In any case, some of the fluids are radioactive or contain radioactive material.

Tube and pipe used in nuclear reactor systems vary in size from 1/16 in. od or less with wall thicknesses from approximately 0.010 to 0.030 in. to piping many feet in diameter with wall thicknesses from about 1/2 in. to several inches depending on the service requirements. The trend in the welding of tubing and piping systems is towards the automation of welding processes. This has been brought about by the extremely high quality of welded joints specified for nuclear systems which has in turn resulted from personnel safety, prevention of the spreading of radioactive contamination and maintaining the highest operating efficiency of the nuclear reactor-electrical generating installation. In addition to these factors, the building volume required for the piping systems has been minimized while space for repair work has been provided.

Gas tungsten-arc welding systems for the automated welding of tubing and piping can be performed by either using two fixed position tungsten electrodes (Fig. 87.38) and no additional filler metal or by using a filler metal addition and welding torch oscillation while having provision for an automatic arc voltage control head. In addition, shielded metal-arc and gas metal-arc welding are used extensively on critical as well as less critical piping.

For additional information on the welding of pipe, see Chapter 85, Industrial Piping.

Fig. 87.36.–Welded deep beam support for heavy core internals.

Fig. 87.37.–Upper support column being hot wire gas tungsten-arc welded.

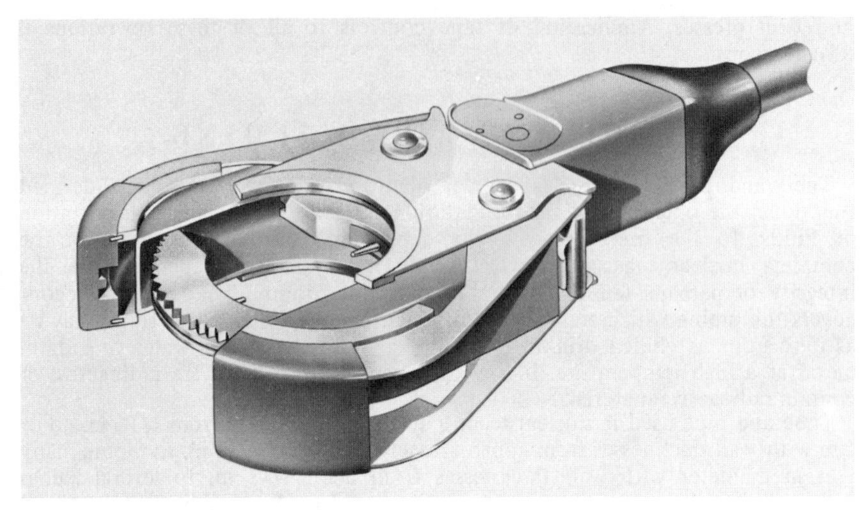

Fig. 87.38.–Automatic tube welding head with two fixed position tungsten electrodes.

CODES

Tube and pipe for nuclear power plants are designed and fabricated in accordance with the rules of Section III of the ASME Boiler and Pressure Vessel Code and other applicable codes such as the ANSI Code for Pressure Piping B31, Section 7, Nuclear Power Piping. Welding procedures are qualified to meet the requirements of Section III and Section IX of the ASME Code and are similar to the procedures used for conventional piping of equivalent material for similar temperature and pressure service conditions.

TUBE WELDING METHODS

The diameter of tubing and its wall thickness dictate that the gas tungsten-arc welding process, preferably automatic, be used to make the reproducible quality of welds required for nuclear reactor use. With the development of the pulsed gas tungsten-arc process for tube welding, the weld quality may be improved.

BASE METALS

Austenitic stainless steel, such as Types 304 and 316, are the most common base metals for construction of tubing systems for nuclear reactors. In specifying tubing base metal, it is important to hold down the impurity and composition ranges as close as economically feasible, for variation in the composition of the base metal will produce variation in the arc voltage, penetration and depth to width ratio of the weld.

Weld Joint Design

Proper joint design and preparation are very important in automatic gas tungsten-arc butt welding applications. There are several requirements and recommendations for a good design of any gas tungsten-arc welded joint.

First, it must be formed by compatible metals. If filler metal is needed from a metallurgical aspect or for weld reinforcement, it should be built into the parts. Ideally, an automatic gas tungsten-arc welding operation should perform without the need for welding operator dexterity and judgment in adding the filler metal to the weld puddle.

A second requirement is to have good fit-up. When two edges are brought together for a butt weld, they must join without gapping. A very common joint is a self-locking joint formed by the insertion of one part into another. The amount of clearance that can be tolerated between the parts is a function of a number of factors. It is important to note, however, that pulsed current techniques have proven to be an important factor permitting greater clearance tolerance between parts. Furthermore, pulsed current techniques help to join parts which have unequal amounts of metal in the immediate weld area. Normally, mass equalization is required in a good joint design.

Finally, a good joint design provides for easy cleaning. A small amount of surface oxides, oil, moisture or dust between two accurately placed parts can cause porosity, for example, in aluminum and titanium welds. The ferrous alloys are much less critical to porosity formation, particularly when the action of the weld puddle is activated by welding current pulses.

Containment of radioactive fluids under pressures and temperatures that exist in nuclear power systems is the principal justification for requiring high quality piping welds. Equally important are safety of personnel and continuity of operation to produce miximum operating efficiencies. In addition, systems utilizing liquid sodium at elevated temperatures produce highly corrosive environments that must be contained. Molten sodium, at any temperature, produces an extreme hazard from its high oxidation rate. If it comes in contact with water, the reaction rate is very rapid.

These fluids may also cause unusually severe thermal shock conditions on the basic components during service. Coolants such as harbor and sea waters, used to dispose of waste heat from the steam condensers, produce a highly corrosive environment in their piping systems. Maintenance of a high degree of cleanliness, design for decontamination, repair and/or replacement for failures and leak tightness are considerations of concern to nuclear power piping designers.

Welding Equipment

The trend in pipe welding of both austenitic and carbon steel is to the use of automated systems. Systems that are used for the welding of tubing may be used directly for the welding of piping with the procurement of pipe welding fixtures. Generally, the precision of control of the welding parameters for pipe welding is virtually the same as for tube welding. The addition of an automatic arc voltage control head, torch oscillator and wire feeder is essential. Hot wire feeders may also be used.

The stringent quality requirements of welds for nuclear reactor piping are producing a trend towards higher precision equipment for gas tungsten-arc, gas metal-arc and submerged arc welding power supplies, wire feeders and rotational equipment. Benefits that come with the use of higher precision equipment are reduced costs per weld by less rework and less time to make a weld.

Welding power supplies for shielded metal-arc welding are fairly well standardized. However, they are showing a trend towards the use of constant current controls to maintain constant current during welding in order to eliminate current fluctuation produced by line voltage variation, temperature variation of the welding power supply, power supply cables and welding current cables.

Mechanized fixtures for the welding of pipe are available from stock for pipe up to 3 ft in diam and 4 in. wall thickness with larger sizes available on special order.

PIPE WELDING METHODS

The welding process to be used for a particular application is determined by many factors. However, with the stringent demands for quality welds in the nuclear industry, the use of automated pipe welding is required in many instances. Because of the economics of automated pipe welding, the trend in the welding of nuclear piping is in this direction.

This is particularly true with the austenitic steel piping, but is also being applied to carbon steel piping. Although the automatic pipe welding systems are expensive compared to manual welding equipment, the joint welding time is very low compared to manual welding and the rework rate approaches zero. The reason for this is the exact process control that is obtained with punched card programmers, precision parameter control and recording of process parameters.

Shielded metal-arc welding is still used extensively on carbon steel piping of all sizes. The gas metal-arc and flux cored arc processes are being used for on-site and shop fabrication. Submerged arc welding is used to a limited extent in shop fabrication where the pipe can be rotated.

Base Metals

Base metals presently used in pressure piping systems for nuclear power are also in use in conventional piping systems with minor modifications. Austenitic stainless steels, such as Type 304, and austenitic stainless steel-clad carbon steels are the most common materials of construction for the primary piping in nuclear reactor systems. The need for minimizing corrosion and corrosion products which may become radioactive makes the use of stainless steels especially desirable.

Filler metals are essentially the same as those used with equivalent base metal in conventional piping. Stainless steels for nuclear service conditions that currently prevail do not require the use of carbide stabilizers for long-time resistance to corrosion in steam, water or sodium service.

It is important to give the welding electrodes proper care. For those with low-hydrogen coverings, moisture content can be mimimized by procuring electrodes in hermetically-sealed containers, dry storing them until the time of use and rebaking in accordance with the manufacturer's instructions when necessary.

In austenitic stainless steels, a controlled delta ferrite content reduces the sensitivity of the weld deposit to hot cracking during welding. Consumable inserts, bare filler wire and covered electrodes usually have their chemistry adjusted to furnish the desired amount of delta ferrite in the weld deposits. Except for this, the chemistry of the weld metal should simulate that of the base metal.

Weld Joint Design

There are a number of factors that should be given consideration when designing a weld joint for nuclear piping. Because of the demands for cleanliness and crevice elimination, it is desirable to use joints with a smooth, crevice-free internal contour. Backing rings are not advisable unless they are later removed. Joints are designed for accessibility to suit the welding process employed, particularly joints which are to be field-welded.

Square groove joints are generally used for light-gage pipe up to 1/8 in. in thickness. Although Vee-groove joints have been successfully used for thick-wall pipe, a wall thickness of 3/8 in. is often considered a desirable point at which to change to a U-groove type joint or a joint with a compound angle.

The most important area in joint design is the weld root area. A sound root pass is essential to maximum joint integrity since defective root beads have been the cause of weld failure. Most root configurations are designed for fusion welding using the gas tungsten-arc welding process with or without filler metal. To avoid oxidation of the internal surface, an inert gas purge should be provided during the first layers of weld deposit.

Weld joints may be classified by the geometry of the joint preparation and the use of consumable insert rings or added filler metal. Joint types and dimensions which have been used successfully are shown in Chapter 84.

DISSIMILAR METAL PIPE JOINTS

The welding of dissimilar metal joints presents special problems because of the differences in properties of the two base metals and the difficulties involved in establishing a filler metal compatible with both base metals and with the service environment. The choice of filler metals will depend on the base metals to be joined and the service environment to which the joint will be subjected. Each application must be considered as a special case, and the choice of filler metal should be made on the basis of the best compromise among such factors as thermal expansion characteristics, metallurgical compatibility, soundness, practicability and economy.

Careful inspection and nondestructive testing of welds are required to ensure that a high degree of soundness is obtained. Dissimilar metal welds are usually radiographed for the full circumference after finishing. Liquid penetrant inspection is applied to the internal and external weld faces and adjacent base metals after finishing. In addition, for severe service applications, a mock-up of the dissimilar metal joint may be subjected to thermal shock or thermal cycling tests to confirm the ability of the connection to withstand service conditions.

INSPECTION AND TESTING

The consequences associated with a failure in a nuclear power plant have made weld joint soundness of the utmost importance. As a means of achieving the necessary weld soundness, several nondestructive testing methods are used. The primary methods are radiography, liquid penetrant, magentic particle and Depending on the class of piping, these nondestructive tests are used singly or in combination.

REPAIR WELDING

Repair welding in relation to nuclear reactor systems may be divided into the following categories:

1. No radiation or chemical hazards.
2. No radiation but chemical hazards.

3. Radiation and no chemical hazards.
4. Radiation and chemical hazards.

Repair welding may be very complex. Without being too inaccurate, it can be said that the major volume of repair welding associated with a nuclear reactor is with the tubing and piping systems.

Generally, the first two of the above categories can be cleared up without much difficulty. They may include such materials as oil, alkalies, acids or liquid metals. With the latter two categories, involving radiation, the problem is much more serious as the radiation level must be reduced to a safe point or the work must be performed with remote manipulators. In some cases, the complexity of the work is such and the radiation level is so high that it would be more economical to dispose of the faulted item in a "burial ground" and fabricate a new one. Sometimes the high level of radioactivity emanating from the surface or subsurface of the weldment prevents effective decontamination. The material itself may have been exposed to intense neutron bombardment and thus became radioactive, making decontamination impossible and disposal unavoidable.

The supply of skilled personnel that may be maintained for performing cutting and welding operations in a high level radiation zone is limited by economics. Thus, when a repair problem occurs in a radiation zone whose radiation level cannot be reduced to a low level, the following alternatives must be evaluated:

1. Can the weldment be repaired with the personnel on hand without exceeding their legal exposure limits?
2. Can it be repaired with existing tools using remote manipulators and auxiliary shielding?
3. Can it be removed and a new one fabricated and installed?

Components which failed or developed trouble after radioactivity levels were built up have been repaired in many reactors. The methods employed in making repairs have invariably been makeshift in terms of equipment, techniques and procedures because standard equipment for remote maintenance is generally not available. There seem to have been two general approaches to reactor system maintenance. Where possible, flooding with water has been used to provide radiation shielding while still allowing visibility and mobility. In other cases, portable or temporary shields, usually lead, have been combined at times.

Repair welding on the thermal shield of one reactor was performed under water with long-handled tools and television viewing. Another reactor system experienced a number of cracks in welds which were repaired by direct work from behind a lead shield with a lead pipe to protect the welder's hand and arm. By having just a small hole in the lead pipe to allow movement of the welding head, workers were protected as much as possible during their working time inside the radiation zone.

The high quality of welds required for piping and tubing used in the nuclear, space and aircraft fields has brought forth a number of automated welding systems for welding tube and pipe. These systems are directly applicable to repair welding of nuclear reactor system piping and may be positioned, once the joint is in place, aligned and tack-welded either with very short exposure or with remote manipulators.

The most difficult task of all, since it must be done virtually by hand contact is to get the replacement piece in place, perfectly aligned, tack-welded and ready for the automated welding fixture. Since a high level radiation zone is being considered, the piece must fit perfectly and not require any fitting in order to minimize personnel exposure to radiation. In spite of all of these seemingly insurmountable problems, repair welding is being performed safely in high level radiation zones.

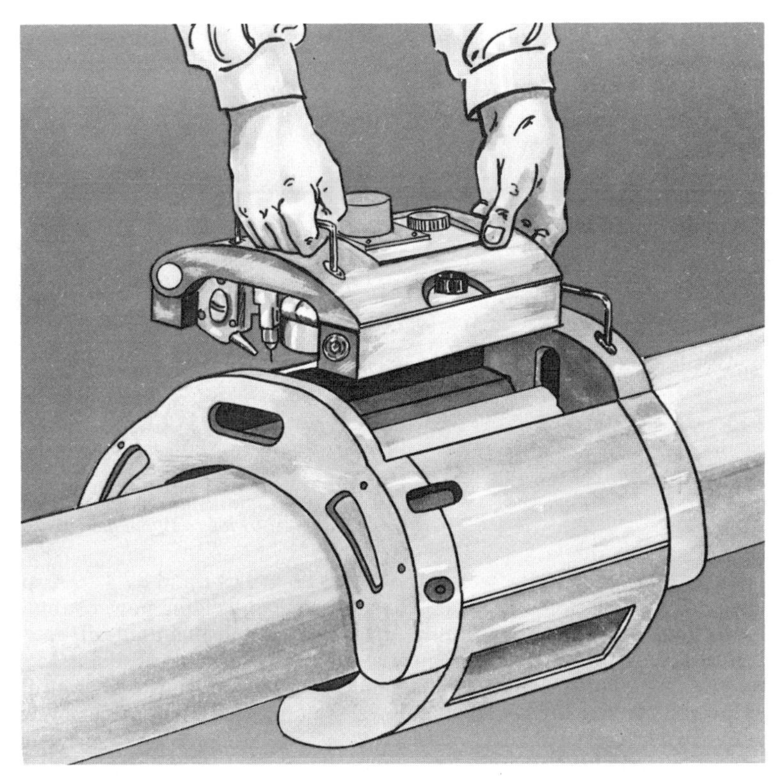

Fig. 87.39.–Sketch of welding module being placed in carriage.

An orbital combination fully automatated precision pipe cutting, milling and welding system has been developed (Fig. 87.39). It was designed to work in close spaces with a minimum of setup time, adjust to the standard dimensional variations of pipe and accurately follow the weld groove. A carriage, the main portion of which is one-half of a cylinder, is equipped with two sets of clamping rollers pivoted axially on the edge of the half cylinder. The other supporting rollers are mounted inside the carriage. Attachment to the pipe is made by retracting the clamping rollers to form a "U" shaped opening, fitting the carriage over the pipe and tightening the clamping rollers against the pipe with an Allen torque wrench. The carriage forms a stable platform which acts as a common carrier for either the milling or welding modules and holds the two ends of the pipe in alignment. Five carriages are available, handling piping diameters from 3 to 24 in. Each carriage has an opening on its periphery into which either a milling or welding module fits, locking into place with quick connectors. The milling and welding modules contain a tractor unit for driving the carriage and module around the pipe.

The milling module (Fig. 87.40) is a double duty unit used for cutting a section of pipe out of a line and/or preparing the remaining end with the desired joint contour for welding. Travel speed, cutter speed, reversing and fine positioning over the joint controls are built into the milling module.

An arc voltage control head, filler wire feeder, welding torch oscillator, rotational drive motor and fine adjustment for positioning the electrode over the

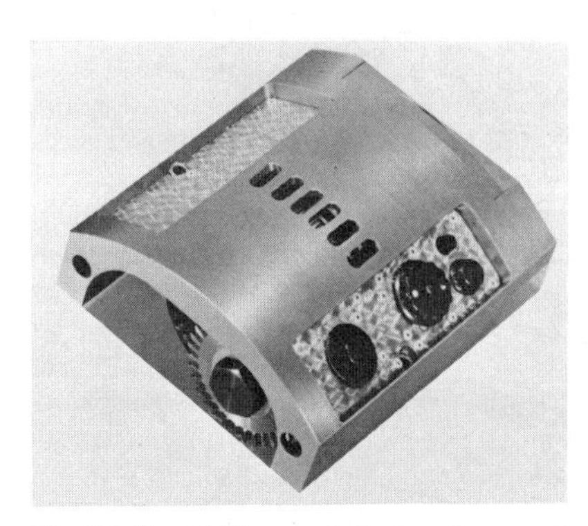

Fig. 87.40.–Milling module showing speed and positioning control knobs.

Fig. 87.41.–Welding module in the open position showing components.

weld joint are included in the welding module (Figs. 87.41 and 87.42). Available are precision combination pulsed and constant current welding power supplies of 100, 300 and 750 A at 100% duty cycle which will maintain all operating parameters within plus or minus 1% of the set point with plus or minus 15% line voltage variation. The usual sequencing and programming controls are available.

For large diameter thick-walled pipe, a welding head has been developed. These are the features of the fixture:

1. Handles 18 to 24 in. diam pipe.
2. Provides all-position welding.
3. Occupies not more than 19 in. along the pipe axis.
4. Operates within a 27 in. radial "envelope" around the pipe.
5. Wire feed, arc voltage and carriage travel rate are automatically controlled. Travel rate is consistent regardless of operational position.
6. The torch has at least 1/2 in. oscillation.

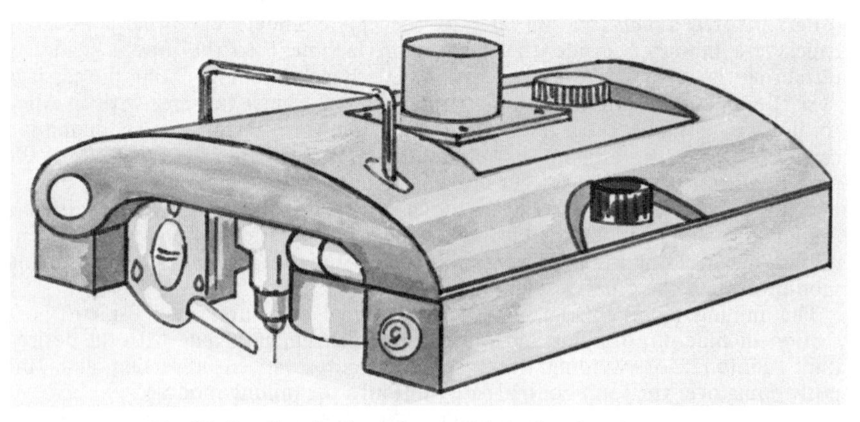

Fig. 87.42.–Sketch of welding module in the closed position.

The pulsed gas tungsten-arc welding process used is ideally suited to the pulling of consumable inserts because of its deep penetrating properties at minimum average heat input to the weld puddle. After completion of the root pass, the joint is filled by using stringer passes for the initial layers followed by oscillated passes further up the joint.

Because the tightly regulated consistency of the automatic process allows a much tighter joint configuration, substantially reduced filler metal is required as compared to a manual operation. The pulsating gas tungsten-arc provides the desirable concave weld beads. Gentle wetting to the joint side walls minimizes undercutting.

SHIPPING CASKS

The movement of fissionable/radioactive material requires that they be contained for one or more of the following reasons:

1. The radiation level does not exceed the limits established for common carrier handling.
2. The container is structurally strong enough to contain the radioactive material in case of a dropping or collision-type accident.
3. The container will prevent damage to the contents during normal shipping.
4. The container will prevent serious consequences in case of an accident.
5. The container shall maintain the maximum allowable temperature either with cooling fins or auxiliary cooling or both.

Thus, the fissionable radioactive material being transported determines the type of container required. Virtually no fissionable/radioactive material may be transported in the liquid state. Liquids, regardless of the radiation level, are placed in a resin to form a jell for transporting. If the material is smearable, then it must be placed in an approved container before placing it in a shipping cask.

The following list of fissionable/radioactive materials is pertinent to the fabrication of shipping containers:

1. New fuel.
2a. Spent fuel, unbroken.
2b. Spent fuel, broken.
3a. High to low level liquid wastes.
3b. Contaminated clothing, rags, paper, etc.
4. Hardware such as piping, core internals, etc.

Because of the nature of each one of these materials and their physical dimensions, shipping containers have been developed for all but one type.

SPENT FUEL SHIPPING CASK

The design of the spent fuel shipping cask is a complex problem for the following reasons:

1. The spent fuel is a high level radiation source with multiple types of emissions.
2. The energy released must be disposed of as heat.
3. The possibility of a critical incident must be prevented.

The casks usually have an inner container which is designed to hold the individual fuel assemblies in place. It fits inside the outer cask which contains

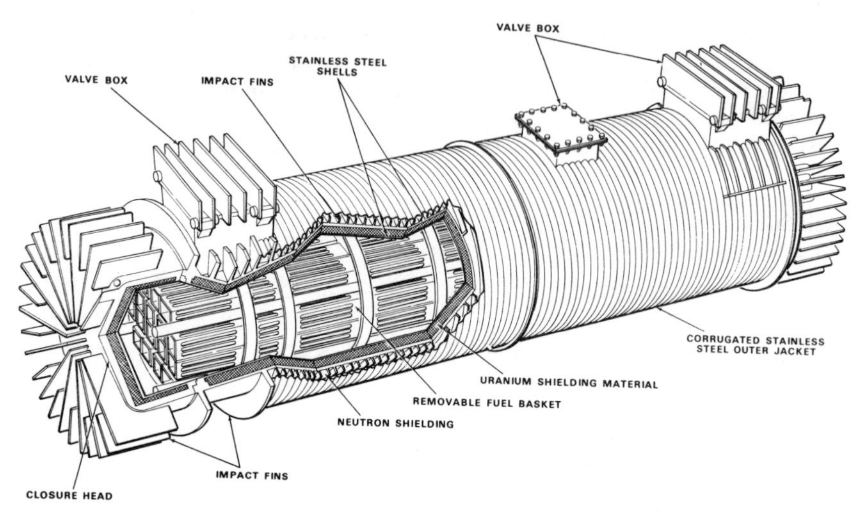

Fig. 87.43.–Spent fuel multi-element shipping cask.

the solid radiation shielding in the form of lead or depleted uranium and the auxiliary cooling systems in the forms of cooling coils and fixed metal fins.

The space between the inner containment and the cask is usually filled with light water to serve both as a neutron absorber and a heat transfer medium. Attached to the outside of the shell of the cask is the mechanical shock absorbing material. It may be a foam-type plastic, foamed-metal, laminated corrugated metal or similar high-density, shock-absorbing material.

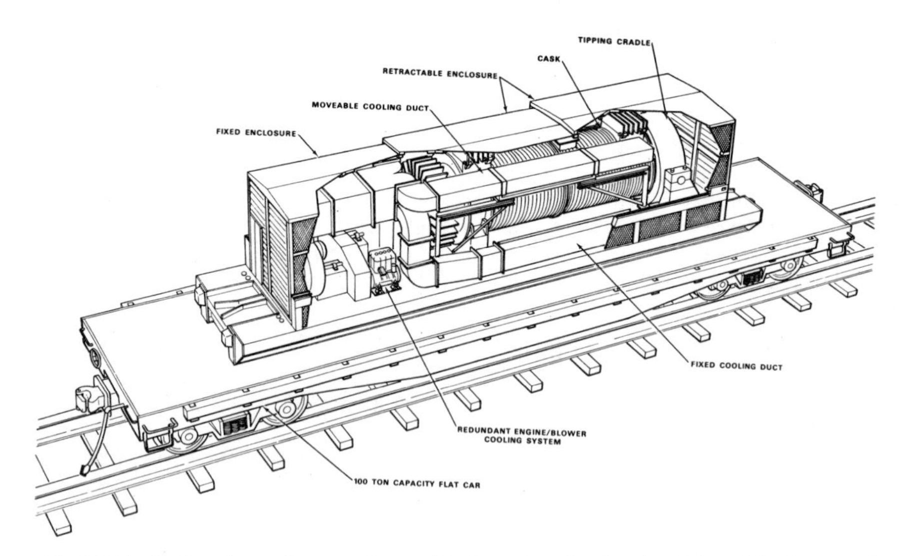

Fig. 87.44.–Railroad car for carrying multi-element cask, showing integral heat removal systems at each end and crash frame.

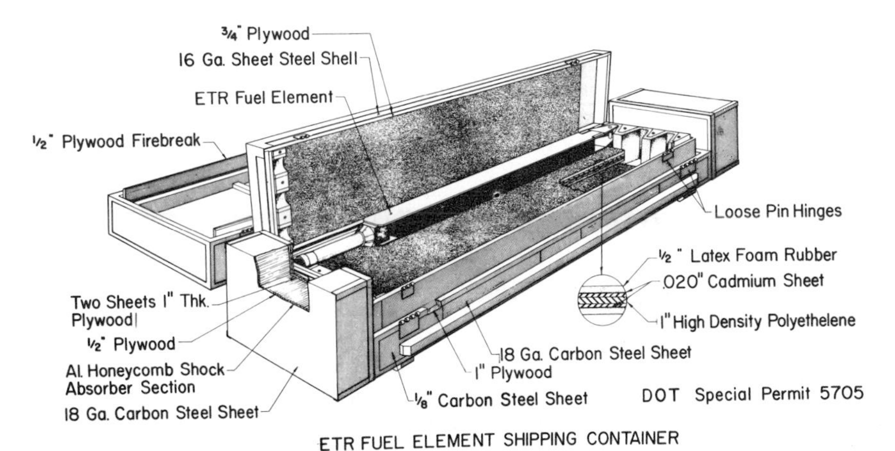

Fig. 87.45.–New fuel shipping container.

Shown in Fig. 87.43 is a multi-element spent fuel handling cask. It is virtually impossible to keep the weight of spent fuel shipping casks below the 50 ton limit set by most states, therefore they must be shipped by railroad. Shown in Fig. 87.44 is a railroad car for handling this type of container. It has the auxiliary heat exchanger system heat disposal equipment and built-in crash frame.

NEW FUEL ELEMENT SHIPPING CONTAINER

The new fuel element shipping container shown in Fig. 87.45 is designed to reduce the radiation level at the surface of the containers to the required level and to protect the fuel from mechanical damage during shipment. The principal problem in this case is to design and fabricate a container with enough mechanical shock-absorbing capacity to absorb the shock of an accident without damaging the fuel or producing a critical incident.

WELD DESIGN

Welds are often particularly vulnerable to serious damage as a result of free-fall accidents and subsequent thermal exposure. This is true due to a variety of factors. In addition, many are located in areas where they are subject to gross deformations.

Perhaps the most vulnerable welds are corner welds or the welds joining the shell to the top and bottom heads. In the free-fall impact, when the line of action passes through or near the center of gravity, these welds are required to bend or rotate through an angle of 90 deg. This places the root of the weld in tension and may result in a gross crack unless the corner is adequately designed. Other structural welds are required to deform a lesser, but significant amount. Generally, such deformation takes the form of bending of the weld and base metal over a significant length; hence, the weld is not placed in jeopardy to the same degree as the corner weld.

The design of a welded joint that is not a part of the cask proper will be left to the discretion of the designer and should be based on good engineering practice.

Consideration should be given to the effects of angular, lateral and end restraints on the weldment when butt welds are made, particularly with respect to material and weld metal having an ultimate strength of 80,000 psi or higher and heavy sections of both low- and high-tensile strength material. The addition of restraints during welding may result in cracks that might not occur otherwise.

QUALITY ASSURANCE

The safety and reliablility of nuclear power plants are of great importance for several reasons. First and foremost is the need to protect the public, plant personnel and the environment against any large scale release of the radioactive materials contained within the plant. Second is the need to provide an adequate and reliable supply of electric power. Finally, because of the high capital cost of these plants, unnecessary down time may cost the utility company an amount approximating $50,000 per day.

To provide assurance that the plants will indeed be safe and reliable, a quality assurance program is required by law for each plant. This program is applicable to the design, procurement, construction, operation and maintenance of the plant, and it lays particular emphasis on those components whose failure could have severe consequences. The welded components falling into this category include those forming pressure boundaries, i.e., the reactor vessel and vessels such as steam generators, pressure piping, valves and pumps, the containment vessel, equipment used for the processing of radioactive wastes, and core structural components which encapsulate the great bulk of the radioactive materials.

The requirements for a quality assurance program are found in Title 10, Chapter 50 Appendix B of the Code of Federal Regulations, and in Sections III, VIII and XI of the ASME Boiler and Pressure Vessel Code. (The B&PV Code is not applicable to fuel.) ANSI is also developing quality assurance standards which are intended to provide more definitive guidance in a number of areas.

The requirements which a quality assurance program typically imposes upon welding and weldments are as follows:

1. Thorough inspection of the base metals for chemistry, integrity, physical properties, etc., and maintenance of pedigree (traceability) of these materials where they are used in critical locations.
2. Formal qualification of welding procedures and welders.
3. Extensive use of analyses and in-process controls to assure that weld filler metals are as specified.
4. Rigid inventory controls on the filler metal and fluxes: controls on baking and holding ovens for low-hydrogen electrodes.
5. Inspection of joint geometries, cleanliness, fit-ups and adherence to other procedural requirements by a welding inspector.
6. Nondestructive testing of the joints by formally qualified NDT examiners, generally using at least two complementary techniques.
7. A record for each weld joint of the procedure, welder, filler metal heat, base metal heat, NDT examiner, inspector and results of inspection.
8. Reviews and inspections by a third party inspector.
9. In accordance with Section XI of the ASME Boiler and Pressure Vessel Code, a planned and systematic program for periodic inspection of critical welds and other components during operation of the plant.

In addition to the above, it is essential that all organizations performing work on nuclear power plant components and their installation and maintenance keep detailed records to provide evidence that established requirements have been met.

BIBLIOGRAPHY

"Russia Could Have Answer to Power Crisis," *Iron Age*, 209 (4) 25, (January 27, 1972).

"Small Component EB Welding," S. M. Robelotto and R. Moose, *Welding Engineer*, 55 (2) 37–39 (1972).

"Shear-Leach Processing of N-Reactor Fuel–Cladding Fires," Wallace W. Schulz, ARH–2351 Atlantic Richfield Hanford Company, Richland, Washington 99352 (February 15, 1972).

"Plasma-Mig–A New Welding Process," *Welding Engineer*, 55 (6) 46 (1972).

"Explosive Welding Plugs into Heat Exchanger Tubes," W. R. Johnson, *Welding Journal*, 50 (1) 22–32 (1971).

"Current BWR Fuel Design and Experience," H. E. Williamson and Dana C. Ditmore, *Reactor Technology*, 14 (1) (Spring 1971).

"Welding Nuclear Power Plants," *Welding Engineer*, 56 (8) 33–39 (1971).

"Only Guidance Systems Give You True Automatic Welding," *Welding Design and Fabrication*, 44 (8) 44–45 (1971).

"Control of Distortion During the Furnace Cycle," C. C. Tennenhouse, *Welding Journal*, 50 (11) 701–711 (1971).

"Pulsed Current and Its Applications," Eugene P. Vilkas, *Welding Journal*, 49 (4) 255–262 (1970).

"Stainless Steel Reactor Pressure Vessels," P. J. Karnoski, Jr., W. J. Fretague, Uldis Potapovs and L. E. Steele, *Nuclear Engineering and Design* 11, 347–367, North-Holland Publishing Co., Amsterdam (1970).

"Fabrication of Swelling Resistant Uranium Metal Fuel Rods for the Canadian Whiteshell Reactor, WR–1", W. E. Gurwell, T. B. Correy and J. O. Vining, A.E.C. Research and Development Report BNWL–1129, UC–25, Battelle-Northwest: Richland, Washington (August, 1969).

"The Joining of Nuclear Energy and the Welding Art," A. M. Weinberg, *Welding Journal*, 48 (11) 867–875 (1969).

"Nuclear Components," D. E. Davis and J. M. Krase, *Nuclear Engineering International* (12) 1085–1090 (1969).

"Modern Metal Joining Techniques," Mel W. Schwartz, *Wiley-Interscience*, New York (1969).

"Acoustic Emission Exposes Cracks During Welding ," W. D. Jolly, *Welding Journal*, 48 (1) 21–27 (1969).

"Welding and Brazing of High Temperature Radiators and Heat Exchangers," G. M. Slaughter, E. A. Franco-Ferreira and P. Patriarca, *Welding Journal*, 47 (1) 15–22 (1968).

"Current Density and Anode Spot Size in Gas Tungsten Arc Welding," H. C. Ludwig, *Welding Journal*, 48 (5), 234s–240s (1968).

"The Effects of Varying Electrode Shape on Arc, Operations, and Quality of Welds in 2014–T6 Aluminum," R. A. Chihoski, *Welding Journal*, 48 (5) 210s–221s (1968).

"Study of Dissimilar Metal Joining by Solid State Welding," C. H. Crane, R. T. Torgerson, D. T. Lovell and W. A. Baginski, NASA CR–82460, Final Report, Huntsville, Alabama (October 1, 1968).

"Furnace Brazing Inconel 718 for Nuclear Reactors," Leopoldo Marti-Balaguer and Marshall E. Rinker, *Metals Progress*, 91 (5) 113 (1967).

"Welded Fuel Elements for a Breeder Reactor," *Welding Design and Fabrication*, 40 (10) (1967).

"Irradiation Embrittlement of Welds and Brazes at Elevated Temperatures," W. R. Martin and G. M. Slaughter, *Welding Journal*, 45 (9) 385s–391s (1966).

"Site Assembly: A New Approach for U. S. Reactor Vessels," T. L. Cramer, *Nucleonics*, 24 (11) (1966).

"Measurement and Control of Weld Chamber Atmospheres," D. R. Stoner and G. G. Lessmann, *Welding Journal*, 44 (8) 337s–346s (1965).

"The Effect of Electrode Geometry in Gas Tungsten-Arc Welding," W. F. Savage, S. S. Strunck and Y. Ishikawa, *Welding Journal*, 44 (11) 489s–496s (1965).

"The Welding of Ferritic Steels to Austenitic Stainless Steels," G. M. Slaughter and T. R. Housley, *Welding Journal*, 43 (10) 454s–460s (1964).

"Wave Shape Effect on Alloying and Arc Stability of A-C Tungsten Inert-Arc Welding," Thomas B. Correy, AIEE *Applications and Industry*, (September 1961).

"Permeation of Water Vapor Through Polymeric Films," J. E. Ayer, D. R. Schmitt and R. M. Mayfield, *Journal of Applied Polymer Science*, 3 (7) 1–10 (1960).

"Joining Zircaloy to Stainless Steel," J. B. McAndrew, R. Necheles and H. Schwartzbart, *Welding Journal*, 529s–534s (1958).

"The Capabilities and Commercial Usage of Explosive Welding," Vonne D. Linse, Douglas Laber and E. G. Smith, Jr., Society of Manufacturing Engineers AD70–572.

"Long Bellows Vacuum Chamber Manipulator," Drawing No. H-3-23204, Sheets 1–4, U.S. Atomic Energy Commission, Richland Operations Office, Richland, Washington 99352.

"HE-103-3 Vacuum Manipulator," Vacuum/Atmospheres Company, 4652 West Rosecrans Avenue, Hawthorne, California 90250.

CHAPTER **88**

SHIPS

PREPARED BY A COMMITTEE CONSISTING OF:

H. G. ACKER, *Chairman*
Bethlehem Steel Corp.

B. L. ALIA
American Bureau of Shipping

T. J. DAWSON
Naval Ship Engineering Center

R. A. MANLEY
Newport News Shipbuilding and Drydock Company

R. S. PARROTT
National Steel and Shipbuilding Company

LT. CDR. R. G. WILLIAMS
U. S. Coast Guard

CHAPTER **88**

SHIPS

INTRODUCTION

The problems involved in the design of floating structures are different from those which confront the designers of bridges, buildings and other land structures. The utmost economy in weight of material is important. Any excess of material over the minimum required for structural strength, including a suitable margin for corrosion and wear, diminishes carrying capacity and speed.

Ships differ widely in type and conditions of service. In a barge of simple box section, engaged in smooth water service, the hull stresses are comparatively simple and readily calculable. Ocean-going vessels, on the other hand, are subject to high stresses due to wind and waves as well as to stresses caused by variable weight distribution of the cargo. The problem of economical structural design of such vessels is complex; solutions have evolved largely on the basis of experience, with due regard to scientific method of analysis and evaluation.

Practically all modern vessels are all-welded. Regulatory agencies usually require a few fore and aft strakes of special notch-tough steel to be placed at strategic locations around the main hull. Generally these strakes are now used in lieu of the alternate requirement of riveted "crack arrestor" seams.

There is a trend toward using the higher strength steels, having yield points ranging from 47,000 to 100,000 psi, in the main hull. These steels are being produced in many types or grades having different strength levels and different chemical compositions; they can be furnished in the as-rolled, normalized or quenched and tempered conditions, depending on the grade of steel, the thickness involved and the properties required. With this trend toward using the higher strength steels must go the warning that the welding of the low-alloy steels is much more exacting and complicated than in the case of welding the ordinary carbon structural steels.

GOVERNING RULES AND REGULATIONS

Merchant vessels are constructed in accordance with the requirements established by the U.S. Coast Guard and by the American Bureau of Shipping. Naval combatant vessels and many merchant-type Naval vessels are constructed in accordance with U.S. Navy specifications. The American Bureau of Shipping (ABS) requirements can be found in its *Rules for Building and Classing Steel Vessels*. ABS also publishes separate "Rules" for river service ships, steel barges, offshore mobile drilling units and manned submersibles.

Coast Guard regulations are a part of the Code of Federal Regulations and come under Title 46, Shipping; Chapter 1, Coast Guard: Inspection and Navigation. The sections shown in Table 88.1 refer to welding.

Table 88.1—U. S. Coast Guard regulations* pertaining to welded ships

Sub-Chapter	Subject	Section
D	Tank Vessel	31.10-1, 33.01-5, 35.01-1, 38.05-10
E	Load Lines	43.15-17
F	Marine Engineering	50.25-20, 50.30, 52.05, 52.15-5, 53.13, 54.01-5, 54.05, 54.22, 56.70, Part 57, 59.10
H	Passenger Vessels	72.01-15, 72.01-20, 72.05-10,
I	Cargo and Miscellaneous Vessels	92.01-10, 92.01-15, 98.15-40, 98.20-30, 98.20-70, 98.25-20, 98.25-40, 98.25-90
T	Small Passenger Vessels	177.01-1, 177.10-1
Q	Specifications	160.032-3(f), 160.035-3(e), 160.035-6(e)
Other Sources		
Navigation and Vessel Inspection Circular No.		
7-68	Notes on Inspection and Repair of Steel Hulls	
2-63	Guide for Inspection and Repair of Lifesaving Equipment, p. 4-3 Par. 7, p. 4-7 Par. 5	
11-63	LSTs as unmanned barges; structural reinforcement and drydocking; hull inspection requirements	
1-66	Requirements for Hull Structural Steel—Structural Continuity	
12-69	Special Examination in lieu of drydocking for large mobile drilling rigs	
Equipment Lists	Items Approved or Accepted under Marine Inspection and Navigation Laws CG-190 (Includes list of approved electrodes)	

*From "Code of Federal Regulations," Title 46, Shipping; Chapter 1, Coast Guard; Inspection and Navigation.

Table 88.2—ABS requirements for ordinary-strength hull structural steel

PROCESS OF MANUFACTURE	FOR ALL GRADES: OPEN-HEARTH, BASIC-OXYGEN OR ELECTRIC-FURNACE						
GRADES	A	B	C	CS	D	E	R
DEOXIDATION	Any method	Semi-killed or killed	Fully killed, fine-grain practice[1]	Fully killed, fine-grain practice[1]	Semi-killed or killed	Fully killed, fine-grain practice[1]	Semi-killed or killed
CHEMICAL COMPOSITION (Ladle analysis)							
Carbon, %		0.21 max	0.23 max[4]	0.18 max	0.21 max[5]	0.18 max.[5]	[5]
Manganese, %		0.80–1.10[2] [6]	0.60–0.90[2]	1.00–1.35	0.60–1.40[5]	0.70–1.50[5]	2.5 x C min[5]
Phosphorus, %	0.05 max	0.05 max	0.05 max	0.05 max	0.05 max	0.05 max	0.05 max
Sulphur, %	0.05 max	0.05 max	0.05 max	0.05 max	0.05 max	0.05 max	0.05 max
Silicon, %		[3]	0.10–0.35	0.10–0.35	0.35 max	0.10–0.35	
HEAT TREATMENT			Normalized over 35.0 mm (1.375 in.) thick.[9]	Normalized		Normalized	
TENSILE TEST							
Tensile strength	For All Grades: 41–50 kgs per sq mm[7] or 58000–71000 lbs per sq in. or 26–32 tons per sq in.						
Elongation	For All Grades: 21% in 200 mm (8 in.)[7] (24% in 50 mm (2 in.) or 22% in 5.65 $\sqrt{A}$) (A equals area of test specimen)						
BEND TEST	For All Grades: 180 deg around a diameter equal to 3 times thickness of the specimen.[8]						

Table 88.2 (continued)—ABS requirements for ordinary-strength hull structural steel

GRADES	A	B	C	CS	D	E	R
IMPACT TEST CHARPY STANDARD V-NOTCH							
Temperature					O C (32F)	−10C (14F)	
Energy, avg., min					4.8 kgm (35 ft-lbs)	6.2 kgm (45 ft-lbs)	
No. of specimens					3 from each 40 tons	3 from each plate	
STAMPING	AB/A	AB/B	AB/C (as rolled) AB/CN (normalized)	AB/CS	AB/D	AB/E	AB/R

Notes

1 The requirement for fine-grain practice may be met by either (a) a McQuaid-Ehn austenite grain size of 5 or finer in accordance with ASTM Designation E 112 for each ladle of each heat; or (b) minimum acid-soluble aluminum content of 0.020% for each ladle of each heat or (c) minimum total aluminum content of 0.025% for each ladle of each heat.

2 Upper limit of manganese may be exceeded, provided carbon content plus 1/6 manganese content does not exceed 0.40%.

3 When the silicon content is 0.10% or more (killed steel) the minimum manganese content may be 0.60%.

4 For normalized Grade C plates, the maximum carbon content may be 0.24%.

5 Carbon content plus 1/6 manganese content shall not exceed 0.40%.

6 Where the use of cold-flanging quality has been specially approved (3.1), the manganese content may be reduced to 0.60–0.90%.

7 The tensile strength of cold-flanging steel shell shall be 39–46 kg/mm^2 (55000–65000 psi) and the elongation 23% min in 200 mm (8 in.).

8 The bend-test requirements for cold flanging steel are:
19.0 mm (0.75 in.) t and under . . 180 deg flat on itself.
Over 19.0 mm (0.75 in.) t to 32.0 mm (1.25 in.) t, 180 deg around diam of 1 x specimen thickness.
Over 32.0 mm (1.25 in.) t, 180 deg around diam of 2 x specimen thickness.

9 Grade C plates over 35.0 mm (1.375 in.) in thickness are to be ordered and produced in the normalized condition when intended for important structural parts.

10 A tensile strength range of 41–56 kg/mm^2 (58000–80000 psi) may be applied to Grade A shapes provided the carbon content does not exceed 0.26% by ladle analysis.

Table 88.3—ABS requirements for higher strength hull structural steel

PROCESS OF MANUFACTURE	FOR ALL GRADES: OPEN HEARTH, BASIC OXYGEN OR ELECTRIC FURNACE		
GRADES[1]	AH 33 or AH 36	DH 33 or DH 36	EH 33 or EH 36
DEOXIDATION	Semi-killed or killed[2]	Killed, fine-grain practice[3]	Killed, fine-grain practice[3]
CHEMICAL COMPOSITION (Ladle Analysis)			
Carbon, %	0.20 max	0.20 max	0.20 max
Manganese, %	1.60 max[2]	0.90–1.60	0.90–1.60
Phosphorus, %	0.04 max	0.04 max	0.04 max
Sulfur, %	0.04 max	0.04 max	0.04 max
Silicon, %	0.50 max[2]	0.15–0.50	0.15–0.50
Nickel, %	0.40 max	0.40 max	0.40 max
Chromium, %	0.25 max	0.25 max	0.25 max
Molybdenum, %	0.08 max	0.08 max	0.08 max
Copper, %	0.35 max	0.35 max	0.35 max
Aluminum, %[4]	0.06 max	0.06 max	0.06 max
Columbium (Niobium), %[4]	0.05 max	0.05 max	0.05 max
Vanadium, %[4]	0.10 max	0.10 max	0.10 max
HEAT TREATMENT	None required[5]	None required 25.5 mm (1.0 in.) and under[5, 6] Normalized over 25.5 mm (1.0 in.)[7]	Normalized
TENSILE TEST			
Tensile strength	For All Grades: 50–63 kg/mm^2; 71,000–90,000 psi; 32–41 tons per sq in.		
Yield point, min	For All Grades: 33 kg/mm^2; 47,000 psi; 21 tons per sq in. or 36 kg/mm^2; 51,000 psi; 23 tons per sq in.		
Elongation	For All Grades: 19% in 200 mm (8 in.); 22% in 50 mm (2 in.); 20% in $5.65\sqrt{A}$ (A equals area of test specimen)		

Table 88.3 (continued)—ABS requirements for higher strength hull structural steel

GRADES	AH 33 or AH 36	DH 33 or DH 36	EH 33 or EH 36
BEND TEST	For All Grades: 180 deg around diameter equal to 3 times thickness of the specimen.		
IMPACT TEST STANDARD CHARPY V-NOTCH			
Temperature	None required	−20C (−4F)[8]	−40C (−40F)
Energy, min avg.		2.8 kgm (20 ft-lbs)	2.8 kgm (20 ft-lbs)
No. of Specimens		3 from each 40 tons	3 from each plate
STAMPING	AB/AH 33 or AB/AH 36	AB/DH 33 or AB/DH 36 AB/DHN 33 or AB/DHN 36[9]	AB/EH 33 or AB/EH 36

Notes

1. The numbers following the Grade designation indicate the yield point to which the steel is ordered and produced in kg/mm². A yield point of 33 kg/mm² is equivalent to 47000 psi and a yield point of 36 kg/mm² is equivalent to 51000 psi.
2. Grade AH to 12.5 mm (0.50 in.) inclusive may be semi-killed or killed unless otherwise specially approved, Grade AH over 12.5 mm (0.50 in.) to be killed with 0.15 to 0.50% Silicon and 0.90 to 1.60% Manganese.
3. The requirement for fine-grain practice may be met by one of the following:
   (a) a McQuaid-Ehn austenite grain size of 5 or finer in accordance with ASTM Designation E112 for each ladle of each heat, or
   (b) minimum acid-soluble aluminum content of 0.020% or minimum total aluminum content of 0.025% for each ladle of each heat, or
   (c) minimum columbium (niobium) content of 0.020% or minimum vanadium content of 0.050% for each ladle of each heat, or
   (d) When vanadium and aluminum are used in combination, minimum vanadium content of 0.030% and minimum acid-soluble aluminum content of 0.010% or minimum total aluminum content of 0.015%.
4. Grades DH and EH are to contain at least one of these grain refining elements in sufficient amount to meet the fine grain practice requirement defined in note (3).
5. Columbium (niobium)- treated steels over 12.5 mm (0.50 in.) in thickness are to be normalized. See Note 7.
6. Grade DH, 19.0 mm (0.75 in.) up to 25.5 mm (1.0 in.) inclusive is to be ordered and produced in the normalized condition when intended for the special applications covered in 43.5.3b.
7. Control rolling of Grade AH and DH may be specially considered as a substitute for normalizing.
8. Impact tests not required for normalized Grade DH.
9. The marking AB/DHN is to be used to denote Grade DH plates which have either been normalized or control rolled in accordance with an approved procedure.

U.S. Navy regulations and requirements governing welding are set forth in five basic documents:

1. NAVSHIPS 0900-000-1000–*Fabrication, Welding and Inspection of Ship Hulls.*
2. NAVSHIPS 0900-060-4010–*Fabrication, Welding and Inspection of Metal Boat and Craft Hulls.*
3. NAVSHIPS 0900-014-5010–*Fabrication, Welding & Inspection of Non-Combatant Ship Hulls.*
4. MIL-STD-278–*Military Standard Fabrication, Welding and Inspection of Machinery, Piping and Pressure Vessels for Ships of the United States Navy,*
5. NAVSHIPS 0900-006-9010–*Fabrication Welding and Inspection of HY-80 Submarine Hulls.*

The detail specifications for each class of ships refer to the basic documents where appropriate. Those engaged in Naval shipbuilding and repair should realize that requirements vary for different ships.

BASE METAL

Industry now recognizes the need for commercial steels that combine conventional mechanical properties with a suitable degree of notch toughness and low hardenability in the heat-affected zone (HAZ). Notch toughness cannot be judged by the conventional tensile and bend tests, and special tests such as the Charpy V-notch are used. However, the degree of notch toughness can be regulated by specifying certain limitations on chemistry and steelmaking practice.

Another factor affecting notch toughness is the thickness of the material. If both thick and thin plates are rolled from the same heat of steel, the steel in the thin plates will usually be more notch tough than the steel in the thick plates. To preserve notch toughness in thick plates, the plates must be made from a more notch-tough chemistry than that needed for thinner plates.

The American Bureau of Shipping specifications for hull steel published since 1948 recognizes variations in notch toughness due to thickness of plates by specifying various grades. The present rules include specifications for high tensile steels of 47,000 to 51,000 psi yield point (Tables 88.2 and 88.3). Applications for these steels are given in Table 88.4.

Various commercially available high-strength steels are now being used in some merchant ship applications. They include as-rolled and normalized steels up to about 50,000 psi yield point, and quenched and tempered steels up to about 100,000 psi yield point. Care must be exercised in selecting such steels to ensure both suitable mechanical properties and adequate weldability.

The U.S. Navy Specification MIL-S-22698A, *Steel Plate, Carbon, Structural, for Ships*, is in substantial agreement with the American Bureau of Shipping specifications for ordinary strength hull structural steel. In heavier thicknesses, both specifications contain specific requirements for normalizing heat treatment to refine the grain and enhance the low-temperature notch toughness. The U.S. Navy Specification for a high-tensile steel is MIL-S-16113C, grade HT. This is a carbon steel whose properties are: maximum ultimate tensile strength, 85,000 to 92,000 psi depending on thickness; minimum yield point, 42,000 to 50,000 psi depending on thickness; elongation, 20% in 8 in. The chemistry of the grade HT is shown in Table 88.5.

HT Type II, an improved HT steel with lower sulfur and phosphorus contents and a range for vanadium, is also contained in the specification. The Type II requires ultrasonic testing for quality assurance and is intended for critical applications, such as submarine construction.

Several strength levels of quenched and tempered steels from 50 to 100 ksi yield strength are covered by Naval Material Procurement Specifications. Development work shows promise of 130 to 150 ksi yield strength material for weldments of high toughness in the near future.

U.S. Navy MIL-S-24113 (Ships) covers carbon-manganese steel plates heat treated by quenching and tempering (QT50). Specification properties are shown in Table 88.6. Residual element control is specified. Mechanical properties for 1 1/4 in. plate are: yield strength 0.2% offset 50,000 to 70,000 psi; minimum elongation in 2 in.–20%; Charpy V-notch impact energy 30 ft-lb average at −10 F. This specification also contains a normalized grade similar to ASTM A-537.

HY-80 is a popular material used principally for critical applications such as submarine hulls and certain areas of naval surface ships. HY-80 and HY-100, Specification MIL-S-16216F, are quenched and tempered, high yield strength low-alloy materials with chemical and mechanical properties as shown in Table 88.7.

WELDING

PROCESSES

Manual shielded metal-arc welding (SMAW) continues to be the major welding process used in ship construction. However, many automatic and semiautomatic

Table 88.4—Thickness limitations for ABS grades (in.)

Type	Grade	Ordinary Applications(2)	Special Applications(2)(3)
Ordinary Strength	A	0.50	----
	B	1.00	0.63
	C	2.00(4)	.89 (1.08 when normalized)
	D	1.375	.89
	CS	2.00	2.00
	E	2.00	2.00
	R	0.75(5)	(6)
Higher Strength	AH	0.75(5)	0.75
	DH	2.00	1.08(7)
	EH	2.00	2.00

(1)Plates over 2 in. in thickness are to be produced to specially approved specifications.
(2)Shapes and bars–any of the grades listed for a given strength level are acceptable except that when bars or shapes are used for plate applications, or as other external effective longitudinal material, restrictions noted in the table apply.
(3)Where special material is required in the rules owing to applications in bilge strake, sheer strake, strength-deck hatch-side strake or stringer plate.
(4)Plates over 1.375 in. are to be normalized when used in important structural parts.
(5)Acceptable up to 2.00 in. except for the bottom, sheer strake, strength-deck plating within the midship portion, and other members which may be subject to comparatively high stresses.
(6)Acceptable to .75 in. max. for bilge strake when rule double bottom is fitted.
(7)Plates over 0.75 in. should be normalized.

Table 88.5—Chemical composition of Navy HT steel (by percentage)

Carbon	0.18 max
Manganese	1.30 max
Sulfur	0.05 max
Phosphorus	0.04 max
Silicon	0.15–0.35
Copper	0.35 max
Nickel	0.25 max
Vanadium	0.02 min
Titanium	0.005 min
Chromium	0.15 max
Molybdenum	0.06 max

Table 88.6—Chemical composition of QT50 (by percentage)

Carbon	0.20 max
Manganese	0.90 to 1.45
Phosphorus	0.035 max
Sulfur	0.040 max
Silicon	0.15 to 0.35

processes are continuing to be developed and adopted for use in ship construction. The increased deposition rate obtained with these processes lowers construction costs. Such processes include submerged arc (SAW), gas metal-arc (GMAW), flux cored arc (FCAW), and special vertical welding processes such as electroslag (EW) and electrogas using GMAW or FCAW electrodes. The EW and electrogas process enable heavy-section welds to be made in one pass. Recently, special submerged arc welding techniques for making one-side welds in long lengths have been used with varying degrees of success.

Selection of the process and filler metal to be used depends on many factors including material to be welded, welding position limitations, sequencing of work, equipment portability and economics. A process and filler metal should be selected so that the weld deposit will be compatible with the base metal and will match the mechanical properties and minimum notch toughness expected of the base metal.

Since many of these processes are new or in the developmental stage, procedure qualification tests should be run to demonstrate the capability of that particular welding procedure to produce satisfactory welds. Procedure qualification and approval for each yard will normally be required by the various regulatory agencies prior to production use.

FILLER METAL

Filler metals are qualified for Naval use by tests in a Navy-approved laboratory. Lists of qualified filler metals are issued at intervals by the Naval Ships Engineering Center.

Filler metals for welding on ships classed by the American Bureau of Shipping (ABS) are qualified by demonstration by the electrode manufacturer, in the presence of a Surveyor of ABS. Lists of approved filler metals are issued annually by ABS. The U.S. Coast Guard accepts filler metals approved by ABS without further testing.

Different grades of filler metals are used for various applications in ordinary and higher strength steel hull construction. Specific filler metal grades are selected to produce welds with strength and toughness comparable to the base metal. Specifically designed low-hydrogen manual electrodes, or processes, are usually specified for all higher strength steel welding, for welding heavy sections and castings, and for welding under restrained conditions or other conditions where hydrogen underbead cracking may be encountered. Several grades of electrodes exhibit low elongation, as compared to other electrodes of the same

Table 88.7—Chemical and mechanical properties of HY-80 and HY-100 steels

Chemical composition (ladle analysis)

Element	Percent, maximum, unless a range is shown	
	HY-80	HY-100
Carbon	0.18	0.20
Manganese	0.10–0.40	0.10–0.40
Phosphorus	0.025	0.025
Sulfur	0.025	0.025
Silicon	0.15–0.35	0.15–0.35
Nickel	2.00–3.25	2.25–3.50
Chromium	1.00–1.80	1.00–1.80
Molybdenum	0.20–0.60	0.20–0.60

Residual elements	Maximum percent permitted	
	HY-80	HY-100
Titanium	0.02	0.02
Vanadium	0.03	0.03
Copper	0.25	0.25

[1] Add 0.02% to the maximum for plate 6 in. thick and over.
[2] The percent of phosphorus and sulfur together should not be more than 0.045.

Mechanical properties

	HY-80	HY-100
Ultimate tensile strength, psi	[1]	[1]
Yield strength, 0.2% offset, psi	80,000 to 95,000	100,000 to 115,000
Elongation in 2 in., min %, Type F2 specimen	...	...
Elongation in 2 in., min %, Type R1 specimen	20	18
Reduction in area, min %, Type R1 specimen		
Longitudinal	55	50
Transverse	50	45

[1] To be recorded for information only.

Impact requirements, Charpy V-Notch

Plate thickness (in.) nominal	Specimen size, millimeters	Ft-lb Average of three tests, min	Test (coolant) temperature, degrees F
Up to 1/2 excl.	10 x 5 (1/2 size)	..	−120 + 3
1/2 to 2, incl.	10 x 10	50	−120 + 3
over 2	10 x 10	30	−120 + 3

strength level. These lower elongation electrodes (E6012, E6013, E7014, and E7024) are therefore not approved for any joints in shell plating, strength decks, tank tops, bulkheads and longitudinal members of large vessels because of their lower ductility. In addition, since their penetration qualities are somewhat low, they are not approved for welding on galvanized material.

The U.S. Navy provides an electrode specification, MIL-E-22200/1 with the following classifications and intended uses :

Type MIL-7018 for welding of medium carbon steels, such as classes A, B and C of Specification MIL-S-22698, and grade HT (under 5/8 in. thickness) of Specification MIL-S-16113.

Type MIL-8018 for welding of grade HT (5/8 in. and thicker) plate of Specification MIL-S-16113 and QT50 of Specification MIL-S-24113.

Types MIL-9018, -10018 and -11018 for welding of low-alloy high yield strength steel, such as grade HY-80 of Specification MIL-S-16216.

Type MIL-11018 and -12018 for welding high yield low-alloy steel, such as grade HY-100 of Specification MIL-S-16216.

In addition to these materials for shielded metal-arc welding, the increasing use of automatic and semiautomatic welding processes in shipbuilding applications requires procurement and stocking of suitable welding expendables. Generally such materials are supplied in coils or spools for gas metal-arc, submerged arc, flux cored arc, electrogas and electroslag welding. Due to the diversity of materials joined by these processes, an extensive selection of filler metal chemistries is available. Due to the relative newness of some of these processes, specification coverage is also not always complete; useful and technically acceptable filler metals are sometimes not yet covered by pertinent specifications. In cases where it is desirable to use such material before it is granted specification coverage, it is the usual practice for the fabrication activity involved to request Navy, ABS or Coast Guard approval for such use on an individual basis. The presentation of substantiating data on tests made by the fabricator is usually required to support such requests.

Filler metals used in shipbuilding are described by a variety of specifications issued by AWS, U.S. Navy and ABS. The following are the specifications most commonly used for steel ship construction.

American Welding Society

A5.1	- Mild Steel Covered Arc Welding Electrodes
A5.5	- Low Alloy Steel Covered Arc Welding Electrodes
A5.17	- Bare Mild Steel Electrodes and Fluxes for Submerged Arc Welding
A5.18	- Mild Steel Electrodes for Gas Metal-Arc Welding
A5.20	- Mild Steel Electrodes for Flux Cored Welding

U.S. Navy

MIL-E-15599	- Electrodes, Welding, Covered, Low-Medium Carbon Steel
MIL-F-18251 and MIL-F-19922	- Fluxes, Welding, Submerged Arc Process, Carbon and Low Alloy Steel Applications
MIL-E-19822	- Electrodes, Welding, Bare High Yield Steel
MIL-E-22200/1	- Electrodes, Welding Mineral Covered, Iron Powder, Low Hydrogen Medium and High Tensile Steel

MIL-E-22200/6	- Electrodes, Welding Mineral Covered, Low Hydrogen, Medium and High Tensile Steel
MIL-E-22749	- Electrodes (Bare) and Fluxes (Granular) Submerged Arc Welding, High Yield Low Alloy Steels
MIL-E-23765	- Electrodes and Rods–Welding, Bare, Solid, Mild Steel

American Bureau of Shipping

Rules for Approval of Electrodes for Manual Arc Welding in Hull Construction
Provisional Rules for the Approval of Filler Metals for Welding Higher Strength Steels
Rules for Approval of Wire-Flux Combinations for Submerged Arc Welding
Provisional Rules for the Approval of Wire-Gas Combinations for Gas Metal Arc Welding

All electrodes, but particularly those having low-hydrogen type coverings, must be carefully packaged, stored and handled. Electrode coverings are by their nature highly hygroscopic, and if excess covering moisture levels prevail during welding, moderate to severe weld flaws may accrue. This is particularly true with respect to filler metal compositions designed to deposit high-strength weld metal when used on high-strength materials. These electrodes are baked during manufacture to provide extremely low covering moisture levels, thus ensuring the absence of dangerously high arc atmosphere hydrogen levels during welding. Since the presence of rather small amounts of hydrogen in high-strength weldments can lead to cracking in the weld and heat-affected zone, it is mandatory that precautions are taken to prevent pick-up of atmospheric moisture. Good practice requires that such a program will, as a minimum:

1. Provide a receipt inspection for detection of damaged packaging.
2. Operate a sampling system to monitor electrode covering moisture contents at receipt, after prolonged storage, and at the point of use.
3. Reject for use or rebake electrodes found to exhibit covering moisture contents in excess of the limits established for the electrode type in question.
4. Provide an administrative system that prevents excessive exposure of electrodes to ambient atmosphere conditions during use.

Moisture contents for type E70XX, E80XX and E90XX to E120XX low-hydrogen type electrodes are generally specified as 0.6%, 0.4% and 0.2% respectively. Experience has shown that this can be achieved and maintained by baking the electrodes at temperatures as high as 800 F prior to use, and holding in ovens at 275 to 300 F until used. However, some brands of electrode coverings will not tolerate 800 F without damaging the covering, and baking procedures should be checked with the electrode manufacturer for adequacy of control without damage to the covering.

A four-hour exposure limitation under normal fabrication conditions under cover has been adequate to maintain a 0.2% limit. It has been shown that effective moisture control must be exercised also for granular flux to be used in the submerged arc welding of high-yield strength steels.

WELDING HIGH-TENSILE CARBON STEELS

The normal high-tensile carbon steels, 45,000 to 60,000 psi yield point, are welded using procedures similar to those used with ordinary mild steel.

Low-hydrogen electrodes should be used for welding the higher strength carbon steels and may be used without preheat except when welding thick plates, restrained structures, or during cold weather.

WELDING HIGH-STRENGTH LOW-ALLOY STEELS

When welding low-alloy quenched and tempered steels having yield points of 80,000 to 100,000 psi, special precautions must be taken. Low-hydrogen electrodes must be used. Since preheating is required, 100 to 300 F maximum, and since high heat inputs are detrimental, welding must be done with small electrodes with control over minimum and maximum interpass temperatures. The Navy specifies 55,000 joules per in. maximum heat input with a minimum preheat of 200 F and a maximum interpass temperature of 300 F for welding HY-80 or HY-100 in thickness above 1 1/8 in. Submerged arc may be used, but only with carefully controlled heat input. It is desirable that welding be done in the shop or at least under controlled conditions.

In critical areas it is recommended that tempering beads be used and that large fillet welds be slightly concave. Tempering beads are effective in lowering the HAZ hardness, thereby reducing the tendency for cracking at the toe of the intermediate passes as well as the finish passes. A particular effort should be made to minimize undercutting and to avoid arc strikes on the base plate.

Generally, an electrode is selected so that the deposited weld metal will match the base plate in strength. However, this may not always be necessary, since lower strength welds such as certain fillet welds may be used in some areas with certain reservations, and still provide adequate strength. These lower strength welds may have less tendency toward cracking. When welding low-alloy steels to mild steel, low-hydrogen electrodes suitable for the alloy steels should be used, although it may only be necessary to match the mechanical properties of the mild steel.

Stress relieving generally should not be performed on the low-alloy steels. When welding is done with certain types of electrodes, stress relieving is prohibited. Where possible, the structural design should be such that a minimum of hot forming followed by reheat-treatment would be required.

As for joint design, it is much more important to locate welded joints away from stress concentrations in the case of low-alloy steels than in the case of mild steel. Full penetration welds are preferred to partial penetration and fillet welds.

Joint preparation is usually accomplished by oxygen cutting. Machining and arc gouging are also employed. With arc gouging, a selected technique should be employed to avoid excessive base plate hardening. Quite often a thin hard surface film is formed upon arc gouging and, generally, should be removed, usually by grinding to bright metal.

In all cases, it is strongly advised that the recommendations of both the steel manufacturer and the electrode manufacturer be followed before any fabrication by welding is performed on any higher strength steel. When welding quenched and tempered steels, it is most important to follow the welding procedures given in the specifications.

DESIGN

In the construction of welded ships, the structure should be designed specifically for welding. In the ideal arrangement, every step of the work is planned in advance, to make the best use of welding processes. Consideration should be given to the yard practices and procedures from the preliminary design stage onward.

Hull structures may be divided into subassemblies, which are usually fabricated clear of the shipways, preferably in a covered shop. These subassemblies can be as large as easy handling permits. Welding in the shop has many obvious advantages over welding aboard ship. The work and equipment are protected from the weather, necessary cleaning is more easily accomplished, the welders and welding operators are less subject to discomfort that would tend to impair work quality, and supervision is usually more efficient. Shorter leads from machines to electrodes and easy access to machines for correct setting reduce the welding variables. By the use of positioning equipment, or by turning whole subassemblies, the bulk of welding may be done in the flat position, resulting in higher quality welds produced at greater speed. If several identical units are to be made, it is often useful to prepare jigs or forms on which the subassemblies are set up. Details of welded joints should be arranged to permit the use of a suitable welding procedure and sequence without difficulty.

Subassemblies that must be joined on the shipways should be planned so that the welds will be in the most accessible positions and arranged, wherever practicable, for flat or machine welding. The root side of such welds should be accessible for back gouging and welding, as ship welding practice generally requires joints to be welded from both sides, except where butts are welded to backing strips or to other structural members. In some cases, lap joints may be used to join subassemblies, so that no final trimming is required. However, this practice is not usually followed in the shell and strength decks, or in other primary strength connections.

FRAMING

Framing of plate panels in ship structures includes shell frames, deck beams and bulkhead stiffeners. This framing, in the smaller sizes, may consist of flat bars and inverted rolled angle shapes. In the larger frame sizes, flanged plates or rolled channels and I-beams with one flange removed are used. In the larger tankers and container ships, the deck stiffeners may be large plate slabs 20 in. x 1 3/8 in. or larger, and some shell and bulkhead stiffeners are required to be so large that they must be fabricated from plate.

Framing members are often intermittently welded to the plating, except at the ends and at intersections with supporting members. This saves welding footage but not necessarily total deposited metal, since the required size of the intermittent weld is greater than the required size of a continuous weld. Distortion is less and removals for repair are easier than if continuous welds are used. Intermittent fillet welding should not be used where the welds might be exposed to water or weather; subject to fatigue, dynamic or ballistic loading; or where galvanizing is used to retard corrosion. All intermittent fillet welding should have continuous welds at the ends of the member being welded. The minimum required length of continuous fillet welds at the end connections is different for each application.

In tanks containing liquid cargo or fresh water, all fillet welding is invariably made continuous.

Where framing can be attached by downhand welding, as in subassemblies, it may be advantageous to use automatic machine welding. Long covered electrodes employing gravity and spring-type feeding mechanisms are also used

Table 88.8—Spacing of welds in merchant ships: size, length and spacing of fillet welds in inches

ITEMS	Length of Fillet weld	1-1/2	2-1/2	3					
	Lesser thickness of members joined, inches	Not over 0.19	Over 0.19 to 0.25	Over 0.25 to 0.32	Over 0.32 to 0.38	Over 0.38 to 0.44	Over 0.44 to 0.50	Over 0.50 to 0.57	Over 0.57 to 0.63
	Nominal size of fillet "w"	1/8	3/16	1/4	1/4	5/16	5/16	3/8	3/8
Single Bottom Floors	To center keelson Note: Connections elsewhere to take same weld as floors in double bottom	In accordance with 30.9.4							
Double Bottom Floors	To shell in aft peaks of vessels having high power and fine form	. . .	. . .	6	5	6	6	7	6
	To shell flat of bottom forward (fore end strengthening) and in peaks	. . .	. . .	10	9	10	10	11	10
	To shell elsewhere	*12	*12	12	11	12	11	12	12
	Solid floors to center vertical keel plate in engine room, under boiler bearers, wide spaced floors with longitudinal frames and in vessels where length exceeds 500 ft	In accordance with 30.9.4							
	Solid floors to center vertical keel plate elsewhere, and open floor brackets to center vertical keel	*10	*10	10	9	10	9	10	9

Table 88.8 (continued)—Spacing of welds in merchant ships: size, length and spacing of fillet welds in inches

Double Bottom Floors	Solid floors and open floor brackets to margin plate	In accordance with 30.9.4							
	To innerbottom in engine room	In accordance with paragraph (5) (c)							
	To innerbottom at forward end (fore end strengthening)	*11	*11	11	10	11	10	11	10
	To innerbottom elsewhere	*12	*12	12	11	12	11	12	12
	Wide spaced with longitudinal framing to shell and innerbottom	In accordance with 30.9.4							
	Solid floor stiffeners at watertight or oiltight boundaries	. . .	12	12	11	12	11	12	12
	Watertight and oiltight periphery connections of floors throughout double bottom	In accordance with 30.9.4							
Center Girder	Nontight to innerbottom or center strake in way of engine and to shell or bar keel	In accordance with 30.9.4							
	Nontight to innerbottom or center strake clear of engine	6	6	6	5	6	5	6	6
	Watertight or oiltight to inner bottom, rider plate, shell or bar keel	In accordance with 30.9.4							
Double Bottom Intercostals	Intercostals and continuous longitudinal girders to shell on flat of bottom forward (fore end strengthening) and to innerbottom in way of engines	. . .	6	6	5	6	5	6	5
	Intercostals and continuous longitudinal girders to shell and innerbottom elsewhere and to floors	*11	*11	11	10	11	10	11	10
	Watertight and oiltight periphery connections of longitudinal girders in double bottom	In accordance with paragraph (5) (c)							

Table 88.8 (continued)—Spacing of welds in merchant ships: size, length and spacing of fillet welds in inches

Frames	To shell in aft peaks of vessels having high power and fine form	. . .	. . .	6	5	6	6	7	6
	To shell for 0.125 L forward and in peaks	. . .	. . .	10	9	10	10	11	10
	To shell elsewhere—See Note A	*12	*12	12	11	12	11	12	12
	Unbracketed to innerbottom	Dbl. Cont.	Dbl. Cont.	Dbl. Cont.	Dbl. Cont.	Dbl. Cont.	Dbl. Cont.	Dbl. Cont.	Dbl. Cont.
	Frame brackets to frames, decks and innerbottom	Dbl. Cont.	Dbl. Cont.	Dbl. Cont.	Dbl. Cont.	Dbl. Cont.	Dbl. Cont.	Dbl. Cont.	Dbl. Cont.
	Longitudinals to shell and innerbottom	*12	*12	12	11	12	11	12	12
	Longitudinals to shell on flat of bottom forward (fore end strengthening)	Dbl. Cont.	Dbl. Cont.	Dbl. Cont.	Dbl. Cont.	Dbl. Cont.	Dbl. Cont.	Dbl. Cont.	Dbl. Cont.
Girders and Webs	To shell and to bulkheads or decks in tanks	. . .	8	9	8	9	8	9	8
	To bulkheads or decks elsewhere	. . .	. . .	10	9	10	9	10	9
	Webs to face plate where area of face plate is 10 sq. in. or less	*10	*10	12	11	12	11	12	12
	Webs to face plate where area of face plate exceeds 10 sq. in.	. . .	. . .	10	9	10	9	10	9

Table 88.8 (continued)—Spacing of welds in merchant ships: size, length and spacing of fillet welds in inches

ITEMS	Length of Fillet weld	1-1/2	2-1/2	3					
	Lesser thickness of members joined, inches	Not over 0.19	Over 0.19 to 0.25	Over 0.25 to 0.32	Over 0.32 to 0.38	Over 0.38 to 0.44	Over 0.44 to 0.50	Over 0.50 to 0.57	Over 0.57 to 0.63
	Nominal size of fillet "w"	1/8	3/16	1/4	1/4	5/16	5/16	3/8	3/8
Bulkheads	Peripheries of swash bulkheads	. . .	8	9	8	9	8	9	8
	Peripheries of nontight structural bulkheads	. . .	9	10	9	10	9	10	9
	Peripheries of oiltight or watertight bulkheads	In accordance with 30.9.4							
	Stiffeners to deep tank bulkheads—See Note A	. . .	*12	12	11	12	11	12	12
	Stiffeners to ordinary watertight bulkheads and deckhouse fronts—See Note A	. . .	*12	12	11	12	11	12	12
	Stiffeners to nontight structural bulkheads; stiffeners on deckhouse sides after ends—See Note B	*12	*12	#*12	12	#12	12	#12	12
	Stiffener brackets to beams, decks, etc.	Dbl. Cont.	Dbl. Cont.	#Dbl. Cont.	Dbl. Cont.	#Dbl. Cont.	Dbl. Cont.	#Dbl. Cont.	Dbl. Cont.
Decks	Peripheries of platform decks and non-tight flats — Upper Weld	Cont.	Cont.	#Cont.	Cont.	#Cont.	Cont.	#Cont.	Cont.
	Peripheries of platform decks and non-tight flats — Lower Weld	12	12	#12	12	#12	12	#12	12
	Peripheries of strength decks as required by Table 1, exposed decks, and all watertight or oiltight decks, tunnels and flats	In accordance with 30.9.4							

Table 88.8 (continued)—Spacing of welds in merchant ships: size, length and spacing of fillet welds in inches

Decks	Beams (transverse or longitudinal) to decks	*12	*12	12	11	12	11	12	12
	Beam knees to beams and frames	Dbl. Cont.	Dbl. Cont.	#Dbl. Cont.	Dbl. Cont.	#Dbl. Cont.	Dbl. Cont.	#Dbl. Cont.	Dbl. Cont.
	Hatch coamings to exposed decks	. . .	. . .	. . .	In accordance with 30.9.4				
	Transverses or deep beams to decks in tanks	. . .	8	9	8	9	8	9	8
	Transverses or deep beams to decks elsewhere	. . .	. . .	10	9	10	9	10	9
Foundations	To top plates, shell or innerbottom for main engines and major auxiliaries	Dbl. Cont.	Dbl. Cont.	Dbl. Cont.	Dbl. Cont.	Dbl. Cont.	Dbl. Cont.	Dbl. Cont.	Dbl. Cont.
	To top plates, shell or innerbottom for boilers and other auxiliaries	In accordance with 30.9.4							

*Fillet welds are to be staggered.
#Nominal size of fillet "w" may be reduced 1/16 in.
NOTE A—Unbracketed stiffeners of shell, watertight and oiltight bulkheads and house fronts are to have double continuous welds for one-tenth of their length at each end.
NOTE B—Unbracketed stiffeners of nontight structural bulkheads, deckhouse sides and after ends are to have a pair of matched intermittent welds at each end.
Where beams, stiffeners, frames, etc., pass through slotted girders, shelves or stringers, there is to be a pair of matched intermittent welds on each side of each such intersection and the beams, stiffeners and frames are to be efficiently attached to the girders, shelves, and stringers.
Intermittent welding of members over 0.63 in. is to be specially considered.

Table 88.8 (continued)—Spacing of welds in merchant ships: size, length and spacing of fillet welds in inches

ADDITIONAL WELDING FOR VESSELS CLASSED "OIL CARRIER"

ITEMS								
	Length of Fillet weld	3						
	Lesser thickness of members joined, inches	Not over 0.25	Over 0.25 to 0.32	Over 0.32 to 0.38	Over 0.38 to 0.44	Over 0.44 to 0.50	Over 0.50 to 0.57	Over 0.57 to 0.63
	Nominal size of fillet "w"	3/16	1/4	1/4	5/16	3/8	3/8	7/16
Transverses	Bottom transverses to shell	Dbl. Cont.	Dbl. Cont.	Dbl. Cont.	Dbl. Cont.	Dbl. Cont.	Dbl. Cont.	Dbl. Cont.
	Side, deck and bulkhead transverses to plating	Dbl. Cont.	Dbl. Cont.	Dbl. Cont.	Dbl. Cont.	#Dbl. Cont.	#Dbl. Cont.	#Dbl. Cont.
	To face plates	6	6	6	6	6	6	6
Girders and Webs	Centerline girder to shell	Dbl. Cont.	Dbl. Cont.	Dbl. Cont.	Dbl. Cont.	Dbl. Cont.	Dbl. Cont.	Dbl. Cont.
	Centerline girder to deck	Dbl. Cont.	Dbl. Cont.	Dbl. Cont.	Dbl. Cont.	#Dbl. Cont.	Dbl. Cont.	#Dbl. Cont.
	Bulkhead webs to plating	Dbl. Cont.	Dbl. Cont.	Dbl. Cont.	Dbl. Cont.	#Dbl. Cont.	#Dbl. Cont.	#Dbl. Cont.
	To face plates	6	6	6	6	6	6	6

#Nominal size of fillet "w" may be reduced 1/16 in.

The welding of longitudinals may be as required under Frames or Decks above. In addition, they are to have double continuous welds at the ends and in way of transverses equal in length to the depth of the longitudinal. For deck longitudinals only a matched pair of welds is required at the transverses.

In barges intermittent welds 3 in. long and spaced at 6 in. centers may be used. The fillet size for face plates may be as required for the member to which it is attached.

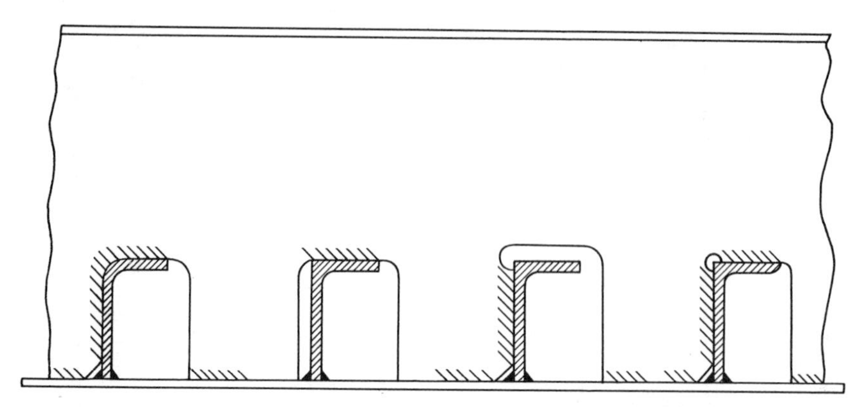

Fig. 88.1.–Typical details for slotting deep-framing members.

for this purpose. Continuous welds of smaller sizes are generally used in lieu of large multipass intermittent welds.

Large web frames or girders intended to support smaller framing members are built up of a web plate and face plate or made of flanged plates. These girders are usually slotted to permit the smaller members, which they support, to pass through. Typical slot cutouts are shown in Fig. 88.1. These slots should be cut with smooth, rounded corners.

Where framing members are welded to plating after completion of the plating joint, such as a butt or seam, no frame scallop in way of this joint is necessary or particularly desirable.

To preclude serious defects in butt joints special care should be exercised in the preweld preparation of butt joints of angles, tees, bulb angles, etc., even where these members appear to be of minor structural importance. Figure 88.2 (a) and (b) shows examples of defective welds.

Standard practice for the arrangement and spacing of fillet welds connecting framing members to plating in merchant ships is included in Table 88.8. Corresponding information for Naval ships is shown in Figs. 88.3 and 88.4.

Fig. 88.2.–Defective welds: (a) bulwark cap rail (b) bilge keel.

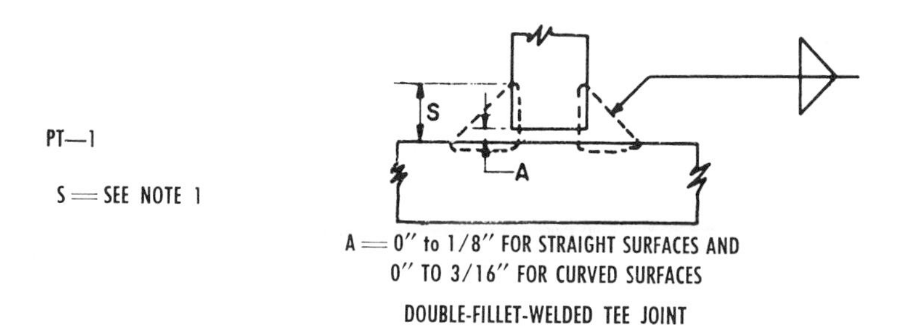

PT—2 (SEE NOTES 2, 3 & 4)

T

S = T MAX.

C

L

S

S

L-C

CHAIN INTERMITTENT—WELDED TEE JOINT

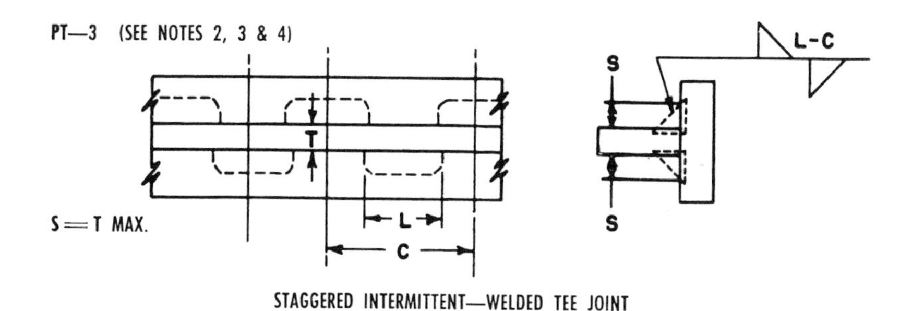

STAGGERED INTERMITTENT—WELDED TEE JOINT

NOTES

1: WHERE A IS GREATER THAN 1/16 INCH, S = THE SIZE GOVERNED BY DESIGN REQUIREMENTS PLUS A.

2: L MIN. = 8 TIMES S, BUT IN NO CASE LESS THAN $1\frac{1}{2}$ INCHES.

3: L MAX. = 16 TIMES THINNER MEMBER, BUT IN NO CASE MORE THAN 6 INCHES.

4: C MAX. = 16 TIMES THINNER MEMBER, BUT IN NO CASE MORE THAN 12 INCHES.

Fig. 88.3.–Welded tee joints (Naval).

DETAILS

The guiding principle in welded ship design is to avoid abrupt discontinuities and to ensure an even stress flow. Good practice has required tying the ends of major strength members gradually into the adjacent structure, as in the case of long superstructures, or changing deck levels. In welded ships, this principle must be applied to the smaller details as well.

EFFICIENCY CHART FOR CONTINUOUS DOUBLE-FILLET WELDED TEE JOINTS MADE BETWEEN MEDIUM STEEL MADE WITH MIL-6011 ELECTRODES. (BASED ON MINIMUM U.T.S. OF 60,000 P.S.I. FOR MEDIUM STEEL AND SHEAR STRENGTH OF 46,400 P.S.I. FOR WELDS.)

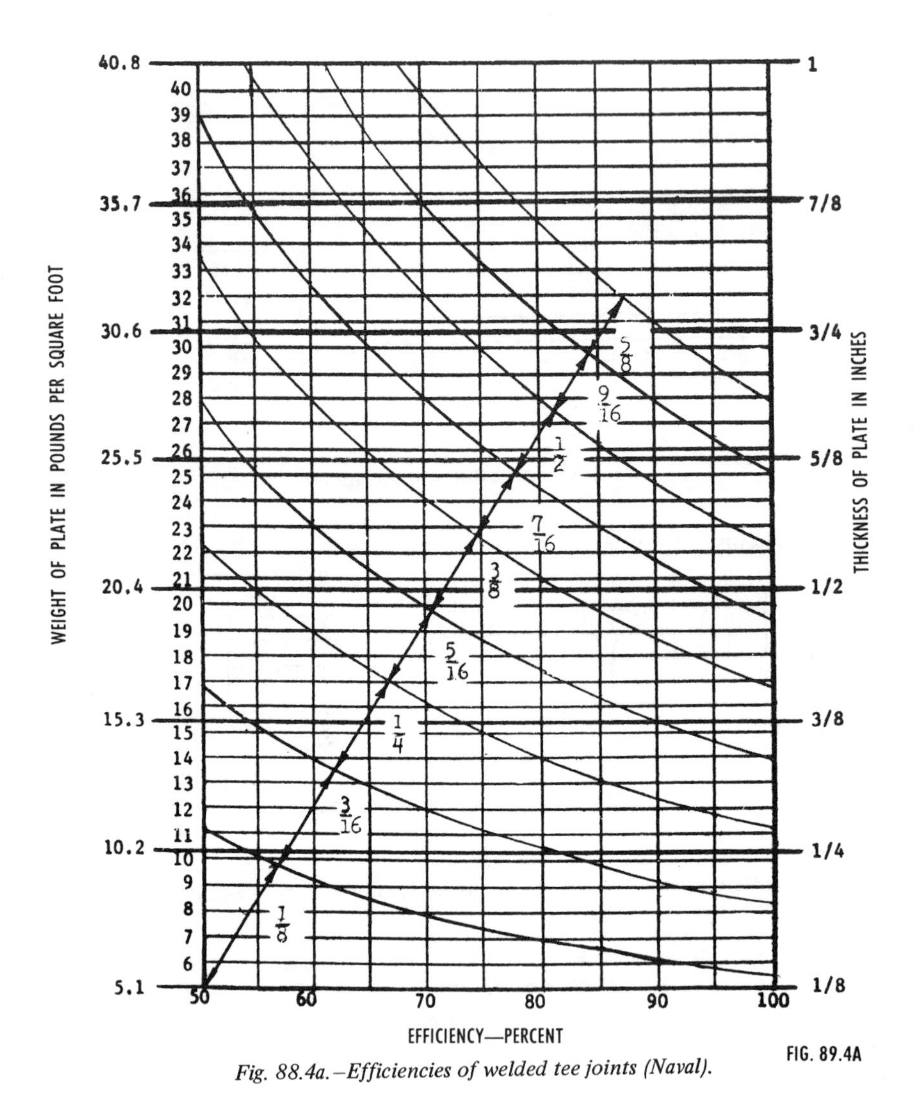

Fig. 88.4a.–Efficiencies of welded tee joints (Naval).

Plug and slot welds should not be used except where absolutely necessary or where it is not practical to use any other type of weld. The generally approved method of making a slot weld is to cut an oval slot big enough so that a fillet weld can easily be made around the inside periphery. Generally, the width of the slot should be at least one and a half times the plate thickness, and the length

EFFICIENCY CHART FOR CONTINUOUS DOUBLE-FILLET WELDED TEE JOINTS MADE BETWEEN MEDIUM STEEL MADE WITH MIL-7016-15 OR MIL-7018 ELECTRODES. (BASED ON MINIMUM U.T.S. OF 60,000 P.S.I. FOR MEDIUM STEEL AND SHEAR STRENGTH OF 58,100 P.S.I. FOR WELDS.)

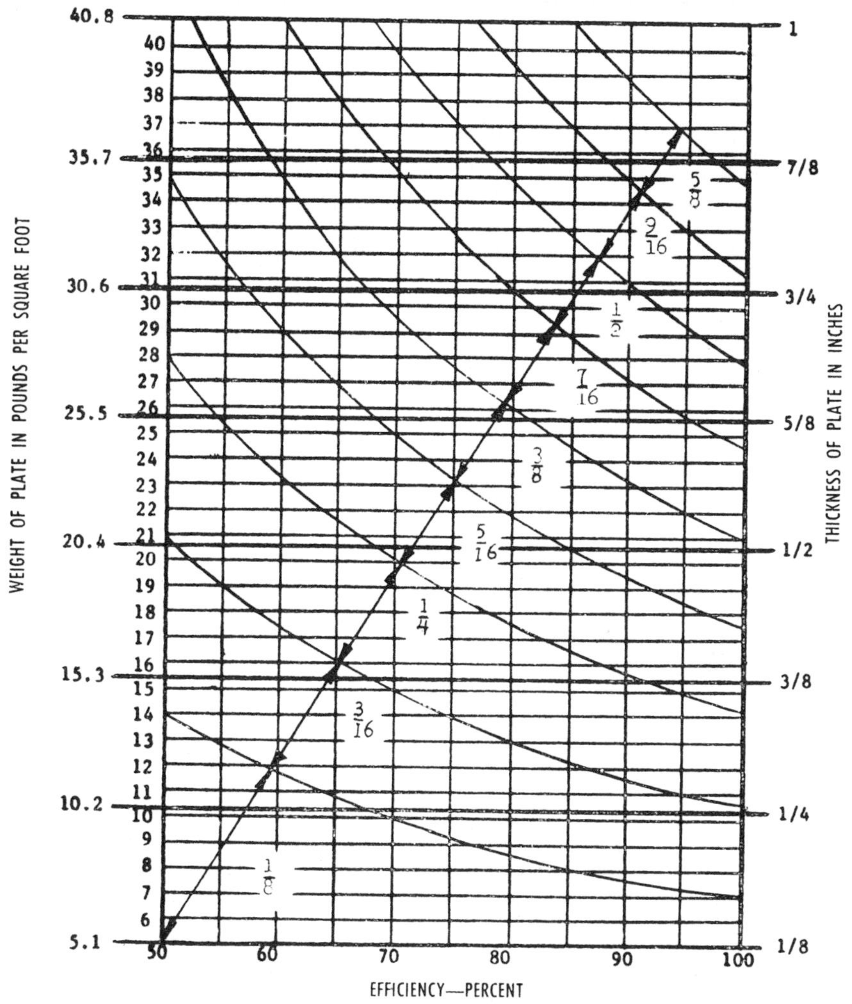

Fig. 88.4b.–Efficiences of welded tee joints (Naval).

should be three times the slot width. A slot weld is sometimes completely filled with weld metal to make the surface flush, as is usually done on rudder closing plates. The oval slot weld is preferred to the circular plug weld.

All openings in any part of the ship should be made with rounded corners. Even small cuts for ventilators, sea chests, etc., should have smooth, rounded

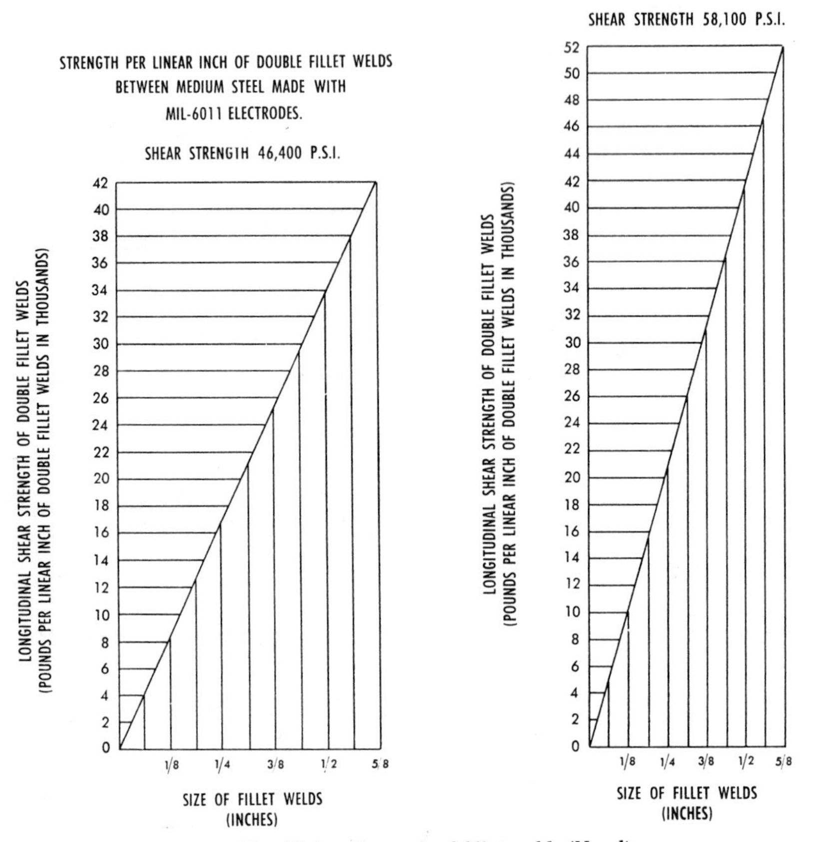

Fig. 88.4c.–Strength of fillet welds (Naval).

corners. At highly stressed points, such as the corners of cargo hatchways, a generous radius, smoothly cut with edges rounded or chamfered, is advisable. The ends of superstructures should be treated similarly by using long, sweeping fashion plates carefully faired into the upper edges of the sheer strake.

Welds in plating should be located at least a short distance away from a geometric change, since even a slight weld defect has the effect of further concentrating stresses at such a location (Fig. 88.5). If the toes of the fillet welds attaching two adjacent members to the surface of a plate are close enough to form a Vee between them, the plate may crack between the welds when under stress. Service records show that care must be taken to avoid even minor discontinuities and to locate welds away from changes in sections, even in seemingly unimportant structural parts. Instances have occurred where a crack started in one part, such as a bulwark, and spread rapidly into the main hull structure from the top of the sheer strake to which the bulwark plate was

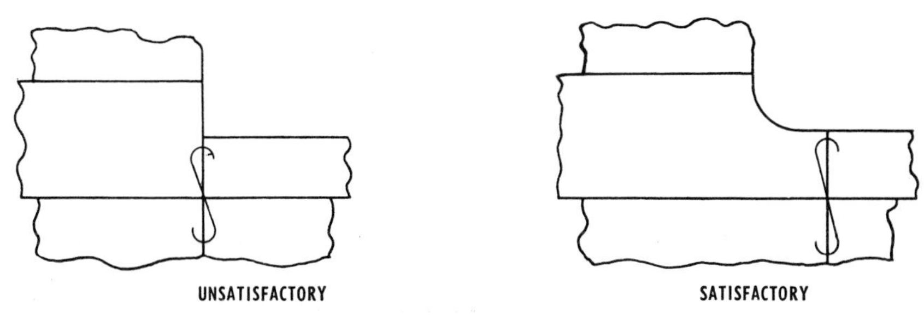

Fig. 88.5.–Hatch corner details.

welded. Many other examples could be cited, but all are basically the same, in that abrupt changes or notch effects are potential sources of trouble.

JOINT DESIGN

The following details of joints are intended as standards to be followed in association with manual arc welding. Other forms of joints may be proposed by individual builders to suit special conditions or special welding processes. In such cases, it may be required that specimen joints be submitted for examination and testing. Such joints are to be prepared by using the same equipment and materials under approximately the same conditions that will prevail in the actual construction. Whenever practicable, all joints should be welded from both sides.

BUTT JOINTS

Manual Welding

Butt joints of plating not more than 1/4 in. thick may ordinarily be manually welded without beveling the plate edges; otherwise, the edges of both plates are to be similarly beveled from one side or both sides to form an included angle of 60 deg. The root face or shoulder should not exceed a 1/8 in. width, and the root opening or gap between plates should not be less than 1/16 in. nor more than 3/16 in., except in the case of single-Vee joints welded in the flat position. Here, tight fits may be used. In all cases, the root of the weld is to be removed to sound metal by an approved method before depositing the closing pass. Where this is impracticable, a substantial backing is to be provided, and the plates are to be beveled to an included angle of not less than 45 deg, and spaced to leave a gap at the root of the weld sufficient to ensure a full-penetration weld. When it is intended to prevent the root of a butt weld in way of a riveted seam lap from fusing to the other plate, separation may be effected by a steel shim or woven glass tape (electrical type). The use of copper for this purpose is not recommended.

Automatic Welding

When an approved automatic machine process is used to weld both sides of a butt joint, the plate edges need not be beveled for thicknesses not greater than 5/8 in. If manual welding is used to make the backing pass for automatic welding, the joint should be beveled for the manual weld to a minimum of 60 degrees. If other details are proposed, such details should be submitted specially for approval to the cognizant regulatory body.

With submerged arc welding, thick plates are beveled from one or both sides to an included angle of 60 deg. The root or land should not exceed 1/4 in. Where the metal is deposited from one side only, plates are usually beveled to an included angle of 30 deg with 1/4 in. opening at the root.

With electroslag and electrogas welding, consideration must be given to the specific application and mechanical properties of the welds and heat-affected zones.

COMBINATION OF RIVETING AND WELDING

If welding is done after riveting, rivets immediately adjacent to a weld may be loosened by heat and should be checked for tightness.

Extensive tests have been made to determine the extent to which the welding of framing and other attachments influences the rivets in adjacent completed joints and the degree to which this practice may affect residual stress. As a result, approval has been given in many cases to the riveting of longitudinal seams or butts in plating prior to the attachment of framing by welding, usually with the following restrictions: rivets immediately adjacent to the weld (usually within 6 in.) are omitted before welding, or when such rivets are driven before welding, they are to be carefully inspected after welding to detect any looseness and redriven if necessary.

In general, reaming and riveting should be done after butt welding has been completed. In the case of a groove weld that crosses a riveted seam, the rivets for about 12 in. on each side of the weld should be driven after the weld has been made.

Although welding a transverse butt joint causes uniform bodily movement of a free plate, it is not always necessary to omit shop-punched holes, if the proper allowance for this shift is made in layout. Following the welding of the butt, the plate may be securely bolted and regular reaming and driving practice carried out.

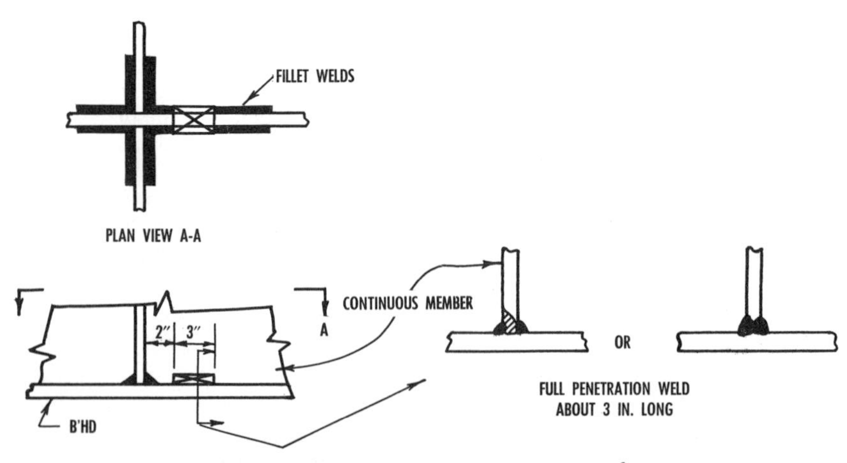

Fig. 88.6.–Welded stop at intersecting members.

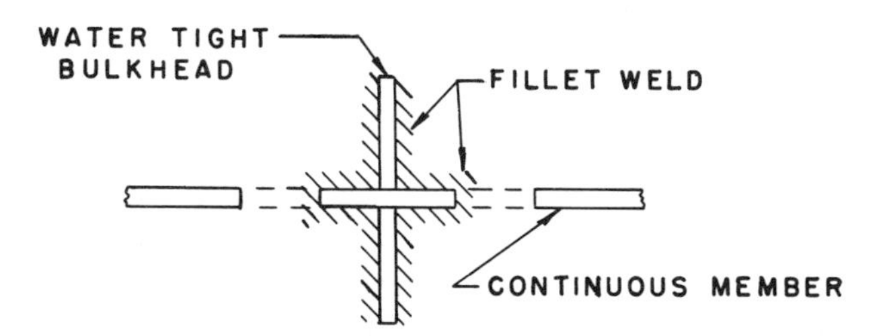

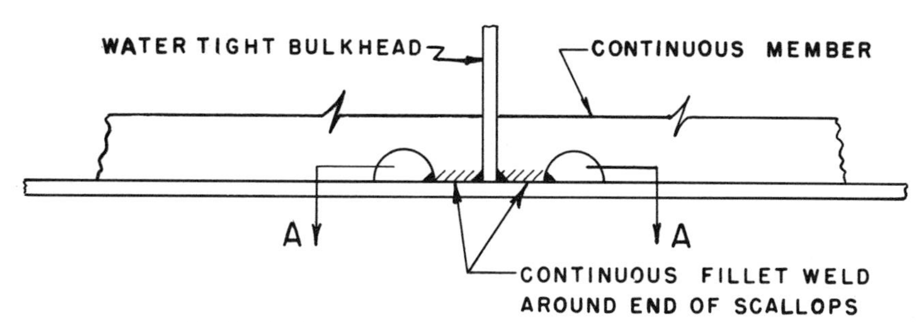

Fig. 88.7.–Alternate welded stops at intersecting members.

Oil and water stops designed to prevent the uncontrolled spread of liquids along fillet-welded or riveted boundaries between the faying surfaces may be of the welded type, although in way of riveted joints, impregnated canvas stops are sometimes considered more dependable. An effective stop at intersecting members is the full penetration welded stop (Fig. 88.6). A common type of stop can be made by simply providing a scallop a short distance from the water-tight bulkhead (Fig. 88.7) and welding around the scallop. In way of riveted or lap welded seams, a full penetration single-Vee stop is generally used (Fig. 88.8).

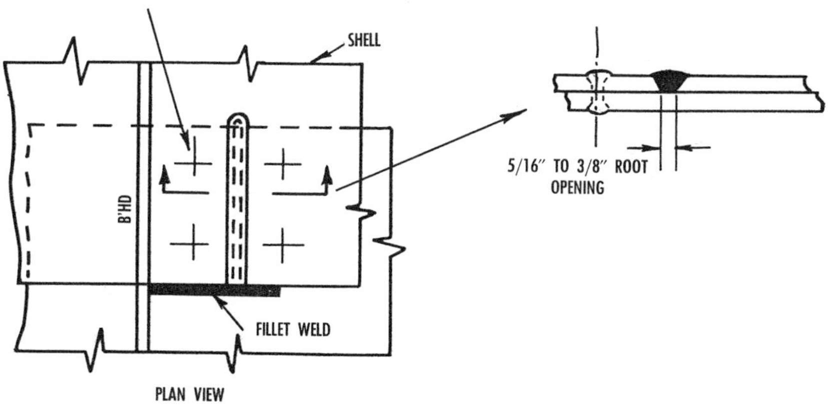

Fig. 88.8.–Welded stop at riveted seam lap.

WORKMANSHIP

Shipbuilding workmanship is affected by so many factors and not just limited to welding. All other operations prior to welding may ultimately affect the quality of the welding.

To avoid difficulty in welding, the molds and templates should be accurate, the material cut or machined to close tolerances and the edges prepared to the correct bevel. Subassemblies and units should be made on well-supported floors or skids that hold the intended shape. When completed, they should be handled and stored with adequate care to prevent damage to prepared edges, which must match adjacent structures.

In all welds made from both sides, the root of the weld deposited from the first side should be ground, chipped or gouged to sound metal prior to deposition on the second side using a technique which provides adequate geometry for further welding (Fig. 88.9).

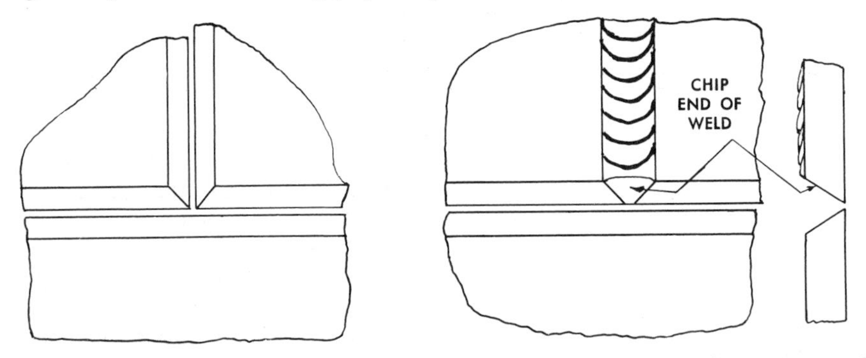

Fig. 88.9.–Welding at the intersections of two joints. Before welding (left). Treatment of first weld (right).

Fitting up the part and assemblies on the shipways necessitates tack welding and various fittings such as clips and strongbacks. These should be arranged not to interfere with welding operations or restrain the parts too greatly against contraction from welding. Some freedom of movement should be allowed (Fig. 88.10). Tack welds should be removed, or the final welds should be properly fused to the tacks; they should be used no more than is absolutely necessary. Cracked tack welds must be removed.

Preheat may be necessary prior to tack welding when the materials to be joined are highly restrained. When welding the higher strength steels, particularly those that are quenched and tempered, the same preheats specified in the welding procedures should be followed when making any type of permanent weld.

Preparation of joints on the shipways must often be done without the use of mechanical equipment, and the quality of the work will depend upon the skill of the cutters and chippers. Guiding devices for manual oxygen cutting are effective if properly used. Errors such as cutting beyond the end of a cut, or chipping an irregular bevel or one too wide at the root, may cause conditions that result in defective welds (Fig. 88.11).

When proper planning and preparation have established a procedure with adequate supervision, the welder himself must be considered. The work should

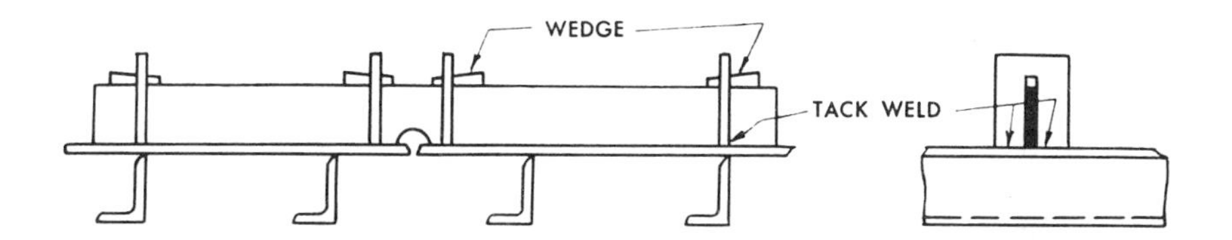

FLOATING STRONGBACK

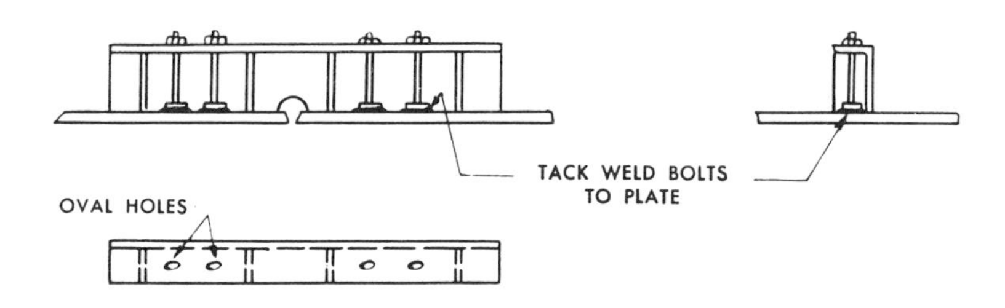

BOLTED ANGLE STRONGBACK

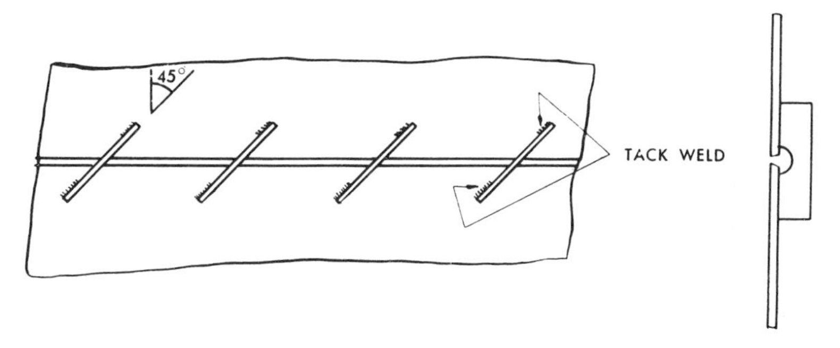

SATISFACTORY ARRANGEMENT OF FLAT BARS

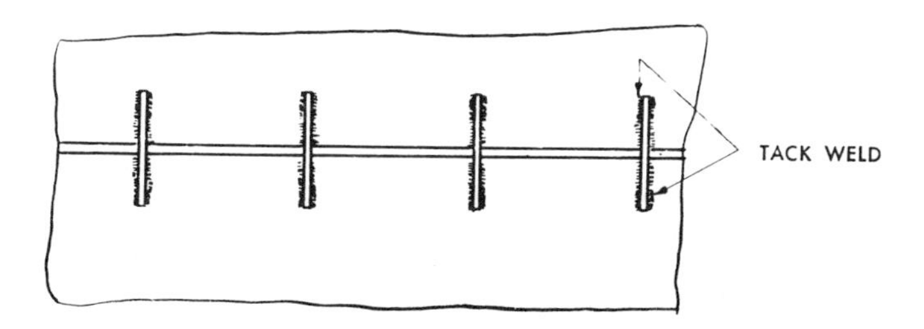

UNSATISFACTORY ARRANGEMENT OF FLAT BARS

Fig. 88.10.—Use of strongbacks to minimize restraint.

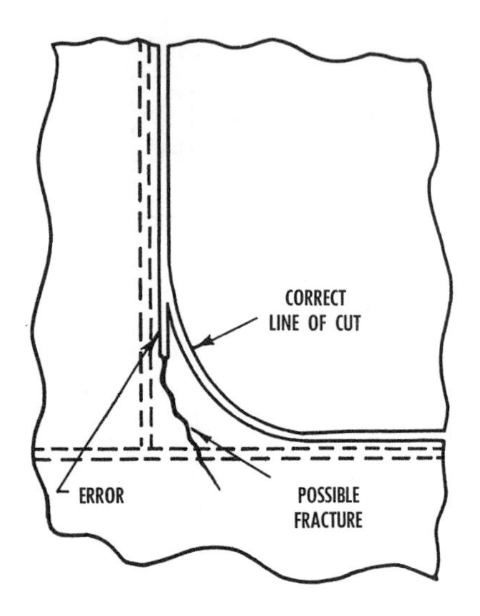

Fig. 88.11.–Notch effect produced by cutting error.

be positioned to make him comfortable. Adequate ventilation, especially in confined spaces, and good accessibility are requisites for comfort and consistently good workmanship by men who must wear shields or helmets and heavy, cumbersome clothing. These factors contribute to increased production as well as good workmanship.

Electrodes should be properly stored, or conditioned if necessary, to minimize moisture pickup. Adequate maintenance of welding equipment is also required to assist the welder in welding to prescribed parameters.

It is imperative that adequate shelter from the elements be provided during welding operations, not only from the standpoint of the physical comfort of the welder but also from the purely mechanical aspects of assuring weld soundness. SMAW, GMAW or GTAW in exposed locations under high wind conditions is almost certain to result in excessive porosity, due to the disturbance of the shielding. Welding under rainy conditions is equally undesirable.

Some leeway for judgment may be allowable in fabricating carbon steels. Since welding of high-strength materials almost always involves the application of preheat and interpass temperature control, it will probably be necessary to provide suitable shelter from atmospheric conditions. In both cases, the edges and surfaces to be welded should be cleaned and dried to minimize the inclusion of foreign matter and entrapped hydrogen which would be injurious in the weld metal.

The most difficult and important joints in the ship should be assigned to highly skilled welders. The supervisor should be familiar with the capabilities of the men under his supervision, to aid in controlling the quality of work. However, a weld attaching even a small fitting to main structural strength members requires the same quality and soundness as joints in the strength members themselves. Failure of an apparently unimportant weld has frequently led to fracture of deck or shell plates.

When errors are made in cutting or fitting, they may be corrected if suitable methods are used. If the edge or end of one member does not meet the surface of a plate to which it should be welded, a liner may be used (Fig. 88.12). However, liners should not be used where the joint transmits tension. In butt

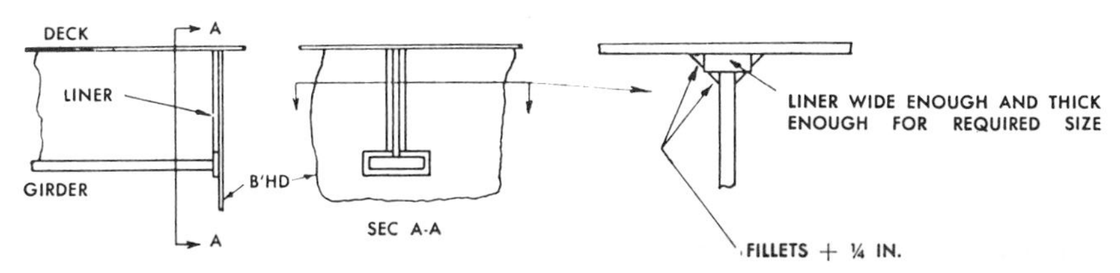

Fig. 88.12.–Use of liners to correct poor fit-up.

joints that cannot be made to fit within specified root opening tolerances, one or both edges of the plates may be built up, using a temporary backing strip if necessary, before attempting to join the plates by making the butt weld. Generally, when the gap is greater than the plate thickness, a plate insert should be considered. Long narrow insert plates should be avoided. Where defective welding is removed, it is essential to prepare the ends of the cavity by chipping or gouging to form a groove suitable for sound welding. Any of the methods of subsurface inspection, although used only as a spot check, is effective in maintaining the quality of workmanship.

PEENING

There have been cases where attempts were made to caulk or peen leaky welds to make them tight. This procedure is not permissible because internal cracks could be covered up by bobbing material over them. Such cracks remain in the structure as potential triggers for subsequent structural failures. Cracks in any weld should be carefully removed and rewelded.

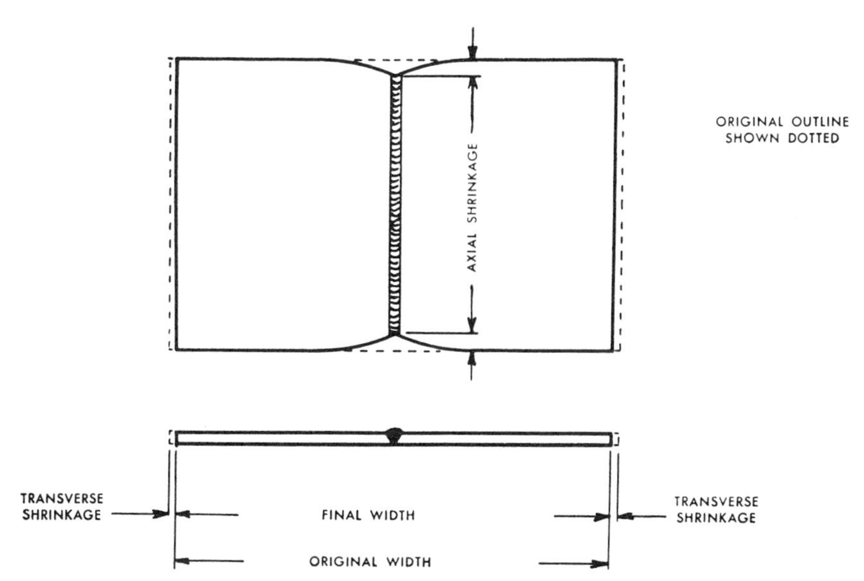

Fig. 88.13.–Shrinkage effects of cooling after welding. The same effects, to a lesser degree, are evident in fillet welding.

Although peening is not recommended, it may, when properly applied, be used to correct distortion or to minimize residual stresses. Peening should not be done on a single bead weld or on the first and last layer of a multiple layer weld. Peening should be done immediately after depositing and cleaning each pass of the weld deposit.

DISTORTION AND CRACKING

Investigations of structural failures in welded ships have indicated that the locked-in stresses did not contribute materially to these failures. Also, longitudinal residual welding stresses cannot be altered by variations in welding procedures. Reaction stresses in the plate perpendicular to the weld can be controlled by adjusting the assembly and welding sequence.

The participation of shrinkage stresses in the phenomena of cracking and distortion is not fully understood. Attention should be focused on control of distortion and cracking. To prevent cracking of the weld metal when welding heavy sections or constrained assemblies, local preheating of the base metal is advisable. On heavy weldments and castings, preheating is always advisable.

A proper sequence of welding is one of the most important factors in welding procedure. The objectives in setting up a welding procedure are: to avoid excessive restraint against weld shrinkage, to minimize distortion and to facilitate the deposition of sound weld metal. These objectives are not always compatible, and the welding engineer may have to seek a satisfactory compromise. No universal rule can be stated.

The weld metal and adjacent heat-affected zone of the base metal contract on cooling and tend to draw closer together (Fig. 88.13). Mass, shape, structural attachments and previous welding may tend to prevent this shrinkage. To minimize the tendency toward distortion in lighter plates, or cracking in heavier ones, some special techniques are employed. Skip, backstep, block and cascade

Table 88.9—Allowances for shrinkage

Butt Welds	
Transverse	1/16 to 3/32 in. for all thicknesses
Longitudinal	
Over 1/2 in. thick	1/32 in. in 10 ft
3/8 to 1/2 in. thick	1/32 to 1/16 in. in 10 ft
1/4 to 3/8 in. thick	1/16 to 3/32 in. in 10 ft
1/4 in. and thinner	1/16 to 1/8 in. in 10 ft
Fillet Welds	
Tucking Allowance*	
Over 1/2 in. thick	No allowance
3/8 to 1/2 in. thick	1/64 in. each stiffener
1/4 to 3/8 in. thick	1/32 in. each stiffener
1/4 in. and thinner	1/16 in. each stiffener

*Developed for use on flat plates with continuously welded stiffeners. Intermittent welding will result in about one-half the tucking tabulated above, when the weld lengths and spacings conform to the requirements for shipbuilding.

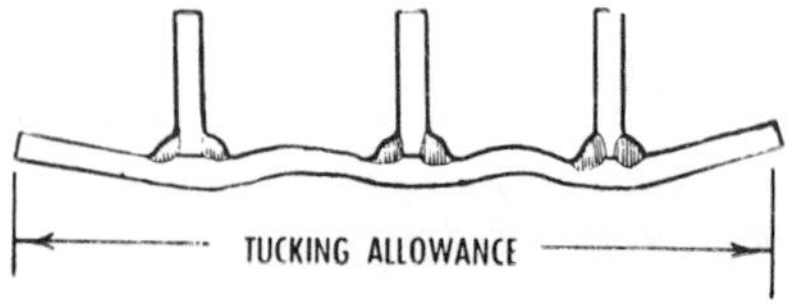

TUCKING ALLOWANCE

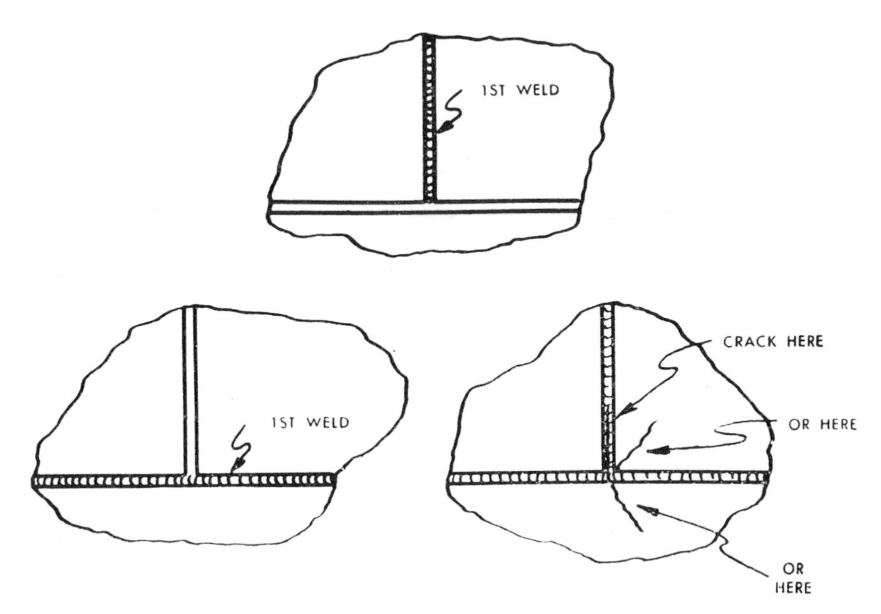

Fig. 88.14.–Sequence for welding junctions of two joints: (a) correct sequence (b) incorrect sequence (c) probable result of incorrect sequence.

welding, intermittent welding, prebending and peening have each proved helpful under certain circumstances.

Since shrinkage cannot be avoided altogether, allowance must be made for loss of dimension. Another practice is to allow excess on the plates and structure at the end away from midship for trimming as work progresses. The scales used by one shipyard are given in Table 88.9. To minimize shrinkage and distortion effects, certain general rules of welding sequence have become accepted on the basis of long experience. These rules are based on the principle of welding toward free ends, or in the direction of greatest freedom. Particular attention must be paid to weld junctions.

The one basic rule that should not be broken in this regard is illustrated in Fig. 88.14. The application of this rule to plate sequence is shown in Fig. 88.15. It may be extended, as shown in Fig. 88.16, almost without limit. When used extensively through the midship body, this sequence results in a marked tendency for the ship to rise from the keel blocks at bow and stern. This tendency may be relieved by welding a complete band around the ship and

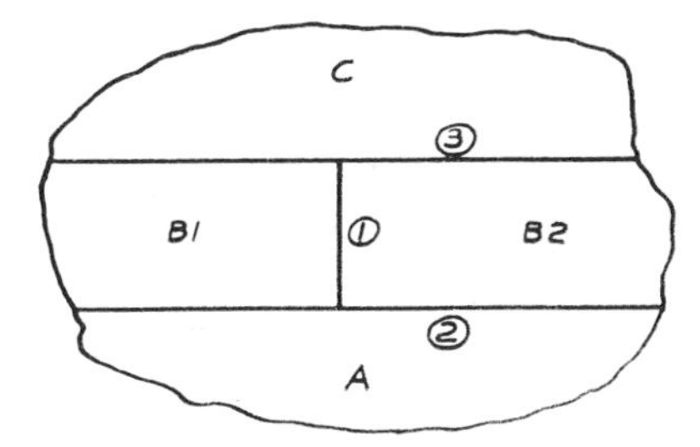

Fig. 88.15.–Proper sequence for welding plate: (1) butt joint complete (2) inboard seam weld complete (3) outboard seam weld complete.

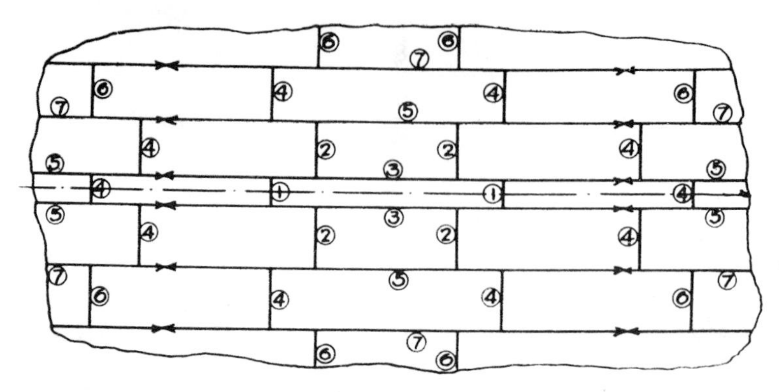

Fig. 88.16.–Welding sequence for bottom shell with staggered butts (butts in line simpler).

proceeding forward and aft, limiting the completed welding by a vertical cross section instead of permitting the bottom to progress ahead of the upper deck. The use of subassemblies and panels facilitates this method.

Nearly all plated structures in shipbuilding are complicated by the attachment of frames, stiffeners, bulkheads, etc., to the plating. The extension of the general sequence in Fig. 88.15 to include this framing is given in Fig. 88.17. The idea is to permit most of the shrinkage to take place before the plates are tied together at their edges. On small plates, where this method becomes less practical, the complete framing may be welded in steps 3 and 7 of Fig. 88.17 instead of just the unwelded crossovers. One advantage of this welding sequence for plates is that the framing may be welded independently of the butt and seam welds, as long as the plates are not welded together. This permits a greater number of welders to be put to work and reduces the overall time of welding.

The illustrations so far have shown the butts staggered. When butts are in line, subassemblies can be welded completely in the shop. This permits trimming to exact size after shrinkage has taken place. These subassemblies are then treated as single plates.

If subassemblies cannot be made on the ground because of shape or other peculiarities, the sections can be welded as panels on the ship using the described sequence. The seams or butt joints can be welded complete, with regard to local sequence, and all the framing can be welded to the panel complete within about 12 in. of all panel edges. This permits many welders to be put to work without damage to the structure. Since there are only a few tie-in welds to be made, the overall structure shrinks very little, thus reducing the tendency of the ship to rise at the bow and stern. This sequence is illustrated in Fig. 88.18.

The basic principle illustrated in Fig. 88.14 should be strictly followed to obtain sound welds and avoid cracking. The illustrations are all extensions of this basic principle, showing simplifications that can be made in complicated structures. Attempts to break welding sequences into small individual steps lead to such complicated practices that they cannot be carried out in the field. The sequence should be simple and practical, with steps large enough to permit a reasonable number of welders to work.

All of the foregoing rules pertaining to welding sequence may be summarized as follows:

1. Do not weld across an open joint, i. e., do not weld a seam across an unwelded butt joint.

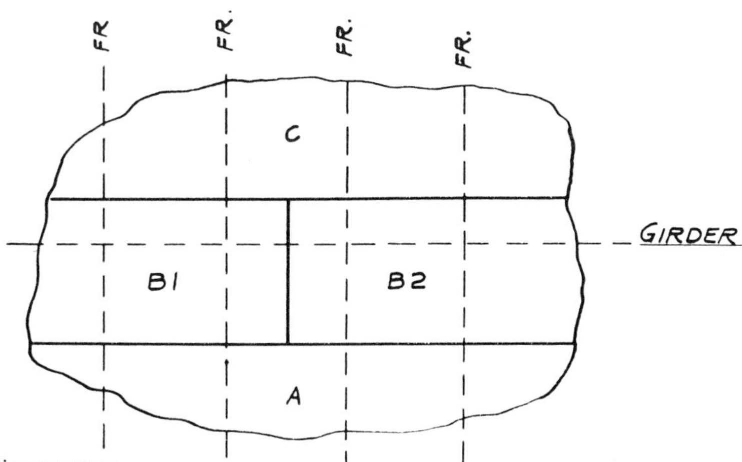

Welding sequence:

1. Weld the four frames and corresponding length of deck edge to side shell plates to within about 12 in. of all unwelded butts and seams
2. Weld butt complete
3. Weld unwelded portion of deck in way of butt
4. Weld inboard or lower seam to point 12 in. from next butt
5. Weld unwelded portion of frames in way of seam
6. Weld outboard seam to point 12 in. from next butt
7. Weld unwelded portion of frames in way of completed seam

Fig. 88.17.—Welding sequence for butts in line (with underframing).

2. Weld as much of the structure as possible in subassemblies before tying them into the ship.
3. Follow a sequence that will minimize distortion and ensure good welds.

FAIRING TOLERANCES

The following standards for the fairing of welded structures are reproduced from NAVSHIPS 0900-000-1000, *Fabrication, Welding and Inspection of Ship Hulls*.

Unfairness of welded plating shall not exceed the tolerances shown in Fig. 88.19. Excessive deformation in alloy steels shall be faired by releasing the joints contiguous to the deformation, fairing by strong backing only and then refastening the joints.

Fig. 88.18.—Welding sequence for butts in line: (1) Weld seams in panels P1 and P2. (2) Weld all to within 12 in. of panel edges. (3) Treat panels as individual plates in over-all sequence.

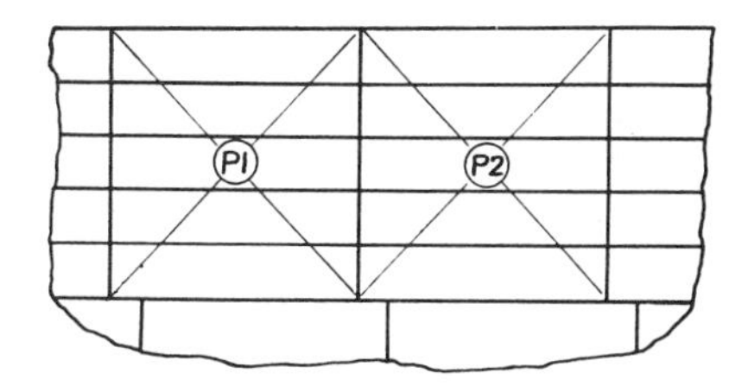

Medium- and high-tensile steel plating may be faired by the application of local heat; however, this procedure on high-tensile steel shall be kept to the absolute minimum due to high local stresses and the possible degradation of the mechanical properties of the material.

From the viewpoint of structural efficiency, plate unfairness is more serious with transverse framing than with longitudinal framing. It is especially important to have minimum distortion of transversely framed strength decks, bottoms and inner bottoms.

The attention of all assembly trades should be drawn to the necessity for ensuring structural fairness, and welding supervision should be cautioned to delay joining of components displaying out-of-tolerance unfairness until corrective measures have been taken.

Additional information on hull construction details can be found in Chapter III of AWS Hull Welding Manual, D3.5.

INSPECTION

Supervision and inspection are necessary to ensure high quality workmanship. In the case of groove welds, a considerable portion of this may be visual inspection of groove preparation before welding. Root openings that are too small to permit complete penetration at the root are likely to result in inadequate gouging, back chipping and incomplete fusion of the finished weld. Root openings that are too wide may also be the cause of unsatisfactory welding if the sides of the groove are not built up with care before bridging across the opening. After welding is completed, visual inspection is effective only in determining acceptability of weld contour, undercut, etc. Under some conditions, a more searching means of inspection may be desirable to check the weld quality throughout the joint.

The principal methods of inspection of finished welding currently employed in shipbuilding are radiography, ultrasonic, magnetic particle and dye penetrant inspection. Utrasonic inspection offers the advantages of no radiation hazard to personnel and an ability to locate the exact position of flaws despite material thickness; the latter factor aids greatly in effecting repairs where needed. The advantage of radiography is that it examines a larger sample of the weld (usual film length is 10 to 17 in.). The magnetic particle inspection is limited to defects close to the surface. The dye penetrant method is only good for surface inspection.

The more serious types of defects likely to be found are cracks, incomplete fusion and excessive slag. These defects are common causes of weld failure. They may also provide seepage channels that destroy water-tightness. Radiographic examination is not always effective in detecting cracks. This should be borne in mind when judging the acceptability of welds by radiography. Magnetic particle or dye penetrant inspection is frequently used to provide surface inspection in conjunction with radiography when the presence of cracks is suspected.

When defects of an unacceptable nature, as determined by the governing standard, are located, they must be removed or rewelded. Removal of the defects to their full extent is usually checked by magnetic particle inspection of the excavation; the adequacy of the final repair is determined by sample radiographs.

The Navy has developed nondestructive test procedures. MIL-STD-271 Military Standard Nondestructive Testing Requirements for Metals, and accep-

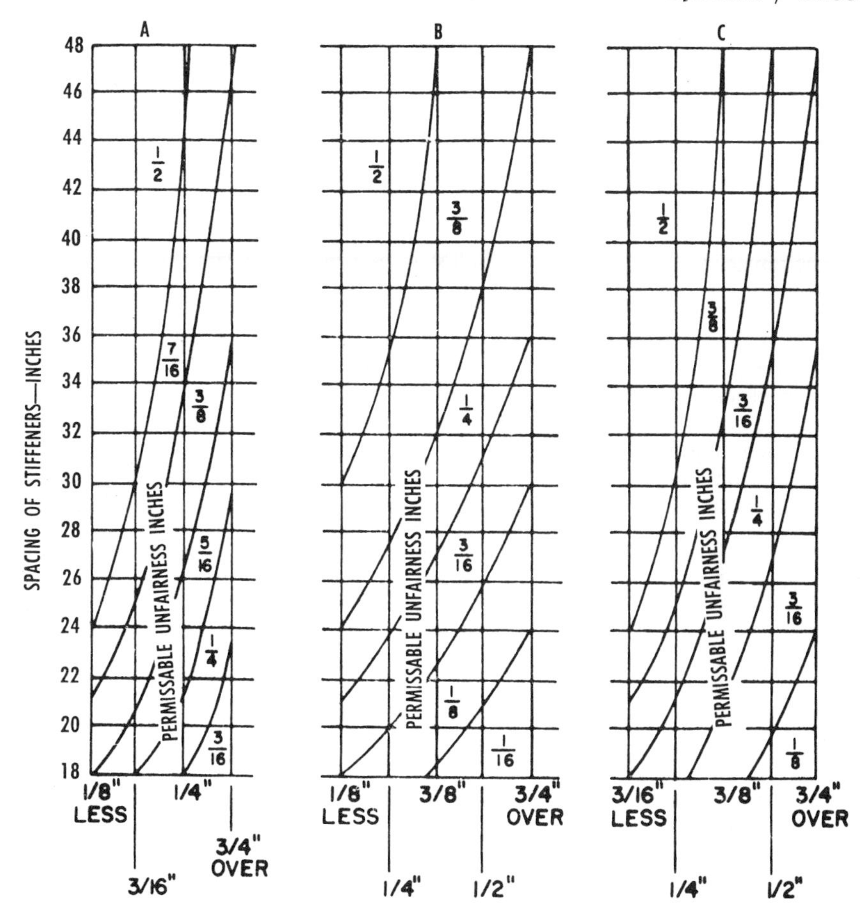

A TO BE USED ON LIVING QUARTERS AND LONGITUDINAL BULKHEAD PLATING IN OTHER BULKHEADS UNFAIRNESS OF 1″ PER PANEL WIDTH WILL BE PERMITTED.

B TO BE USED ON DECK PLATING.

C TO BE USED ON SHELL PLATING.

(IN APPLYING THE ABOVE TOLERANCES. THE UNFAIRNESS OF THE PLATING SHALL BE MEASURED ACROSS THE MINOR DIMENSION OF THE PANEL)

Fig. 88.19.–Tolerance for fairness of welded structure. (Permissible unfairness in inches is indicated within curves.)

tance/rejection standards for nondestructive tests such as NAVSHIPS 0900-003-9000 Radiographic Standards for Production and Repair Welds. There are several grades of which Grade III is intended for hull structural welding of surface ships. The Navy has also developed a standard for ultrasonic weld inspection that is acceptable as an alternate method for radiography: NAVSHIPS 0900-066-3010 Ultrasonic Inspection Procedure and Acceptance Standards for Hull Structure Production and Repair Welds. The American Bureau of Shipping has established radiographic standards that are in their publication *Requirements for Radiographic Inspection of Hull Welds.*

REPAIR INSTRUCTIONS

The repair work to be done on a hull structure should be carefully studied, and a procedure for doing the work laid out in advance. In making repairs, a number of points should be kept in mind.

1. The fractures in hull plates, etc., usually start at localized concentrations of stress. In the preliminary inspection, the first thing to be determined is whether or not the fracture started in a notch or sharp angle (stress raiser), and, if it did, to provide for the elimination of these factors.
2. When large breaks have occurred, considerable material should be removed and new plates or frames inserted; the repairs may therefore involve appreciably more restraint with less latitude for proper welding and direction sequences than in new construction. In cases where welding is done under unusual restraint, such as in welding heavy plates or castings, it is recommended that low-hydrogen electrodes be used, with preheat if necessary. Careful planning is absolutely essential.
3. In the repair of cracks in the plating of vessels where it is not intended to add new material, somewhat different conditions prevail and the following should be observed:
   a. Locate the ends of the crack. At a point not less than one plate thickness beyond the apparent end of the crack, drill a hole of diameter about equal to the thickness of the plate.
   b. After Veeing out, if the joint has a root opening too wide for closing with the first bead, build up the sides of the joint with light beads until a standard joint is prepared. Then weld as usual.
   c. In general, large pieces of plating should not be removed unless the crack has opened up to a large extent or there is definite evidence of deterioration or poor quality in the fractured plate. The plate cracked should not be taken as prima facie evidence that the steel is inferior in quality to other parts of the structure.
4. When possible, welding should be avoided in cold weather, particularly when the temperature is below freezing. However, the effects of cold weather may be overcome by preheating the part to about 150 F before welding. Heavy sections should be preheated prior to making a repair, regardless of the ambient temperature. A tent or shelter should be used to protect the welder and the work from bad weather, and the arc from strong winds.

BIBLIOGRAPHY

"Aluminum Shipbuilding Demands Extra Attention to Welding," *Welding Journal*, 49, 281 (April, 1970).

"Underwater Welding Method for the Construction of Outsized Ships," T. Kumose, K. Yamada and H. Onone, *Welding Journal*, 47, 194–206 (March, 1968).

"Performance Tests of a High-Yield Strength Steel for Ships," W. P. Benton, Jr., W. D. Doty and R. D. Manning, *Welding Journal*, 47, 534s–542s (December, 1968).

Hull Welding Manual, American Welding Society, D3.5 (1962).

CHAPTER 89

RAILROADS

PREPARED BY A COMMITTEE CONSISTING OF:

A. J. PALMER, *Chairman*
Locomotive Products Department–General Electric Company

W. S. ADAMS
Electro Motive Division–General Motors Corporation

W. O. BROWN
Southern Pacific Transportation Company

J. D. CASE
Penn Central Company

F. A. DANAHY
Association of American Railroads

M. F. HENGEL
Missouri Pacific Railroad Company
The Texas and Pacific Railway Company
Chicago and Eastern Illinois Railroad Company

CHAPTER 89

RAILROADS

INTRODUCTION

The transporting of freight and passengers by railroads in the United States involves 330,000 miles of track, 1,780,000 freight cars, 29,000 locomotives and 11,000 passenger cars. In addition, thousands of vehicles are used daily in intercity rapid transit systems. Between 1966 and 1970, acquisition of freight cars varied annually from about 56,000 to nearly 90,000 units.

Several trends are noteworthy in the design and construction of freight cars. These include increased carrying capacity, increased use of welding to replace rivets, design of cars for carrying a specific commodity, growing popularity of car-cushioning devices and the use of high-strength low-alloy steels. The increased carrying capacity requires larger trucks with higher and longer car bodies. This in turn raises the center-of-gravity of cars resulting in higher static and dynamic forces acting on most parts.

Welding is the major process used in the fabrication of all railroad facilities and structures and in the repair and maintenance of equipment. Practically all known welding processes are being used or have been applied to components of the industry.

LOCOMOTIVE WELDING APPLICATIONS

The manufacture of diesel-electric locomotives and component assemblies utilizes the full range of welding processes. The basic welded components are the underframe, cab, front and rear hoods, engine, generator, trucks, traction motors, electrical cabinet and equipment components, such as fuel tanks and air reservoirs. Many of these components are visible on the completed locomotives shown in Fig. 89.1.

Fig. 89.1.–Two 3600 horsepower, 6 axle, diesel-electric locomotives.

UNDERFRAME

A widely used locomotive underframe design currently being built is a self-supporting, rigid type, utilizing "T" center sills and a bottom plate of from 1 to 3 in. in thickness (see Figs. 89.2 and 89.3). The underframe is designed to withstand operating loadings, depending on the service for which it is intended.

Fig. 89.2.–Typical underframe fabrication facility.

In some cases the maximum is a buff load of 1,000,000 lb through the fabricated draft gear pocket, bottom plate and center sills. A design concept sometimes used and evident in older locomotives utilizes the truss principle, wherein loads are transmitted through the underframe, side frames and roof trusses.

The underframe assembly is fabricated and welded with the aid of jigs and fixtures used to accurately locate the various component parts and subassemblies and build camber into the underframe. The entire underframe assembly is placed in a trunnion to position weld joints so that most welds can be made in the flat and horizontal positions. The underframe structure is an all-welded design, composed of steels containing less than 0.26% carbon with and without added alloys. Critical welded subassembly components such as bolsters and center plates are stress relieved prior to assembly into the underframe. Some bolster

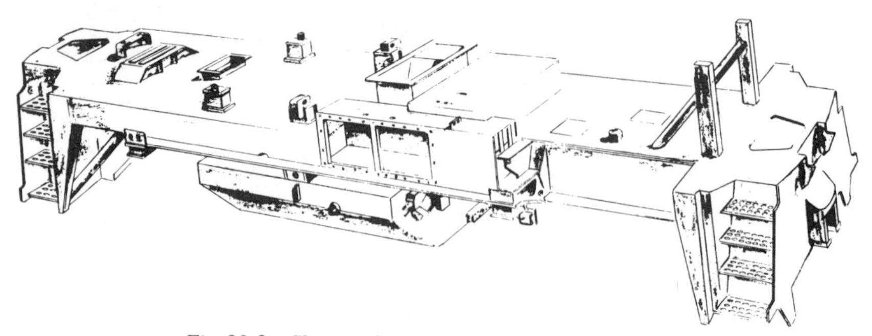

Fig. 89.3.–Sketch of a completed underframe assembly.

welds are magnetic particle or ultrasonically tested before assembly to the underframe, and all weld terminations are dressed by grinding to provide smooth contours to minimize stress concentration sites.

Welding is performed on the underframe and components with the shielded metal-arc process, using electrodes of the E60XX and E70XX classification. Submerged arc welding and flux cored arc welding with CO_2 shielding gas are also employed. The choice of the process is dependent on the weld size, weld joint accessibility, welding position and length of weld. All tack welds made prior to welding with the flux cored wire process are made with low-hydrogen electrodes.

In the underframe design, the majority of welds are fillet welds. Chamfers on pieces 3/8 in. or less are 45 deg, while heavier sections require a 30 deg chamfer on adjoining pieces. This results in a 60 deg included angle for welding. Critical structural joints are welded from both sides to obtain full penetration welds.

CAB AND HOODS

Cab and hoods are all-welded constructions consisting of an inner framework of angles, zees or rectangular tubing with a sheet metal cover. Shielded metal-arc welding is used with E60XX and E70XX electrodes. Gas metal-arc welding is used extensively for the sheet metal cover. Filler metals for this process with larger amounts of deoxidizers such as E70S-2, E70S-5 and E70S-6 per AWS A5.18-69, are used to compensate for the rimmed condition of the base metal. In addition; latches, hinges and a variety of other attachments are made by resistance welding during the fabrication phase. Completed cab and hood structures are shown in Figs. 89.4–89.6.

DIESEL ENGINE

Extensive welding is utilized on various components of the diesel engine. The engine frame or crankcase is either a casting or a welded fabrication. The welded

Fig. 89.4.–Front cab assembly.

Fig. 89.5.–Rear cab assembly.

Fig. 89.6.–Engine cab assembly.

assembly consists of forgings, special rolled channels, formed sheets and plate stock. The materials are low-carbon steels with less than 0.26% carbon. Firebox or flange quality grades of steel are specified in some areas. The design is such that 75% of the fabrication is done by submerged arc welding. Two other commonly used processes are shielded metal-arc and flux cored arc welding. All groove weld joints made with manual welding processes are designed for full penetration welds to obtain 100% joint efficiency. In order that all joints are accessible for welding, the fabrication is made progressively in subassemblies.

Fig. 89.7.–All-welded, 20 cylinder turbocharged diesel engine.

Inspection stations are coordinated with each subassembly weld to achieve 100% inspection coverage. Stress relieving, shot blast cleaning and machining are performed after welding to complete the fabrication. An all-welded turbocharged diesel engine is shown in Fig. 89.7.

In the cast engine frame design, a hardfacing application is used on the cast steel cylinder head valve seats. The gas metal-arc welding process is utilized for this operation with a ferritic stainless steel alloy filler metal. Preheating and postweld heat treating techniques are used to achieve the desired properties. In addition, various structural welds are made with the same process with low-alloy filler metal. A turbocharged diesel engine with a cast frame is shown in Fig. 89.8.

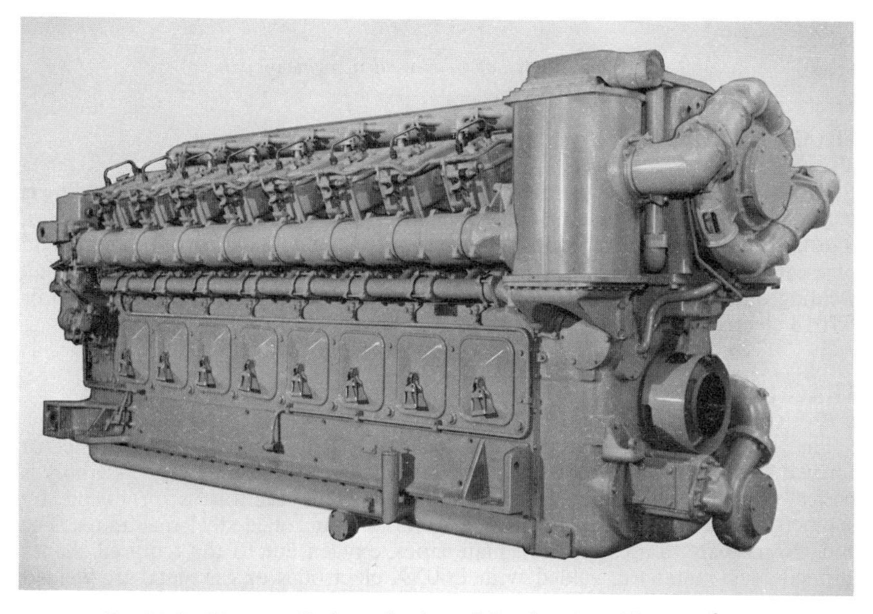

Fig. 89.8.–Sixteen cylinder turbocharged diesel engine with a cast frame.

The engine oil pans are also welded fabrications consisting of forgings, sheet or plate of low-carbon steel. Shielded metal-arc and gas metal-arc welding processes are used. After welding, oil pans are stress relieved, cleaned and machined. A completed oil pan is shown in Fig. 89.9.

GENERATOR FRAMES

The generator frame and structural attachments are usually fabricated from a special low-carbon steel to facilitate welding, sizing and the proper magnetic qualities. The ring is joined by the electroslag welding process in which the joint cross section is equal to the base metal cross section. The supporting base and brackets are shielded metal-arc welded with E60XX electrodes. The frame is mechanically sized with an expanding mandrel. The plastic deformation produced serves as a checkpoint for the weld joint.

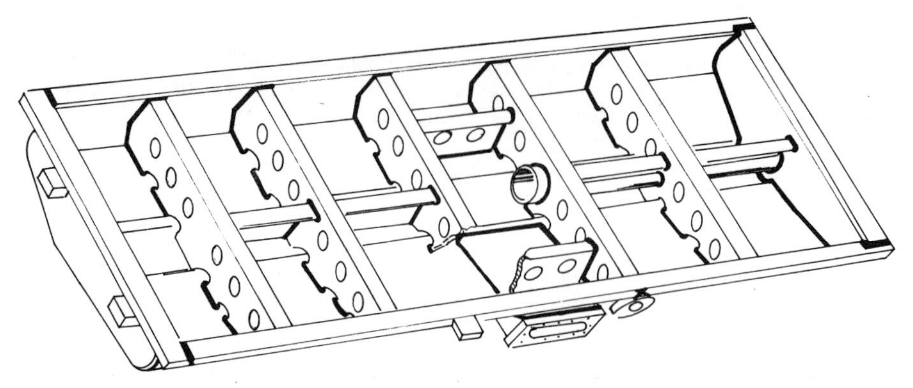

Fig. 89.9.–Sketch of an all-welded engine oil pan.

TRUCK FRAMES

The truck frames on most diesel locomotives are made of cast steel. Welding is confined to casting repair or the application of pipe brackets, chafing plates, journal box liners, bolster stops, etc. Except for the application of medium- and high-carbon steel wear plates (welded with E310-16 stainless steel electrodes), all welding on cast steel truck frames is restricted to low-hydrogen type E7016 or E7018 electrodes.

TRACTION MOTOR FRAME

The traction motor frame is a welded assembly consisting of a special rolled or formed section of SAE 1020 steel and fabricated end closures. The assembly is basically composed of a magnetic section. This section is produced by submerged arc welding or electroslag welding four rolled-steel quadrants. The end closures are fabricated from plate stock oxygen cut to the required shapes and shielded metal-arc welded with E60XX electrodes or gas metal-arc welded with flux cored filler metal.

All traction motor frames are stress relieved and shot blasted prior to machining. A completed traction motor frame is shown in Fig. 89.10.

ELECTRICAL CABINET

The electrical cabinet structure is a welded assembly of rectangular tubing, channels, angles and formed sheet steel. The outer sheets must be sealed to provide a leak-tight cabinet. The seal is made either by continuous welding or with the use of a sealing compound applied to the faying surfaces of sheet-to-structure and subsequent intermittent welding. All structural materials are mild steel welded with shielded metal-arc, gas metal-arc and/or resistance-welding processes.

ACCESSORY EQUIPMENT

Fuel and water tanks are fabricated from rolled plate. They are designed to meet requirements for required performance, track clearance and underframe

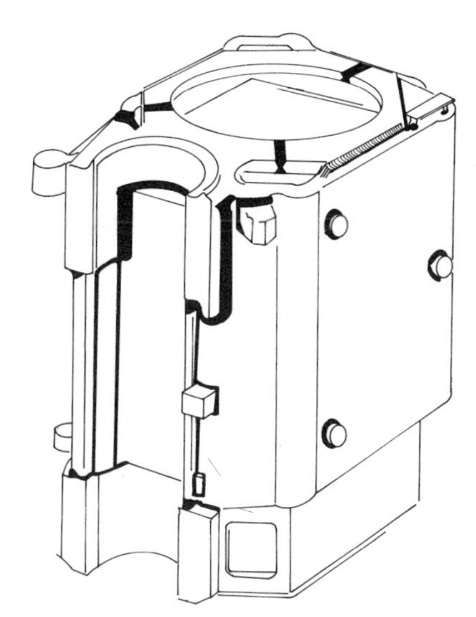

Fig. 89.10.–Sketch of an all-welded traction motor frame.

mounting. The plate thickness varies, depending upon the capacity of the tank and the distance between tank hangers. The design is such that welding can be accomplished on the inside of the tank, as well as on the outside, except for the top closing plates. The plate normally consists of semi-killed low-carbon rolled stock with a 27,000 psi yield strength and 22% elongation. The fuel tanks are welded by the shielded metal-arc process with E60XX electrodes and the flux cored arc. All fuel and water tanks are pressure tested. All exposed weld joints are soap tested while the tank is pressurized. A second test utilizes a gas leak sensing device.

Main air reservoirs are manufactured in accordance with the ASME Boiler and Pressure Vessel Code, Sections VIII and IX. The material is ASTM firebox quality Grade C steel, with a 55,000 psi tensile strength and 30,000 psi min yield strength. The shell portion is submerged arc welded with a flux-filled copper backup chill bar. The ends are also submerged arc welded with permanent backup strips. All flanges and permanent fittings are welded with the shielded metal-arc welding process using E6010 or E6011 electrodes.

Locomotive piping consists of compressed air lines, steam lines, boiler water lines, fuel lines and lubricating oil distribution lines. The steam and compressed air lines are fabricated from extra heavy pipe and are shielded metal-arc welded with E6010 and E6011 electrodes, including backup rings designed to minimize flow restrictions. Socket-type fittings are used wherever directional changes or multiple outlets are required. Single outlets on pipe runs are applied in the form of weld-o-lets or thread-o-lets where takeoffs do not exceed one-half the diameter of the run pipe. The carbon content of these fittings is limited to a maximum of 0.26%.

Fuel and lubricating oil lines are generally copper tubes which are used to combat corrosion, expedite cleaning and form easily, to circumvent obstacles. They are generally brazed with a silver brazing alloy where attachments are necessary.

Tubes and fittings made from copper or copper alloys may be joined with

BCuP-5 (copper, silver, phosphorus) brazing filler metal. Copper tubes or fittings which are joined to steel or malleable iron fittings are usually brazed with BAg-1 (silver, copper, zinc and cadmium) alloy.

Malleable iron fittings can be precleaned electrolytically in molten salt prior to brazing and after machining. Manual torch brazing and automatic gas heating are used with preplaced filler metal rings, nuggets or foil stock. All assemblies are pressure tested, either as assemblies or in the final application.

Exhaust mufflers and manifolds are manufactured from firebox quality steel or high-temperature heat-resistant steels of chromium-molybdenum compositions. These materials are welded with the shielded metal-arc welding process, with E8018-B2 electrodes and the flux cored arc welding process utilizing a filler metal electrode of comparable analysis.

Welding of stainless steel assemblies is almost always performed with a welding electrode of similar chemical composition to that of the base metal. However, when corrosion or operation at critical temperatures is expected, the columbium-stabilized E347 type electrode is used.

MAINTENANCE AND REPAIR OF COMPONENTS

Because of its welded design, the diesel locomotive and many of its components are readily maintained and repaired by welding. Damaged or worn sections can be replaced by means of oxygen cutting or related methods and reclaimed by welding.

Almost all the materials are weldable and with the proper selection of welding processes, filler metals and joint designs, almost any type of repair can be made.

Care should be taken to prevent over-or-underwelding. Overwelding is just as detrimental on critical joints as is underwelding. An oversized weld tends to increase the heat input, causing excessive distortion and thus concentrating stresses at the joint. Underwelding is undesirable for obvious reasons.

Diesel Engine Valves

Diesel engine valves can be reclaimed at a fraction of the cost of new valves by surfacing (hardfacing) the valve seats. This can be done by using the oxyacetylene welding process with a nonferrous precision cast alloy rod, containing cobalt, chromium and tungsten. In addition to resistance to corrosion and abrasion at elevated temperatures, these valves will withstand medium impact. Deposits result in a hardness of 45 Rockwell C. The valve seats should be machined in a lathe with a roundnose tool ground to a 3/8 in. radius. Premachining the seat area will allow for a deeper weld deposit after final machining and grinding. The valve is then placed in a positioner and preheated to about 800 F. The positioner should be constructed as shown in Fig. 89.11. Care must be taken to protect the valve from cold drafts during welding and cooling.

When surfacing a valve, a continuous operation should be employed. If the surfacing is interrupted, the deposit may contain porosity. When completing the deposit, overlap the starting point by about 3/16 to 1/4 in. The welding rod should not be applied so as to be diluted with the base metal but should be applied with a braze welding technique. After the surfacing has been applied and postheated to 700 F, the valve should be submerged in hydrated lime with the stem upright, allowing it to cool to room temperature. The valve can then be machined to original dimensions by using a sintered carbide tool.

Fig. 89.11.–Application of oxyacetylene welding to reclaim damaged engine valves.

Cylinder Head Valve Seats

Valve seats can be reclaimed or repaired on the cast steel cylinder heads by surfacing. This can be done by following essentially the same procedures used for the initial welding.

Diesel Cylinder Heads

Before welding the cast iron, the head is carefully examined under hydrostatic pressure for cracks. Then it is preheated over a four-hour period to 1000 F for visual inspection. The purpose of pretesting the head is to determine any internal cracks or defects which might develop. Any cylinder head with defects which cannot be repaired economically is immediately scrapped at this stage. If the head can be reclaimed, it is preheated to 1200 to 1400 F with preheat furnaces using charcoal. More elaborate ones use natural gas or electricity with pyrometric control. The cylinder head is welded with the oxyacetylene welding process using flux and a cast iron welding rod. When the welding has been completed, the head and furnaces are completely covered with asbestos sheeting and allowed to cool for approximately 24 hours. After the head is control-cooled, it is taken to a vertical boring mill where it is rough machined in the valve seat areas. It is then hydrostatically tested and if no leaks appear, it is finish-machined. The cylinder head is then placed on a surface grinder and the face is ground.

Diesel Locomotive "A" Frame

Frames with oversized crankshaft-bearing areas can be restored to standard size by building up with the gas metal-arc process using a 0.035 in. diam wire and CO_2 shielding gas. After the bores have been built up, they can be machined to original dimensions.

Eddy Current Clutches

As a permanent repair for eddy current clutches, a copper overlay is built up on the steel bore of the clutch, thus producing a permanent bond. This increases the service life of the clutch many times, at a fraction of the cost of new clutches. A 1/16 in. diam deoxidized copper welding wire and an argon-shielded gun are generally employed when applying overlays on these clutches. After the overlay is applied, the copper overlay is machined, and the clutch is ready for service.

Traction Motors

End frames with deformed bores can be restored for about one-half the cost of new ones. The bore may be built up by the use of either shielded metal-arc with E6012 or the gas metal-arc process and CO_2 gas. The heat input during welding is controlled to protect the frame from distortion. After the frame bores have been built up, they can then be machined and ground to original dimensions (Fig. 89.12).

Frames with oversized axle bores can be restored to standard size by building up with the gas metal-arc process and welding vertical down with 0.035 in. diam wire. The heat input during welding is greatly reduced with the gas metal-arc process which minimizes distortion. This process reduces the cost of reclaiming by 35% over shielded metal-arc welding.

Reverser Contacts

Reverser contacts can be reclaimed by welding with either copper or aluminum bronze wire. The contactors are placed in a jig and, in order to minimize distortion, skip welding techniques are used. After welding, the contactor jig is mounted in a milling machine, or lathe, and milled or turned to shape. The aluminum bronze deposits average 77 to 81 Rockwell B hardness,

Fig. 89.12.–Traction motor end frame buildup for finishing. These end frames can be reclaimed for one-half the cost of replacement.

Fig. 89.13.–Applying wear liner to journal box with the gas metal-arc welding process using 18 Cr-Ni austenitic stainless steel 0.035 in. diam wire and argon gas.

while the deoxidized copper will average from 51 to 71 Rockwell B. The hardness of the original base metal will average from 75 to 80 Rockwell B. While the electrical conductivity of copper is higher than aluminum bronze, the abrasion resistance of aluminum bronze is greater.

Journal Box

Figure 89.13 shows the application of manganese steel liner to journal boxes with the gas metal-arc process. The wire used is 18 Cr-18 Ni austenitic stainless steel 0.035 in. diam, with argon gas shielding. Liners can be applied by the gas metal-arc process for one-half the cost of shielded metal-arc welding.

In general, when maintenance and repair requirements are evaluated on system components, welding can be performed if the proper discretion is given to such factors as base and filler metal compatibility, joint design and preparation. In many cases the dimensional integrity of the fabrication will also require consideration.

WELDER TRAINING AND CERTIFICATION

Welder training and certification is an important prerequisite to quality and cost control of welded fabrications. The welding skills required to obtain optimum results can only be achieved through adequate training and certification programs. A recent trend in welder training involves the utilization of a programmed course. Audio-visual units coupled with workbooks are integrated with the practical welding lessons. The trainee is introduced to welding terminology and concepts in small bits of information. This information is supplemented with a series of practical welding lessons that results in a steady progression to the qualification test. The qualification test encompasses the essential welding variables of most industry-wide codes such as Section IX of the ASME Boiler and Pressure Vessel Code. Upon being certified the trainee is assigned to a production-type welding job where additional training is required. This training involves the implementation of welding skills into products through proper interpretation of job instructions and requirements. This is accomplished through close supervision and diligent surveilance by quality control.

In a large manufacturing facility, it is essential that the proper welding specifications be transmitted from the engineering department to the production and inspection departments. This is accomplished with the aid of welding symbols on drawings in accordance with AWS A2.0-68 or equivalent. Supplementary welding specifications are also used to provide general or detailed information to production and inspection as required.

RAILWAY/COMMUTER CAR WELDING APPLICATIONS

MATERIALS

The materials usually employed in the fabrication of stainless steel railway cars are AISI 430 and type 201 or 301 or both.

Type 430 is a corrosion- and heat-resisting 17% chromium steel and is used in nonstructural areas.

Type 201 is an austenitic chromium-nickel manganese steel and is used in the cold rolled or annealed condition. Type 301 is an austenitic chromium-nickel steel capable of obtaining high strength and ductility by moderate or severe cold working. Both these stainless steels are used in structural areas.

RAILWAY CAR COMPONENTS

Car Floor Assembly

The floor assembly consists of two end underframe assemblies of high-strength low-alloy steel joined by manual and semiautomatic arc-welding processes. These units are plug-welded or riveted or both to the center sill which is made up of stainless steel drawbench sections which are resistance-welded together. The transverse stainless steel floor beams and floor pans are, in turn, resistance-welded to the center sill. The side sills enclose the beams, floor pans and end underframes. Following structural assembly, the floor is inverted for installation of underfloor equipment (see Fig. 89.14).

Car Side Assembly

The sideframe is a semimonocoque type structure with shear panels at the ends and is constructed with a 1 1/4 in. camber in 85 ft. It is a completely resistance-welded unit with the exception of the butt joints between the deadlight panels in which the gas tungsten-arc welding process is utilized. The outer exposed surfaces below the deadlight panels may be of light gage corrugation-type construction resistance-welded directly to the structure. Fluted panels may be attached later at car assembly. The gas metal-arc welding process is used to attach heavy stainless steel tie-in members at the floor level in door areas.

Car End Assembly

The end frame (Fig. 89.15) consists of end sheets, angles for connection to roof and side-frames and collision post assemblies which connect the longitudinal roof members (purlins) and the end underframe. This assembly is resistance-welded.

Car Roof Assembly

Zee-shaped, stretch-formed roof carlines, hat-shaped roof purlins and side plate assemblies are put together and welded in a cambered fixture. Corrugated mat assemblies, running the entire roof length, are then welded to this frame. Resistance welding is employed throughout. This completed roof is inverted for installation of piping, wiring, insulation, air ducts and air conditioning. Ceiling support members are also resistance-welded in place at this time.

Final Assembly

Structural general assembly (Fig. 89.16) is divided into two stations: assembly of sides and ends to the floor and installation of the roof. Resistance welding with portable tools is the major joining method with shielded metal-arc welding used to plug weld the collision post to the end underframe using stainless steel electrode E308-XX.

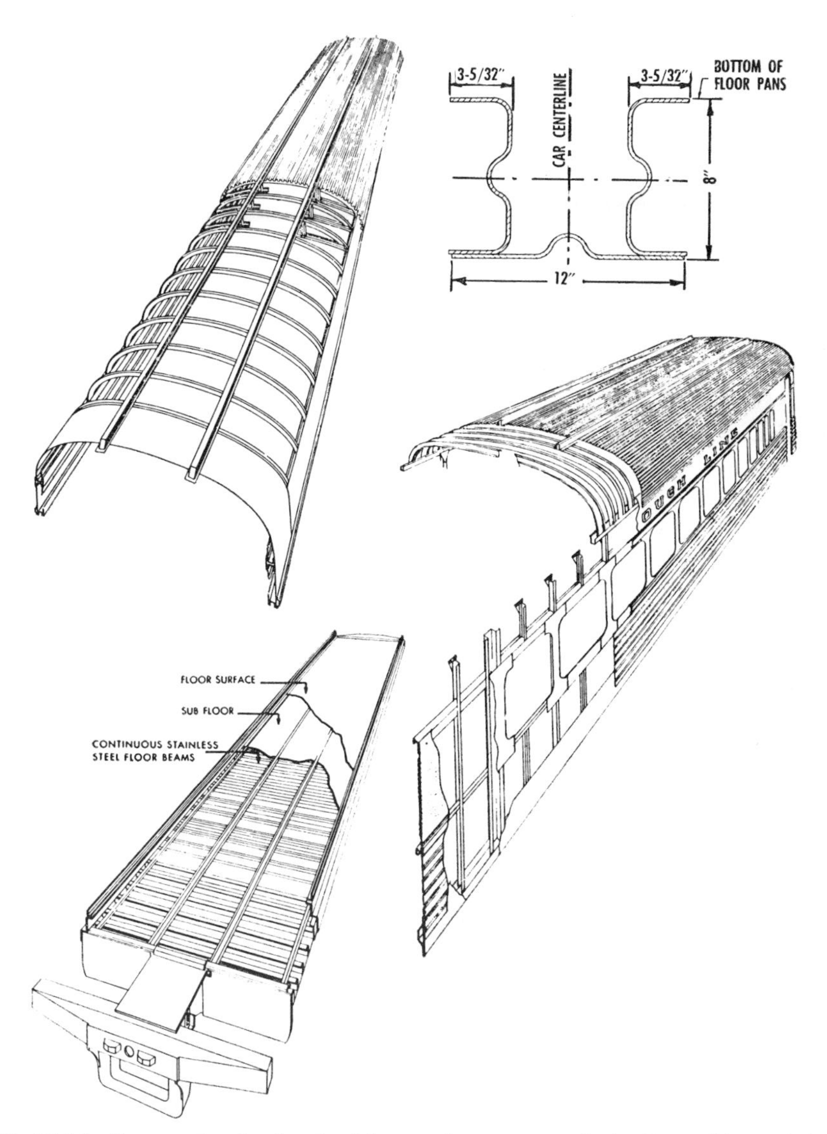

Fig. 89.14.–Construction details of stainless steel passenger car body. Car roof (upper left), center sill cross section (upper right), side frame assembly (lower right), floor construction (lower left).

Interior trim, floor, seats, windows, partitions and hardware are added following water testing. Arc-spot welding has been used to weld stainless steel corrugation floors to the floor pans when this type of structure is specified. Interior aluminum panels and subassemblies are joined by resistance welding and gas tungsten-arc welding before installation in the car. Car trucks may be of the

Fig. 89.15.–A resistance welded car end frame assembly.

cast type or a completely welded structure of low-carbon alloy steel employing semiautomatic and automatic welding equipment.

The lightweight stainless steel resistance welded structure is employed in all types of mainline cars, such as coaches, sleepers, vista domes, double deck, baggage and dining cars. The self-propelled railway diesel car (RDC), as well as subway cars, are also fabricated by these methods.

Fabrication

Jigs and fixtures are designed to hold the parts in alignment with emphasis on welding tool accessibility as well as the control of shrinkage and distortion where

Fig. 89.16.–Resistance-spot welding of car side components.

procedure and sequence cannot eliminate this condition. Large assembly fixtures are combined with motorized structures to permit convenient travel and weld tool suspension. Lifting and turnover devices are designed to expedite loading and unloading of large assemblies. Portable jigs are limited to approximately 15 lb including parts, and, as a result, most jigs are of the stationary type.

Due to their flexibility, portable resistance welding installations are desirable when large assemblies are involved. Lightweight tool construction and good tool balance are important for the ease of operation necessary to produce quality welds of good appearance and even spacing. Air operated single- or double-acting cylinders of various pressure ranges are normally used. Hydraulic actuation is limited to tools requiring electrode force in excess of 2500 lb. "C" type and "X" type action tools with throat depths up to 126 in. are used. Portable seam welding equipment is used to attach stiffener members to flat panels, and resistance-seam welding is done on long sections by stationary seam welding equipment. Press-type welding machines are limited to lightweight sub-assemblies.

Special base metal chemistry analysis and design techniques have been established by testing programs which make possible the use of shielded metal-arc welding in these structures. Under these conditions, the unstabilized stainless steel can be fabricated without danger of intergranular corrosion and fatigue failure.

The use of automatic resistance- and arc-welding machines is considered when their initial cost can be justified. The present engineering trend is to design structures that will permit utilization of such equipment to a greater degree.

COMMUTER CARS

Underframe Design and Construction

The underframe for the commuter car is shown in Fig. 89.17. Basically, this design has not changed during the past fifteen years. One minor change was the use of pressed shapes, eliminating many small sheared pieces which required welding. All of the structural parts, including the center sill, are low-alloy steel.

The use of large positioners for the complete underframe assembly and the major subassemblies has contributed to the speed and quality of the welding. These assemblies are principally welded with either the automatic submerged arc process or shielded metal-arc process using E6020 electrodes. At critical locations, either E7016 or E7018 low-hydrogen electrodes are used, and copes have been added to improve accessibility and reduce locked-up stresses at these locations. The welding procedures and fixtures have been adapted to minimize residual stresses during welding.

Wherever possible, alloy steel castings are being welded instead of riveted. Also, the false floor sheets, which are either 0.030 in. stainless or 0.040 in. thick low-alloy steel, are spot welded to the completed underframe instead of being riveted. Both of these changes have contributed substantially to the reduction in weight of the completed car.

The center sill subassembly is the longitudinal section down the center of the car shown in Fig. 89.17. This member is designed to withstand a buffing load of 800,000 lb in compression. The center sill is made up of two Z-26 sections which are submerged arc welded together for the full 54 ft 6 in. length. The sill sections are clamped in a fixture which maintains a camber of 4 1/2 in. while welding.

Fig. 89.17.–Center sill subassembly for commuter car underframe, usually assembled by automatic submerged arc welding or the shielded metal-arc welding process.

There are two side sills per car. These are the other longitudinal members that are visible in Fig. 89.17. The side sill is made by spotwelding a rolled angle and a pressed angle. The crossbearers and floor beams are the transverse members that support the floor. These members are pressed channels and rolled Zee-sections and are welded with the manual shielded metal-arc process to the center sill and the side sill.

There are two bolster subassemblies per car where the trucks are mounted. These are of box-section design with 5/8 in. thick top and bottom cover plates and 1/2 in. thick web plates. The top cover plates, which are 16 in. wide, are continuous for the full width of the car. These are welded with an automatic submerged arc welding machine mounted on a portable gantry frame. For these automatic welds, web plates are beveled 45 deg on the outside edge in order to obtain the required penetration. An additional fillet weld is made on the inside of the web plates at the bottom cover plate. Copes are located in the web plates at the intersections of fillet welds.

The draft sill subassembly is a continuation of the center sill to the end of the underframe. It has approximately the same cross-section dimensions as the center sill, but is constructed of web and cover plates. The draft lugs and reinforcing plates are welded on the inside. The draft sill is butt welded to the center sill during the final assembly of the underframe. The bolster slides on and is welded in place, and the other end of the draft sill is butt welded to the center end sill subassembly. The latter flairs out to provide clearance for the coupler.

The center end sill, draft sill and bolster are fitted together and welded to make the end platform assembly. After straightening, this assembly is fitted into the underframe fixture with the other underframe parts. These are then welded in a large fixture which positions the joints properly for welding.

Side Design and Construction

In order to use existing passenger car equipment and still be able to accommodate the increased height of the double-deck cars, the sides are constructed as two separate and complete portions. These separate portions, which are joined together, are 7 ft 5 3/8 in. high and 2 ft 11 15/16 in. high. The method used for construction is practically the same as that used for standard height cars. Each side portion is divided into two complete subassemblies, each the full length of the car, consisting of the side sheets and the framing members, which are subsequently spot-welded together. By completing all of the arc-welding operations prior to spot welding the subassemblies together, it is possible to better control the shrinkage from welding and prevent the formation of buckles.

The 14-gage low-alloy steel side sheets are assembled from a number of smaller separate sheets. The window openings are blanked originally and trimmed to size after welding. The corrugated stiffeners, which cover almost the entire side sheet, are spot-welded in place. The sheets are then clamped rigidly in place in a fixture and groove welded together with the submerged arc. A grooved copper backup controls the penetration of these welds. The three horizontal seams are welded first in order to make side sheet sections of proper height. Then the five vertical seams join the sections into a complete side portion. The side sections are moved freely from position to position on rollers mounted on the welding tables and at the transfer locations. With the exception of the side sill angle, the framing members are all Zee-sections of low-alloy, 3/16 in. thick

Fig. 89.18.–Assembly of passenger car sides. Windows are provided for by the framed blank areas.

steel. There are four continuous longitudinal members which are visible in Fig. 89.18. The vertical posts are coped and arc welded to the horizontal members, so that one leg of each Zee-section fits flat for spot welding to the side sheets. Since camber must be built into passenger car sides, it is provided for in the locating fixture for the framing members. This is a vertical fixture, the full size of the side portion, for locating, clamping and arc welding the frame.

The side sheet subassembly is placed against the copper backing of the side spot-welding fixture. This continuous 1/2 in. copper backing supports the entire car side portion and permits welding pressure to be applied to any point over an area 8 ft 6 in. high by 85 ft long. After the approved primers and sealers are applied, the side frame is located on the fixture. A definite welding procedure is followed to ensure that the camber of the completed side is maintained.

The side spot-welding machine travels in a pit on a track parallel to the fixture, and an external water spray may be used around the electrodes in addition to the normal internal cooling water. An impulse spot weld is usually required because of the different thicknesses of the side sheet and framing members.

Roof Design and Construction

The roof design has been standardized for a number of years. The smooth, oval appearance of the completed roofs is shown in Fig. 89.19. The 14-gage low-alloy steel roof sheets are 2 ft 9 1/2 in. wide. The longitudinal purlins, used for stiffeners inside the sheet, are spot-welded in place with standard press-type resistance welding machines. The carlines are transverse hat-sections that are pressed and drawn to shape.

The carlines are first fitted on the matching copper backup bars on the fixture. The roof sheets are then located on top of the carlines with enough gap allowance for welding the sheets to the carlines. Straps are used to hold the sheets down during the manual arc-welding operation. The roof requires over

Fig. 89.19.–Car roofs of 14 gage low-alloy steel. All welds must be proven watertight.

336 ft of welding, which must be sequenced properly to eliminate any buckles in the completed assembly. To ensure a leak-proof construction, all of the welded seams are water-tested.

End Design and Construction

The elimination of conventional fasteners from the construction of the ends has improved the appearance. The collision posts are moved out, allowing more room in the interior. Vertical stiffeners are spot-welded to the inside of the end sheets, and the end sheets are welded at the edges.

Final Assembly

Car piping is either silver brazed or gas welded. Figure 89.20 indicates the amount of pipe welding required on the underframe. The underframe, sides, ends and roof are brought together with turnbuckles and braces in preparation for welding. The sides are then joined to the side sills on the underframe. The roof carlines are arc welded to the side plate, and the roof sheets are joined to the side sheets. Finally, the ends are welded to the sides and roof with lap, butt and fillet welds.

After the braces are removed, the seat supports, inside finish, heating and plumbing facilities are welded in place. This completes the car for finish welding of items such as studs for the support of insulation, and for the diaphragms on the ends. Many of the interior appointments, which may be aluminum, monel, stainless steel or copper, are fabricated by shielded metal-arc welding, gas tungsten-arc welding or brazing.

MAINTENANCE AND REPAIR OF COMPONENTS

Passenger Cars

There are in operation today on American railroads two distinct types of passenger train cars: heavyweight and lightweight. The heavyweight is the older of the two types and now obsolete in design, although many are still in operation. The heavyweight type is usually of riveted carbon steel construction.

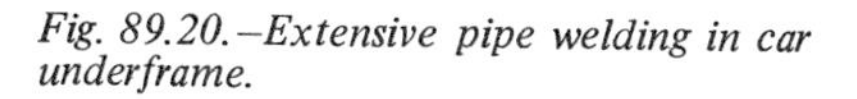
Fig. 89.20.–Extensive pipe welding in car underframe.

The lightweight type is fabricated from either low-alloy, high-tensile or stainless steel, with the stainless steel type employing either arc welding or resistance-spot welding for the fabrication. The high-tensile low-alloy car is usually fabricated by one of the several arc-welding processes. The car fabricated by arc welding will have an assist from both spot welding and riveting, while the car fabricated by spot welding will have a similar assist from arc welding and rivets. Both types of cars confront the repairman with the usual problems of induced stress, shrinkage, distortion and camber maintenance. The modern passenger train car is an involved unit. Its primary purpose is to furnish a safe and comfortable ride for its passengers over long distances. It may contain its own electric generating plant and lighting system, air conditioning system, running hot and cold water systems, plumbing and bathroom facilities, and often telephones, radio, communication systems, sleeping quarters and dining facilities. These components, along with the car structure and running gear, will contain most of the metals utilized by modern engineering. The repair processes commonly employed are oxyacetylene, shielded metal-arc, gas metal-arc, gas tungsten-arc and the resistance-welding processes.

Repairs to passenger train cars are usually governed by the [illegible]nstruction engineering department of the individual railroad, but regulations and specifica-

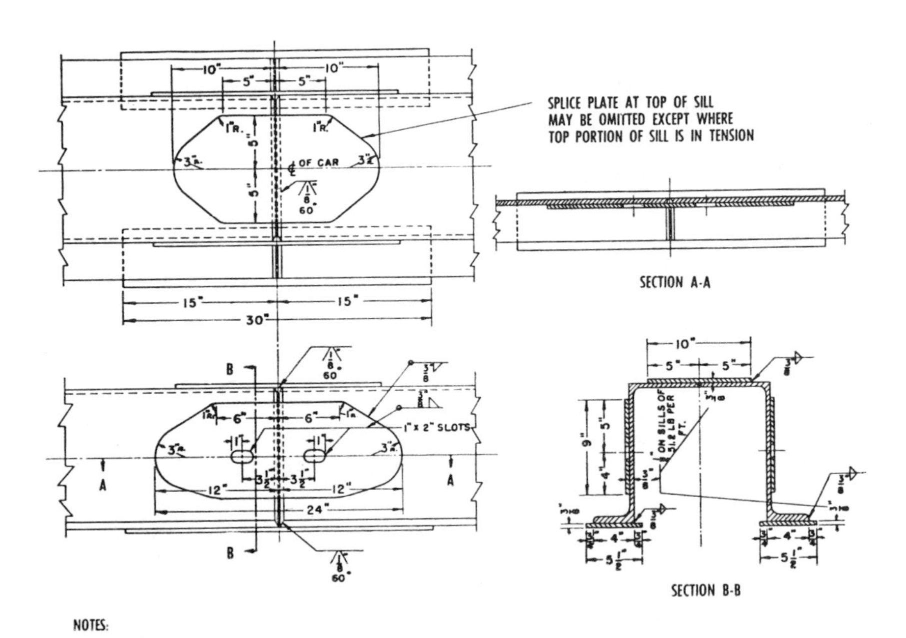

NOTES:

1) REPAIR TO BE LOCATED AT ANY POINT WHERE CONSTRUCTION OF CAR WILL PERMIT AND DIMENSIONS OF REINFORCEMENT AS SHOWN CAN BE ADHERED TO

2) REINFORCING PLATES 3/8 IN. THICK LOCATED ON OUTSIDE OF ZEE BAR WEB AND BOTTOM OF ZEE BAR FLANGE TO BE OF DIMENSIONS SHOWN

3) ALL WELDING TO BE DONE AS SHOWN

Fig. 89.21.–Requirements of the Association of American Railroads governing center sill splicing.

tions, of such governing bodies as The Association of American Railroads, The Interstate Commerce Commission, The United States Postal Department and The United States Department of Health, Education and Welfare must be considered and complied with.

Defective parts of passenger train cars are prepared for welding either by arc or oxygen cutting, or by mechanical sawing. Engineering calculations are made to ensure that repair replacements (including weld seams in either patching splicing or alterations) will maintain calculated structure strength.

Underframes

Underframes in modern passenger train cars are fabricated from steel plates and shapes joined by arc welding. In designing the underframe, the car builder recognizes The Association of American Railroads' requirements for safe operation of the car. Consequently, the repairman must always, in replacing parts or sections, splicing or compensating for wear, maintain the designer's calculated strength and camber, avoiding repair design that will promote the eccentric loading of a repaired member. Figure 89.21 shows the Association of American Railroads' requirements for center sill splicing.

Car Trucks

Presently, extensive welding repairs are made on all truck parts, either for welding defects, fabricating repair sections or to compensate for wear. Such components as the truck frame, air brake equipment (Fig. 89.22), levers, rods, equalizers, bolsters, swing hangers, crossbars, center plate pedestal jaws and spring planks receive welding repairs for defects and wear when the car is in the repair shop. Since these parts are either cast or forged steel, one of the arc-welding processes is employed; heat treating of parts after welding is mandatory in certain classes of steel.

Fig. 89.22.–Welding repairs being made to fabricated brake seam assembly with the shielded metal-arc welding process.

Side Framing

Side framing repairs usually involve the repair of defects or the replacing of sections damaged by accidents or corrosion. In replacing a section, each individual job must be evaluated so that side frame strength and camber will be maintained and distortion avoided. Welding process, filler metals and welding procedure should be based on the base metal in the original frame, so that framing will be brought to its original dimensions (Fig. 89.23).

Roofs

Repairs to passenger train car roofs are generally made to correct damage, leaky conditions or alterations. In making repairs to roof framing and skin plates, weld seams and splices are designed to maintain roof strength and to be

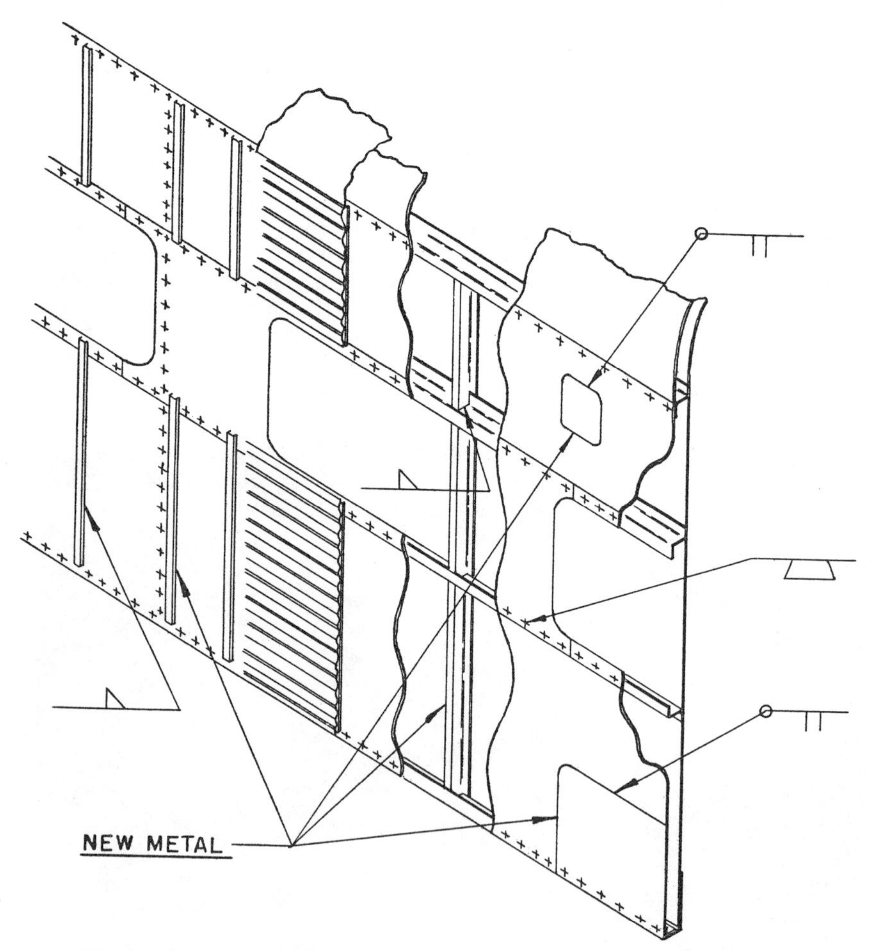

Fig. 89.23.–Typical sheathing and side framing repairs to passenger train cars.

weatherproof. On arc-welded roofs, one of the arc-welding processes may be employed for repair. On resistance-spot welded roofs, either spot welding or arc welding may be used. The car shown in Fig. 89.24 had been involved in a fire and holes had been chopped in the original plates. Repair plates were applied as an overlay welded to the car structure and seal welded around all edges for weatherproofing.

Fig. 89.24.–Repairs to stainless steel roof.

Miscellaneous Car Parts

The couplers shown in Fig. 89.25 are reconditioned for defects and wear. The Association of American Railroads' specifications are followed which cover extent of welding repairs allowable, type of filler metal used, heat treatment after welding and coupler finish. Other coupler parts, such as yokes and locks, may be reconditioned in a similar manner.

Soldering and brazing are used in and around passenger train cars for repairs in dining car kitchens and air-conditioning and refrigeration equipment. Filler metals of this type should be cadmium-free when used where food is served or processed.

FREIGHT CAR CONSTRUCTION

GENERAL

Freight cars generally are made up of five components: trucks, brakes, underframes, draft gears and bodies. The following paragraphs describe the

Fig. 89.25.–High-tensile couplers after reconditioning by welding.

design and construction of each component. Special features and details, however, are included within the general descriptions of the car types.

FREIGHT CAR TRUCKS

The principle members of freight car trucks are either cast or wrought steel. Shielded metal-arc and gas metal-arc welding processes are used for repairs of casting defects and to attach replaceable wear plated rings. Also minor brackets are sometimes attached by welding.

BRAKES

Air brake auxiliary reservoirs are often constructed by welding in accordance with applicable pressure vessel and railroad codes. Many brake parts such as brackets and levers are cut to shape by means of single or multiple head oxyacetylene torches mounted on tracing machines.

UNDERFRAME

The underframes of most freight cars are similar, with variations to suit the type of car body for which it is designed. Underframes are usually an all-welded skeleton structure comprised of center sill, side sills, end sills, body boltsters and crossbearers that provide intermediate connections between the center and side sills. Cars with heavy floor loadings, e.g., box cars and gondolas, have longitudinal floor stringers spaced between center and side sills to carry and help distribute loads (see Fig. 89.26).

An important variation in underframe construction occurs when the sliding sill type of underframe cushioning is used. The sliding sill is a one-piece member running through the entire length of the car and having a coupler attached to

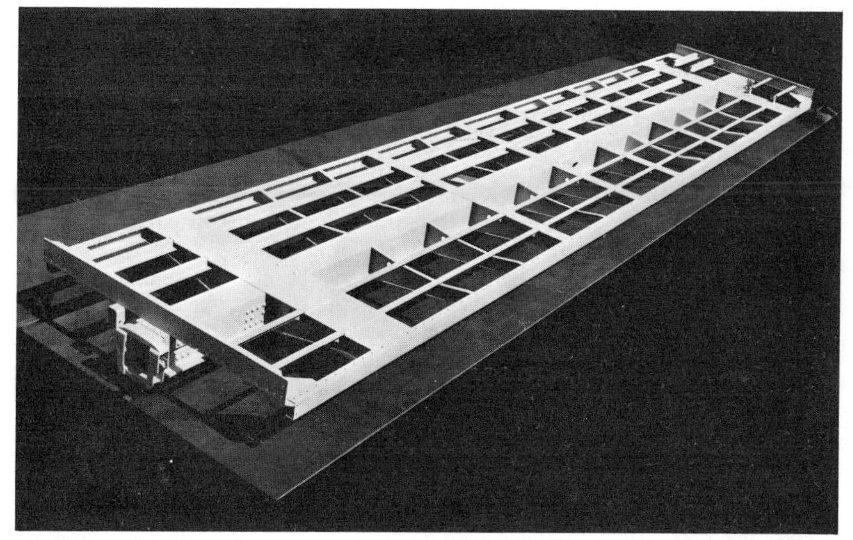

Fig. 89.26.–Refrigerator car underframing similar to that used in many types of freight cars.

each end. Coupler forces are transmitted into the main underframe through a hydraulically-controlled dampening device. The sliding sill is re-centered after each shock by a large helical spring or a gas-operated piston.

When underframe cushioning is not used, mounting, attaching and restraining means for end-of-car cushioning or conventional draft gears are applied at the ends of the center sill.

Construction of the sliding sill of a cushion underframe car and the center sill of a conventional car are nearly identical. They usually consist of two rolled section Zee bars placed back-to-back and welded along the edges of the two upper flanges to form the sill. The two Zee bars are held in place on a jig by a series of clamps which maintain the inside width dimension. Camber is built into the jig so that the sill will be straight after welding. Welding is accomplished by means of a traveling gantry which straddles the Zee section, carrying the power sources, flux-recovery units, reels of welding wire, automatic welding heads and secondary power requirements (see Fig. 89.27). Submerged arc welding is employed, and during the welding process the underside of the weld is usually backed up with welding flux contained in a canvas retainer and held tightly

Fig. 89.27.–Two Zee bars being welded with the submerged arc welding process. Hydraulic clamps maintain the bars in alignment for welding.

against the weld joint by an inflated air hose. As the traveling gantry moves along the weld, flux is deposited on top of the sill ahead of the weld and is picked up by a vacuum flux-recovery device behind the weld.

A triple welding electrode arrangement is generally used. The first two electrodes deposit a weld that penetrates approximately halfway through the flanges of the Zee sections. The third wire trails the first two by approximately 3 1/2 in. and increases the penetration to approximately three-quarters of the way through the flanges over the greater part of the weld. For a distance of approximately 2 ft on either side of the bolster center, the current on the third wire is raised so that the penetration is 100%.

The control panel is mounted on the carriage beside the welding head. It indicates amperage, voltage and travel speed to the welding operator permitting him to make adjustments as the work progresses. Average welding speeds are 50 to 80 ipm.

An alternate method may be employed to obtain full fusion. The technique is to first deposit filler metal from one side as described above. The sill section is then turned upside down and an additional pass deposited from the root side generally using automatic submerged arc techniques. This method is used where controlled joint fit-up cannot be maintained and when long lengths of full joint penetration are required.

After the center sill is removed from the welding fixtures, a striker is fitted to each end. Front and rear draft lugs and other members are applied to tie the lower portions of the sill together. They are fitted and welded manually by the shielded metal-arc welding process.

When sliding sill underframe cushioning is used, a "window" must be provided through the underframe structure. Usually in this construction the center sill is comprised of a pair of carbuilder's channels, one located on either side of the sliding sill.

Typical body bolsters (two per car) are a welded box section. The top cover plate is continuous for the full width of the car, and is attached to the top of the center sill with intermittent fillet welds, slot welds or mechanical fasteners. The ends of the top cover plate often lap under the side sill flanges and are attached with fillet welds. One end of the bottom cover plate laps the bottom flange of the side sill and in some designs an access hole permits the welds inside of the box section to be tied in. The other end is welded to the bottom of the center sill flange.

Underframe parts and subassemblies are brought together in an underframe assembly jig. The underframe is assembled in an upside-down position with the center sill and bolster assemblies placed against their respective guides and stops. Floor stringers, bulkhead bottom track (where used), crossbearers, crossties and end sills are also placed in the jig. Where used, the sliding sill assembly is placed in the underframe jig while it is in the inverted position. Related cushioning mechanism may be applied at this stage.

The body center casting used with the sliding sill construction or the center filler and center plate used with conventional underframe construction are applied ready for fastening with mechanical fasteners, slot welds, fillet welds or a combination of all three.

Underframe components are usually held in place in the jig by means of air-operated clamps. When assembled, the entire underframe is tack welded together ready for final welding. The assembled underframe is then removed from the jig and placed in a rotatable positioner. The positioner rotates the frame about its longitudinal axis so that all welds can be made in the flat position.

When a steel floor is used (for example in a gondola car), it is applied and tack welded while the underframe is still in the positioner. After tack welding, the positioner is revolved to the inverted position and the floor-attaching welds are completed from the underframe side.

The all-fabricated underframe described above is typical for modern freight cars; however, some cars have cast draft sills (replacing the center sill from the body bolster to the end of the car and including draft lugs, center plate, and striker) or cast underframe ends in which the draft sill and the body bolster are cast in one piece for incorporation into the underframe weldment. Most underframe structural members are rolled steel sections, and many are car and shipbuilders' special shapes. Materials are carbon or low-alloy high-strength steel or combinations thereof, ranging from 30,000 to 60,000 psi yield strength. Cast steel is usually used for miscellaneous components including truck central bearing and coupler horn strikers.

CAR SIDES

Most freight car sides are of similar construction whether used for box cars, hopper cars or gondolas. There are a few notable exceptions such as the teardrop-shaped covered and open-top hopper cars, stock cars, tank cars, etc.

Generally, side construction consists of a framework composed of a series of vertical posts, end posts, top and bottom chord members and sheets or plates of appropriate thickness covering the area between the framework members. These are sometimes pressed to provide pockets for load-restraint locating rails. In practically all cases, the sides are made up as an all-welded subassembly.

The side posts are usually either a Zee section (e. g., double-sheathed box cars), an "I" or a hat section (single-sheathed boxcars, hopper cars, gondolas etc.). Top chord members (side plates on box and refrigerator cars) are selected to meet requirements for roof connections (if used), structural forces and operating conditions such as heavy battering by loading buckets, vibrators,

Fig. 89.28.–Typical boxcar weighing approximately 23 tons having a capacity of 50 tons.

magnets or scrap ladings. The bottom structural side member is sometimes built as part of the underframe and the side subassembly is joined to it during final assembly. A typical double-sheathed box car side is shown in Fig. 89.28.

Box car sides have openings for doors, usually centered longitudinally and with heavy framing all around. Openings generally are from 8 to 16 ft in width and full height of the side from floor level to underside of the top chord member. Auxiliary door components such as guides, tracks and stops are usually attached to the side subassembly while it is in its welding jig.

Figure 89.29 is a cross-section view of a typical side showing the principal framing members and welding details. The 0.10 in. thick side sheets are automatically butt-welded to the outer leg of the Zee-section posts. The side sheets lap the side plate on the top and the side sill on the bottom, and are

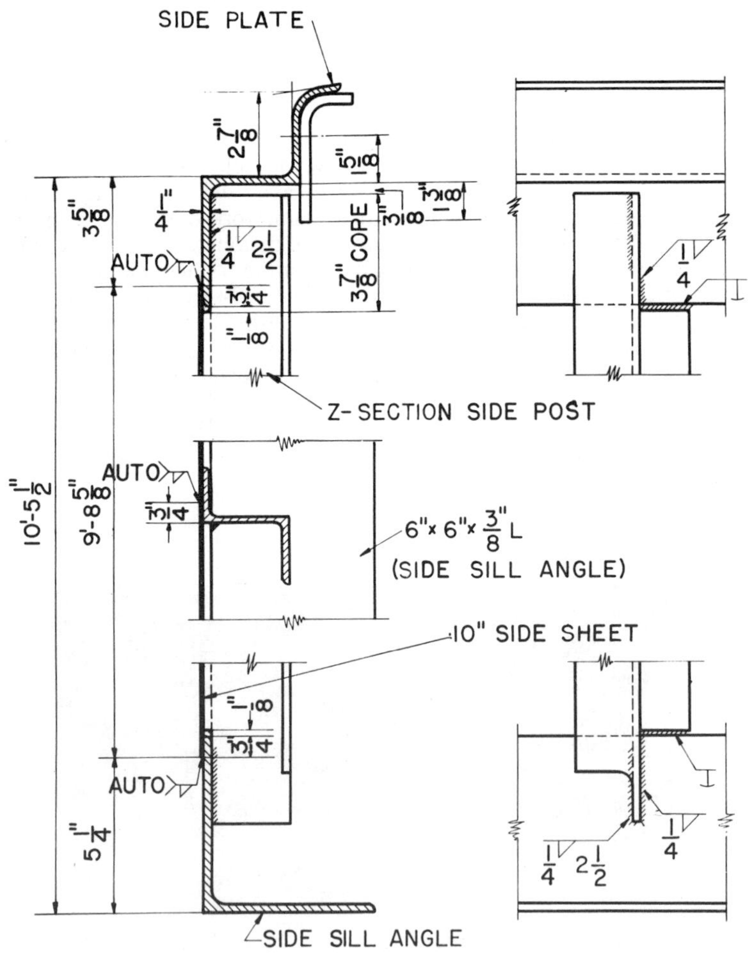

Fig. 89.29.–Cross section of a side showing principal framing members and welding details.

welded automatically by the submerged arc or gas metal-arc process for the full length of the car. The side sheets may be riveted or welded to the corner posts. The side sill is usually reinforced beneath the door opening.

Automatic submerged arc welding of the side sheets was the first high-speed double-head welding operation used on freight cars. Many carbuilders now use gas metal-arc welding; however, the procedures are similar with either process. The framing members are assembled and tack welded in a horizontal locating fixture. The side frame is then transferred to the automatic welding fixture, where the side sheets are located and tack welded in place. The fixtures are straddled by individually powered welding gantries used for welding the vertical seams. Submerged arc systems use two direct-current control and wire feed units mounted on the carriage; the power sources are mounted on a platform adjacent to the fixture. A guide wheel is used for locating the electrodes over the seam, and welding speed is 130 to 140 ipm.

The gas metal-arc high-speed welding process with shielding gas for attaching the side sheets to the Zee-section sideposts is similar to the above. After the side sheets are tack-welded to the side frame, the side is moved into the transverse welding position where it is tilted approximately 6 deg. Welds are made downhill at speeds of 150 ipm. Constant current or constant potential power sources are used with triple deoxidized solid welding wire or cored wire. The downhill welding helps to feather-in the weld bead as well as minimize excessive penetration into the side posts. One automatic welding machine operator will weld over 2500 ft of butt welds in eight hours.

CAR ENDS

Box car, refrigerator car, stock car and gondola car ends are often made from one or two pieces of steel with transverse corrugations hot-pressed in. The vertical edges are usually curved 90 deg for connection to the side assemblies.

When the ends are relatively tall, e.g., in box cars, the top and bottom portions of the pressed end are usually welded together by either gas metal-arc processes. Connection to the sides may be either by welding or by mechanical fasteners.

A comparatively recent alternate end configuration is one in which the end is an all-welded subassembly. In this case, the end is comprised of two side members, generally having a rectangular tube cross section, connected by a corrugated end sheet.

The corrugations may have a sinesoidal cross section; however, they more usually take a trapezoidal shape. The end assembly is tacked and placed in a positioner so that the serpentine weld connecting the corrugated sheet to the side member on each side can be made downhand. This end configuration has been used on both box and gondola cars.

IMPORTANT DESIGN DETAILS OF CERTAIN FREIGHT CAR TYPES: OPEN AND COVERED HOPPER CARS

A conventional open top or covered hopper car is constructed with either two, three or four compartments, each with its loading hatch and discharge outlet. When welded, it will contain from 2500 to 3800 linear ft of groove and fillet welds for cars of 2100 cu ft capacities respectively (see Fig. 89.30).

When these cars are made in quantity they are built on a production line where each welding position or station is required to complete a predetermined

Fig. 89.30.–One hundred-ton hopper car, L&N Series 190200–190369.

amount of welding in an alloted time. To maintain quality standards, control shrinkage and distortion all at the same time, it is desirable to prefabricate or subassemble as many parts as possible and weld them into principal assemblies limited only by size. In the construction of twin, triple and quadruple hopper cars, four principal assemblies are used.

Underframe

The underframe usually is contructed from railroad-type Zee sections welded together to form a center sill to which are assembled and welded the strikers, center fillers, separators and loading bolsters, as described above.

Car Ends

The body bolsters, upper and lower floor sheets with connecting gussets, form the principal car end assemblies.

Partitions

The cross ridge, center partition sheets and gussets are assembled in a similar manner forming the center partitions–one for twin, two for triple and three for quadruple hoppers.

Sides

Sides are generally built as described above.

The underframe, ends and partitions are jig assembled. This section of the body is lifted and placed on the assembly line and the sides are applied. The car then progresses down this line where all flat welding is completed and the carlines are attached. The car is then removed from this track and secured in a positioning device for the completion of welding. Returning the car to the track, the roof sheets are applied and the car is passed under traveling gantries where the roof joints are welded through into the carlines and longitudinal lap welding is completed. The loading hatches are then positioned, tack welded and welded, and preassembled running boards are secured.

Covered hopper cars have been used to transport food products and chemicals which must be protected from the elements and contamination by chemical reaction with bare steel on the interior surfaces of the car. This has required car interiors to be lined with an epoxy paint primer and a urethane finish coat.

These sprayed-on linings require the interior surface of the car to be free from sharp corners, ledges and rough areas. Weld joints must be smooth and well blended into the adjacent plate. Undercuts, craters, discontinuities, weld spatter, excessive convexity or concavity, porosity and slag are not permitted. The sprayed lining has a thickness of 5 to 8 mils, and after spraying, the entire surface is checked with a meter for uniform thickness. Blasting, touch-up, grinding, disking and blending, followed by thorough washing and vacuuming, are required to prepare the interior surface for application of the lining metal.

SPECIAL-PURPOSE FLAT CARS

Since about 1960, a whole family of uses has developed based on extra long flat cars. The basic flat car has evolved into one 89 ft 4 in. long over end sills and with two alternative deck heights. The standard-deck car has a deck height of 41 in. and the low-deck car has a deck height of 31 in. The low-deck car uses special 28 in. diam wheels in achieving its low-deck height.

In the late 1950's use of these cars for piggyback highway truck trailers and special racks for new automobiles began to rise. This car type has also been used increasingly for transporting intermodel-type shipping containers.

More recent developments include modification by the addition of a superstructure to make a Vert-A-Pac car. This car has five large doors on each side that fold down and rest on the ground. Three subcompact autos are loaded aboard each door and when the door is closed, the autos hang nose-down on special hooks. Thirty autos are hauled in one Vert-A-Pac compared with a maximum of 18 similar autos on a conventional tri-level rack-equipped car.

Another recent use of the long special-purpose type car is shipment of three-tiered containers, each handling three standard automobiles. Four containers are carried in pedestals on each flat car.

Fig. 89.31.–A low-level piggyback car, 89 ft long, weighing 56,300 lb with a 130,000 lb carrying capacity.

Several manufacturers have standard 89 ft flat car designs that are manufactured for stock and are available "off-the-shelf." A typical example is the car shown in Fig. 89.31 which is of all-welded design. It contains 36,000 lb of steel, most of which is high-tensile strength low-alloy material, and is 89 ft 7 in. in overall length. It weighs 56,300 lb with trucks and trailer hitches, and it can transport 130,000 lb.

The body structure is made up of 3/8 or 7/16 in. thick steel plate which forms the car deck or floor. The floor is formed by automatic submerged arc welding sections into one piece 85 ft long. These butt welds are made from one side with a flux backup which results in 100% penetration with reinforced top and bottom. The completed floor plate is moved to a fixture in which pressed channels are fitted and tack welded to the underside. These give the floor longitudinal stiffness. They are automatically fillet welded to the floor by submerged arc welding heads mounted on a gantry. The center sill section consisting of the web plates and bottom cover are built up in a subassembly using automatic and semiautomatic submerged arc welding. The end sections of web plates, bottom covers, bolsters, center filler and center plate are also constructed into a subassembly using both semiautomatic submerged arc welding and manual shielded metal-arc welding.

A typical side sill assembly consisting of groove-welded seamless tubing and formed side plate is submerged arc welded using tandem welding heads at speeds of 100 ipm (Fig. 89.32). Other side sill designs incorporate angles, channels and rolled sections. These individual assemblies are fitted together in a master jig, and all flat and horizontal welding is completed using both the semiautomatic submerged arc, gas metal-arc, and manual shielded metal-arc welding process predominately with E7024 electrodes.

Fig. 89.32.–Side sill assembly fabricated by submerged arc welding at speeds of 100 ipm.

The entire structure is then placed in a positioning device for final welding. The shorter welds are made with covered electrodes while the long runs are made with semiautomatic or automatic welding processes.

Upon completion of all welding, the car structure is then placed on trucks and the trailer hitches are welded to the car deck. The car then moves for application of brakes, couplers, bridge plates and safety appliances followed by final cleaning and painting.

This car contains over 3200 linear ft of welding, 1200 ft of which is applied by submerged arc welding. A total of approximately 185 lb of coiled wire and 475 lb of covered electrodes is used on each car. Welding machine operators average 1450 ft of welding in eight hours. Semiautomatic averages 380 ft and manual welders average 200 ft.

TANK CARS

Railroad tank cars came into being around 1860 and consisted of wooden barrels mounted on flat cars. They were used primarily by the petroleum industry. Cars of this type were largely fabricated by carpenters.

Most tank cars are now fabricated from carbon steel; however, other metals sometimes used include weldable aluminum alloys and stainless steel alloys such as type 304, type 304 ELC, type 316 and type 430 chromium iron. Nickel steel is used for cryogenic liquids.

Ladings transported by tank cars include liquids, gases, chemicals, slurries meltable solids and emulsions.

The Interstate Commerce Commission and the Association of American Railroads regulate the design, specifications and codes for the manufacture of tank car underframes and vessels. These tanks or vessels are designed for a specific use and, consequently, there are many cases where tank cars designed for a certain type of product cannot carry another product because of the nature of the base metal from which the vessel is fabricated. In all cases, tank cars are designed around a specific testing pressure, and the operation of the tank car must remain within the limitations of this design. The choice of base metal is made in accordance with the commodity that is to be transported or by a given line of commodities from a standpoint of corrosion. There are also cases where the choice of base metal is made to take advantage of the added strength of a certain base metal, thereby enabling the design engineer to use a thinner plate and effect a weight saving.

A large percentage of modern tank cars is of monocoque (center sill-less) design. In this instance, no separate underframe is used and all train forces are carried through the shell of the tank. Elaborate gussets and bracing are required to support the shell on stub underframes at each end. The stub underframes connect to the car's trucks, couplers and braking gear (see Fig. 89.33).

Codes

The codes under which tank car vessels are built are almost identical to the ASME Boiler and Pressure Vessel Code. Design of each new car type must be approved by the Committee on Tank Cars of the AAR. The requirements are the same as ASME for the qualification of welding procedures and welders. The specifications for radiographic inspection are the same as for any unfired pressure vessel. However, the AAR Code permits what is known as a

Fig. 89.33.—Welded tank car 61 ft long with a capacity of 32,900 gallons. The car is used to transport liquefied petroleum gas.

"qualification" of a straight run of similar tanks by having a minimum specified number of defects in the first group of vessels; this indicates that the ability of the welding machine operators and the procedures established for that particular base metal and type of joint are compatible, and will produce welds that meet specifications. The Code then permits the welding of the next numbered group of tanks as long as they are of the same design, class and type of material; they are produced with only X-ray inspection at the junctions of the seams. This is approved and permitted only on low-pressure tank cars. All other tank cars are required to be 100% radiographed, and the radiographs are required to be kept for a period of five years. Radiography is required only on the butt joints of the vessel and not on any fillet welds or on any part of the underframe.

Other things that govern the classification of a tank car include the types of safety valves and vents, the type of dome and manway, the fittings in the dome or in the manway, whether unloading is bottom outlet or by pressure from the top through the manway or dome, and whether the tank is riveted or welded, insulated or uninsulated. If it is insulated, it would have an outer jacket to hold and protect the insulation against the tank shell. Some tanks are lined with rubber, paint, or plastic linings or through a nickel cladding process. As many as six compartments have been incorporated in the tank car design.

Tank car heads are either pressed or spun and the edges of the plate are usually prepared with a bevel or, if the thickness is great enough, a double bevel and small nose will be either planed (in the case of aluminum or stainless steel) or (in the case of carbon steel) cut with multitip oxyacetylene cutting torches. This is done by placing the head on a rotating turntable with the cutting torch assembly riding against the head, and cutting both bevels and nose or land at the same time.

The plate edges are prepared in the same manner. For steel, this operation is effected with a bridge type of fixture on which the torch assembly floats on the plate. This is done in order to follow any buckling of the plate that might be present. The sheets or plates are then rolled on large mechanical rolls to the diameter necessary for the size of tank. The plates are fitted and tack welded into position; they are then welded on one side completely, then, when necessary, they are gouged on the opposite side either by mechanical means, or by oxygen or arc cutting. The last side is then welded with the automatic welding process. The tank is then X-rayed and the radiographs are interpreted by trained personnel. After any defects are removed, repairs are made and re-X-rayed.

After the tank has passed radiographic and dimensional inspection and all required appendages have been welded, it is ready to be stress-relieved. Carbon steel tank cars are generally stress-relieved at 1150 F and tank cars of stainless steel are stress-relieved at a temperature suitable for the specific grade of stainless steel. Aluminum tank cars are not stress-relieved. Some aluminum tank cars are built of heat-treatable aluminum which is heat-treated prior to fitting for welding. When this aluminum type is used, sections are designed around the weakest portion of the joint, which is the heat-affected zone of the weld.

After stress relieving, the tank car is insulated and the jacket applied. If it is an uninsulated tank car, other appendages such as grab irons and safety appliances are applied.

The repairing of tank cars is controlled by the Interstate Commerce Commission and the Association of American Railroads. Welders should be qualified in the same manner as welders performing work on new products. Repairs should be radiographed as are welds on new tanks. In many cases, local stress relieving is permitted when the repair is not too large. Occasionally, however, the repair is so large that the tank must be removed from the underframe and stress-relieved as a whole unit in a large stress-relief furnace. Unauthorized repairs by unqualified persons are prevented by referring such qualifications and specifications, as well as the execution of an application for approval of welded tanks, to the Association of American Railroads.

ALUMINUM FREIGHT CARS

The strength, light weight and corrosion resistance of aluminum are characteristics which are desirable for railroad rolling stock. Some of the most rewarding applications are for the captive cars that complete a carrying job without leaving the railroad's own system. Typical are the covered hopper cars used for cement, salt, fuels and food products and gondolas for coal, ores and similar commodities.

The ease and speed with which aluminum can be welded have been prime factors in the successful use of this metal.

Presently, aluminum cars are more expensive to build than those of steel because of the current five-to-one differential in base metal cost. However, the light weight of the aluminum car greatly increases the load-carrying capacity, reducing the number of cars needed to transport the payload. In addition, these factors are also valuable assets of aluminum: lower maintenance costs, resistance to weathering in the unpainted condition, resistance to many of the corrosive cargoes, smooth interior that aids in cleaning when interchanging cargoes and an ultimate high scrap value.

Design

Advantage is taken of the large single plates available in aluminum. An example is the side plates for the quadruple hopper car which is 49 ft long by 9 ft 4 in. wide. This eliminates many groove welds necessary with smaller plate sizes. Inexpensive special extruded shapes, some with built-in edge preparation, which have been developed to simplify construction and give design freedom are shown in Fig. 89.34.

Care should be taken by the designer to give good joint accessibility with the gas metal-arc welding process and to use convenient connections between the

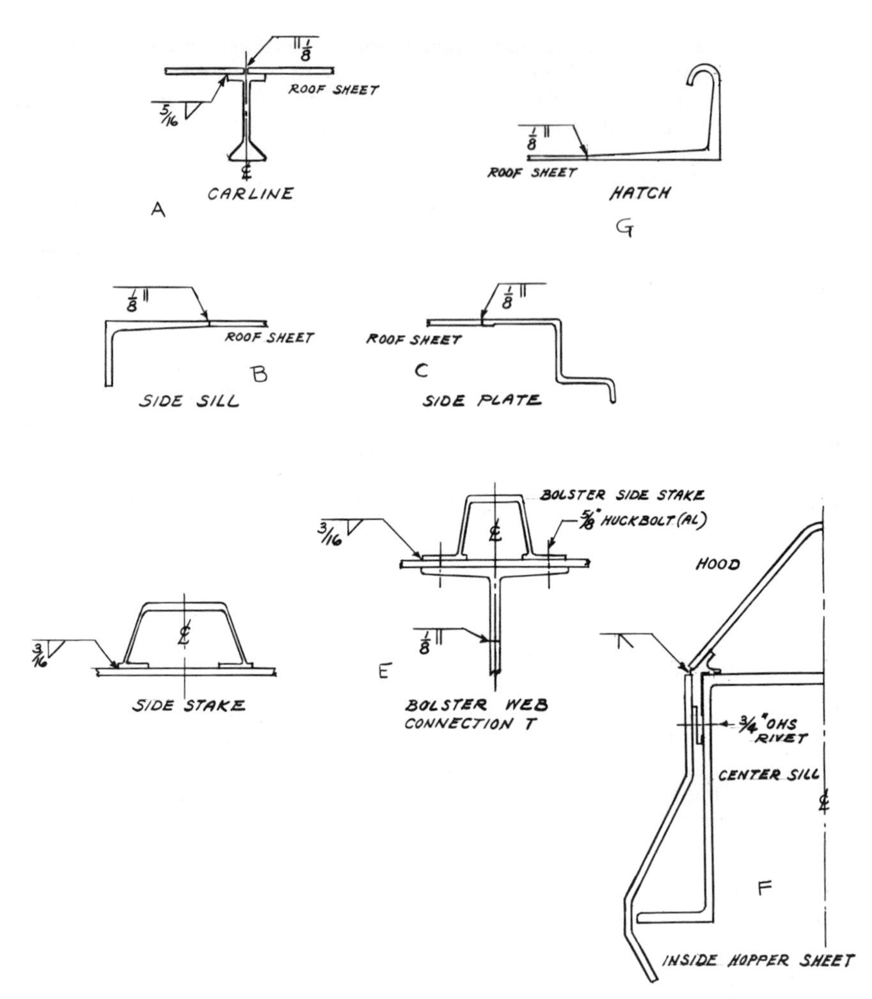

Fig. 89.34.–(A) Carline–extruded shape for roof support. Roof sheets 3/16 in. thick, groove welded into carline. (B) Side sill–extruded shape welded to side sheet without additional edge preparation. (C) Roof sill–extruded shape welded to side sheet. (D) Side stake–reinforcement for side of car. (E) Chute extrusion and bolster web connection showing edge preparation for groove weld. (F) Bolster stake–extrusion riveted to center sill with groove weld tying in inside hopper sheet and hood over center sill. (G) Hatch.

principal assemblies joined at the main assembly line to compensate for minor dimensional changes caused by weld shrinkage.

Base Metals

The aluminum cars are generally made of 5083-H112 or 5083-H321 alloy for plate and sheet varying in thickness from 3/16 to 5/8 in. Where severe forming is required, 5083-0 alloy is used. The extrusions are 5083-H111 alloy of various special shapes developed to simplify construction. The welding electrode is usually 5356 alloy, in 3/64, 1/16 and 3/32 in. diam.

Cleaning

Surface preparation of aluminum prior to welding is required to hold weld porosity to acceptable limits. Components are immersed in a heated (160 F) dilute (15%) solution of phosphoric acid for a period of five minutes followed by a water rinse to remove dirt, grease and all contaminants accumulated during fabrication. Larger members, such as side sheets, are stainless steel power wire-brushed within the welding areas and washed in a similar manner.

Welding

The gas metal-arc welding process is used with both fully automatic and semiautomatic wire feeders coupled to push and pull type guns usually powered by constant-voltage type, d-c reverse polarity power source. The shielding gas is piped to the welding stations from a central source, usually a liquid converter.

Fabrication

A conventional aluminum covered hopper car is usually made with two, three or four compartments, each with its loading hatch and hopper outlet. When welded it will contain from 2500 to 3800 linear ft of groove and fillet welding. It is constructed much the same way as a conventional covered hopper car. In the construction of twin, triple and quadruple hopper cars, five principal assemblies are used.

PRINCIPAL ASSEMBLIES

Underframe

The underframe is usually constructed from two special railroad-type mill rolled steel Zee sections welded together to form a center sill to which are assembled and welded the strikers, center fillers, separators and loadbearing bolsters. The aluminum body connecting members are then riveted to and insulated from the steel Zee sections by a coating of nonhardening butyl rubber compound to retard electrolytic action.

Car Ends

The body bolsters and upper and lower floor sheets with connecting gussets form the principal car end assemblies.

Partitions

The cross ridge, center partition sheets and gussets are assembled in a similar manner forming the center partitions, one for twin, two for triple and three for quadruple hoppers.

Sides

Car sides are made on an adjacent assembly line where the bottom sill extrusion, side sheet and top extruded plate are jig assembled with side stake stiffeners. All of these are completely welded with automatic welding heads mounted on overhead traveling gantries.

Roofs

Two types of roofs are available for aluminum covered hopper cars. A trough hatch roof is used where the shipper desires a continuous opening over all hoppers for fast loading. Covered by interlocking hinged covers the trough hatch roof permits opening of one or all hoppers as desired. This roof is made from two full length 3/16 in. thick aluminum sheets subassembled and square-butt welded to the extruded hatch frame (Fig. 89.34G) forming the roof sides; two end closure sheets with a similar extrusion completes the car roof and rectangular hatch.

Round hatch roofs are usually made from several transverse sheets prepunched for the application of 30 in. diam cast aluminum hatches. These are automatically clamped and fillet lap welded to the roof sheet; the transverse sheets are square-butt welded into the carlines (Fig. 89.34A).

The underframe, ends, partitions and outside hopper sheets are now jig assembled and tack welded. This body section is lifted from the jig and placed on trucks in the first position of the assembly line. Here the sides are applied. The car moves progressively along the track past several welding stations where all flat welding to the underside is completed. The cars are lifted from the track, secured into rotating head and tail stocks, then turned 90 deg in both directions for the application of roof carlines and the completion of all welding that would otherwise be out-of-position work. Returning the car to the track, the roof is applied and passed under traveling welding gantries for longitudinal lap and transverse butt welding.

MAINTENANCE AND REPAIR TO FREIGHT CARS

Heavy repairs to freight cars, having sustained extensive deterioration from continued use and/or wreck damage, are generally carried out under a program governed by the car construction engineering department of the individual railroad. Repairs of defective components are generally made through replacement in kind, including section size and grade of steel. Because prevalent amounts of high-strength low-alloy (HTLA) steel are being used in principal load-carrying members of the more recently built freight cars, the grade of steel used for the repair is particularly important in order to preserve structural integrity. Nevertheless, engineering calculations are made to assure that the replacement repair will maintain the design structural strength and encompass any upgrading to meet the requirements specified in the A. A. R. Specifications for Design, Fabrication and Construction of Freight Cars. It is quite evident that a close liaison between the engineering department and the repair shop is required to properly repair freight cars.

Quality workmanship for the repair is expected and is assured through careful visual inspection and the testing and qualification of welders. Welders are tested and qualified by individual railroads in accordance with test procedures set forth in Section 6 of the Electrical Section of the Manual of Standards and Recommended Practices. In addition, all welders are generally qualified carmen and hence familiar with the A.A.R. Interchange Rules.

Because the cars of all railroads are offered and accepted in interchange, the periodic inspection, maintenance and repairs to the cars are governed by the Association of American Railroads and the Department of Transportation (viz., safety requirements). Repairs are made in accordance with the A.A.R.

Interchange Rules which were formulated as a guide for properly handling all matters concerned in the interchange of freight traffic with regard to determining responsibility and providing an equitable basis for charging such repairs and damages.

The Interchange Rules are quite specific with regard to the inspection, gaging and wear limits of various car components (running gear) which, if exceeded, are sufficient cause for removal. Also, the rules are specific with regard to welding car components: the welding of some components such as wheels, axles and safety appliances which can be unsafe and often disastrous are prohibited while the welding and salvage of other components such as truck sides, bolsters, center plates, couplers, knuckles and yokes are permitted.

Draft Components

Coupler, knuckle and yoke castings are currently obtained in three grades of steel, i.e., Grade B, Grade C-high tensile and Grade E-high tensile. Selection is made on the severity of service: general service or unit train service.

Grade B (70,000 psi T.S.) castings are salvaged by welding cracks and building up worn areas with E7015, E7016, E7018 low-hydrogen covered electrodes or equivalent flux cored wires, followed by normalizing.

Grade C-high tensile (90,000 psi T.S.) castings are similarly repaired using E9015, E9016, E9018 low-hydrogen covered electrodes or equivalent flux cored wires, followed by normalizing and tempering heat treatment.

Grade E-high tensile (120,000 psi T.S.) castings are salvaged with E12015, E12016 and E12018 low-hydrogen covered electrodes, followed by a normalizing or quenching and tempering heat treatment.

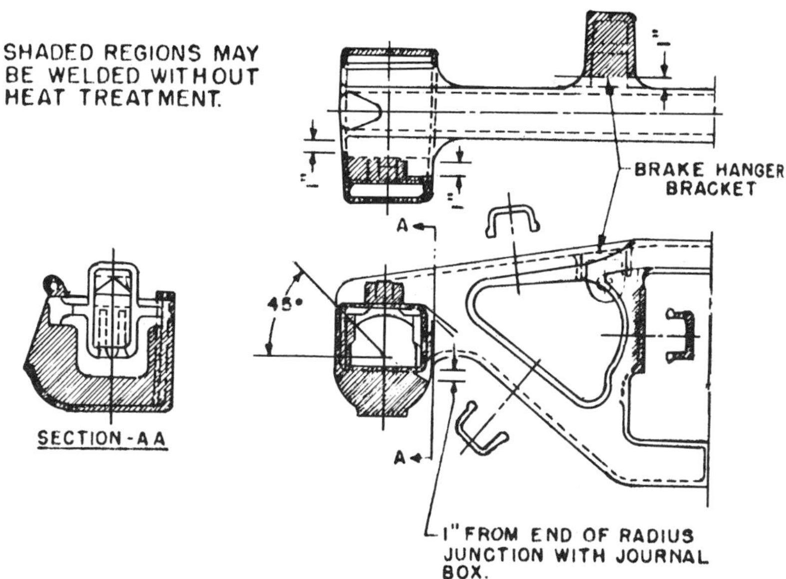

Fig. 89.35.–Reproduction of interchange rules for repair welding cast journal boxes without heat treatment.

To reduce coupler shank wear, hardened wear plates are often attached by welding, using low-hydrogen electrodes.

Truck Components

Truck sides and bolsters are cast in Grade B and Grade C-high tensile steels and are similarly salvaged and/or repaired as comparable grade draft castings mentioned above. The Interchange Rules are specific with regard to areas of the castings in which welding is permitted without heat treatment, as well as those areas where welding is prohibited unless followed by heat treatment (see Fig. 89.35 and 89.36).

To reduce wear of truck bolster center plate bowl, horizontal and vertical 11–14% manganese steel wear liners are applied with the vertical ring wear liner secured to the bolster by intermittent or continuous welding as shown in Fig. 89.37 and 89.38. AWS Type E308-16, E309-16, E310-16, (or equivalent) stainless steel electrodes are used.

WELDING OF RAILS

The two basic systems used for pressure welding of railroad rails are the pressure gas and the flash welding processes. Both types of welding have been qualified for the American Railway Engineering Association's standard two million cycle rolling load tests for railroad rails and have given long term satisfactory main line service under fast and heavy density traffic.

Since most continuous rail is in one-quarter mile lengths, the number of bolted-and-bonded joints is reduced considerably and substantial savings are effected by decreasing joint maintenance expense. Although reduction of

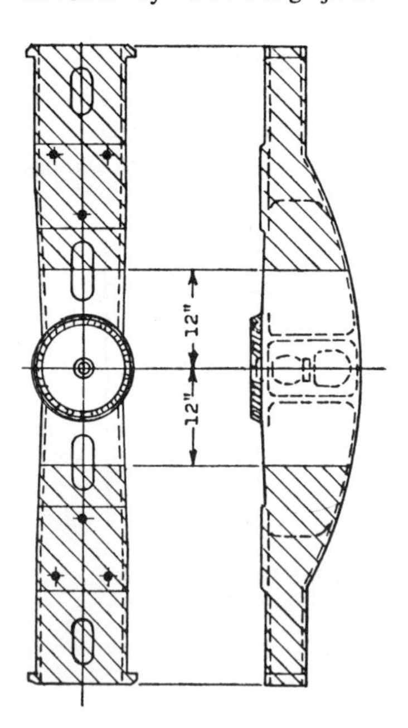

Fig. 89.36.–Reproduction of interchange rules for repair welding cast steel bolsters which require postweld heat-treating.

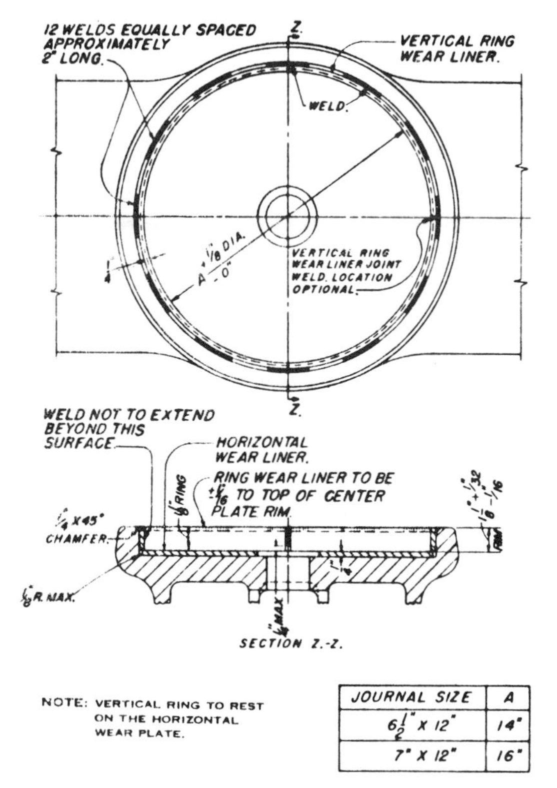

JOURNAL SIZE	A
6½" X 12"	14"
7" X 12"	16"

Fig. 89.37.–Application of two-piece horizontal and vertical wear liners to the truck bolster center plate with intermittent weld per A.A.R.

maintenance offers immediate savings, welded rail promises greater savings from longer service life since most jointed rail is removed from the track because of worn joint conditions. Most long welded rails are joined by joint bars which require joint maintenance and additional rail anchors. However, some rails are joined by thermit welding processes to be continuous between insulated rail joints.

In either system of rail welding, the rail must be moved initially from railroad cars to stockpiles or to simple or automatic transfer racks or conveyor lines. This can be either a simple or sophisticated operation depending on the manpower savings possible, available money and the permanence of the welding site.

GAS PRESSURE WELDING

This method of welding continuous rail requires perfectly parallel and meticulously cleaned rail end faces. This is accomplished by making a saw cut with a thick blade between rail ends butted together or by cropping each rail end with a precision type abrasive cutoff machine. When there is a delay between cleaning and welding, the ends are protected with a thin oil coating to prevent corrosion. These ends are then carefully cleaned with a solvent such as chlorasol immediately preceding the welding operation.

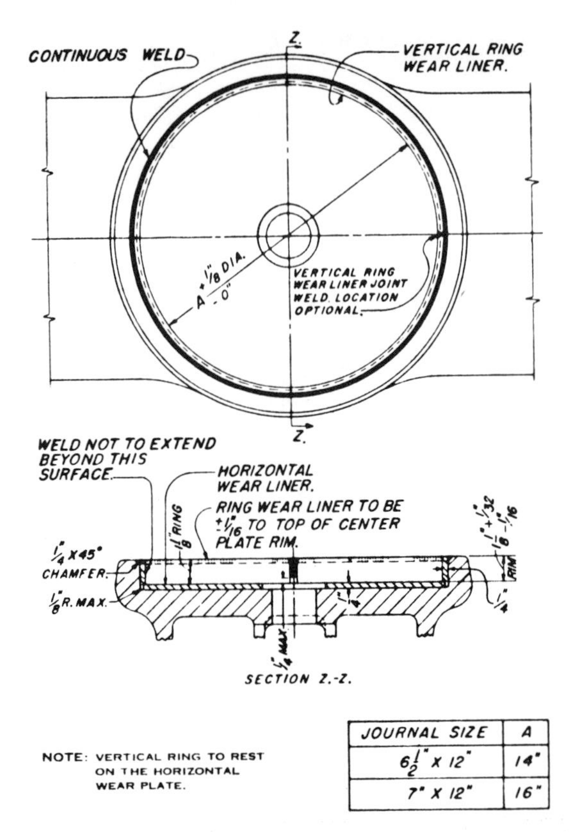

JOURNAL SIZE	A
$6\frac{1}{2}$" X 12"	14"
7" X 12"	16"

Fig. 89.38.–Application of two-piece horizontal and vertical wear liners to the truck bolster center plate with continuous weld per A.A.R.

After cleaning, the ends are abutted and the rails are gripped by cam-acting clamps to force the rail ends together under a hydraulically imposed pressure of 3000 psi of the rail cross section. Heat is applied by an atmosphere-excluding envelope of flame from four rail contoured heads containing a myriad of small heating tips. These heads are oscillated automatically through a desired stroke to bring the ends to a uniformly heated short column plastic condition. As the ends approach the proper state of plasticity, the empirically determined end pressure of approximately 3000 psi will upset the ends slowly. When a 7/8 in. upset is obtained, the cohesive temperature will have been crossed, and the flame and pressure are then turned off to complete the welding process.

After welding, the rail is moved to the next station to be ground. The entire upset bulge is removed on the base and head and partially in the web zone by abrasive belt grinding machinery while the weld is still hot. Formerly. this bulge was partially removed while red-hot by oxygen cutting and was finished by solid abrasive wheel machinery.

The gas pressure welding process is a relatively simple one not requiring the massive and complex equipment used in the flash welding process. The system, however, though functionally simple, required much development work to evolve an effective process. Such items as pressure regulators, quick-shutoff valves, provision for fine adjustment of the flame ratio, ease of setting up rail ends for welding and determination of proper relationship between temperature and upsetting pressure are required for optimum weld quality. In addition, much thought has gone into the design for rigidity of the clamping and upsetting system to maintain precise alignment of the rail ends. Also, the simplicity of the system allows a good deal of welding operator integrity and care to be reflected in the weld quality.

FLASH WELDING

This method of welding continuous rail demands little if any rail end face preparation, but it does require that all mill scale, corrosion, dirt or other foreign material be removed from the tops and bottoms of the rail ends where the electrodes make contact with the rail. Cleaning is necessary to provide a good distribution of welding current over the electrode contact surface to avoid areas of excessive current density which could cause metallurgical damage to the surface of the high-carbon rails. This cleaning is generally done with abrasive belt grinding machinery or with manually-held grinding equipment. After cleaning, the rails are clamped and aligned for welding.

There are several different makes of rail welding machines, all of which have a means of hydraulically clamping and aligning the rails, with automatic controls for the welding process. All machines have welding transformers closely connected to electrodes which produce high currents at low voltages through the rail ends. These machines do equivalent jobs of welding but employ slightly varying techniques and mechanisms to accomplish the same ends.

Usually some initial flashing is used to match rail end configurations, but all machines use a short duration off-on pulse type short circuiting of the rail ends to preheat them to a red temperature preceding final flashing. This final flashing may or may not be accelerated and is terminated by upsetting with the current shutoff. A timer action, a limit switch controlling the power supply or a magnetic circuit breaker can effect the current shutoff.

Electrode voltages vary with the size of rail to be welded and with the frequency of the power supply. For example, on 136 lb per yd rail (13.35 square in. area), the open circuit electrode voltage will vary from a low of 6 to 7 V at 10 Hz, 8 to 9 V at 25 Hz and 10 to 13 V at 50 to 60 Hz. Because of less reactance to voltage drops at the lower frequencies, lower electrode voltages can be used for preheating and, in turn, provide this lower voltage for flashing without further switching or additional power supply control circuitry. Maintaining the lowest voltage consistent with easy flashing is considered a desirable flashing technique. Whether the low or high frequencies are used depends upon the design considerations preferred by the manufacturer.

A 600 kVA commercial three-phase power installation is considered adequate for all railroad rail, although the single-phase instantaneous pulse loads at the welding machine may reach 750 kVA on larger rail such as a 136 lb per yd section. Load balancing is generally accomplished with motor-generator sets or with electronic converting equipment. Some portable welding plants use diesel-generator sets for power generation.

After the weld is made, the bulk of the upset may be removed by a number of means. One system used a burr-removing shear device which closely surrounds the rail section and is positioned in the welding machine outside the clamping zone while the weld is being made. After the weld is completed, the shearing device is moved adjacent to the weld where it is clamped and pushed across the weld to shear off the burr to 1/16 in. of the base section. Other systems use a stripping-broaching technique whereby the plastic weld is pushed or pulled through a series of cutters each of which strips a short section of the upset until the entire periphery is progressively removed. Another system removes the upset by grinding while the weld is still red hot.

Finishing of the welds is accomplished with solid grinding wheel machinery or with automatic or semiautomatic belt grinding equipment.

Unloading the long strings is accomplished by anchoring one end of the long strings to the ground, two at a time, and then pulling or pushing the railroad cars out from under the long rails for transfer to the roadbed to await the rail laying crew.

THERMIT WELDING

The thermit method of field welding of rails dates back to the turn of the century. Its basis is the alumino-thermic reduction of iron from iron oxide, Fe_2O_3. The reaction $Fe_2O_3 + 2\ Al$ produces heat plus slag plus iron. Of course, various additives produce the mechanical properties desired for a particular rail stock. The temperature of the reaction is approximately 2450 C, including losses by radiation and conduction.

In the United States, these two processes are now being used for making molds for thermit welding of rail:

1. The Croning Process.—A mold is formed from thermosetting resin-bonded sand mixtures and brought in contact with preheat (300 to 500 F) metal patterns. This results in a firm shell with a cavity corresponding to the rail section to be welded.
2. The CO_2 Mold.—A mold is formed out of clay binding or pure quartz sand with water glass binding by passing CO_2 through the mold for hardening.

There are basically three types of thermit welding applications in the rail industry:

1. Welding with a collar—preheating.
2. Welding without a collar—preheating.
3. Welding without a collar—no preheating.

Preheating is done by a torch or a flammable material coated to the walls of the mold.

Endurance limit testing of various rail samples has shown that the method of preheating without a collar is superior. Obviously, the collar design produces stress concentrations resulting in lower fatigue. For many years it was believed that the collar added strength.

Thermit welding can be applied to rail joints between ties, rail joints on ties, composite joints with different rail profiles and different degrees of wear, and special joints such as switches, etc.

These are the typical mechanical properties obtained from the joining of two steel rails (U.S.) 136 lb per yd: weld hardness: 284 BHN, rail hardness: 254 BHN, deflection: 15/16 in.,and breaking load (39 in. center): 173,000 lb.

The weld stages are similar for both the Croning and the CO_2 mold processes. The six stages (with process differences) are:

1. The welding gap is obtained allowing some super-elevation to accommodate the cooling of a weld. The ends are usually straight.
2. The molds are applied with luting sand. The Croning shells sometime require certain sealing pastes or asbestos rope.

3. Preheating is applied by an external burner, or an external burner is used to ignite a flammable section of the mold. The preheating time is from 6 to 12 minutes depending on the rail section and method of heating.
4. The welding compounds are poured into the crucible, ignited and tapped. The CO_2 mold process uses an independent crucible and a trained welding operator; tapping is based on the reaction. The Croning process has the crucible built as part of the mold and uses the melting of steel disks as its tapping control.
5. After three or four minutes, the CO_2 mold process allows mold removal; however, longer time periods are required for the shell molds. Removal of the protruding metal is accomplished in various ways.
6. After rough removal of the excess metal, the head and edges of the base are ground to match the section profiles. All other excess metal should be ground down to within 3/16 in. of the rail profile.

Research is continuously progressing toward improving the mechanical properties of welds and, in the last ten years, it has provided the railways with a much improved thermit method.

RAIL INSPECTION AND HANDLING

Testing at the welding plant consists mostly of some form of magnetic particle inspection, because this system lends itself readily to testing at the elevated temperatures encountered on this type of production line. After the rails have been laid in track, the welds are generally tested by ultrasonic methods although several railroads have used radiography to some extent.

The anticorrosive application is most easily done at the welding site while the weld zone is still hot. The anticorrosive material is applied with a brush or sprayed on by aspirating the anticorrosive from a container with high pressure air.

The long strings of continuous welded rail are generally pushed onto a string of railroad flat cars each equipped with special roller assemblies to convey and support several strings. The earlier roller assemblies supported 12 strings of rail while later versions carry as many as 24, 30, 40 and 54 strings (Fig. 89.39). Each string of rail is carried on an individual roller of the multiroller assembly mounted on each car. Although most strings are progressively loaded on flat cars one rail at a time as each weld is made, sometimes the long strings are stored on

Fig. 89.39.—View of roller fixture mounted on flat cars for conveying and shipping continuous welded rail from welding plant to laying site.

racks for shipment. Before shipping, the long strings are secured with special clamps which tie the rails to a holddown fixture securely anchored to the center flat car. This center car anchoring of the long strings divides both the slack action and the apparent shortening and lengthening of the long strings as they bend around the outside and inside of curves while in transit.

RAILWAY MAINTENANCE BY WELDING

General Yard and Trackage

Arc and oxyacetylene welding have been used for many years in the maintenance of railroad way departments for the repair and reclamation of many metals previously removed from the track and replaced. Many items can be repaired in place for less than the replacement cost. Very often these repaired items will outlast new parts. This part of the chapter outlines, in a general way, most of the major applications of welding in the maintenance of way departments.

Rail Ends

One of the most widely applied welding operations is the building-up of battered rail ends.

The service life of rail, joint bar and joint ties is materially increased. This, plus a lengthened cycle for ballast cleaning, track surfacing and less damage to rolling stock, provides a worthwhile saving.

The welding program should be governed as follows:

1. Repair should be undertaken where 75% of the rail ends have a batter of 0.025 to 0.080 in.
2. Generally the length of track scheduled should be one mile or more but shorter lengths that can be coordinated with other programs can be scheduled.
3. No out-of-face welding should be scheduled on rail that has an average head wear of 3/16 in. or more.

Rail ends having a batter of more than 0.080 in. or requiring a weld of 16 in. or more should be removed and cropped, but an occasional joint of this type may be taken care of by out-of-face welders.

Rails requiring welding and having more than four driver burns should be removed from the track prior to the arrival of the welding crew.

Three different patterns for rail end welding are used. They are the strip weld as shown in Fig. 89.40A and B, the 3/4 ball shown in Fig. 89.40C and the full ball in Fig. 89.40D.

The pattern chosen should be dependent on such things as the amount of batter and gross tons of traffic. The weld limits should be marked out as shown in Fig. 89.41, but the weld limit must not go beyond the last angle bar hole.

The three processes generally used for rebuilding rail end are oxyacetylene, shielded metal-arc and gas metal-arc welding.

The most satisfactory and economical results can be obtained if the following is done prior to welding:

1. Replace all rails having excess damage.
2. Renew worn joint bars or tighten all bolts if bars are not worn.
3. Surface the track properly.

The welding must not precede but should follow, as soon as possible, after this

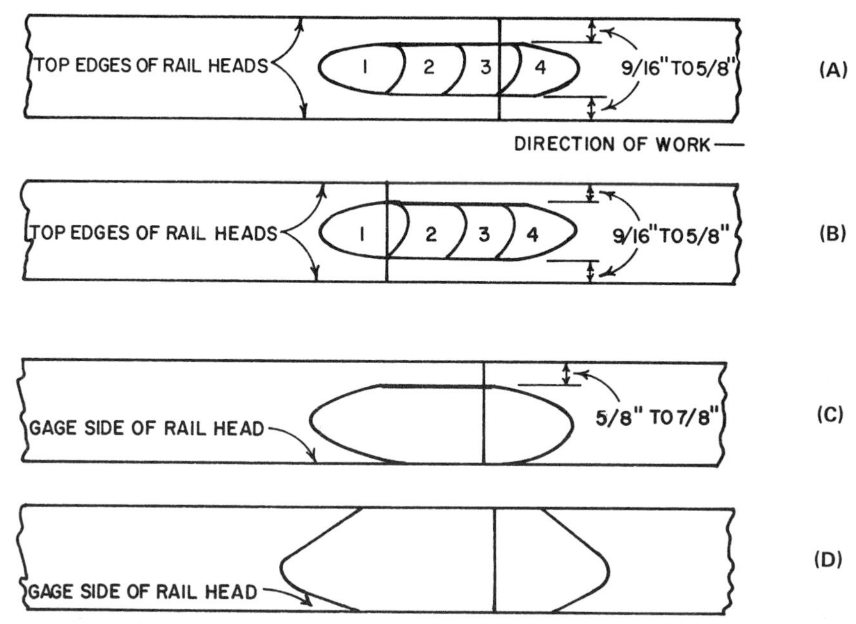

Fig. 89.40.–Typical methods of rebuilding rail end batter. For light tonnage tracks, use sequence shown in (A) or (B). For medium to heavy tonnage tracks and joints with batter of 0.080 in. or more, use sequence shown in (C) or (D).

sequence. The welding must be programmed to take advantage of dead track whenever possible.

Oxyacetylene Welding

The size of the welding tip necessary may vary from one which consumes approximately 60 cfh of oxygen for lightweight rail to one which consumes 80 cfh on heavier weight rail.

A neutral flame should be used for preheating in preparation for application of the weld metal, for melting out loose or laminated metal and for reheating to smooth with a flatter or to make a cut with a hot-cut chisel. Generally, joints are surface ground and cross-slotted by use of grinding machines. Instructions for cross-grinding and cross-slotting rail ends are shown in Fig. 89.42.

In applying metal to rail ends the welder must work in small patches, thoroughly hammer-working each patch before welding the next patch. Weld metal must extend out into the gap on open joints so that when cross-ground, a full rail end exists. On tight joints, the weld metal should be carried full width across the joint gap to approximately 1 to 2 in. past the joint. After peening, reheat an approximately 1/2 in. wide area directly above the gap to a dull red color. This will ensure a square break and will complete the weld.

Grinding

Surface grinding should be done after cooling in order to make the welded area conform to the existing (not original) section of rail. Tolerance for rail

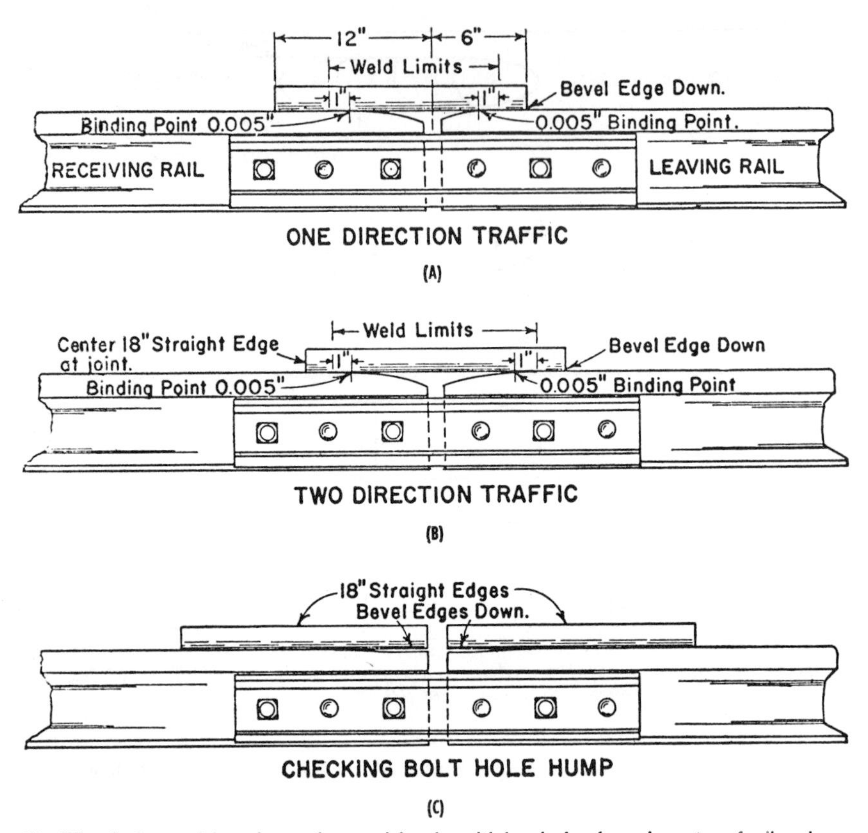

(1) When laying out joints always place straight edge with bevel edge down, in center of rail and perpendicular to the rail head.

(2) Place straight edge as shown in (a) for track having one-direction traffic and (b) for track having traffic in both directions and then proceed as follows:
1—Check if 0.015 inch feeler gage will go between straight edge and rail head (See (a) and (b).
2—If 0.015 inch feeler gage goes under straight edge, mark joint for welding. To determine length of weld, use 0.005 inch feeler gage and mark each rail 1 inch past point where 0.005 inch feeler gage binds under straight edge.

(3) If 0.015 inch feeler gage will not go under straight edge, check each rail as shown by (c) for possible "bolt hole hump." If rail shows a "bolt hole hump," mark for grinding. If rail does not show this hump cross grind only.

Fig. 89.41.–Method of measuring rail end batter.

surface grinding should be no more than 0.010 in. high or 0.005 in. low. Surface grinding and cross-grinding of rail ends which have cooled should be completed by the close of work each day.

Welding Rod Composition

The typical analysis of a suitable welding rod is carbon, 0.45%; manganese, 1.09%; silicon, 0.52%; chromium, 1.10%. A maximum Brinell hardness of 375 for deposited metal without heat treatment is most desirable. Harder deposits generally spall at weld ends. Welding rods of 1/4 in. diam by 36 in. long are generally the most suitable, but 3/16 in. diam by 36 in. may be used for light deposits.

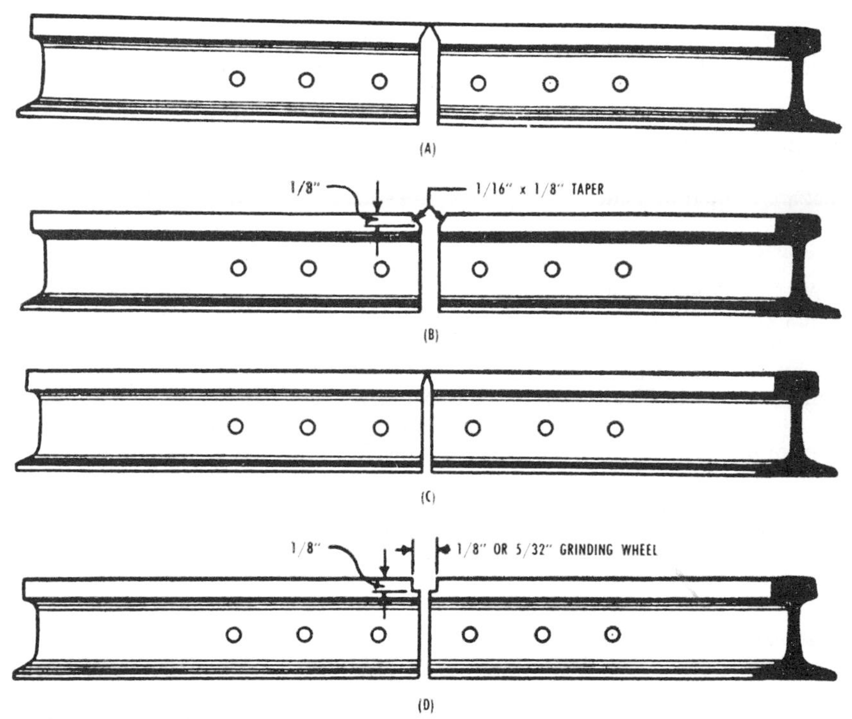

Fig. 89.42.–Typical instructions for cross grinding of rail ends. On joints where expansion is 1/16 in. or greater, the 5/32 in. wide grinding wheel should be employed to clean out all excess metal in the expansion area (A). The joint will then be as shown in (B). On joints where expansion is 1/16 in. or less (C), 1/8 in. or 5/32 in. wide grinding wheel should be used. Grinding should be done to a depth of 1/8 in. as indicated in (D).

Shielded Metal-Arc Welding

The following is a typical welding procedure employed for the arc welding of rail ends:

1. Preheat rail ends to 700 F.
2. Weld rail ends.
3. Postheat rail end to 1100 F.
4. Surface grind and cross-slot.
5. Blanket welds for three minutes or longer (applicable only when temperature is 25 F or less).

Preparation

Remove all flaws or defective metal by grinding, cutting electrodes or air carbon-arc cutting prior to welding. If the flawed metal is removed by electrode cutting, the rail end must be preheated to 300 F prior to metal removal.

Preheating

All rail ends should be preheated to 700 F before welding with an approved preheating device. The preheater should be positioned so that approximately two-thirds of the crucible length is placed beyond the weld limit mark, off the weld limit entirely. Where weld length is less than four in. the crucible is placed so that the inside end (end nearest joint) is approximately 1 in. from the rail ends. Preheating and welding must be coordinated so that welding starts within one minute after removal of the preheater from the joint. Reheat all joints that have been allowed to cool for more than one minute. This also applies to joints that are to be rewelded.

Welding

Welding should begin on the rail end requiring the shortest weld. Stringer bead welding is not recommended as this can cause deep crater cracks. Start the first pass about 1 in. back of the rail end and 1 in. in from the weld limit. Progress the weld to the rail end across to the outside weld limit, then down the outside limit to the weld length desired, and back along this pass to the starting point. All additional passes should be made in this manner, starting and ending at the same point. Use a 1/4 in. electrode and weave the electrode about twice its diam. This will produce a bead approximately 1 to 1 1/4 in. in width. Taper the weld ends out on the gage side so that there will be a gradual transition of the wheel load from the weld to parent metal. Welds must be carried back at least 2 1/2 in. from the rail end, regardless of the actual batter, to avoid excess hardness in the weld.

Postheating

Immediately after welding is completed, all welds should be postheated to approximately 1100 F. The postheater should be positioned in the same manner as the preheater. Length and correct amount of postheat will be determined by hardness tests. When welding is performed at temperatures of 25 F or less, all welds must be covered by approved rail blankets for a minimum of three minutes.

Surface Grinding

Rail surfaces should be ground so that the welded area will conform to the existing (not original) section of rail. Tolerance for rail surface after grinding should be no more than 0.010 in. high or 0.005 in. low.

Hardness Limits

The hardness limits at the surface of welds over the entire weld area must be within 330 and 390 BHN. The welder should check the hardness of at least ten joints each day.

Gas Metal-Arc Welding

The typical welding procedure consists of these five steps:

1. Preheat rail ends to 500 F.
2. Weld rail ends.
3. Postheat rail end to 700 F.
4. Surface grind and cross-slot.
5. Blanket welds for three minutes or longer (applicable only when temperature is 25 F or less).

Preheating

Use the same technique as outlined under shielded metal-arc welding, but set the preheat at 500 F instead of 700 F.

Welding

Run the stringer bead about 1 in. in width; run the wire back into the buildup when completing a pass to avoid deep craters. Start welding at the rail ends and work back to avoid overheating the rail end. Be sure the ends on open joints are welded out far enough so that when cross-slotted, a full end section is obtained. Taper the weld ends out on the gage side so that there will be a gradual transition of the wheel load from the weld to base metal.

Postheating

Use the same technique as outlined under shielded metal-arc welding, but set the postheat at 700 F instead of 1100 F.

Surface Grinding

Use the same technique as outlined in shielded metal-arc welding.

Hardness Limits

Use the same technique as outlined in shielded metal-arc welding.

FLAME STRAIGHTENING OF JOINT BARS

General

Joint bars are flame straightened when the bars are bent and will prevent proper surfacing of the track. The signal supervisor must be notified in advance of all flame straightening and should coordinate the work. Flame straightening of joint bars should not be performed while it is raining or snowing.

Preparation

Cover all bond wires with protective covers. Remove the two center bolts and raise the track spikes for three ties each side of the joint center 3/4 in. to allow the joint to come up as the bars cool. The amount of straightening is controlled by the size of the area heated and this size can readily be determined by making a few test heats.

Procedure

Using a neutral flame, apply heat with a large-consumption heating tip (125–150 cfh) to both joint bars simultaneously. Start at the center line of the bolt holes directly below the rail ends, and heat an area, extending down to the bottom of the bar, which increases from zero at the bolt hole to a width at the base of from 3 to 5 in (see Fig. 89.43). Heat this area to between 1400 and 1450 F (cherry red). The width of the lower section of the heated area will vary according to the amount of lift required: the greater the width, the greater the lift.

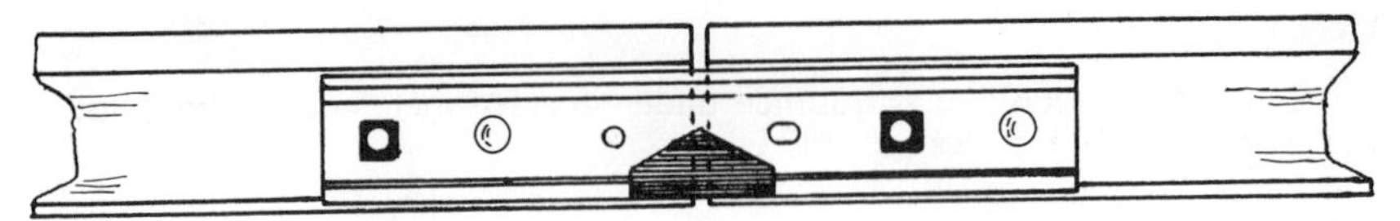

Fig. 89.43.–Sketch of heating pattern for bars.

Joints must be allowed to cool one minute before trains pass over them. All bolts must be tightened, and the ties under and on each side of the joint tamped as soon as possible after joint bars have cooled.

SWITCH POINT WELDING

Instructions for Building Up Worn Switch Points: Oxyacetylene Process

As a general rule, switch points may be repaired in the field if the following conditions exist:

1. Traffic conditions permit at least five hours of productive labor in an eight hour work day.
2. Switch point and stock rail have less than 3/8 in. head wear.
3. Overall weld length on switch point does not exceed 24 in.

Preparation

Grind off all overflow on the gage side of the stock rail opposite the switch point contact area, extending the ground area four in. beyond each end of the contact area. Grind off the overflow on the adjacent side of the switch point and all fatigued, laminated, chipped or otherwise defective metal. House the stock rail approximately 1/8 in. at the point end and 4 in. past the point end. Taper this housing back to zero at approximately 20 in. from the point end to protect it; then properly adjust the switch point.

Welding

The length of the weld must be governed by the existing conditions. The welding contour must be kept uniform so that it will permit the safe passage of trains at all times in either direction.

Oxyacetylene Welding Procedure

Use a 3/16 in. welding rod and a welding tip (point end) that will draw 20 to 30 cfh. For heavier sections, increase the tip size as the point widens. Select a tip that allows proper control of molten metal but does not overheat the weld metal.

Weld with the point in open position and start welding from the point end to avoid possible derailments. Before starting to weld each first pass, preheat the area (generally 4 to 6 in. in length) and check the soundness of base metal. If any imperfections were missed, they can easily be seen during preheating and should be ground and flushed out.

Build the point up by welding successive passes equal to the preheated zone until the desired height is obtained. Then peen each pass as welded.

The point must be closed against the stock rail after each pass in order to shape and fit the switch point. Be sure all overflowed weld metal between the point and stock rail is ground off prior to closing or it will be impossible to correctly fit the point to the stock rail. The end of the welded point should be about 1/4 in. thick at the bottom and tapered to 1/8 in. at the top, with the top about 1/4 in. below the stock rail surface (see Fig. 89.44).

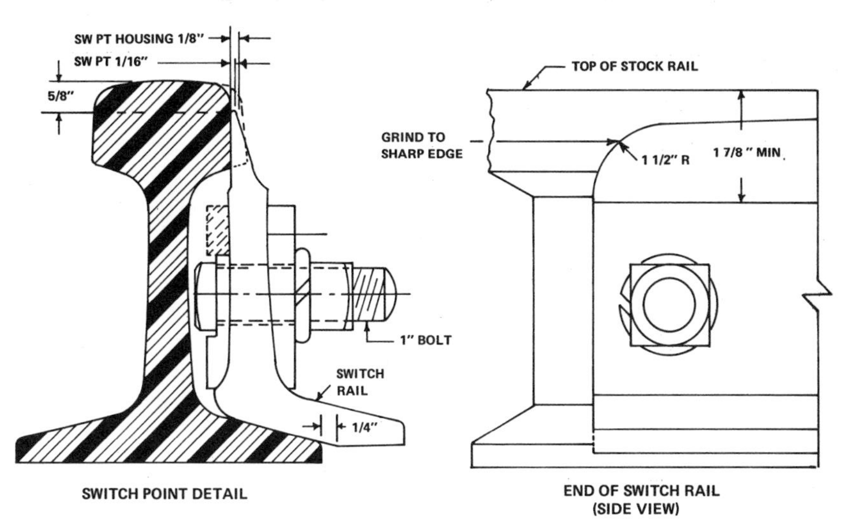

Fig. 89.44.–Sketch of switch point rail.

Buildup is continued in this manner (4 to 6 in. passes) until height is obtained. If additional width is desired, the weld metal should be added by running horizontal beads on the gage side of the point, the first bead on the top edge and additional beads as needed below each other. The end of the bead must be tapered to prevent wheel pick if the point is used before welding is completed.

When the desired height and width have been obtained, the newly completed area should be heated to a forging temperature (bright red, approximately 1500 F). Then the switch point should be closed against the stock rail and shaped to the desired contour with the flat side of a hammer (see Fig. 89.45). Continue welding and hammer forging in this manner until the switch point is completed.

The contraction of the weld metal will cause the switch point to warp. This can be prevented by applying heat to the base immediately after each pass is completed. Heat the base of the switch point to a bright red color. Apply slightly more heat to the inside thin base section than to the outside full base section to avoid side warpage.

When welding, hammer forging and flame straightening are completed, close the switch point against the stock rail. Preheat the first 10 to 12 in. of the end of the point to a cherry red color, leaving the point against the stock rail to assure slow cooling. This slow cooling will prevent excessive hardness in thin point sections.

Fig. 89.45.–Welding and hammer forging of a switch point along the track.

Grinding

The finished ground switch point should closely conform to new switch point specifications.

Identifying Welded Switch Points

With a 3/4 or 1 in. flat cold chisel, mark YD on reinforcing strap between the first and second clips.

Final Inspection

Open and close the switch to be sure the switch stand is operating correctly and that both switch points fit correctly.

Shielded Metal-Arc Welding Process Preparation

Use the same technique as outlined under the oxyacetylene process.

Welding

Use the same technique as outlined under the oxyacetylene process.

Welding Procedure

Preheat the weld area to 500 F before applying the weld metal. If for any reason this temperature drops to 300 F or lower, reheat again to 500 F. Start the buildup at the point end and work back toward the heavier section. Peen each bead lightly–first at the weld end, and then from the starting point toward the finish. After completion of welding, the switch point must be ground to the correct contour and the switch checked for the correct working condition.

WELDING OF FROGS AND CROSSINGS

The repair of frogs and crossings by welding is now common practice on all railroads. The most economical and satisfactory results can be obtained by careful inspection, qualified supervision and setting up a definite repair schedule based on this inspection. The factors determining whether or not to weld frogs are: the general condition of the frog; the kind, amount and speed of traffic over it; the probable condition at the next scheduled welding date; and the probable life of the frog at the present location.

In setting up a time schedule for this type of work, the following should be considered: the type and amount of traffic, the size of the frog (one No. 20 will consume about twice the welding time of a similarly worn No. 10), the distance from headquarters, the availability of the location and the amount of welding necessary.

Frogs and crossings should be surface ground after four to six weeks' service and reground whenever overflow causes the flange-way to decrease 1 5/8 in. This can be easily checked with flange-way gages. When the flange-way clearance is reduced to less than 1 5/8 in. the car wheels passing through the reduced flange-way act as a wedge, thus causing flange-way cracks. Frogs cannot be repaired in the field when transverse cracks appear. This cracking of flange-ways occurs in manganese steel frogs and crossings only (see Fig. 89.46).

Marking Weld Limits

Mashed out or fractured areas should be welded to conform to existing heights on either side to provide a good ride. On badly worn frogs, check the point on both wings and frog point where the worn area begins, mark these points and build up the wings and points accordingly.

On frogs or crossings that have heavy traffic on one side and light traffic on the other, use the heavy traffic side as a basis for correct height and taper the weld back on the light traffic side. Use a 24 or 36 in. straight edge to mark out weld limits.

Manganese Steel Frogs and Crossings

Frogs and crossings of 12 to 14% manganese steel must be repaired by arc welding. Field welding should be attempted if traffic conditions will permit five hours productive labor in any eight hour day, head wear is less than 3/16 in. on wing rails and cracks can be completely welded from running surface. The damaged area should be small enough so that when ground out, an unsafe track condition does not occur.

All frogs which need a complete overhaul or those with longitudinal cracks in the filler section or transverse cracks that extend into the filler area should be welded in the shop.

Preparation

Remove defective metal by grinding, cutting electrode or air carbon-arc cutting. If the metal is removed by electrode cutting, all heat-affected metal must be removed by grinding. This is not required, however, when metal is removed by air carbon-arc gouging. Caution must be used when grinding out deep flaws or the casting may be overheated.

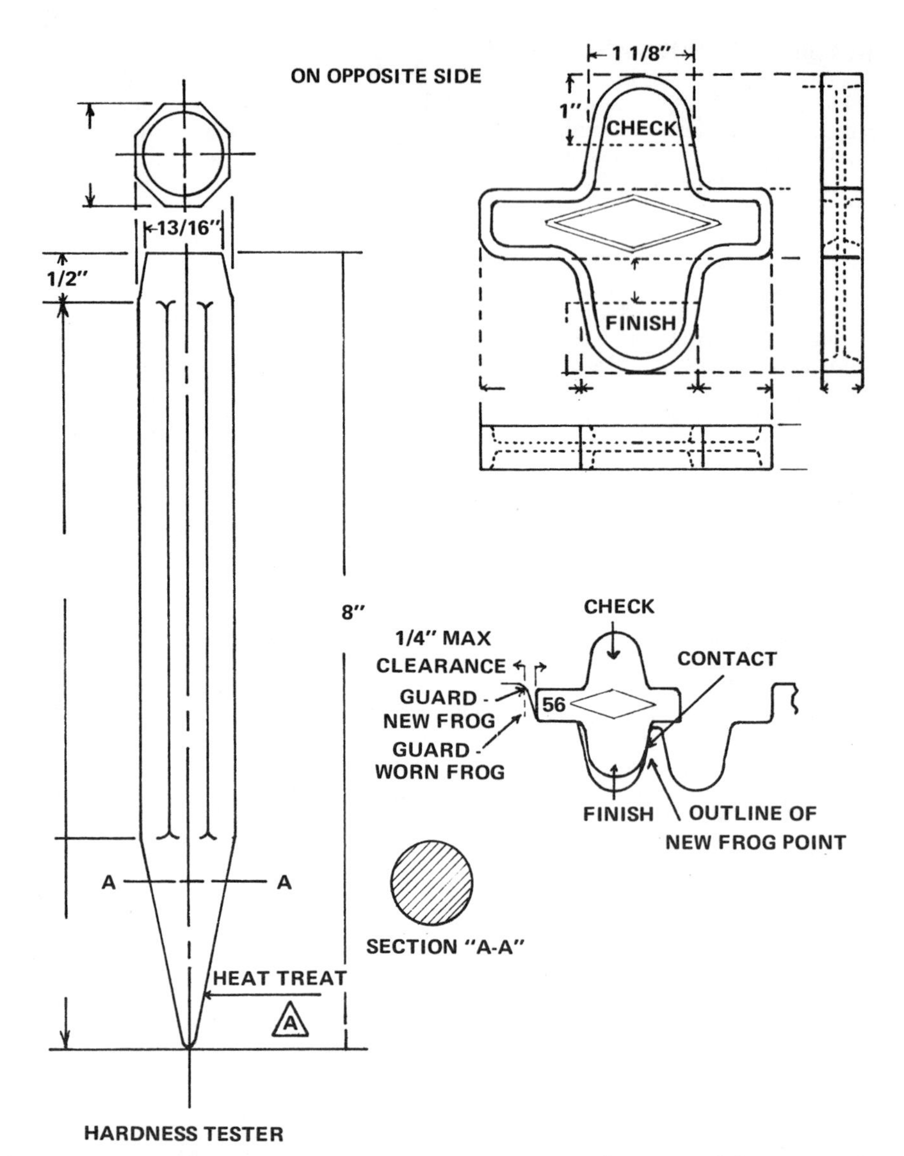

Fig. 89.46.–Flangeway gage check: apply after grinding flangeway of frog or crossing. Guards on self-guarded frogs (for checking maximum wear of guards on self-guarded frogs): apply at actual 5/8 in. point as illustrated above and restore surface of guard when clearance is greater than 1/4 in.

Grind out the work-hardened surface in any area that will require welding even if the area is free of flaws. This may require a depth of 3/16 in. in areas that receive the greatest impact. To determine when the work-hardened metal has been sufficiently removed, take the center punch (Fig. 89.46), place it in the bottom of the flange-way (give it a good rap), then do the same on the running surface being ground. When equal impressions are obtained, sufficient metal has been removed.

Grind off the sharp edges along the flange-way and round them slightly prior to welding.

Shielded Metal-Arc Welding

Weld with dc, reverse polarity and, when possible, weld at alternate locations to avoid overheating the casting. Use a short arc, approximately the diameter of the electrode, and weave the electrode about twice its diameter. This will produce a bead about 1/2 to 5/8 in. in width. Limit the length of the bead to about 4 in. or the amount produced by one electrode. Ensure that the heat in the casting does not exceed 600 F when tested with a 600 F temperature indicating crayon 1/2 in. from the weld area.

When possible on frogs, weld alternately on wings and point, so that they remain at about the same height. Hold the electrode at about a 20–25 deg lead angle in the direction of the weld bead travel. At the end of each bead, run the electrode back into the built-up metal to avoid crater cracks, then immediately peen the bead with the flat side of a hammer. First peen at the finishing end, then start at the beginning of the pass and peen in the direction of the welding. On the next bead the arc is restruck ahead of the crater, then drawn back to the crater and the weld bead continued. Cross-hatch each layer to reduce stresses in weld metal. Be sure to apply sufficient weld metal to the casting so that when it is finish ground, the surface, gage and field sides provide full sections free of voids and low spots.

Recommended Welding Current Range

Electrode Diameter (in.)	Current (A)
3/16	140–230
1/4	200–300

(The midpoint of the current range is suggested.)

Self-Guarded Frogs Under Traffic

If the raised guard is worn more than 3/8 in., it must be completely welded before the point is rebuilt. When rebuilding the raised guard section, lay the first bead in the bottom corner, with additional beads above this. Ensure that the weld is tapered so that it will not cause wheel pick and possible derailment.

Gas Metal-Arc Welding

Many railroads are now building up 12 to 14% manganese steel frogs using the gas metal-arc welding process with excellent results. Heat input into the base metal is less than the shielded metal-arc application. Weld metal is deposited three to four times faster, therefore both time and material are reduced. Metal is applied by stringer bead welding and very little peening is required. Overheating the casting must be avoided as described under Shielded Metal-Arc Welding.

WELDING OF OPEN-HEARTH FROGS

The general procedure for inspection, marking, measuring batter and weld preparation is much the same as that outlined for manganese steel frogs.

Frogs that have been poorly maintained will have a point rail section considerably lower than wing rails. When this condition exists it is necessary to build a ramp on the back or heel end of the point to allow wheels to rise to the height of the wings. Generally, frogs in this condition should be shop repaired.

Open-hearth frogs can be welded by either the oxyacetylene or arc-welding processes.

Oxyacetylene Welding

After the area to be built up has been determined, remove fatigued, laminated, chipped or otherwise defective metal by grinding or by flushing with a welding torch. Grinding is preferred because it provides a cleaner surface for weld applications and does not warp the frog.

The welding tip used should be one that consumes 60 to 70 cfh of oxygen on wing rails and where the point is 2 in. or wider. A tip that consumes 50 cfh should be used on the balance of the point.

The type of flame used should be the same as outlined for oxyacetylene rail end welding. Generally a 1/4 in. diam by 36 in. welding rod is used on heavy sections and a 3/16 in. diam by 36 in. on a thin point section.

In the application of weld metal a small flame is used and the metal is applied in longitudinal passes 3/4 to 1 in. in width and 4 to 6 in. in length. Each pass is peened before starting the next pass. The best results are obtained when a deep weld puddle is used.

Welding is started on the heel end of the frog at the point where the wheels transfer to the wing rail. The first bead of each pass is applied along the inside edge of the wing rail and the bead is carried toward the toe end of the frog. The welder then works back to the starting point allowing the weld metal to overhang on the inside edge. On frogs with normal wear, an experienced welder should apply sufficient metal in one pass to reach the correct height. Each pass should be peened after weld application. The welder should work alternately on wings and point so that they will remain at about the same height. When possible, the frog is completed by grinding, but if the frog is to be finished with a hot cut and flatter, each pass should be smoothed with a flatter and trimmed with a hot cut as each pass is built to correct height. Welding on frog points is started at the heel end and worked toward the thin end. Warping of frogs can be corrected by applying heat along the base of frog wing rails to counteract welding heat and weld contraction.

Heat Treated–Bolted–Open-Hearth Frogs and Crossings

Preheat prior to grinding. Preheat the repair area with either a propane preheater or an oxyacetylene torch to 300 F prior to grinding out the work-hardened, deformed, spalled, chipped or cracked surface. Keep the welding tip approximately 3 in. from the rail, and oscillate the welding torch so that an even soaking heat is obtained. Extend the heated area to approximately 4 in. beyond the repair area.

Grind out the repair area carefully to avoid overheating the rail and thus causing cracks. Grind down the area of greatest impact to approximately 3/16 in. below surface. Taper grind back on the gage side at about a 45 deg angle so that when welded, there will be a gradual transition of the wheel load from base metal to weld metal.

Shielded Metal-Arc Welding

Reheat the repair area to 500 F, using the same technique and covering the same area as outlined under Preparation.

Build up the welding area by running stringer beads that are approximately twice the diameter of the welding electrode. Immediately peen each bead with the flat side of a hammer. Cross-hatch each layer of metal to reduce weld stresses. Build up the weld metal approximately 1/8 in. more than desired. Weld the height or width so that the full rail section will be obtained after finish grinding.

Frequently check the heat at about 2 in. from the repair area and if the temperature drops to 300 F, reheat again to 500 F.

Postheating

Postheat the weld area to approximately 700 F immediately after weld completion, again using the same technique as outlined under Preparation. After postheating, the weld area must be protected against rain, snow, etc., and cooled as slowly as possible, preferably by covering with an insulating blanket.

Repair Sequence

	Reheat when rail cools below
1. Preheat to 300 F.	200 F
2. Grind out repair area.	
3. Preheat to 500 F.	300 F
4. Weld.	
5. Postheat to 700 F.	
6. Finish grind.	

Special

Use a temperature indicating crayon to check temperature.
Do not heat the rail above 800 F.
Do not repair this type frog or crossing by oxyacetylene welding.

Care of Electrodes

The low-hydrogen electrode which is used should be baked at 350 F to 400 F for four hours. Electrodes that have been exposed to atmospheric conditions longer than two hours should be baked again at 350 F to 400 F for four hours before using. Do not use electrodes that have come in contact with water, oil or grease. Purchase in 10 lb hermetically-sealed containers to avoid loss from moisture pickup.

Shielded Metal-Arc Welding (Standard Open-Hearth Frogs and Crossings)

After the area to be built up has been determined, all fatigued, laminated, chipped or otherwise defective metal and overhang should be removed by grinding. Under certain conditions, some welders may prefer to use scarfing by oxyacetylene or electric arc methods to remove defective metal. Before scarfing is performed, all the material surrounding the area to be scarfed must be preheated. Burned metal, slag and similar material left by scarfing must be removed by grinding.

Preheating

Before any weld metal is applied, the area to be built up should be preheated to a temperature of 500 F min. This minimum temperature should be maintained throughout the welding operation. This can be accomplished by confining the welding to one wing or point in cold weather, or one wing and point in mild weather. During very hot weather, or if the work is being done indoors, it is permissible to build up both wings and point simultaneously by using a rotating welding sequence. The 500 F temperature must be maintained by additional heating if train movements or other causes interrupt the welding routine.

Welding

For an efficient deposition rate, 3/16 or 1/4 in. diam electrodes should be used to rebuild frogs and crossing corners. The electrode should deposit successive beads 3/4 to 1 in. wide applied parallel to the wing or point until the necessary buildup is obtained. Each bead must be thoroughly peened and all slag removed before depositing the next layer of weld metal.

Postheating

Immediately after welding is completed, the built-up areas must be postheated to a temperature of approximately 1000 F. This will result in a BHN of 275 to 325 for the as-deposited weld metal. Cold working under traffic will raise the BHN to 350-390, which is a satisfactory range for this service.

Finish Grinding–All Frogs and Crossings

Grind the running surface to a good riding surface on frogs, drop the point 3/16 in. below the wing rails and raise to full height 8 in. back from the point. Top gage corners of the wing rails and the point should be ground to a 5/8 in. radius. Avoid square corners or deep grinding marks.

When completed, the flange-ways must be in good alignment and 1 7/8 in. wide (check this with Gage No. 1). Grind out all overflow between casting and binder rails on rail bound frogs with a 1/8 in. cross-slotting wheel.

When possible, regrind all frogs or crossings within four to five weeks after welding.

SYSTEM FROG REBUILDING

The rebuilding of frogs at a centralized shop has many advantages and will prove economical and feasible if worked in conjunction with field welding. Frogs worn or damaged beyond economic field repair, or requiring new binder rail or bolts, can be reclaimed at a centralized shop without excessive cost. This is due to the increase in productive labor, availability of correct facilities and production line output.

The welding procedures used are the same as those used when field welding except that filler bars are frequently used and the welding procedure should avoid setting up local stresses.

Grinding should be done by a reciprocating bed grinder and wet grinding must be used to avoid overheating the manganese steel castings.

FIELD RAIL END HARDENING

All joints in new rail, standard insulated heels of switch points and frog connections, which are not to be field welded, must be field hardened. Field rail end hardening is accomplished by rapidly heating rail head surface and then permitting it to air cool quickly.

When possible, use the field end hardening box shown in Fig. 89.47. If the box cannot be used, place tie plates across rail heads 3 1/2 in. back from the joint to help localize the heat.

The reasons for using a flame hardening box include the following:

1. It protects the flame from wind.
2. It is much easier to see correct color which is almost impossible in bright sunlight.
3. It helps prevent heat spread in rail head and thereby provides faster cooling and resultant harder rail ends.

Do not end harden fit rail. The rail head of fit rail is already, nearly as hard as obtainable by this method and occasionally the rail head will uncap if an attempt is made to harden it further.

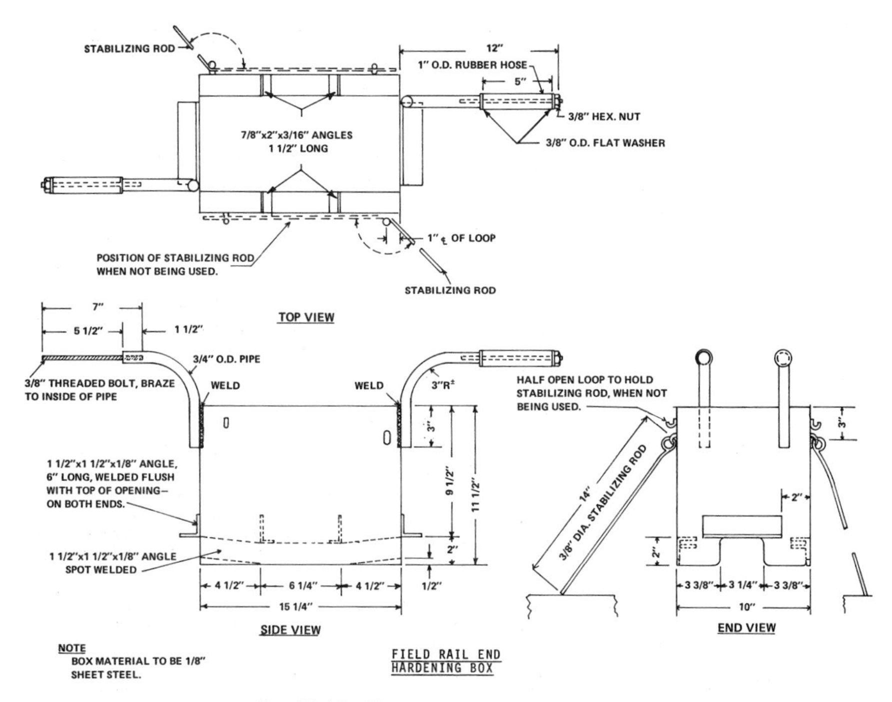

Fig. 89.47.–Field-rail end hardening box.

Use a welding tip that will consume 150–250 cfh of oxygen. Use a single flame tip. Before starting to field harden, make sure the welding tip is in good condition. Check the flame carefully and if necessary, clean the tip with the correct size tip cleaner. When using heating tips of this size, use a two-cylinder acetylene manifold to obtain optimum heat.

Method

Chamfer the rail head 1/16 in. as shown in the cross-slotting instruction. Measure back 1 3/4 in. from the joint on both rail ends and mark with a soap stone. Light the torch and adjust it to a maximum neutral flame. Start heating in the center of the rail head at the 1 3/4 in. mark, rotate the flame in a small circular motion until a spot about 1 in. in diameter shows a bright red color, then heat the opposite end in the same manner (see Fig. 89.48 A and B). During the entire heating operation, be sure the white cone of the heating flame is about 1/4 in. above the rail as this provides the greatest heat with a minumum chance of melting the rail surface.

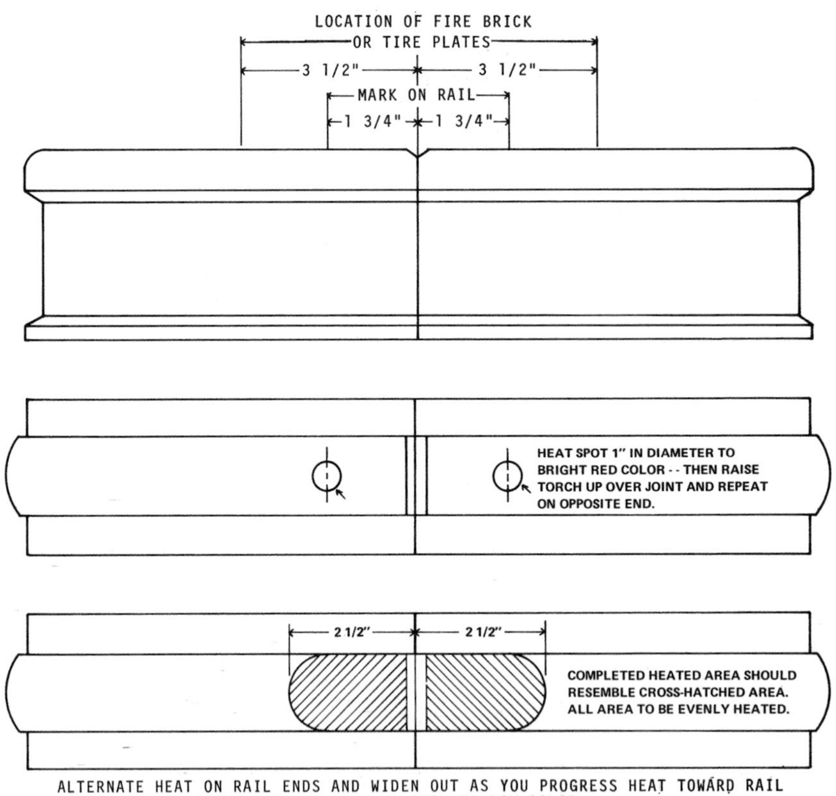

Fig. 89.48.–Field-rail end hardening.

Return the heating torch to the original end and increase the heated area. Now continue heating the rail ends by alternating the heat from one rail end to the other, meanwhile increasing the width and moving the heated area toward the rail ends until the entire width of both rail ends are heated to an even yellow color. Be careful to lift the torch when moving from rail to rail to avoid melting the rail ends, especially on open or recently cross-ground joints. The heated area should be as shown in Fig. 89.48 C.

Remove heating torch and check temperature with a 1550 F temperature indicating crayon. This should melt readily. Immediately remove box or tie plates to assure fast cooling of rail ends.

The entire heating time should not exceed 50 seconds to reach a temperature of 1550 F on both rail ends or 30 seconds when heating one rail end. The hardness of field hardened rail ends should be 40 to 60 points higher than the Brinell hardness of unhardened rail.

Special Conditions

When field hardening a rail end adjacent to known plant-hardened rail ends such as fully heat-treated railroad crossings, open up the joint 1 in. or more (when practical) and harden in a normal manner on the untreated rail end.

If the rail ends must be kept tight, protect the hardened end by placing a tie plate across the rail head with about 1/8 in. of the tie plate protruding past the end of the hardened rail end. Proceed in a normal manner except that as the heating is completed, keep the white cone of the flame back about 1/4 in. from the edge of the tie plate and let the yellow heat run out to the rail end. Immediately remove the tie plate from the track to permit fast cooling.

Rail ends must be hardened prior to joining with the frog since manganese steel will transform to the "As-Cast" condition when heated to over 600 F and it is impossible to flame harden the rail end and not overheat the adjacent frog metal. "As-Cast" condition manganese steel is a very brittle material and will quickly break out under traffic.

BIBLIOGRAPHY

"Fixtures and Manipulators for Mechanized Arc Welding," Spencer Payne, *Welding Journal*, 48(12)942–949 (1969).

"Weldability of High-Carbon Steel Using the Gas Metal-Arc Process," N. R. Braton, D.A.G. Stegner and S.M. Wu, *Welding Journal* 46(7) 329–336 (1967).

Car and Locomotive Cyclopedia (1966).

A. A. R. Yearbook of Railroad Facts (1965).

"So_2 Gas-Welds Long Rail," Edward T. Myers, *Modern Railroads*, 16, 63–66 (1961).

"Production Welding of Aluminum Tank-Type Covered Hopper Cars," H. Bertrand et al, *Welding Journal*, 42(7) 561–566 (1963).

"Welding Standardization Cuts Railroad Costs," R. L. Thomas and F. J. Graham, *Welding Journal*, 41(3) 238–242 (1962).

"Welding of Aluminum Gondola Cars," B. Karnisky and W. G. Boese, *Welding Journal*, 40(11) 1137–1153 (1961).

"A New Concept–Push-Pull Suburban Cars," W. Van Der Sluys, Paper No. 60-WA-264, published by ASME (August, 1960).

"Research and Development of Continuous Welded Rail," G. M. Magee, *Welding Journal* 39(9) (1960).

"Expand New Rail Welding Plant," *Modern Railroads*, 14, 85–86 (1959).

"L & N Cuts Continuous Rail Costs," Edward T. Myers, *Modern Railroads*, 14, 97–100 (1959).

Railroad Car Facts–American Railway Car Institute.

"Laboratory Tests of Continuous Welded Rails," R. E. Cramer, *Proceedings American Railway Engineering Association*, 59, 896–904 (1958).

"Studies of Upset Variables in the Flash Welding of Steels," E. F. Nippes, W. F. Savage, G. Grotke and S. M. Robelotto, *Welding Journal*, 36(8) 192s to 216s (1957).

"All-Welded Construction in Freight Cars," W. G. Boese, *Modern Railroads* (June, 1956).

"An Application of Pressure Welding to Fabricate Continuous Welded Rail," D. C. Hastings, *Welding Journal*, 34(11) 1065–1069 (1955).

"Tests of Electric Flash Butt Welded Rails," R. E. Cramer and R. S. Jensen, *Proceedings American Railway Engineering Association*, 55, 684–694 (1954).

"Engineering, Fabrication and Welding of Aluminum Railroad Equipment," Kaiser Aluminum Technical Services.

CHAPTER **90**

AUTOMOTIVE PRODUCTS

PREPARED BY A COMMITTEE CONSISTING OF:

T. W. SHEARER, JR., *Chairman*
Fisher Body General Offices
General Motors Corporation

D. W. SHAW
International Harvester

J. W. MITCHELL
Mfg. Eng. and Dev. Office
Ford Motor Company

L. J. ROSE
Fisher Body
General Motors Corporation

P. SHAUGHNESSY
Chrysler Corporation

A. J. DEARING
A. O. Smith Corporation

W. E. MUMFORD
Motor Wheel Corporation

N. E. O'CONNOR
Fisher Body Division
General Motors Corporation

CHAPTER 90

AUTOMOTIVE PRODUCTS

INTRODUCTION

The ability to join metals by the various welding processes has much to do with the overall concept of automotive product design. The choice of the specific welding process to be applied on any given product is directed by the metals to be joined, product design, type of service requirements, production rate and economics.

Many types of metal are welded in the automotive field, but low carbon cold-rolled sheet steel and hot-rolled plate, usually pickled and oiled, predominate. The metals to be joined, whether ferrous metals (plain carbon or alloyed steel) or nonferrous metals, influence the choice of the welding process to be used. The trend to achieve greater corrosion protection has emphasized the use of steels which have been coated for corrosion protection. The corrosion-resistant metallic and paint coatings have a definite effect on the welding processes, and necessitate modifications in equipment and production techniques. Typical metals being welded in the automotive industry are shown in Table 90.1.

Product design obviously determines to some extent which welding process can be used. Joint accessibility, sequence of assembly operations and finished apperance of the part must all be considered in selecting a welding process.

The intended service of any product has much to do with the product design and choice of metals for the component parts. Also related to service is the joining process used to fabricate the product. In many instances, more than one process is applicable, and a choice must be made. Each of the processes has requirements of joint description and tolerances best suited to its application,

Table 90.1—Typical metals used for automotive parts

Parts	Metal Used
Engine blocks and heads	Alloy cast iron and aluminum
Door hinges and chassis brackets	Malleable iron
Frames	Low carbon (0.25% max) steel
Body components	Low carbon, cold rolled steel sheet Low carbon, hot rolled picked and oiled Galvanized cold rolled sheet steel Zinc-rich paint primed cold rolled steel Occasionally, low-alloy, high tensile sheet steel
Rear axle housings	Low or medium carbon steel stampings Cast steel or malleable iron
Propeller shafts	Low carbon steel tubing welded to steel forgings
Wheels	Low carbon rimmed or semi-killed steel
Fuel tanks	Terne plate
Trim and hardware	Stainless steel, aluminum and zinc die cast
Commercial bus and coach components	Aluminum sheet, angle, channel and extrusions
Commercial trailers (cargo and liquid carriers)	Aluminum sheet, angle, channel and extrusions

and it is important that these items be considered in the design stage. The application of a process to a product whose joint description is not in accord with the process requirements will lead to poor quality, higher unit cost, lower production rate or any combination of these.

The type of service requirements that welded joints of automotive products must meet can generally be indicated as:

1. Structural characteristics
   a. Mechanical strength
   b. Fatigue resistance
   c. Rigidity in bending or torsion or both
2. Corrosion resistance
3. Liquid-tight characteristics
4. Appearance characteristics

The high production rates encountered in the automotive industry have led to a high degree of automation and associated special equipment. In the case of automobile body components, multispot welding machines make as many as 100 to 125 spot welds simultaneously, at production rates up to 600 components per hour. Other associated components, smaller in size and requiring fewer welds, are produced at typical rates of 800 per hour. Obviously, the use of specialized equipment for either resistance welding or automatic fusion welding requires a high initial investment and must be justified by the volume to be produced. In cases of lower volume production, more versatile manually operated standardized equipment is used.

Economics is an important factor in selecting the metal-joining process for the product. In the case of high volume production, a fraction of a cent saving per part in either consumable material or labor can add up to a considerable amount of money over a year's production.

ENGINE APPLICATIONS

CAST IRON ENGINE BLOCKS AND HEADS

Engine blocks and heads are usually cast iron castings. Relatively large numbers of these castings are produced with imperfections which can be repaired by welding. These types of defects include misruns, laps, blowholes, cracks, porosity, tool gouges, machining errors and broken-off portions.

These major defects are found by visual inspection of the as-cast components after sandblasting. Other defects are recognized by hydrostatic testing, magnetic powder and fluorescent penetrant methods.

Preparation of these defective areas for repair by welding may require removal of metal to eliminate sand inclusions and the defective region. This metal removal may be accomplished by chipping, grinding or gouging with the oxygen or air carbon-arc processes.

In the as-cast condition the majority of repairs are made by the oxyacetylene process using a cast iron filler metal. Preheating of the entire casting is usually done in a furnace when a major defect in a restrained portion of the casting is to be repaired. Edges, corners and bosses can be repaired with local preheat procedure. If a color match is not required some repairs can be made by braze welding. This requires a lower preheat, less heat input to the casting in making the repair and no melting of the base metal.

Defects which are found after machining are repaired using special arc-welding techniques. In the case of sand inclusions in the bore, two basic methods have been successfully used: (1) *Low Voltage-High Current Transformer* (sometimes referred to as "cold welding" or "battery welding") in which bare nickel electrode wire is used at a low enough voltage to prevent an arc to be maintained; however, the high current is sufficient to melt the electrode wire and it is deposited by a scratching technique. Considerable skill is required to secure a reliable bond and porosity-free deposit. (2) *Intermittent Welding* in which the welding current obtained from a special transformer is interrupted on a programmed basis by a contactor turning the arc on and off, preventing accumulation of heat in the workpiece. This method is faster and yields a better bond and sounder deposit.

For repair of small defects and machining errors on finished or semifinished castings, the shielded metal-arc process using pure nickel or 55% nickel electrodes has proven very satisfactory. The weld deposit is machinable; however, a heat-affected zone does exist that is harder than the deposited metal but is still able to be machined. The accepted technique of welding involves depositing the weld bead in short increments and peening the deposit while still at a dull red heat, then permitting the deposit to cool before the next increment of weld is made. This procedure is followed until the repair is completed.

ALUMINUM ALLOY

Aluminum engine blocks and heads are presently fabricated by the permanent mold casting process, primarily from the silicon-aluminum alloys (Alloy 356).

Casting defects that may be repaired by welding are porosity or gas inclusions (generally at steel core locations). These defects are recognized by visual inspection and hydrostatic testing.

Preparation of defects for repair by chipping or grinding is not satisfactory due to the possible shear closure of adjacent defects. The preferred method is to puddle melt the area and scrape out the defective area with a steel rod.

The gas tungsten-arc welding process using an a-c transformer, high frequency stabilized, and 1% thoriated tungsten with argon shielding gas yields satisfactory repairs. The repair requires considerable welder skill and must be performed prior to porosity sealer treatment, heat treatment and machining operations. A general preheat of 500 F is necessary to permit reasonable time for repair. Postweld heat treatment does not appear to be necessary; however, the casting should be slow-cooled.

Localized repairs, such as tool gouges and machining errors can be made using the gas metal-arc welding process with an argon shielding media.

VALVE MECHANISMS

Rocker arms, although usually made of forgings, are welded in some instances. Fabrication involves the resistance welding and brazing of stampings and screw machine parts to form the complete rocker arms.

Valves have a wafer of hardenable metal (for example, SAE 8640) on the bottom face, which rubs against the cam. This disk is projection-welded, with the proper weld program to minimize embrittlement, using a dome embossment, to the remainder of the tappet, which is generally made as a screw machine part.

Valves designed for heavy-duty truck service frequently have their seating surfaces improved by means of a surfacing metal. Still higher temperature service requires valves of the sodium-cooled type. Such valves are forged or machined hollow. The top cap or bottom of the stem contains a small hole in its center through which the proper amount of sodium is introduced. This hole is subsequently closed by resistance or arc welding.

RADIATORS

The major processes used for radiator fabrication are soldering and resistance welding. The lock seam joints of the core tub are sealed with solder on the tube forming mill. The core-to-fin and header joints are furnace-soldered. The hose fittings-to-water tank joints and the water tank-to-header joints are soldered by hand or by automation using torch-soldering or flow-soldering techniques. Projection welding or spot welding is sometimes used to replace soldering for attaching the support brackets and baffles.

EXHAUST SYSTEMS

Exhaust and tail pipes are made of low-carbon aluminized steel strips formed and resistance upset seam welded in one continuous operation. Double wall exhaust pipe tubing is also formed and upset seam welded in one continuous operation. The requirements are gas tightness and easy formability.

Exhaust mufflers are made of light gage sheet parts rolled and stamped to shape. The side seams are usually spot or seam welded. End joints are rolled as well as seam welded. Baffles and other interior parts are spot welded in place. Aluminized sheet steel is used by some manufacturers for greater corrosion resistance on exhaust systems. Figure 90.1 shows a cross section of one type of muffler.

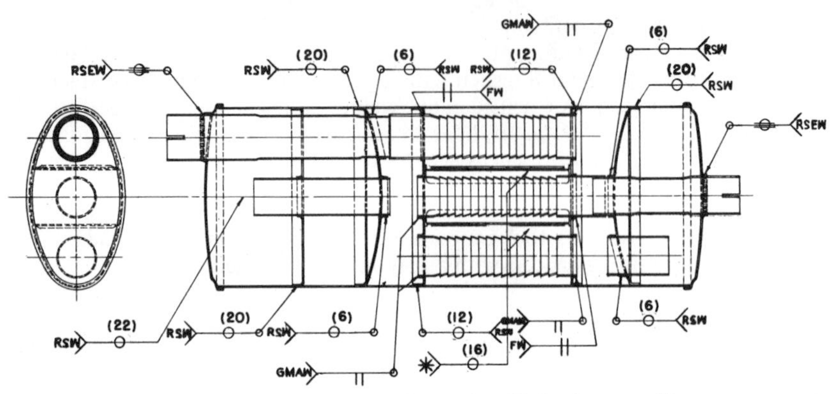

Fig. 90.1.–Cross section of typical welded exhaust muffler.

FUEL LINES

Considerable quantities of small diameter tubing are used in such automotive applications as fuel, oil, hydraulic and vacuum lines. This tubing is of two types–single or double wrapped (Fig. 90.2).

Double-wrapped tubing is made by roll-forming a copper-plated strip on a tube mill and brazing it in an atmosphere furnace in one continuous operation. The copper-brazed joint extends around the full circumference of the double wrap. The double-wrapped tubing is used extensively for brake lines.

The single wrap is rolled in a single thickness and resistance upset welded in a continuous tube mill operation. The tubing is reduced in diameter after welding in a separate drawing operation. Although this tubing is rated lower than the double-wrapped brazed tubing it is used in many applications where the requirements are freedom from leaks and easy formability, such as oil and gasoline lines. Welded tubing is not used for brake lines.

OIL PANS

The oil pan body consists basically of a deep-drawn stamping with several baffles and reinforcements spot welded and/or projection welded to it.

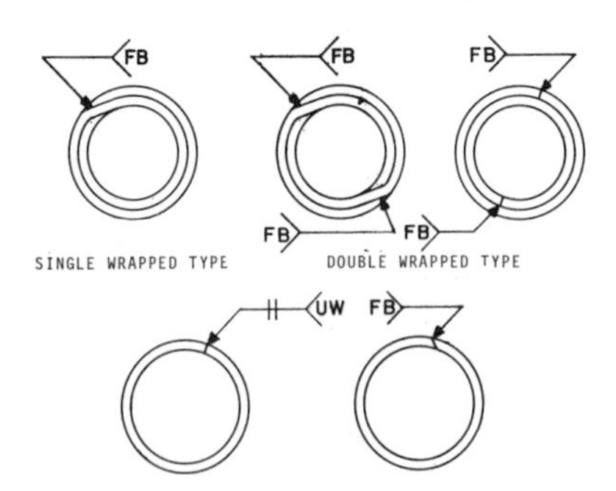

Fig. 90.2.–Single- and double-wrapped tubing used in the automotive industry.

Where production requirements are high, oil pans are usually welded on a series of high-speed, multiple-point, multiple-transformer-type press welding machines. These machines are placed in a straight line so that the pan passes from one machine to the other with a direct flow of material until completion.

Oil pan drain plug bosses are usually projection welded. Some are made of screw machine parts joined by projection welds using annular embossment to the pan at a level low enough to permit good drainage. This boss is ordinarily located on the rear face to be safe from impact.

Still other bosses are stamped from flat sheet and are dome projection welded on the inside in a depression to obtain full drainage. The drain plug gasket seals against the outside surface of the pan. Impacts from road obstructions do not rupture this weld. Figure 90.3 illustrates typical projection welded drain plug bosses.

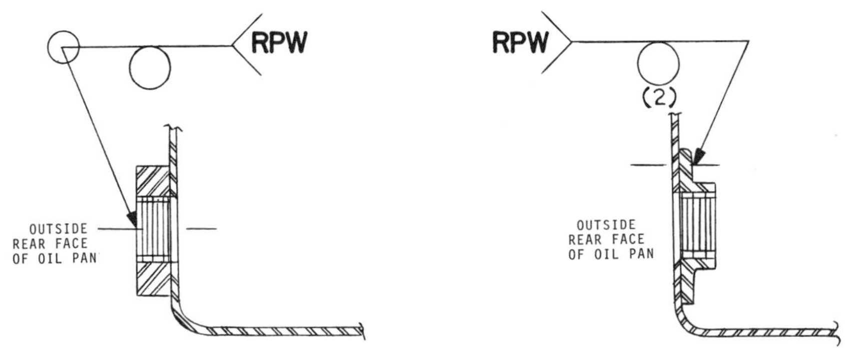

Fig. 90.3.–Typical oil pan drain plug bosses. At left is shown a ring projection welded boss and at right, a dome projection welded boss.

OIL FILTERS

Oil filters are made of deep drawn stampings, screw machine fittings, etc. Some manufacturers use resistance welding entirely for joining.

CHASSIS APPLICATIONS

FRAMES

The frame of our modern automobile, because of its complexity of shapes, high volume of production and stringent demands for low-cost fabrication, requires a variety of welding processes. The desired product is a frame which has structural integrity at economical cost. This is true of a frame for the heavy luxury car, a compact car or the components used in lieu of a frame in the unitized body constructed automobile (Figs. 90.4 and 90.5).

Due to the large number of repetitive parts (Fig. 90.6), there has been a continuing growth in the use of automatic and semiautomatic welding processes,

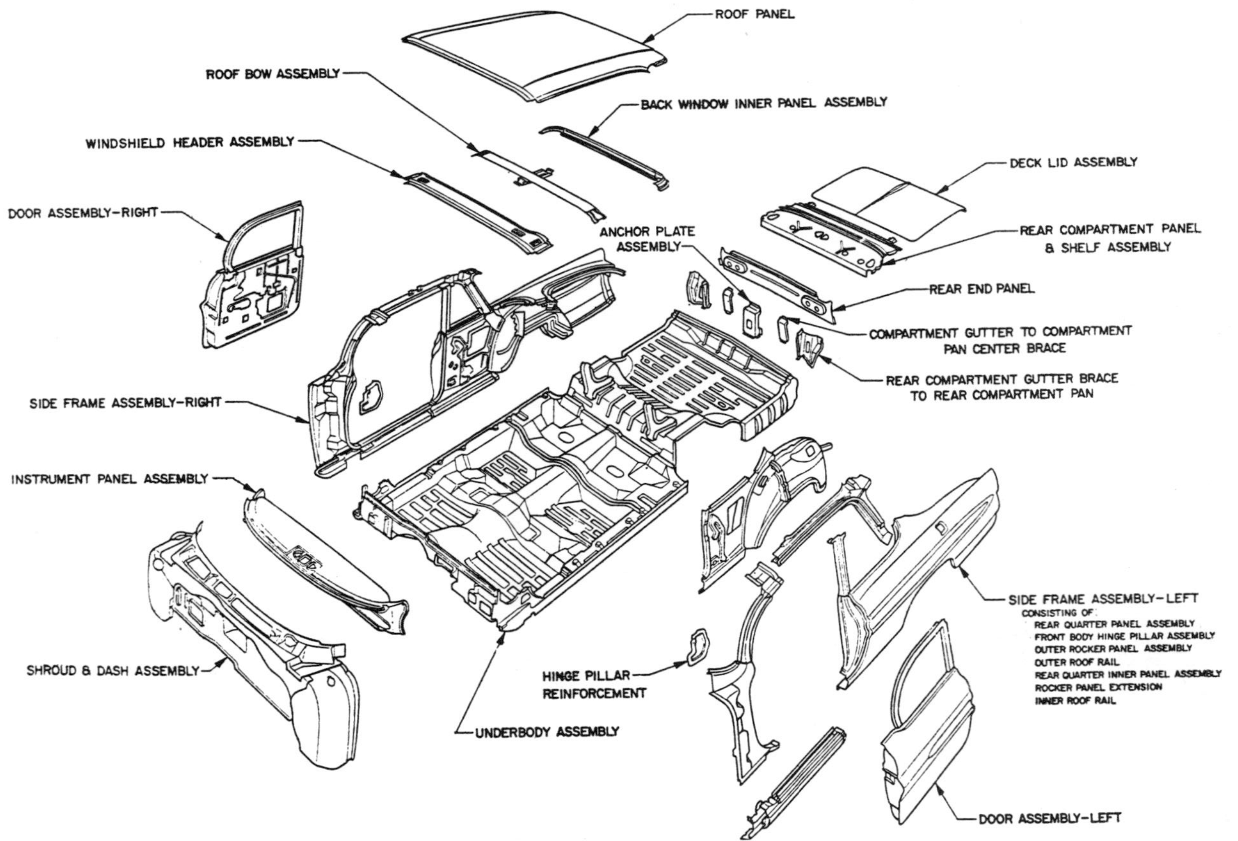

Fig. 90.4.–Typical example of conventional automobile body construction where the body is carried on a frame.

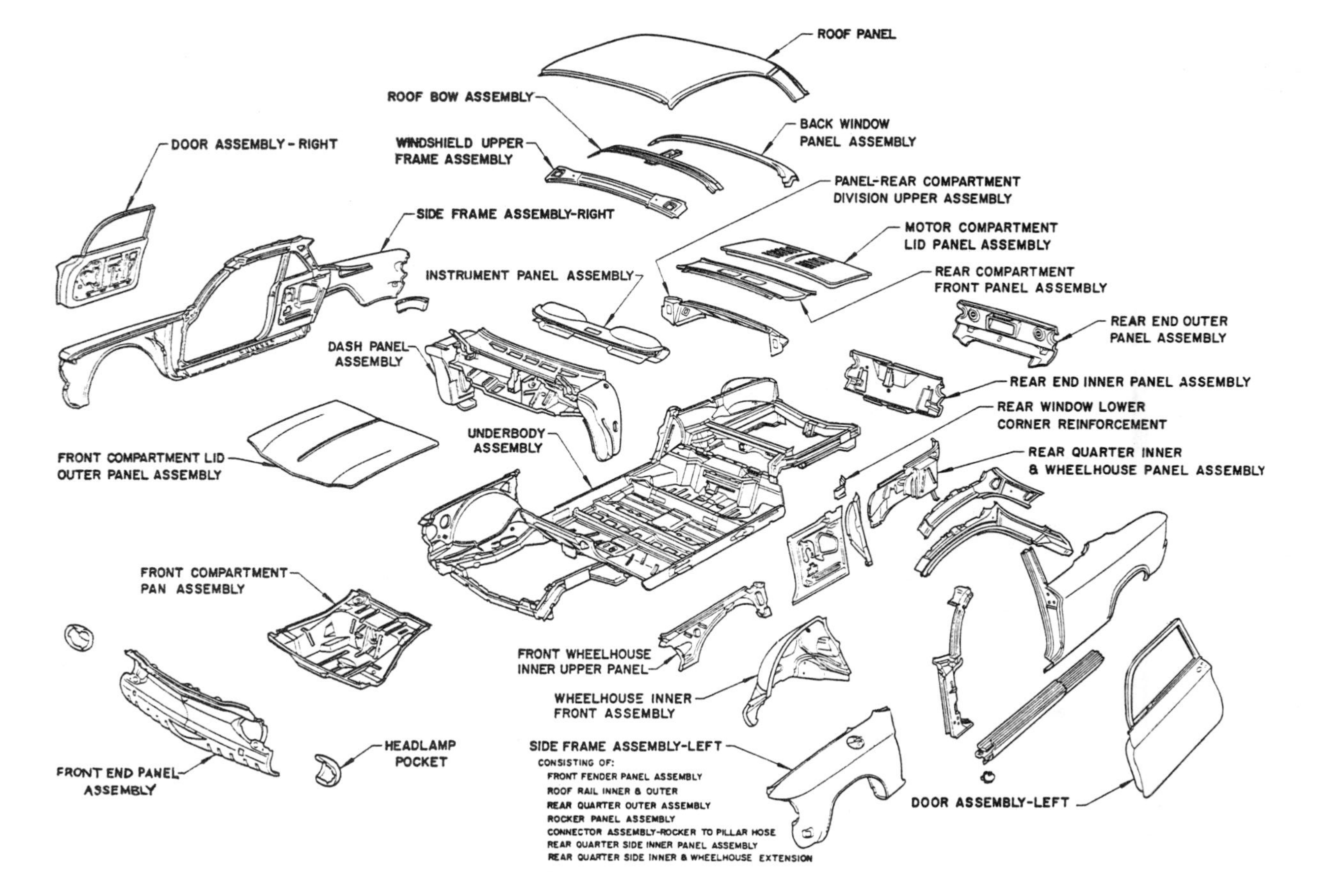

Fig. 90.5.–Typical example of unitized or integral automobile body construction.

Fig. 90.6.–View of a frame storage area showing a large number of repetitive pieces.

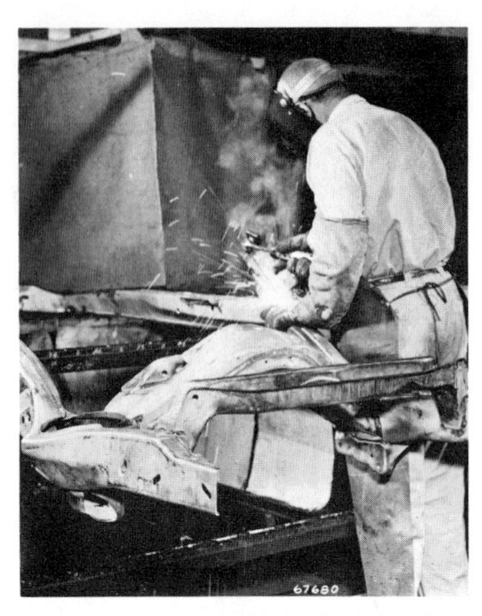

Fig. 90.7.–Example of shielded metal-arc welding of frame brackets.

particularly in the gas shielded metal-arc, and resistance welding processes. While manual stick electrode is still used to a great exent, its volume is declining.

Since the economics of welding play such an important part in the total fabrication cost of an automobile frame, the welding engineer and the design engineer must carefully analyze the welding processes that are to be applied to manufacture today's automobile frame.

The engineers should be certain that the design and the welding process are compatible. Also, such items as capital investment, labor costs and efficiency, the duty cycle of welding and number and rate of pieces to be made must be carefully considered before specifying the various welding processes.

It is also important that the susceptibility to design change be carefully considered as this could have a decided influence on the degree of automation or tooling costs which are practical for the specific frame.

Some general guidelines can be set for the selection of the following welding processes for frame welding: manual shielded metal-arc welding (SMAW), semiautomatic welding with the gas metal-arc process (GMAW–SA) and the fully automatic gas metal-arc process (GMAW–AU). (The use of the flux cored automatic and semiautomatic process is expanding.)

Manual shielded metal-arc welding with covered electrodes, while one of our oldest processes, is still used in the fabrication of frames, and a number of special electrodes have been formulated specifically for frame welding.

This process requires the lowest capital investment per welding unit and must be considered where the volume of repetitive parts is low or where the duty cycle of actual arc time is a low percentage of the time the part is in this particular operation.

For example, welding a bracket on a frame using a covered electrode (Fig. 90.7) may be the most economical method since the weld is short and the time of actual welding is very low, compared to the handling time in and out of the fixture.

The manual shielded metal-arc welding process is also used where there is a high susceptibility to model change. This is due to the lower cost of the welding equipment as well as low cost jigs and fixtures.

The advantages of frame fabrication by means of manual welding with covered electrodes include lowest capital investment in equipment, jigs and fixtures; most flexible welding process to accommodate poor fit-up and out-of-position welding; low maintenance cost of welding equipment. Against these advantages must be weighed the higher cost of welding labor or lower productivity, waste through stub loss and loss of time to change electrodes.

Perhaps the most popular welding process used in the fabrication of frames is the semiautomatic gas metal-arc process. This process has several modifications using solid wires with CO_2 shielding gas as well as argon mixtures. Flux cored wires are also gaining in acceptance in frame fabrication using the semiautomatic welding equipment (FCAW). CO_2 shielding gas is used on occasion with this process when increased penetration is desired.

The semiautomatic gas metal-arc process is specified to take advantage of its high welding speed when the production volume is not sufficient to justify fully automatic welding. It has replaced covered electrode welding on many of the short and out-of-position welds.

The capital investment for welding equipment and jig and fixture costs are somewhat higher for this process than manual SMAW, but the savings in labor must be measured against these costs. The rapid growth of this process in frame welding indicates that this process has had a good economic return over a wide scope of operations. This process allows the choice of using either solid filler metal electrodes with CO_2 or argon mixture shielding gasses or flux cored electrodes, with or without gas shielding. In general, the use of solid electrodes with the gas metal-arc process requires more skill by the welder, better fit-up of parts and more repair due to burn-thru than when using flux cored electrodes. However, there has been a considerable increase in the use of the FCAW process in frame fabrication because of the better appearance of the welds, more tolerance of part fit-up, somewhat less penetration and better usability characteristics for the welder.

The advantages of semiautomatic welding are higher welding speeds thus lower labor costs, excellent weld metal quality, relatively low-cost jigs and fixtures (Fig. 90.8) and adaptability of equipment to model changes.

Factors that must be considered to justify the semiautomatic equipment over manual SMAW are larger capital and maintenance costs.

As labor costs increase and fatigue and labor efficiencies affect frame fabrication costs, there is a growing trend toward the fully automatic welding processes, particularly the gas metal-arc process using either solid or flux cored electrodes.

When utilizing this process, which involves the greatest outlay of capital for fixtures and welding equipment (Fig. 90.9), justification must be obtained by having a large number of identical parts. Then the increased speed of welding can amortize the cost of equipment before a major model change.

Paramount in the utilization of the fully automatic welding process is exacting control of part fit-up, cleanliness of parts and a method of accurately tracing the joint to be welded. The process requires minimum skill from the machine operator although there is a requirement for skilled technicians to maintain and service the equipment.

Fig. 90.8.–Typical fixture for semiautomatic welding of automobile frames.

Fig. 90.9.–Complexity of fixtures needed for automatic welding of automobile frames.

Resistance welding is utilized in some cases for bracket attachment as well as to hold parts in position until they are arc-welded.

In addition, the upset welding processes are applied as follows:

1. Upset butt welding of parts end-to-end for blank welding.
2. Continuous butt seam welding, both low and high frequency, for use in formed tubular and box structures.

Fabrication of automobile frames, by the various welding processes, has contributed greatly to the quality, safety and low cost of our modern automobile frames.

SIDE RAILS

Some frame side rails are intermittently welded with manual arc welding processes. It has been established that more rigidity from thinner material is obtained from sections that have been continuously welded. For this reason most manufacturers employ semiautomatic and fully automatic processes for this application. Frame side rail welding clearly shows how other factors, such as tooling, product design or production requirements, influence the selection of a welding process.

REAR AXLE HOUSINGS

The modern pressed steel rear axle housing assembly (Fig. 90.10) usually requires the use of several welding processes in its fabrication. The halves can be formed and automatically arc welded along the four seams. Either all four seams may be welded at once in the 3 and 9 o'clock positions, or two at a time in the 12 o'clock position. The flanged bearing ends can be either hot-formed from the tubing section or formed separately and automatic arc welded onto the housing. The housing can also be made by arc welding the sheet metal banjo center section to formed tubular end pieces. The center section usually has four arc welded seams.

The cover is also welded to the housing. The various suspension brackets that make up the assembly may be projection welded or arc welded to the housing. The brackets themselves are often welded subassemblies. Small tube or cable retaining clips may be projection welded to the housing. They are usually placed so that several may be welded at one time.

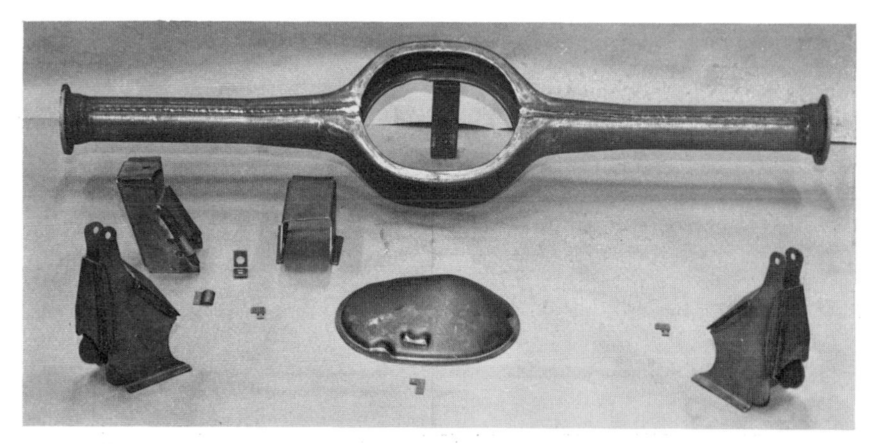

Fig. 90.10.–Pressed steel rear axle housing assembly fabricated by welding.

The automatic arc welding processes most commonly used are submerged arc welding and gas metal-arc welding using both solid and flux cored wires and a variety of gases and gas mixtures such as CO_2, CO_2-argon and argon-oxygen. The joint design, production rate and welding machine layout will dictate the process used as well as the economy of the process itself.

The "Salisbury" type of axle housing commonly used today consists of a cast iron (either gray or malleable) carrier and steel tubing assembly. The various sheet metal suspension brackets are automatic arc welded to the resistance welded steel tubing. The wheel flange can be either hot-formed from the tube or a separate piece formed and welded to the tube. The carrier and tubes are assembled by pressing the tubes into the carrier, using a 0.002 to 0.005 in. press fit. Additional resistance to torque is obtained by one of two methods. The most common method is to automatic arc weld the tube through several holes drilled in the carrier thus forming plug welds. Gas metal-arc welding is used with CO_2, CO_2-argon or argon-oxgen mixture shielding gases. To secure sufficient penetration without burn-thru, the wire is rotated in a circle whose diameter is slightly less than that of the hole being filled. As the penetration is mainly determined by the first pass, a dual welding heat may be used with high current for penetration at first, switching to a lower value to fill. The holes may be either partially filled (2/3 minimum) or completely filled and capped if oil seepage is a problem (Fig. 90.11).

FUEL TANKS

The use of terne plate for fuel tanks is standard, since it is corrosion-resistant and readily soldered and welded by the resistance-seam welding process producing an economical, leakproof unit (Fig. 90.12).

Through the years, refinement in welding terne-plated steel for use in fuel tanks has resulted in welding speeds of 300 ipm for 20- to 22-gage fuel tank material. This work is done with tandem machines or seam welders facing each other. Welding is done on a straight line across one side or end of the fuel tank, with automatic equipment handling the tanks. The indicated speeds require

Fig. 90.11.–Automatic plug welds in a Salisbury type axle assembly.

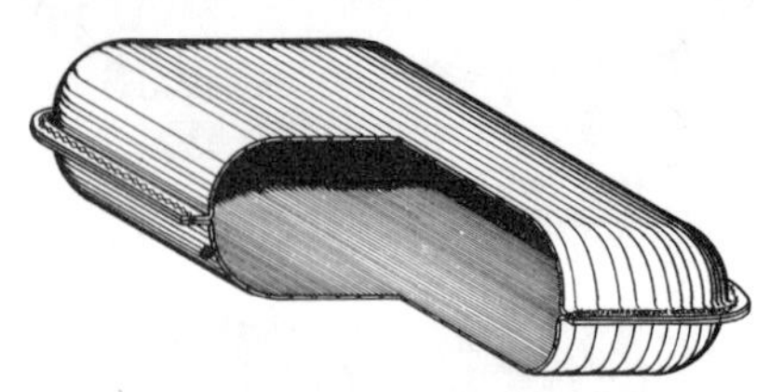

Fig. 90.12.–Seam welded fuel tank.

heavy duty equipment, utilizing either one cycle on and one cycle off schedules or similar timing techniques on a-c machines, or continuous schedules on d-c machines to obtain consistently sound welds. Careful attention must be paid to the knurl design and shape of the wheel-type electrodes to give consistent production quality by controlling the buildup of tin and lead on the wheels. Pretacking is required, and care must be exercised in the crossing of the welds at the corners in order to prevent leaks or blowholes.

Some of the exterior-mounted fuel tanks for commercial vehicles are fabricated from heavier (3/32 to 1/4 in.) bare steel or floor plate, using manual, submerged arc, or gas metal-arc welding processes. Fuel tanks are also made of aluminum in 3/32 or 1/8 in. thickness, and may be fabricated by the gas tungsten-arc or gas metal-arc welding process.

PROPELLER AND DRIVE SHAFTS

Propeller and drive shafts are commonly made from resistance welded tubing with the end forgings seam welded, flash welded or arc welded in place. To avoid the formation of cracks, special attention has to be given to joint design, the type of material welded, the shape of the forgings and the welding wire composition when arc welding is used.

The submerged arc and gas metal-arc welding processes with solid and flux cored wires with or without CO_2 shielding gas are among those used. The welded assemblies are balanced by means of resistance welded balance weights. As the torque requirements are high, quality welds have to be made both in the tubing and end forging weldments.

Some development has been made using friction or inertia welding to join the component parts. In this case the tube is rotated and the two end forgings are held stationary. The forgings are machined to obtain the necessary equal cross sections. Satisfactory weld joints are obtained.

WHEELS

The welding of passenger car and truck wheels was made possible by extensive research and development in wheel design and welding technique. Because of the cyclic nature of the load on a wheel when in service, special attention must be given to wheel design, shape of the weld nugget, weld location, control of process parameters, and the fit of the parts to obtain maximum resistance to fatigue.

A passenger car wheel basically consists of a rim and spider (disk). The spider is a steel stamping, which is pressed into the rim and forms the center section and mounting means of the wheel assembly. The spider can be joined to the rim by either resistance spot welding (spot or overlapping roll spot) or arc welding. At present, considerable development work is being done to use electron beam welding for fabricating the spider to the rim and for replacing flash welding in the rim joint. Geometry of the assembly, service factors and production requirements generally determine which of the welding processes will be used.

The spot welded assembly usually has eight elongated spot welds, approximately 0.40 by 0.60 in., the length of the nugget exceeding the width by 0.20 in. for maximum resistance to fatigue. When production requirements are low, single electrode press type welders are put to use; for volume production, special high-speed welding equipment is used.

A typical high-speed wheel welding line will have a pressing station for assembly, two four-gun welding stations and a valve hole-piercing station. If required, special stamping stations, inspection units or devices to "true" the mounting with respect to the tire seats can be included in the line.

The weld test procedure for production is a push-out test. The wheel assembly is placed in a press and the spider is pushed through the rim. This causes the metal to tear out around the eight spot welds and the size of the weld is easily determined (Fig. 90.13).In addition to the weld tests, equipment performance is monitored; in some instances, continuously, and in others, on a scheduled basis. Characteristics monitored may include time, nugget expansion, acoustic emission, voltage between tips, current, force, electrode size, shape and alignment, resistance between electrodes, weld energy level, electrode cooling and power supply condition.

The procedure for assembling arc-welded passenger car wheels is similar to that used for resistance welded wheels; the spider is pressed into the rim and joined by a continuous fillet weld or with intermittent fillets. The length of the spider leg must be controlled to within 0.10 in. to ensure proper positioning of the welding wire in each assembly.

A high production arc-welding line usually contains a pressing station for assembly and staking, a welding station (or more, depending on production

Fig. 90.13.–Welded wheel after undergoing push-out test. Spider (center of wheel) is pushed through rim causing metal to tear around spot welds.

requirements), equipped with four automatic arc welding heads and a station for piercing the valve hole, plus units for stamping identification, etc. (Fig. 90.14). Some passenger car wheels are now being welded by arc spot welding.

Fig. 90.14.–Arc welding machine for fabricating automobile wheels.

The center section of a truck wheel assembly may be a stamping similar to the passenger car spider or a tapered rolled disk. The edge of the disk is flanged over to provide means for attaching the disk to the rim. The disk is usually welded to the rim by a lap fillet weld or an arc seam (burn-thru weld); the edge must be fairly uniform for a fillet weld. In some instances a trimming operation is required to provide a suitable edge for welding. While the arc seam weld requires no special edge preparation, very good fit-up between the rim and the disk is necessary to maintain consistent penetration of the arc through the disk and into the rim. Submerged arc, GMAW-CO_2 and some of the flux cored processes have been used to weld truck wheel assemblies.

The passenger car rim is made from a strip of low-carbon mild steel. The strip stock is cut to length, coiled and flash welded to form a circular band. Approximately 5/16 to 3/8 in. of metal is used for flashing and upset. After the upset metal is removed, the band is formed into a rim by various pressing and rolling operations and expanded to size. The diameter of the finished rim is 13 to 15% greater than the diameter of the flash welded band. The flash butt weld must be free from defects caused by defective metal and improper welding techniques to withstand the subsequent die forming operations. Flash welding machines used for passenger car rims range in size from 300 to 750 kVA. Flash welders used for truck rims are generally 750 kVA machines. Truck rims are usually made from hot-rolled steel sections.

Construction equipment wheels utilize rims fabricated from hot-rolled steel sections coiled, welded and sized. These sections, with "as-rolled" joint preparation or machined joints, are combined with bands made from flat steel bars to make a rim of the desired width.

These rims are used as "demountable" rims in conjunction with cast wheels, or are fabricated into disk wheels by welding a mounting ring to the inside of the rim. Some are fabricated into "integral" wheels by welding the rims to cast, forged, fabricated or combined hub and strut (or disk). Usually it is necessary to do all finish machining on the wheels after all fabricating in order to maintain the required dimensional tolerances. Wheels used for construction equipment range in size from a nominal 15 in. diam to a 57 in. diam in production, with low volume specials as large as 84 in. diam.

WELDING PROCESSES

RESISTANCE-SPOT WELDING

Spot welding is the most commonly used resistance welding process in automotive body assembly. The general types of spot welding equipment are stationary and multiple spot welding machines and portable gun equipment.

Stationary Spot Welding Machines

These machines are generally used for small assemblies which can be handled easily and which contain relatively few welds, or for miscellaneous low production runs which require the flexibility obtainable with this type of equipment.

Resistance welding machine efficiency is the ratio of the usable production time to the total shift time. Welding machines can be operated at an efficiency in excess of 80%, which is the same as other machine tools. This increased efficiency has been brought about by improvements in the mechanical and electrical performance of the machines, sequence controls and weld timers.

A single-point welding machine can operate at a speed which is limited only by the rate of loading and unloading the parts. The total production time for each part includes the loading, welding and unloading time. If the loading, welding and unloading time occurs simultaneously, the total welding time per part is the longest of the three times. A dial welding machine accomplishes this savings in cost, except that the index time is lost unless the welding operator needs time to orient the parts before loading. A dial welding machine may be constructed in several forms: one of these is a direct series weld in which the only moving electrical connections are the two replaceable upper electrodes. Since the welding machine produces two parts each time it is energized, the duty cycle is one-half that of a single machine running at the same production rate; the unproductive indexing time is also halved.

The use of standard or in-line welding machines effects a labor saving since the parts are loaded once and several operations are performed without further handling. During their cycle, these welding machines often perform other operations, such as drilling, tapping, forming, pressing and assembling. A standard welding machine may be equipped with an automatic feeding and unloading device which will allow one welding operator to run a long multi-section machine.

When high production rates are required, the choice of electrode materials and contours is especially important, since the total output of a machine is reduced by the time necessary to replace or redress the electrodes. The best materials must often be selected by trial runs.

Multiple Spot Welding Machines

A typical automobile body is fabricated from sheet steel components or subassemblies which, for the most part, are spot welded together. An automobile body may contain as many as 6000 structural spot welds, of which half may be made by means of multiple spot welding machines, often referred to as press welding.

Press welding is a multispot welding method wherein any number of welding heads or welding guns containing spot welding electrodes simultaneously engage and apply pressure to the workpieces, after which they are immediately initiated electrically, either as a single large group or successively in smaller groups to produce the required spot welds.

Although the name press welding implies the use of a press, it is not necessary to have the welding electrodes mounted in a press. The mechanical press, when used, merely raises and lowers the workpieces into welding position and does not perform the functions usually associated with a power press. Simplified setups may be constructed utilizing a fixed platen welding machine. Regardless of the type of setup or welding machine used in the multiple spot-welding process, the welding heads or welding guns, to which the welding electrodes are attached, are moved into welding position and retracted by means of either air or hydraulically powered cylinders. Figure 90.15 shows a typical press welding machine.

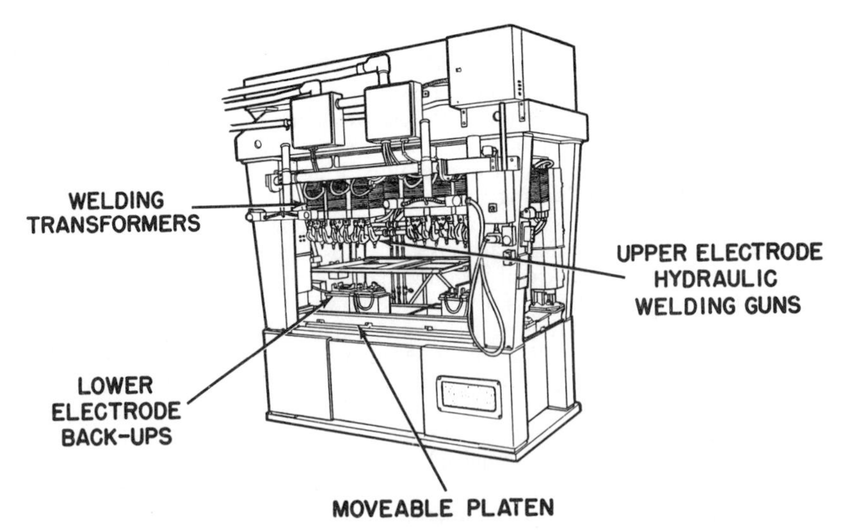

Fig. 90.15.–Typical press welding machine used in the automotive industry.

Multiple spot welding machines may be of several different types including a four-post press; an open, two-column gap press; an alligator or hinged platen press and a table top and other special arrangements. Regardless of the type of welding machine, the main functions of multiple spot welding machines are to act as framework on which to mount multiple electrodes and welding transformers, to bring the work into welding position and to resist the forces applied to the work by the pressure devices which generate the required electrode force.

The electrodes which are mounted in the press welder are generally made up of a framework on which are mounted the welding components: the welding guns for applying pressure, the welding transformers for supplying the welding current, the backup electrode for supporting and completing the welding circuit and miscellaneous items such as manifolds, cables, holders, tips and water cooling hoses. A typical press welding gun assembly for use in multiple spot welding machines is shown in Fig. 90.16.

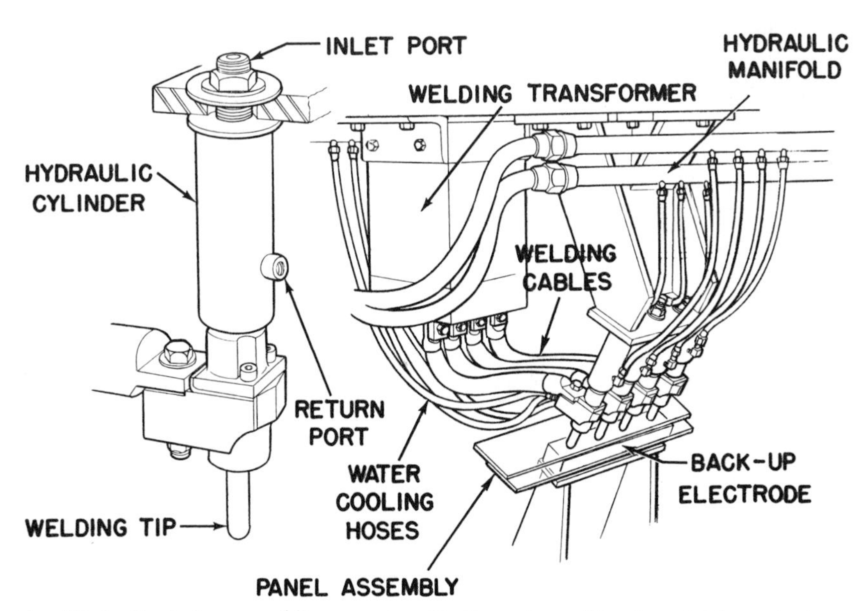

Fig. 90.16.–Typical press welding gun assembly for series welding in multiple-spot welding machines.

Multiple spot welding machines are often set up in a progressive series of three or more machines, with some form of automatic or mechanical handling conveyor systems feeding and connecting them. This permits high-speed production of complicated assemblies, such as underbodies, doors, seats or other automobile parts.

Many multiple spot welding machines are designed with interchangeable welding fixtures so that the same press welder may be used for welding similar assemblies for different car models.

Generally, multiple welds may be made by one or more different methods of welding, as described in Chapter 30, Section 2. Examples of types of welds are series welds, push-pull welds, direct welds or indirect welds. When the series weld circuit must be used because of design, the welds should be located sufficiently far apart to reduce current shunting to a minimum. This method is usually limited to thicknesses of stock .041 in. or less. While applications of single and multiple spot welds follow the general theory as outlined in Chapter 30, Section 2, the individual, specific arrangements may vary.

Portable Spot Welding Equipment

In the assembly of component sheet metal parts where the physical size and shape make it prohibitive to move the workpiece relative to the welding machine (such as with completely assembled car bodies), it becomes most practical to process these assemblies using portable resistance spot welding guns (Figs. 90.17 and 90.18). These guns come in a variety of physical shapes and sizes; the shape and throat area are dictated by the geometry of the component parts that the gun must accommodate. The force and current capacity are a factor of the metal thickness combinations that are to be welded. The air-operated guns are associated with the lower force requirements of the lighter metal gages while the hydraulic-operated guns are applied to heavier metal, higher force applications. Current is supplied by a transformer, rated to supply the magnitude of current required at the duty cycle intended, and transmitted to the gun by means of a low reactance water-cooled secondary cable. The guns are suspended on balancers and hanger devices to facilitate gun handling and manipulation. Welding guns are sequenced by NEMA Type 3B timing devices which meters squeeze, weld, hold and off-time increments. Welding rates will vary with the metal thickness to be joined with typical rates being as high as 200

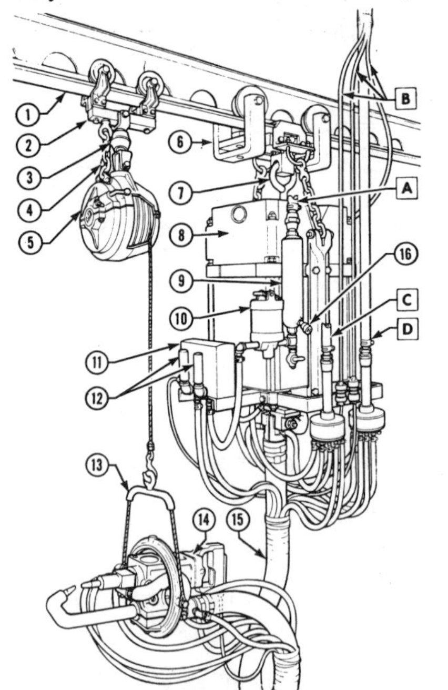

1. OVERHEAD SUPPORTING TRACK
2. BALANCER CARRIAGE
3. BALANCER INSULATOR
4. BALANCER SAFETY CHAIN
5. SPRING BALANCER
6. TRANSFORMER CARRIAGE
7. TRANSFORMER SUPPORT & SAFETY DEVICE
8. WELDING TRANSFORMER
9. AIRLINE FILTER & WATER TRAP
10. AIRLINE LUBRICATOR
11. FOUR WAY AIR SOLENOID VALVE
12. AIR SILENCER
13. PORTABLE GUN HANGER (RING TYPE SHOWN)
14. AIR ACTUATED PORTABLE WELDING GUN
15. WATER COOLED KICKLESS TYPE WELDING CABLE
16. AUTOMATIC SEDIMENT DISCHARGE VALVE

A. TO AIR LINE
B. TO ELECTRICAL CONTROL PANEL
C. TO WATER RETURN LINE
D. TO WATER FEED LINE

Fig. 90.17.–Typical installation of an air-actuated portable resistance spot weld gun.

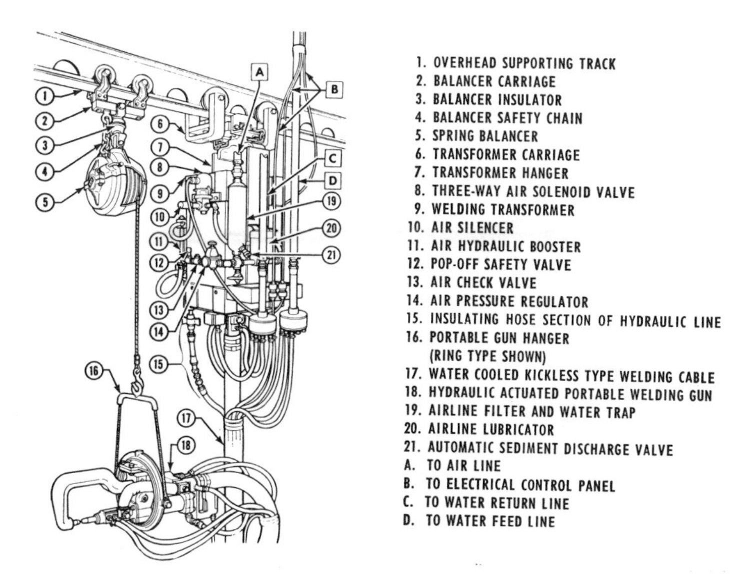

Fig. 90.18.–Typical installation of a hydraulic-actuated portable resistance-spot weld gun.

spots per minute for light gage (0.035 in.) metal and 60 spots per minute for heavy gages (0.105 in. and above).

Since metal fit-up is often poor, it is recommended that guns of sufficient force capacity be utilized to meet the schedules on the high side of the AWS recommended practices. Accuracy of controls, proper equipment and electrode maintenance must be emphasized for portable gun resistance spot welding operators to maintain weld quality.

PROJECTION WELDING

Projection welding is used to obtain multiple closely spaced welds simultaneously in an assembly, to make leakproof welds in areas too small to seam weld, and to weld metals of widely different thicknesses. Projections in sheet metal are formed by means of a punch and die in round and elongated descriptions. Solid metal projections are made by cold heading, forming and machining procedures. Other applications utilize the geometry of the components to be welded to obtain localization of welding force and current, such as cross-wire welding, chamfer screw head geometries to a hole periphery and radiused heads on a flat surface. Much of the projection welding done in the automotive industry is on drawn sheet metal, using commercial cold-headed and formed products such as weld nuts, weld bolts, tapping plates, etc.

Projections must be welded simultaneously. These welding operations are performed in multiple electrode presses or on stand-type welding set-ups utilizing special platen type electrodes to distribute evenly the welding force and current to all the projections concerned. The welding force applying mechanisms used for projection welding should be low-inertia devices so that good follow-up, on projection collapse, can be provided. Figure 90.19 shows a projection welded brake cable bracket on an automobile front floor pan.

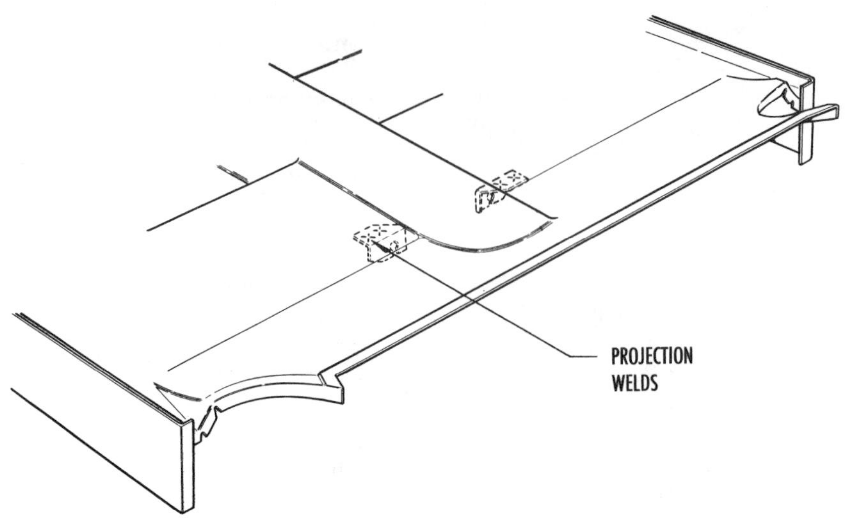

Fig. 90.19.–Brake cable bracket projection-welded to an automobile front floor pan.

Flash Welding

Flash welding can be used for joining of body panels involving short joints of relatively simple configuration. Although it was used quite extensively years ago in joining major exterior panels, the accurate electrode alignment that was required caused excessive die maintenance. It has been used with reasonable effectiveness, on occasion, for joining unexposed parts, such as floor support members for radiators and floor pans.

Seam Welding

In order to assure weathertight joints on roof rails, portable seam welding equipment has been employed. These units are mounted in a manner similar to portable gun spot welding units, with proper counterbalancing and with low reactance cables to the transformers. This type of equipment has ensured weathertight joints, and, in addition, has been successful in lengthening electrode life, thereby giving a much more consistent product. While the welds are not overlapping in the case of the roof joint, they are close enough together so that no moisture can be allowed through the joint after welding. Speeds of approximately 250 ipm are obtained on the roof rail.

Extension to roof panel blanks are now being foil-butt seam welded together by special welding machines. This application consists of butting the edges of the two blanks and introducing a thin, narrow strip of steel foil (approximately 0.010 in. thick) above and below the joint as it is passed between conventional seam welding wheels. The resultant seam weld is strong and will withstand normal drawing operations. In some instances the foil is only used on one side of the joint. Some metal finishing is required on the show surface of the joint. The production rate of this process can be as high as 240 units per hour when properly tooled (welding rate up to 20 ft per minute).

Mash seam welding may be used on similar applications although the resultant weld may not finish as satisfactorily as the foil butt weld.

ARC WELDING

The utilization of arc welding processes has increased for various subassembly and final assembly of automotive components. Equipment and technique improvements have resulted in a trend for semiautomatic and fully automatic procedures to supersede many previous manual welding operations.

Manual shielded metal-arc welding is now generally employed on relatively low production components, coated steels—where weld deposit metallurgy is required to match base metal—and where semi- or fully automatic arc procedures are not applicable or economically justified. The ultimate reliance on the skill of the welder and the comparatively lower production rate are further reasons for the manual process being replaced with newer developed semiautomatic and automatic processes.

Gas metal-arc welding is being utilized in the automotive industry for the fabrication of aluminum, stainless steels, low-alloy steels and light gage and heavy plate low-carbon steel. The filler metal electrode and shielding gas media have to be selected with regard to the metal to be joined, metal thickness, joint design and specific joint requirements.

The semiautomatic gas metal-arc welding process in which limited heat input lessens burn-thru on light gage sheet metal and where arc length is not as dependent on the individual welder has gained wide acceptance in sheet metal fabrication.

Fully automatic welding applications employing the gas metal-arc procedures have also been increasingly used for components that were formerly joined by shielded metal-arc and submerged arc welding.

In the automotive industry, gas metal-arc spot welding is applied through the techniques of edge welding and through welding. Gas metal-arc spot welding is predominately used for the thinner metal gages of steel, stainless steel and aluminum used for component automotive parts.

Submerged arc welding is used in many applications; however, the newer gas metal-arc welding processes have replaced this procedure where high deposition rate was not the primary prerequisite for its application. Processing heavy fabrication in the flat position is particularly suited to this process.

Gas Metal-Arc Spot Welding

Gas metal-arc spot welding is being applied in two ways in the automotive industry: both the weld-through and the edge-type techniques are used. Major advantages of this process include accessibility to only one side of the joint for a localized spot-type fusion weld; welds are of a consistently high level of quality; and a minimum of welding operator skill is required. The process is applicable to steel, aluminum and stainless steel over a wide range of metal thicknesses varying from light gage 0.035 in. sheet metal to perhaps as heavy as 0.188 in. The type and diameter of wire electrode and the shielding medium have to be selected to suit the metal, its thickness and technique of the application. This process is presently being used to fabricate body and frame components for automobiles, trucks and trailers.

STUD WELDING

Stud welding using the stored-arc process is a relatively new concept for attaching moldings and wiring harness clip and hard trim hardware to light gage sheet metal. "T" shaped studs are welded to a sheet-metal panel which mechanically holds a plastic or metal molding retainer for the retention of the hard molding. The primary advantage of this concept is that it eliminates drilling and punching holes in the body outer panels where water leakage and corrosion present difficulties. Inventory reduction of the door is also achieved.

The stud welding systems presently used in the automotive industry are composed of the following:

1. Pistol-type gun.
2. Stored-arc control and power supply—providing voltage regulation and synchronization of the mechanical and electrical functions.
3. Electro-pneumatic automatic feed system—transferring stud from hopper to gun by means of blowing an individual stud through a tube connected between the hopper and a receiver on the welding gun.

The system allows a welding rate of approximately 60 studs per minute. Studs used in automotive trim attaching applications are of a flat-head rivet description and are a cold-headed product made of 304 type stainless steel.

Correlation of the mechanical and electrical functions of the stored-arc process are critical and must be maintained. Total operational time per weld is 57 milliseconds or approximately 3 1/2 Hz with the main energy discharge being only 1/2 Hz or less. The stud is physically lifted off the panel by a solenoid to establish an arc approximately 1/32 in. long. Coil de-energization is correlated with the main pulse so as to terminate the plunge just below the halfway point on the trailing slope of the power pulse.

Stud location is accomplished with bushing-type locators associated with the nozzle geometry of the gun for side molding applications (Fig. 90.20). Location of the stud in window reveal is maintained by fixing the desired spacing between the nozzle outer surface to the center of the stud. Placement of the nozzle on the flange spaces the stud at the proper elevation on the vertical surface which is normal to the pinch weld flange or nozzle rest plane.

GAS WELDING

Gas welding with the oxyacetylene flame is the oldest metal joining process used in production welding of lighter gage sheet steel for automotive components having designed gaps for ease of assembly. The high degree of flexibility and versatility of the process makes it particularly adaptable in automotive sheet metal construction and for repair operations (Fig. 90.21). The inability of the process to localize heat with a flame creates distortion problems, especially in gas welding light gage metal where surface finish is required. The development of newer processes, such as gas metal-arc welding has replaced gas

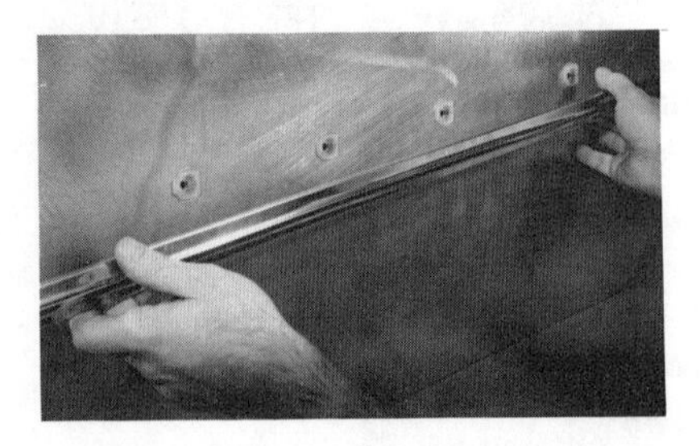

Fig. 90.20.–Studs welded in place and associated with plastic clips for hard trim molding retention.

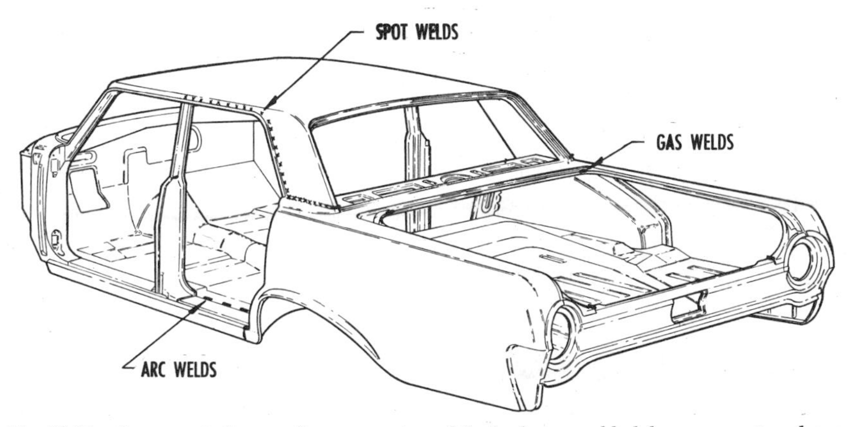

Fig. 90.21.–Representative sections on automobile body assembled by arc, spot and gas welding.

welding in many instances where the joint design and metal fit-up is more compatible with the newer processes.

Limited application of the oxygen-hydrogen flame is used in gas welding aluminum. Here too, gas tungsten-arc and gas metal-arc welding have replaced gas welding in aluminum fabrication. When joint fit-up conditions are not consistent with the newer processes, and when repair is necessary, gas welding is used.

BRAZING

Brazing replaces gas welding in areas where better flow characteristics are required to produce finished joints with minimum distortion and less heat input. Under these circumstances the increased cost of the brazing materials is more than offset by savings effected in the cost of metal finishing. A typical example is the brazing of the upper door frames to lower panels on two-piece door designs. This area is then painted after brazing and remains visible as an exposed joint. Brazing is also applicable to those joints where sealing is prime and the structure secondary.

SOLDERING

Solder processing with the tin-lead alloys is accomplished by numerous procedures in the automotive industry. Procedure and technique are dictated by the category of components to be soldered.

The dip soldering process using solder in the melted or liquid form is utilized for the simultaneous solder assembly of numerous joints in radiator-type cores, motor commuter assemblies and some wiring harness connections. Variations of the dip process are also used for printed-circuit components in radio and instrument panel electrical circuits.

The most extensive use of solder alloys in body assembly is as a filling material to conceal outer surface panel sections with depressed weld joints. This solder is supplied in bar form and is heated to a partially melted state before application.

For economy reasons the tin contents of these fill solders are lower than the alloys used in the dip processes. The joints are normally fluxed, blowpipe torch-heated and wipe-coated with a solder containing at least 15% tin. Additional blowpipe heating is then supplied and the low-tin alloy applied with various shaped wood paddles or blocks to approximate the surface contour. This filling material is subsequently mechanically finished with abrasives to final contours and painted.

Additional applications of generally higher tin content solder in wire form is made to seal gas tank fittings and coach-type sheet-metal joints and as a repair metal for various surface discrepancies.

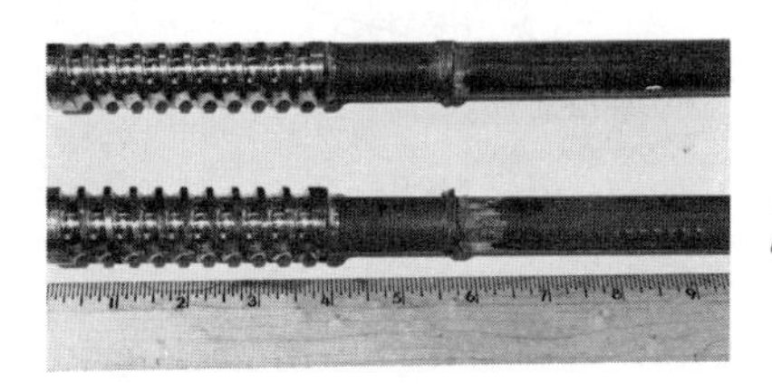

Fig. 90.22.–Automotive steering worm-to-shaft assembly by friction welding.

FRICTION WELDING

Production applications of friction welding in the automotive industry have included steering worm to shaft (Fig. 90.22), valve stem to head, yoke to drive shaft, rear axle housing end flange to tube, diesel engine precombustion chambers, transmission input and output shafts, steering sectors, radius rods, cams for actuating brake shoes and others.

Currently the highest volume American automotive application of the process is the attaching of wear-resisting stems to heat-resisting exhaust engine valve heads. The United States trails England and Japan in the use of friction welding with the latter being the dominant user. This process is capable of producing high quality weld joints between most common and dissimilar metals. One exception is its application to cast iron parts, where to date it cannot be applied. At least one of the components to be joined must be circular or nearly so at the weld face and one part must be rotatable at high speeds.

In friction welding, the rotating mechanism is disengaged and braked to a stop during the upset portion of the weld cycle. A similar process known as inertia welding brings the part up to rotational speed and the energy stored in a flywheel is used to rotate the one part when the two pieces are brought together. This stored energy is dissipated into friction, forming a weld at the end of part rotation.

ELECTRON BEAM WELDING

Electron beam welding in the automotive industry is done primarily on medium (soft) vacuum and out-of-vacuum types of equipment. Automobile components welded on soft vacuum equipment include distributor cams, flywheel ring gears and copper ring blanks for starter motor commutators.

Frame cross members, steering column outer tubes and ball joint assemblies are currently being welded on out-of-vacuum electron beam welders.

The highly concentrated energy input weld permits the welding of the finish machined and heat-treated components of the distributor cam. The butt weld of the copper ring blank is not economically feasible by any welding process other than electron beam. The 14 in. diameter automotive flywheel consists of a ring gear (SAE 1050) joined to a rimmed steel hub (SAE 1010) at welding speeds in excess of 300 ipm.

The most productive medium vacuum welders are of the dial feed type having vacuum chambers just large enough to accept the parts to be welded. Production rates of 1,700 pieces per hour have been reported.

Out-of-vacuum electron beam welding is limited to parts whose configurations allow close access to the weld area by the welding head. Part fit-up is essential on all EBW assemblies. Rimmed steel must be treated with aluminum in the form of paint or metallic particles in the weld area. Porosity found even in the best castings causes problems.

MISCELLANEOUS APPLICATIONS

Torque Converters

Automatic welding involving numerous processes is used in the manufacture of torque converters for automatic transmissions. Submerged arc, gas metal-arc, electron beam, resistance-spot, projection, arc-spot, and submerged arc-spot welding are all in current use for fabricating this component (Fig. 90.23).

The torque converter is actually an oil-filled pressure vessel which houses a stator and impeller through which the engine torque is selectively transmitted to the driving wheels. The hub is welded to the bowl by either submerged arc or gas metal-arc welding. The cover is likewise welded to the bowl. Both of these welds must not leak at operating oil pressures which may reach several hundred pounds per square inch. Internal vanes are attached in place by either resistance welding

Fig. 90.23.–Torque converter assembled with either submerged arc welding or gas metal-arc welding with carbon dioxide shielding gas.

alone or in combination with arc spot welds. The starter ring gear is attached to the cover either directly by gas metal-arc welding or bolted through projection welded studs or nuts. If a flywheel is used, the starter ring gear is attached to it by either electron beam, gas metal-arc or submerged arc-spot welding.

Spring Hangers

Automotive spring hanger fabrication consists of welding a tube into a channel bracket (Fig. 90.24). As is common with many automotive components, high volume requirements permit the use of highly automated automatic welding equipment. Gas metal-arc welding with CO_2 shielding gas is most commonly employed for this assembly.

Starter Frames

Starter and generator frames are produced by forming a piece of blank stock into a cylinder and then butt welding the ends (Fig. 90.25). Both automatic submerged arc welding and gas metal-arc welding with CO_2 shielding gas are used for this application.

SPECIAL VEHICLES

COMMERCIAL BUSES AND COACHES

Many of the items being fabricated on coaches are similar to those currently being used for passenger automobiles. Since the production quotas for this type of vehicle are drastically lower, special purpose equipment generally is not financially justified. The spot, seam and arc-welding fixtures and equipment in coach plants are usually much simpler and more adaptable to change than those in automobile plants.

Aside from the types of equipment involved, the design of parts is similar to that in automobiles, except that the service demands are more severe, and,

Fig. 90.24.–Typical welded spring hanger.

Fig. 90.25.–Starter frame assembled by welding.

therefore, the construction is correspondingly more rugged. For this reason, the metals are necessarily thicker, and equipment must be designed accordingly.

The gas metal-arc welding process using CO_2 or argon shielding gas has been adapted to the welding requirements and has extended the use of welding in this field. The high welding speeds, coupled with the minimum tooling requirements and ease in adaptability, have resulted in this welding process being extensively used.

The stud welding process is used to make removable attachments to aluminum or steel without drilling holes for bolts (see Table 90.2).

Table 90.2–Welding processes commonly used for coach assemblies

Assembly	Welding Process
Jack-knife doors	Spot Welding
Engine Cradle	Gas Metal-Arc Welding (CO_2 shielding gas)
Carlines	Gas Metal-Arc Welding (CO_2 shielding gas)
Window and Windshield Panels	Spot Welding
Doors	Spot Welding
Stepwells	Spot Welding
Fuel Tanks and Surge Tanks	Spot and Seam Welding
Axle Housings	Submerged Arc Welding
Torque Rods	Flash Welding
Coach Structural Beam	Gas Metal-Arc Welding (CO_2 shielding gas)

TRUCKS

Passenger-carrying compartments of cold-rolled sheet steel construction, whether truck cab or automotive body, are manufactured similarly. Methods include multiple spot welding machines, portable resistance welding equipment and semiautomatic and automatic gas metal-arc welding equipment. The same metal gages are used in both assemblies; consequently, the gas welding, brazing and arc welding operations on these components are also similar. In some instances, due to weight consideration, truck cabs are constructed of aluminum, utilizing the gas metal-arc and resistance-spot welding processes.

The frames of trucks are required to withstand high structural loads. The type and degree of load are dictated by the vehicle's size and its intended use. While many truck frames have riveted assemblies, resistance-spot welding, arc welding and arc-spot welding are being used increasingly in today's production. The semiautomatic gas metal-arc welding process has been successful in the fabrication of truck frames.

The fuel tanks used on smaller trucks are very similar to those used in automobiles, being constructed of terne plate and produced by the seam welding process. Fuel tanks for the heavier trucks are much larger and of diverse shapes. The design and size of these tanks do not lend themselves readily to the seam welding process. Being made of heavier gage uncoated metal, these tanks are usually fabricated with the gas metal-arc welding process.

Processing of other accessory items–generator, starter motors, propeller shafts, mufflers, brake and clutch pedals, etc.–is essentially the same as that used in automotive practice. Figure 90.26 shows a specially designed automatic gas metal-arc welding machine used for producing brake and clutch pedals.

Fig. 90.26.–Automatic gas metal-arc welding machine for welding clutch and brake pedals.

TRAILERS

The manufacture of highway freight-hauling trailers is similar to that of trucks in many ways. Like the manufacturing of trucks, the trailer industry is faced with heavier and larger components than automotive, with much less production quantity. However, trailer demand is increasing to the point where methods are being employed more for economic reasons than by the dictates of engineering design.

Where the steel trade can afford to supply sheets to suit the largest unit of an automobile, the trailer industry has had to improvise. An example would be 0.032 in. thick stainless steel roof assemblies where corrugations must be rolled in to stiffen the sheet. Nine such sheets, having 320 ft of seam welding, are welded into one assembly 8 by 40 ft which must be processed with special fixturing in conjunction with the welding equipment. The fact that eye appeal is important in the choice of stainless steel adds additional concern to the welding operation.

The side walls and front wall of this unit are processed from 40 ft long mill-rolled corrugated strips spot welded together as a mat by a gantry-type multiple press spot welding machine (Fig. 90.27), progressing over the fixtured 40 ft long assembly. There are over 15,000 spot welds in the walls of this trailer which places the equipment need beyond that of single spot gun welding that would require long throats where pressure and power loss could be a problem.

Other trailers are being constructed of spot-welded carbon steel or low-carbon, high-tensile strength steel. This allows less weight in the trailer for a greater payload, but demands better control of pressure and heat to ensure good structural quality spot welds, especially where corrosion protection of weld through primer is required.

The gas metal-arc spot weld has been found to be of use on rear posts and

Fig. 90.27.–Trailer side wall welded by a gantry-type multiple-spot welding machine.

headers where arc welding, employing a puddling technique, would be necessary because of the boxed-in section type of design.

Items such as axles, brakes, spring hangers and wheels, that, in the past, have always been considered as forgings or castings, are today welded assemblies with tubes and stampings playing an important part. Special positioning fixtures are developed to handle large assemblies.

The shielded metal-arc method of machine welding has been employed in trailer building for years in long welds similar to rear box posts, in heavy-stressed members like under-construction box cross members, sandwich-type kingpin framing assemblies and in platform trailers, the long 40 ft fabricated I-beam member. Under ideal conditions, a weld of uniform penetration, free of porosity, will be produced.

For better controlled welding, the gas metal-arc welding process is generally used on stainless steel, aluminum and, at times, on carbon steel. This method has been found to be efficient and ensures a good quality production weld.

Recently the gas tungsten-arc welding process has been given careful consideration by the trailer industry for fusing base metal extrusions, rather than applying filler metal. With proper fit-up, good structural welds up to 40 ft long have been experienced.

The aforementioned methods have found application in the specialized trailer of today where product handling, such as bulk hauling of flour, cement and other commodities, has been given special consideration.

Trailers for hauling liquids have the same general manufacturing problems as do standard trailers, with the additional problem of load-shifting and leaks. The industry has gone to machine welding for better quality control wherever possible, and many of the welds are made when the metal is in the flat position, on tank sheets before contour is formed. Some of these type units come under the classification of pressure vessels and must be manufactured and tested under the Boiler and Pressure Vessel Code as set up by ASME (see Chapter 84).

TRACTORS

In the manufacture of farm tractors, the use of welding has become increasingly important. The processes being used include spot welding, seam welding, projection welding and flash welding, as well as gas and arc welding. In certain instances, special machines for resistance-welded assemblies are justified although, in general, standard equipment is being used since the production of these tractors does not warrant large expenditures for highly specialized equipment.

The gas metal-arc welding process, both semiautomatic and fully automatic, has found increased application in the heavy plate fabrication in this category of the automotive industry. The basic reason for the increased use is an economic factor. The high penetration characteristic of the process allows the minimizing of joint preparation and the making of small size fillet welds that have strength equivalent to larger weldments.

Fuel tanks for farm-type tractors are generally seam welded in a manner similar to that used in automobiles, with the exception that the shape of the tank is often altered to suit the lines of the tractor, since the tank is completely exposed in this application. In typical farm tractors, the transmission and differential housing are of cast iron. The framework of the tractor, steering mechanism and miscellaneous items such as hoods, grilles and cowl assemblies are fabricated by welding. Heavier parts, such as the frame and drawbar assemblies, are generally arc welded and the sheet metal parts spot welded, using either portable or pedestal-type welding machines or, in the case of some high production items, special purpose multiple-electrode welding machines. Many of the round shafts employed in the various assemblies of the tractors are flash welded to forged parts, such as universal joints or worm gears. In the manufacture of tractors, there are also many small assemblies which are brazed by using torches, furnaces or induction coils for heating. The metals used for this brazing depend upon the details of the assembly, but most of the common metals are employed.

Manufacturers of heavy, track-laying tractors used for road work employ welding operations which are radically different. The assemblies used in these tractors are subjected to very severe service and, because the loads are heavier, it is necessary to have extremely large parts to avoid breakage. Arc welding is usually used and automatic arc-welding procedures are employed wherever applicable. In some of the heavy, track-laying tractors, the transmission and differential housings are completely fabricated rather than being cast. This involves a large amount of welding on plates ranging from 1/2 to 1 1/2 in. thick. Also, the tractor frames are normally of arc-welded construction. In most cases, these are completely machined after fabrication. Most of the arc welding is done in special positioning fixtures rather than on full positioners, because it is often necessary to rotate the parts in only one plane. It is common to tack weld the component parts of these large assemblies in a locating fixture which is not capable of positioning, and then reload the assembly in positioning fixtures for final welding. Due to the size of many of these assemblies, it is often possible to break them down into small component assemblies in order to simplify the operations and eliminate the necessity of positioning the entire assembly for a large portion of the welding.

BIBLIOGRAPHY

"Designing and Welding Fork Truck Lift Assemblies," D. Wiedenheft, *Welding Journal*, 50(7)483–490 (1971).

"Spreadability of Automotive Solders," H. W. Kerr, *Welding Journal*, 50(10) 441s–444s (1971).

"Gas Tungsten-Arc Hole Piercing," T. W. Shearer, Jr., *Welding Journal* 44 (12) 20–28 (1965).

"Butt Welding Steel Sheet by the Foil-Seam Process," L. W. Mecklenborg, *Welding Journal* 39 (1), 19–28 (1960).

"Recommended Practices for Automotive Welding Design," American Welding Society, D8.4 (1961).

CHAPTER 91

AIRCRAFT

REVIEWED BY THE WELDING HANDBOOK COMMITTEE

Chapter 91, as it appears here, is essentially a reprint from the Fifth Edition. Although the chapter contents have been reviewed with respect to typography, the general text and references to standard specifications have not been updated. The decision not to present a complete revision was based on current circumstances in the aerospace industry (see Preface) and the feeling on the part of the Welding Handbook Committee that the Fifth Edition chapter is still a valuable source of information for anyone seeking useful and relevant data on aircraft.

CHAPTER 91

AIRCRAFT

INTRODUCTION

Welding is universally accepted by the aircraft, missile and rocket industries. A large number of joining processes have found wide application in the fabrication of all types of aerospace vehicles. The advent of supersonic aircraft and missiles has placed greater emphasis on the role that welding techniques must play in the production of new and advanced vehicles. Increased stresses, higher temperatures and other environmental factors associated with supersonic flight have necessitated the development of new materials, improved welding techniques and refinements of presently used welding processes.

MATERIALS

Materials with high strength-to-weight ratios are of the utmost importance to the aircraft industry. Aluminum and magnesium alloys meet this requirement; they are the most commonly used structural materials in the aircraft industry. Low-alloy steels, corrosion-resistant steels and cobalt-base, nickel-base and titanium alloys are also employed where temperature requirements dictate their use.

Rapid advances in technology have made a number of new alloys available to the aerospace industries. Precipitation-hardening stainless steels, improved nickel- and cobalt-base heat-resistant alloys, new iron-base alloys and refractory metals are being adopted. Composite materials such as honeycomb or sandwich structures also are finding wide acceptance in the airframe industry.

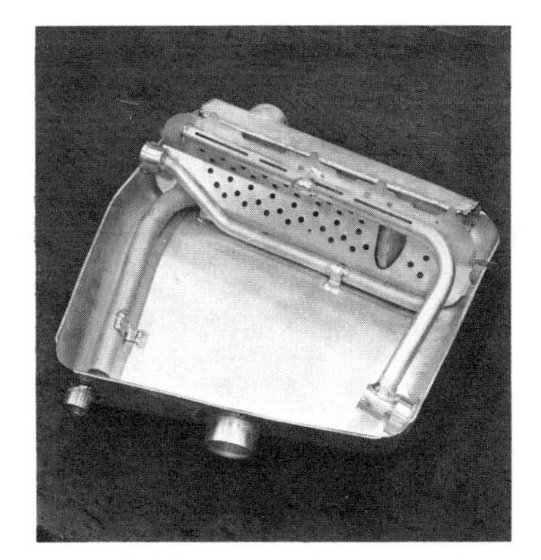

Fig. 91.1.–Oil tank of 5052 aluminum which has been assembled with the gas tungsten-arc welding process. Note confined areas.

ALUMINUM ALLOYS

Welded aluminum alloys are widely used in the aircraft industry. Figure 91.1 shows an aluminum assembly fabricated by gas tungsten-arc welding. Details of the various processes are discussed later in the chapter, and a summary of the weldable aluminum alloys for aircraft use is given in Table 91.1. Reference should also be made to Chapter 69 in Section 4.

MAGNESIUM ALLOYS

Magnesium alloys occupy an important position among the structural metals in everyday use. Aircraft structural uses include applications for fuselage, wing

Fig. 91.2.–Magnesium wing cap being welded with the gas tungsten-arc welding process. The wing cap is stress-relieved in the same fixture.

Table 91.1—Aluminum and aluminum alloys commonly used in the aircraft industry*†

Description	Government Specifications Pertaining to Materials	Spot and Seam Welding‡	Gas Metal- and Gas Tungsten-Arc	Gas Welding	Brazing
Clad plate, sheet, strip					
2014	QQ-A-255	S	L	U	U
2024	QQ-A-362	S	U	U	U
7075	QQ-A-287	S	U	U	U
7178	MIL-A-9183	S	L	U	U
Rolled bars, rods and shapes					
2014	QQ-A-266	L	L	U	U
2017	QQ-A-351	L	L	U	U
2024	QQ-A-268	L	U	U	U
6061	QQ-A-325	S	S	S	S
7075	QQ-A-282	L	U	U	U
1100	QQ-A-411	L	S	S	S
Extruded bars, rods and shapes					
2014	QQ-A-261	S	L	U	U
2024	QQ-A-267	S	U	U	U
6061	QQ-A-270	S	S	S	S
7075	QQ-A-277	S	U	U	U
1100	MIL-A-2545	S	S	S	S
3003	QQ-A-357	S	S	S	S
Forgings					
2014	QQ-A-367	L	L	U	U
2219	QQ-A-367	L	S	U	U
7075	QQ-A-367	L	U	U	U
Sheet and plate					
1100	QQ-A-561	S	S	S	S
3003	QQ-A-359	S	S	S	S
2014	AMS-4029	L	L	U	U
2024	QQ-A-355	L	U	U	U
2219	MIL-A-8920	S	S	S	U
5052	QQ-A-318	S	S	S	S
5083	MIL-A-17358	S	S	S	L
5086	MIL-A-19070	S	S	S	L
5456	MIL-A-79842	S	S	S	L
6061	QQ-A-327	S	S	S	S
7075	QQ-A-283	L	U	U	U
Tubing					
1100	WW-T-783	S	S	S	S
3003	WW-T-788	S	S	S	S
2024	WW-T-785	L	L	U	U
5052	WW-T-787	S	S	S	S
6061	WW-T-789	S	S	S	S
Castings					
356 Perm. mold	QQ-A-596	S	S	S	L
356 Sand	QQ-A-601	S	S	S	L
355 Perm. Mold	QQ-A-596	S	S	S	L
355 Sand	QQ-A-601	S	S	S	L
195 Sand	QQ-A-601	L	L	L	U
220 Sand	QQ-A-601	U	U	U	S
43 Perm. mold	QQ-A-596	S	S	S	U
43 Die	QQ-A-591	S	S	S	U
13 Die	QQ-A-596	"	"	"	"

Code

for all tables: S—for Satisfactory. U—for Unsatisfactory. L—for Limited Weldability (crack sensitivity or loss in corrosion resistance or poor weld properties).

*Refer to text for detailed discussion on welding of aluminum.

†Applications: general airframe structure, cowling, fairing, ducts, tanks, etc.

‡Due to the possibility of corrosion, restrictions have been placed on the spot welding of bars of 7075, 2024 and 2014 alloys to themselves or to each other. See MIL-W-6858.

Table 91.2—Magnesium alloys commonly used in the aircraft industry*

Description		Government Specifications Pertaining to Materials	Spot and Seam Welding	Gas Metal- and Gas Tungsten-Arc	Gas Welding
Rolled bars, rods and shapes					
M1A		QQ-M-31	S	S	S
Extrusions					
AZ31B		QQ-M-31	S	S	S
AZ61A		QQ-M-31	S	S	S
AZ80A		QQ-M-31	S	S	S
ZK60A		QQ-M-31	S	L	L
Sheet and plate					
AZ31B		QQ-M-44	S	S	S
HK31A		MIL-M-26075	S	S	S
Tubing					
AZ31B		WW-T-825	S	S	S
AZ61A		WW-T-825	S	S	S
M1A		WW-T-825	S	S	S
ZK60A		WW-T-825	S	L	L
Forgings					
AZ61A		QQ-M-40	S	S	S
AZ80A		QQ-M-40	S	S	S
M1A		QQ-M-40	S	S	S
ZK60A		WW-T-825	S	L	L
Castings					
AZ63A	Sand	QQ-M-56	Not used	S	S
AZ63A	Perm. mold	QQ-M-55		S	S
AZ81A	Sand	QQ-M-56		S	S
AZ81A	Perm. mold	QQ-M-55		S	S
AZ91C	Sand	QQ-M-56		S	S
AZ91C	Perm. mold	QQ-M-55		S	S
AZ92A	Sand	QQ-M-56		S	S
AZ92A	Perm. mold	QQ-M-55		S	S
EZ33A	Sand	QQ-M-56		S	S
EZ33A	Perm. mold	QQ-M-55		S	S
HK31A	Sand	QQ-M-56		S	S
HZ32A	Sand	QQ-M-56		S	S
ZK51A	Sand	QQ-M-56		L	L
AM100A	Perm. mold	QQ-M-55		S	S
ZH62A		QQ-M-56		L	L
HZ32A		QQ-M-56		S	S

*Refer to Chapter 70, Section 4, for detailed discussion on welding of magnesium. Code same as for Table 91.1.

and empennage parts, landing wheels, engine parts and accessories. Magnesium can be joined by most of the fusion and resistance welding methods. The relative weldability of each magnesium alloy for each process indicated is listed in Table 91.2.

As is the case with other metals, no single factor predominates when a method of joining is to be selected. The joining method is usually dictated by the type of assembly, the equipment and personnel available and the end use to which the assembly will be put. However, because of the susceptibility to stress-corrosion cracking in welds of some magnesium alloys, it is essential that a thermal treatment be employed to remove the residual stress that causes this phenomenon. A gas tungsten-arc welded magnesium wing cap which is welded and stress-relieved in a dual purpose fixture is shown in Fig 91.2.

Reference should be made to Chapter 70, Section 4, for detailed information on the welding of magnesium. Welded joints between magnesium and other

Table 91.3—Fabrication processes commonly employed for aircraft applications involving titanium alloys

Welding Process	Alpha and Alpha-Lean Beta	Sheet Alloy Type†	
		Alpha-Beta	Beta
Fusion			
Gas tungsten-arc	S	U	S
Gas metal-arc	S	U	S
Gas	U	U	U
Arc spot	S	"	S
Electron beam	S	U	S
Shielded metal-arc	U	U	U
Resistance			
Spot, seam, stitch	S	U	S
Flash	S	U	S
Foil butt	S	U	"
Brazing			
Ag-A1	S	S	S
Ag-Li	S	S	S
Miscellaneous			
Pressure gas welding	S	S	S

*Code same as for Table 91.1.
†Sheet alloy types:

Alpha	Alpha-Lean Beta	Alpha-Beta
Commercially pure	6A1-4V	8Mn
5A12.5Sn		4A1-3 Mo-1V
8A1-1Mo-1V	Beta	2-1/2 A1-16V
8A1-2Cb-1Ta	13V-11Cr-3A1	5A1-1.25Fe-2.75Cr
		2Fe-2Cr-2Mo

metals are generally not recommended because of the resulting galvanic action, which promotes rapid corrosion in the weld area.

TITANIUM ALLOYS

Titanium was developed primarily for use in the airframe industry. As a result, welded fabrication techniques have been developed by this industry. Developmental efforts by both producers and users have made titanium welding a routine operation when proper precautions are observed.

Titanium alloys commonly used in the aircraft industry are listed in Table 91.3. Normal gas-shielded arc welding techniques are directly applicable to

Fig. 91.3.—Twenty-five inch diameter helium pressure vessel fabricated by welding.

Fig. 91.4.–Typical fusion zone in 6A1-4V titanium alloy (5X).

titanium and its alloys, but, because of its reactive nature with air and other foreign material, greater emphasis must be placed on preventing contamination. Chapter 73, Section 4 contains detailed information on titanium and titanium alloys and their fabrication by welding.

The fusion welding of heat-treatable titanium alloys in the fabrication of pressure vessels has been a successful application. The 25 in. diam pressure vessel shown in Fig. 91.3 is made of the Ti-6A1-4V alloy. This pressure vessel was fabricated by fusion welding two heat-treated hemispherical forgings together with a girth weld.

A typical fusion weld microstructure of Ti-6A1-4V is shown in Fig. 91.4. It should be noted that this structure is completely different from that normally observed in the base metal (Fig. 91.5), and it generally consists of very large equiaxed grains. This structure should be regarded as normal for any of the titanium alloys that have undergone fusion welding.

One of the newer titanium alloys, Ti-13V-11Cr-3A1, has been fabricated into solid propellant rocket motor cases. The only welding required was two circumferential gas tungsten-arc welds to join the main cylinder to the forward and aft heads.

AUSTENITIC STAINLESS STEEL

The austenitic chromium-nickel stainless steels are used extensively in aircraft. Fusion welded parts subject to corrosive media applications are not made from

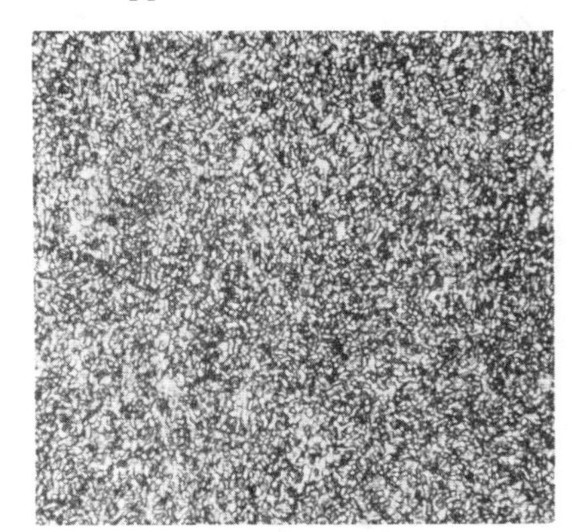

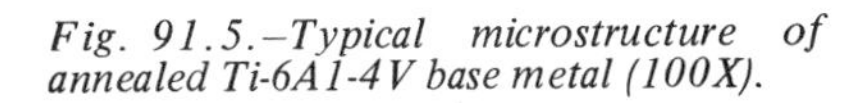

Fig. 91.5.–Typical microstructure of annealed Ti-6A1-4V base metal (100X).

Table 91.4—Austenitic chromium-nickel stainless steels commonly used in aircraft applications[a]

Description	Strength[c]				Projection, Spot and Seam Welding	Flash Welding	Arc Welding		Gas Welding	Brazing
	Hard		Soft							
	Ultimate	Yield	Ultimate	Yield			Gas Shielded-	Shielded Metal-		
18% Chromium-8% Nickel Steel[b]										
Sheet										
AISI 301	185	140	75	30	S	L[d,e]	L[d]	L[d]	L[d]	L[f]
AISI 302	185	140	75	30	S	L[d,e]	L[d]	L[d]	L[d]	L[d]
AISI 321	100	–	75	–	S	L[e]	S	S	S	L
AISI 347	100	–	75	–	S	L[e]	S	S	S	L
Bar										
AISI 302[f]	115	85	75	–	S[g]	L[d,e]	L[d]	L[d]	L[d]	L[d]
AISI 304 Grade 1A	–	–	75	–	S[g]	L[d,e]	L[d]	L[d]	L[d]	L[d]
AISI 321 Class 8	–	–	75	–	S[g]	S[e]	S	S	S	L
AISI 347 Class 8	–	–	75	–	S[g]	S[e]	S	S	S	L
Tubing										
AISI 302	120	75	75	30	S[h]	L[d,e]	L[d]	L[d]	L[f]	L[d]
AISI 321	–	–	75	–	S[h]	L[e]	S	S	S	L
AISI 321	100	–	75	–	S[h]	L[e]	S	S	S	L
AISI 347	–	–	75	–	S[h]	L[e]	S	S	S	L
AISI 347	100	–	75	–	S[h]	L[e]	S	S	S	L

[a] Corrosion-resistant and heat-resistant applications; exhaust stacks, fire walls and seals, surfaces exposed to exhaust gases, high-pressure lines, lavatory equipment and miscellaneous furniture.

[b] Refer to Chapter 65, Section 4, for detailed discussion of welding austenitic chromium-nickel stainless steels.

[c] Typical values, given in 1000 psi.

[d] Solution anneal postheat treatment required to dissolve precipitated carbides unless reduced corrosion resistance is acceptable.

[e] Requires specialized technique.

[f] Except free machining bar.

[g] May be satisfactorily welded in thin, smooth sections.

[h] Poor accessibility and fit-up restrict the use of tubular forms.

Code same as for Table 91.1.

unstabilized grades, such as AISI 301 and 302. Martensitic and ferritic stainless steels are rarely used.

The weldable grades of corrosion-resistant steels listed in Table 91.4 can be fusion welded by normal welding techniques similar to those used for other metals. Reference should be made to Chapter 65, Section 4 for detailed information on the austenitic chromium-nickel stainless steels.

PRECIPITATION-HARDENING CORROSION-RESISTANT STEEL

The commercial grades commonly classed as precipitation-hardening corrosion-resistant steel are 17-4PH, 17-7PH, 350, 355, PH15-7Mo and PH14-8Mo. Other compositions which could fall into the same material category have not found significant application in aircraft or missile manufacture. The steels listed may be readily welded by welding processes used on austenitic stainless steels (Table 91.5). When various aspects of the welding procedure have been adjusted to correspond to the alloy being welded and its heat-treated condition, all of these materials can be successfully welded. These materials do not air harden during welding. Preheating, interpass temperature control and slow cooling are not generally necessary to prevent cracking. Commercial filler metal compositions are available for each of these alloys.

Although welding in the annealed condition prior to the transformation and hardening treatments will result in higher weld metal tensile strengths, this is not always the best sequence to secure the required joint properties. The welding process applied is a major factor in planning a welding and hardening procedure. One helpful feature of working with a steel of this kind is that the same heat-treating steps and temperatures are employed whether the material is thin foil or heavy plate. The grades which contain aluminum as one of the alloying

Table 91.5—Fabrication processes commonly employed for aircraft applications involving precipitation-hardening corrosion-resistant steel

	Sheet Alloy		
Welding Process	350, 355	17–4PH	17–7PH, PH14–8Mo
Fusion			
Gas tungsten-arc	S	S	S
Gas metal-arc	S	S	S
Gas	U	U	U
Arc spot	S	S	S
Electron beam	S	S	S
Shielded metal-arc	S	S	S*
Resistance			
Spot, seam	S	S	S
Flash	S	S	S
Foil butt	S†	S†	S†
Brazing			
Ag-Li	S	S	S

* Must use 17–4 or other dissimilar electrode.
† Thicker filler strips are recommended.
Code same as for Table 91.1.

elements tend to lose this element from the molten weld by oxidation. Therefore, if the weld metal is to have the same composition as the base metal, the use of processes other than the gas shielded-arc techniques is not appropriate. However, if weld metal of a dissimilar composition is to be deposited, any process suitable for stainless steel is applicable.

The grades which contain aluminum exhibit what is commonly called "aluminum effect," a floating oxide on the surface of the weld puddle. The molten weld metal appears to have a hazy film on the surface. The film is particularly noticeable in gas tungsten-arc welding. A certain amount of welder experience is required to gage proper conditions for complete penetration and weld bead formation. Different welding conditions will result in a variation in the density of the film. Although an appreciable oxide-like film can appear on the molten metal under certain welding conditions, the amount of aluminum removed from the weld metal is usually negligible.

NICKEL-BASE ALLOYS

This section describes only those nickel-base alloys which are used for elevated-temperature service in the aerospace industry. In general, the variations of nickel-base alloys consist of matrix-strengthening elements taken into solid solution by the base element, nickel. These alloys are generally used in applications requiring moderate high-temperature strengths combined with excellent corrosion- and oxidation-resistant characteristics.

Chromium, cobalt and iron are the major alloying elements in nickel-base materials. Chromium primarily increases oxidation resistance; cobalt and iron tend to improve high-temperature creep resistance. Molybdenum is added to several matrix-strengthened nickel-base alloys, but usually in small amounts when chromium is present to avoid the formation of brittle intermetallic compounds. Some of the typical matrix-strengthened nickel-base alloys used in the aircraft industry are Inconel,[1] Hastelloy "B,"[2] Hastelloy "W,"[2] Hastelloy "X"[2] and Hastelloy "C"[2]

Welding of Nickel-Base Alloys

Fusion welding of matrix-strengthened nickel-base alloys in the aircraft industry is generally accomplished by either the gas tungsten-arc or shielded metal-arc welding process. The welding characteristics of these materials are similar to the austenitic stainless steels. The following precautions should be observed during welding:

1. The filler metal should be at least equal to the base metal in corrosion- and oxidation-resistant properties.
2. Joints should be properly cleaned prior to welding to remove all traces of grease, paint and other foreign matter which would cause excessive sulfur pickup.
3. Multiple-pass welds should be performed in a string-beading pattern to avoid the occurrence of hot-cracking; care should be taken to clean thoroughly between passes.
4. With the gas tungsten-arc welding process, root openings in sheet metal material should not be necessary, since complete penetration is obtained.

---

[1] Registered trademark of the International Nickel Co., Inc.
[2] Registered trademark of the Stellite Div., Cabot Corp.

5. With the gas tungsten-arc welding process, copper back-up bars with slotted grooves and inert-gas backing should be used when accessibility to the joint makes them feasible.

Applications requiring resistance welding of matrix-strengthened nickel-base alloys are accomplished with little difficulty if the variables involved are carefully controlled. Thus, a specific machine certification is generally required for each alloy, gage and process. The conditions which determine a sound resistance welded joint are current, pressure, time of applied pressure and welding current, electrode tip or wheel contour and surface cleanliness. These conditions should be established on sample coupons; the coupons should be sectioned and evaluated by qualified personnel, prior to welding an actual production part. Spot, seam and overlap spot welding are most commonly used.

Chapter 67, Section 4 contains detailed information on nickel alloys and their weldability.

SUPER-ALLOYS

Super-alloys are those iron-base, nickel-base and cobalt-base materials which, because of their age-hardening characteristics, exhibit excellent tensile strength, fatigue and stress-rupture properties at elevated temperatures. These alloys are particularly applicable to the fabrication of jet engines and hypersonic vehicles because of their high strength-to-weight ratio and their excellent creep resistance compared to other materials.

When aluminum, titanium or columbium is added to iron-, nickel- or cobalt-base alloys, a complex precipitate constituent is obtained which, when dispersed throughout the matrix, hardens the alloy and makes it stronger. Carbide precipitates which dissolve into the matrix at solutioning temperatures also add to the high-temperature properties of the alloy. However, the carbide precipitate formed is usually an embrittling phase in the grain boundaries and is generally undesirable. Measures must be taken to prevent carbide formation in harmful amounts in order to obtain good welding characteristics.

The super-alloys are usually welded in the stress-relieved, solution-annealed condition. After welding, the fabricated part is again solution-annealed and age-hardened. Once the material is in the age-hardened condition, it is difficult to prevent cracking from occurring during fusion welding of parts which have high built-in restraint.

Some of the commonly used iron-, nickel- and cobalt-base alloys are listed in Table 91.6.

Table 91.6—Super-alloys used in the aircraft industry

IRON-BASE	NICKEL-BASE
Incoloy*	Inconel W
A-286	Inconel X
N-155*	Inconel 702
	Inconel 700
COBALT-BASE	Inconel 901
	Hastelloy R-235
S-8 16	M-252
L-605*	René '41
V36	Udimet 500
	Udimet 700
	Inconel 718
	Astroloy

*Nonage-hardenable alloys.

Fig. 91.6.–Machined ring turbine frame for jet engine showing an A286 strut sheet welded to an inner and outer A286 strut end casting.

Welding of Super-Alloys

Welding is generally performed with the gas tungsten-arc welding process. There are a few minor applications where the shielded metal-arc process is used. All welding precautions previously mentioned under the Nickel-Base Alloys section apply and some additional precautions that are inherent in the welding of super-alloys are:

1. Parts which have been worked or deformed are given a stress-relief anneal prior to welding.
2. Welded parts are rapidly heated and cooled through the aging range to avoid the harmful effects of embrittling carbides in the grain boundaries.
3. Since a matrix-strengthened filler metal is generally used, yielding between 80-90% joint efficiency, fusion welds are designed where possible in low stress areas.

Figure 91.6 shows a machine ring turbine frame for a jet engine in which A286 struts are welded into inner and outer strut-end A286 castings.

The resistance welding performed on age-hardenable super-alloys for aircraft applications generally consists of spot, overlap spot or seam welding. Certification of welders and welding operators and equipment is required prior to joining component parts.

ULTRA-HIGH STRENGTH STEELS

The ultra-high strength steels listed in Table 91.7 may be used for frames, engine mounts, brackets, control rods, high pressure hydraulic components and

Table 91.7–Fabrication processes commonly employed for ultra-high-strength steels in aircraft applications (SAE 4340, 4130, 4140, H-11, 300M, D6AC)

Process	Remarks
Fusion welding	Preheat and postheat are recommended for gas tungsten-arc, gas metal-arc and shielded metal-arc welding. Other fusion welding processes are not recommended
Resistance welding	Flash welding is acceptable; other resistance welding processes are not recommended.
Brazing	Both copper and silver brazing are acceptable.

other structures. With a suitable protective coating these materials show a high strength-to-weight ratio in the 800 to 1000 F service temperature range.

The type H-11 (5% chromium die steel) steels require special processing to prevent cracking and serious distortion of weldments. Accepted procedure is to preheat 250 to 700 F prior to welding, depending on metal thickness, and to maintain this temperature during welding. The temperature of the weldment should be equalized to the preheat temperature after welding and then cooled uniformly to room temperature. Parts are stress-relieved at 1200 F for tempering of the welds. Standard heat treatment can then be performed on the welded parts to obtain the desired properties. Heat-treated parts cannot be satisfactorily welded. Type H-11 welding wire and electrodes are available for gas shielded-arc welding and covered electrode welding.

The frame in Fig. 91.7 is made up of H-11 steel, heat-treated to 280–300 ksi. The frame is made of all sheet metal details, except for the two machined corner fittings. Welds require complete joint penetration and radiographic inspection.

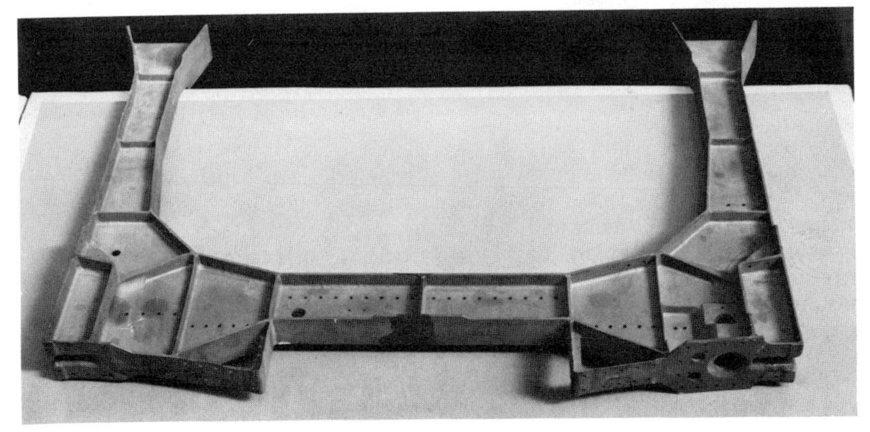

Fig. 91.7.–Jet engine structural frame constructed of H-11 steel heat-treated to 280-300 ksi.

WELDING PROCESSES EMPLOYED IN THE AEROSPACE INDUSTRY

Welding processes now employed in the aerospace industries include fusion and solid state welding, resistance welding, brazing and soldering. Each of these metal joining processes offers a number of variations, some of which are discussed in this section.

The ability of the aforementioned joining processes to meet the service requirements of a wide variety of materials is recognized. However, proper selection of the welding process to be employed depends upon the following factors:

1. Engineering design and service requirements of the product.
2. Inherent characteristic differentiating the various welding processes.
3. Characteristics of welds made by the various processes.
4. Composition of metals to be welded.
5. Weld joint design.
6. Economics.

The decision as to what constitutes adequate quality in a weldment for specific service requirements is sometimes difficult to determine. In some cases, this may require extensive investigation by destructive and nondestructive testing, corrosion testing, proof loading and fatigue and impact testing.

The reliability of weldments in aircraft is of prime importance to assure safety of flight. The penalty for failure of structural components is probably greater for airborne vehicles than in any other industry. In view of the critical nature of aircraft welding, high standards of quality are necessary, and close quality control becomes mandatory.

The major processes to be considered are fusion welding, resistance welding, brazing, brazed honeycomb and bonded honeycomb. Each process is defined and subdivided when necessary, with special emphasis upon process equipment and application to the aircraft industry. Manual fusion welding welders in the aircraft industry are qualified and certified in accordance with MIL-T-5021 to assure quality welds. The applicable Military Process Specifications are discussed in the latter part of this chapter.

ARC WELDING

During the past several years, extensive research and developmental efforts have been made to establish methods of welded fabrication which will satisfy the service demands of supersonic aircraft and missiles. The aircraft industry welding equipment manufacturers and the materials industries are cooperating to advance the art and science of welding to meet these requirements. Whatever welding process is used, it has been found necessary to establish and maintain absolute control over the possible welding variables. Some of the factors and major variables involved in welding are:

a. Material composition
b. Material condition
c. Material toughness
d. Welded joint configuration
e. Filler metal
f. Position of welding
g. Hold-down tooling
h. Backup tooling
i. Weld contamination
j. Gas shield on back (root)
k. Gas shield on top (face)
l. Power supply
m. Current
n. Voltage
o. Weld travel speed
p. Filler wire feed and diameter
q. Electrode

PROCEDURE

The following procedures and requirements are representative of the aircraft and missile industry.

Welder Qualification

Aircraft welders performing manual welding operations are required to have active qualification status on the metal to be welded and the process used, as required by Specification MIL-T-5021. In addition, procedure certifications are often required for each different weld configuration when mechanized welding is used.

Quality Requirements

The following weld quality requirements apply to the fusion welding of critical components in guided missiles and airframes: weld deposits which are flush with the base metal surfaces or weld deposit reinforcement must have a surface appearance portraying an even, repetitive ripple pattern or one resembling the surface of the adjacent base metal. The following types of defects in a weld deposit or adjacent area are cause for rejection of the part: surface roughness, irregular lineal configuration of the weld bead in relation to the weld joint, porosity, inadequate joint penetration, inclusions, nonfused areas, cracks and undercut.

Weld deposits may be machined, ground or roll-planished to conform to design size requirements. After dressing operations, the base metal and weld deposit thickness in the dressed area should not be less than the nominal sheet thickness as set forth in the applicable material specification. Such weld defects in the bead reinforcement as "drop-through" or burn-thru deposits may be removed by grinding, and the part may be considered acceptable without rewelding. Weldment details or sections which do not meet drawing dimensions or contain other defects may not be reworked by weld deposit build-up without responsible authorization.

Cross-section misalignment (offset) of sheet metal or tubular surfaces adjacent to a butt weld must conform to close tolerance requirements. For instance, when a flush surface is required by the drawing, the surfaces must be flush within 0.010 in. or 10% of the minimum thickness of the thinner material in the joint, whichever is less. Subsequent to tacking or tooling fit-up operations, the width of gaps in weld joints should not exceed 1/32 in., or 50% of the thickness of the thinnest section in the joint, whichever is less. However, tolerances vary with material, process and application.

Welding Requirements

Weld joints and the areas immediately adjacent to them must be free of all foreign substances, such as oxides, dirt, grease, oil and marking ink. This also applies to weld tooling in close proximity to the weld area. Localized mechanical cleaning with garnet cloth or a wire brush is used. The residue created by wire brushing or the use of garnet cloth is removed by thoroughly washing the cleaned area with naphtha, methyl-ethyl ketone or ethyl acetate. The use of backup flux is held to a minimum, and cannot be used on joints that are inaccessible for postweld cleaning.

It is necessary that machine welding wire be "level wound" on standard spools, which are adaptable to commercial welding wire feed machines. Manual and machine welding filler metals must be suitably packaged to prevent contamination of the wire in storage or during shipping. All filler metals must be free from surface or internal contamination that would affect weldability or the quality of the weld deposits. The chemical analysis of all filler metals is checked to applicable specification requirements prior to production use.

Weldments that will be subsequently spun, planished or subjected to other forming operations are usually machine welded. Tack welds are used when required to maintain joint alignment and preweld joint spacing. Applied tack welds which would result in unsatisfactory weld quality, such as tacks that contain defects or cracks, must be removed and replaced. When multipass welding is used, all slag, loose scale, flux and defective weld deposits present on a weld pass must be removed before the next weld pass is made, in order to meet quality requirements. Upon completion of each welding operation, the welder or welding machine operator marks each weldment legibly with an assigned identification number. Steel stamps are used for welder identification, if they can be used without causing structural injury or deformation of the weldment configuration. Otherwise, rubber stamps, electric pencils or other approved marking procedures are used for welder identification.

Preheating and Stress Relieving

Steel containing more than 0.33% nominal carbon is preheated 250F to 600 F prior to welding, depending upon the material thickness, the hardenability of the material and the welding process being used. Preheating operations are performed with equipment having close temperature control. The initial preheat temperatures are maintained throughout the welding operation. Upon completion of the welding operation, the part is normally heated to 1200 F for stress relief. Stress-relief temperatures must be compatible with engineering requirements for the material and application.

GAS TUNGSTEN-ARC WELDING

From the standpoint of man-hours, the gas tungsten-arc welding process is the major welding method employed in the airframe industry.

The principles of operation, equipment used, shielding gas, materials, filler metal and various applications of the process are discussed in Chapter 23, Section 2. Additional information on welding equipment is contained in Chapter 28, Section 2.

MANUAL GAS TUNGSTEN-ARC WELDING

Conventional welded aircraft assemblies, such as ducts, fittings, fairing and cowling components do not lend themselves to automatic welding because the particular configurations contain drastic contour and section changes. Manual welding equipment currently in use is generally of the conventional transformer-rectifier type of power supply, usually with a saturable reactor or magnetic amplifier type of control. The electric control of the machine enables the welder to vary the current as required when welding by means of a foot control. This type of control provides up-and-down slope for starts and stops as well as variations required for welding nonuniform sections to minimize cracking and porosity and to obtain the desired weld bead contour

Standard manual gas tungsten-arc welding torches are used with argon or mixtures of argon and helium as a shielding gas. The use of inert gas as a backup during manual welding is increasing as weld tooling techniques improve.

AUTOMATIC GAS TUNGSTEN-ARC WELDING

Significant advances in gas tungsten-arc welding equipment have been made to meet the exacting precision required in the aircraft industry. The greatest progress has been made in the control of the various welding parameters. Present requirements dictate controlled up-and-down slope of current, delayed weld travel, delayed wire start and advanced wire stop. In addition, weld parameter changes must be programmed to meet the requirements of advanced design configurations. General practice is to instrument the machine parameters of current, voltage, weld speed and wire feed. Recording volt and ampere meters are often employed to monitor weld settings.

Prior to welding a production part, the welding procedure to be used for the section is established and the resultant weld quality verified by destructive or nondestructive testing.

Hold-down clamps and backup mandrels must be free of all foreign substances prior to "loading" a fixture with the assembly details. All shielding gas lines are adequately purged prior to starting welding operations, and the dewpoint, as well as other contamination of the gas, is determined.

When a butt joint is to be produced in a weld fixture, the joint should be in metal-to-metal contact at the start of the welding operation. The width of gaps in any location on the remainder of the joint, due to irregular sheet edge planes, should not exceed 1/32 in. or 50% of the thinner base metal thickness in the joint, whichever is less. Sheet, tube or other components forming groove welds must be in cross-sectional alignment except where two thicknesses are involved, in which case the alignment is usually held to one side.

Where design permits, starting and run-out tabs are used on all longitudinal butt joints unless a minimum of 1 in. is to be trimmed from each end of the weldment upon completion of the welding operation. The starting and run-out tabs should be of the same material and thickness as the weldment. These tabs may be an integral part of the weldment, or they may be attached separately at the starting and run-out point of the welding operation.

On circumferential or circular butt joints, the completion point of the welding operation may have to overlap the weld starting point. Tabs have been used to start and stop circumferential welds.

Automatic gas tungsten-arc welding is applicable to material as thin as 0.003 in. foil. For thicknesses up to 0.020 in. it is not generally practical to add filler metal by means of wire. Therefore, a flange of about twice the metal thickness may be turned up on each piece to be groove welded. This flange is fused to provide the desired filler metal for the joint. This procedure has proved to be successful for welding of face sheets used as skins in the fabrication of stainless steel brazed honeycomb aircraft panels. These welds are usually roll planished to improve their mechanical properties and reduce distortion, and to bring the weld zone down to the same thickness as the adjacent material. When butt welding material over 0.020 in. thick, it may be necessary to add filler metal.

Automatic gas tungsten-arc welding is especially suitable for girth and longitudinal groove welds that require a minimum of distortion, narrow beads and heat-affected zones and consistent resutls. Automatic welding also saves considerable time per welded piece because of its ability to operate at regulated high speeds. Its disadvantages are that consistent alignment must be maintained and that additional tooling costs are incurred.

Figure 91.8 illustrates an automatic gas tungsten-arc welding machine with optical tracking and independent voltage, current and filler metal feed controls and trace recordings of the significant welding parameters.

Thin gage materials can be welded with automatic equipment but at considerable expense, since they generally require special accessories and fixturing. Stainless steel foil as thin as 0.003 in. has been welded with this type of equipment.

SHIELDED METAL-ARC WELDING

Shielded metal-arc welding is used on aircraft parts; its use has declined in recent years because of the substitution of gas tungsten-arc welding and other

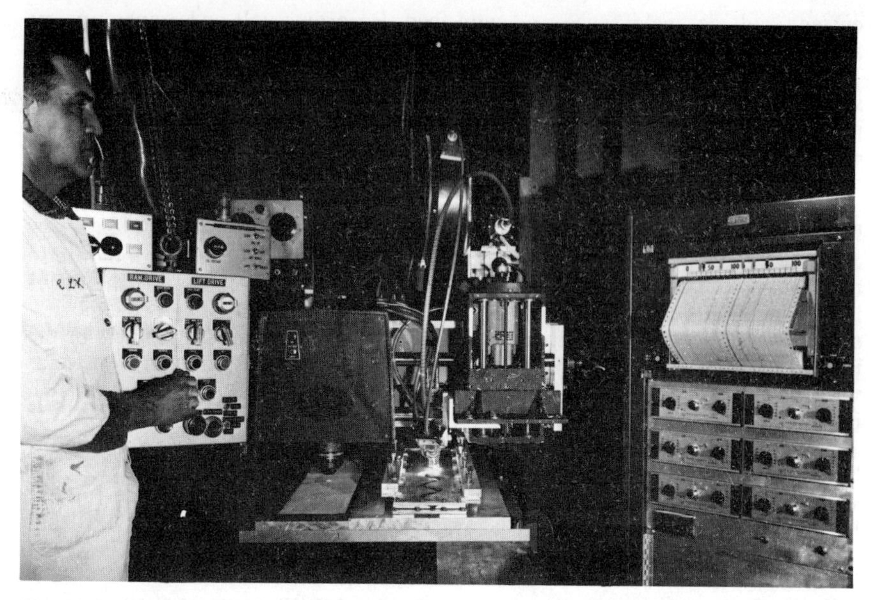

Fig. 91.8.–Electric follower controlling gas tungsten-arc welding on titanium, burn-thru weld. Trace recording is used to continuously record the significant welding parameters.

processes. It is a manual process suitable for 0.125 in. or thicker metal. In comparison with manual gas tungsten-arc, it will give lower distortion; on crack-prone alloys it will often give less hot-cracking. Covered electrodes are available for most stainless steels, super-alloys and high-strength steels. The aircraft industry uses the covered electrode in the fabrication of ground support equipment, which is usually made of low-carbon and low-alloy steels. Detailed information on the shielded metal-arc welding process is contained in Section 2.

GAS METAL-ARC WELDING

The heavier thicknesses of steel and aluminum require the use of the gas metal-arc welding process to obtain maximum joint efficiency. The gas metal-arc welding process is used to minimize weld distortion and the heat-affected zone adjacent to the weld. The advantages obtained by using this method have justified its extensive development. However, limitations as to the minimum metal thickness which may be satisfactorily welded, position of weld and control have limited its uses. Detailed information on the gas shielded-arc welding processes and allied equipment is available in Chapters 23 and 28 in Section 2.

Gas metal-arc welding is used extensively in the fabrication of aluminum rocket motor fuel and oxidizer tanks, as well as the fabrication of in-flight refueling tanks for jet aircraft. These welding operations are primarily mechanized, using stake-type fixtures for the longitudinal weld of the tank cylinders. Spiral internal expanding mandrels are used to facilitate producing the circumferential tank welds. The materials in current use are 6061-T6 aluminum, 5456 aluminum, 5083 aluminum, 7079 aluminum, 2219 aluminum and other alloys in thicknesses ranging from about 0.070 to 0.375 in.

To maintain the precise mechanical alignment necessary for aerodynamic and structural applications of these aluminum tanks, extensive internal and external tooling is used to maintain alignment of the parts and control the weld penetration bead. Austenitic stainless steel is sometimes used for the backup as well as the hold-down tooling because it is nonmagnetic and easily cleaned. Stainless steel does not give as rapid chill as does copper, but has desirable thermodynamic properties of uniform weld heat dissipation. The backup tooling can be segmented and forced mechanically or pneumatically against the weldment. Backup gas is used where beneficial. Weld head-to-work distance and alignment of weld head is usually maintained mechanically.

To consistently produce aluminum welds of radiographic quality, the metal and the tooling must be absolutely clean. It may be necessary to scrape the interfaces of the joints for welds of this type. The front and back sides of the metal should also be scraped to a depth of 0.001 to 0.005 in. and to a distance of 0.500 in. from the edge of the butt joint. This scraping should be done immediately before welding to minimize oxidation of the aluminum. The filler metal must be of high quality with respect to purity, cleanliness and surface condition. The filler metal generally has a very short shelf life, of about one week to three months after the package is opened, depending on the conditions under which the wire is stored. Vacuum tin containers and properly packaged plastic containers considerably prolong the storage time without deterioration.

To meet aircraft and missile specifications consistently, the constant potential welding power supply should be free of fluctuating input voltages. This may be accomplished within the welding power supply or with stabilization transformers in the input supply line.

The pull-type wire feeders are most successful, due to their ability to handle small wire smoothly. The constant speed wire feed gas metal-arc welding equipment is in general use. These wire-drive motors are electronically regulated to provide constant speed under varying loads. For repeatability, the weld parameters, wire feed (which controls amperage), welding voltage, weld travel and distance of welding head from work should be adjustable to the desired values.

The use of manually operated gas metal-arc welding equipment has been limited in general to noncritical aircraft components, tooling and ground support equipment. However, recent developments employing small wires have been used successfully in special applications. Lighter gages can also be welded using relief backing bars or straps. There is an overall reduction in heat input which results in less shrinkage and distortion.

GAS WELDING

Oxyacetylene welding is being replaced with the gas tungsten-arc and gas metal-arc welding processes, but is still used to some extent in aircraft factories. Its inherent disadvantages, principally the reactive atmosphere in the weld zone and excessive heat input to the weldment, have precluded the use of gas welding for space age materials.

ARC-SPOT WELDING

There are two types of arc-spot welding: gas tungsten-arc spot welding and gas metal-arc spot welding. Generally an arc-spot weld is made by applying a welding arc of sufficient heat to penetrate through the top sheet and cause local fusion of two or more sheets of material.

There is no existing military specification for gas tungsten-arc spot welding, but some aircraft companies have had their specifications approved by the military services. The most practical application to date is the spot welding of stainless steel and mild steel.

Aircraft quality arc-spot welding is accomplished by following exacting equipment settings, welding procedures and metal preparation. For consistent results, the faying surfaces of the lap joint to be welded must be in intimate contact. No particular pressure is required other than that necessary to maintain metal-to-metal contact.

The arc can be initated by the touch start system, which advances and retracts the tungsten electrode by the use of high frequency or by the use of the pilot arc start system where a low current arc is maintained between the tungsten electrode and nozzle. With either system, the weld cycle timing should begin with the actual flow of welding current. The welding of a very thin metal (0.005 in. thick) will require very short welding times. Such a weld in stainless steel would require a current rise to welding current in 1/2 cycle (measured on a 60 cycle per second time basis), hold for 4 cycles and decay within 1/2 cycle. To meet these requirements, it is necessary that the welding power supply have a fast response to current change and that the timers be very accurate and consistent. Electronic d-c timers and cold cathode counting tubes are in current use in lieu of the conventional resistance-capacitor timing system.

The usual arc-spot weld control consists of current and time control of the preweld, welding and postweld operations. For material thicknesses over 0.050 in., filler metal should be added where crater cracking exists and where nugget configuration needs to be improved. This may be accomplished by special wire feeders employing rate of feed control and sequence timing so that the wire may be injected into the weld puddle as required for any given weld setting. Reheat treating, quenching and tempering operations may also be accomplished to improve the as-welded properties of heat-treatable steels. This is accomplished by means of a special nozzle attachment providing an oxyhydrogen flame for the heat-treating cycle. The gas flame heat treat cycles are sequenced in with the arc weld cycles to provide an optimum nugget for any given requirement.

An application of arc-spot welding is the welding of extruded sections to the face sheet of brazed honeycomb panels. Gas metal-arc spot welding is being used on many materials and on commercial products.

Figure 91.9 shows a typical arc-spot welded stainless steel corrugated sandwich construction.

STUD WELDING

In general, percussion stud welding has greater application in the aircraft industry than arc stud welding (Chapter 30, Section 2). However, both processes are used to a limited extent. Their chief use is to supply attachment points to bulkheads and other structural members for support of equipment. The selection of welding process depends on the thickness of the material to be welded and the required stud diameter. Smaller studs can be welded to thinner sheet by the percussion welding process than by the arc welding process.

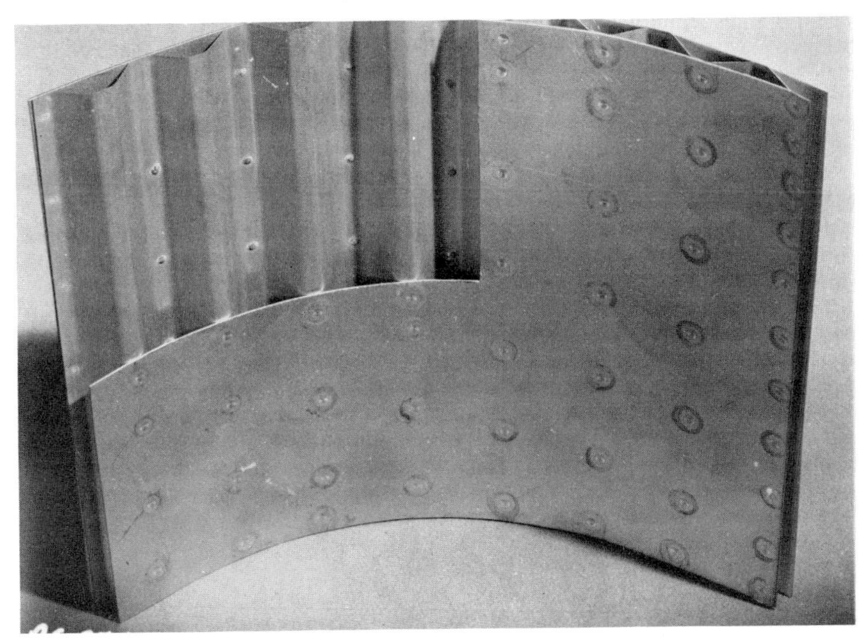

Fig. 91.9.–Typical arc-spot welded stainless steel sandwich structure.

PLASMA-ARC WELDING

A miniature plasma-arc torch has been developed for metal from 0.002 in. gage foil up to 0.035 in. It has advantages in quality and ease of welding over gas tungsten-arc and should supplant it in many aircraft applications. Heat is supplied by a thin 1/16 in. plasma jet extending over 1/2 in. from the torch. The intense heat is uniform for over 1/4 in. so that close control of work-to-torch distance is not as neccessary as in gas tungsten-arc. The torch uses an argon or 95Ar-H_2 atmosphere which also serves to protect the work. The presently available power supply has a range of 0.1 to 10.0 amperes. The torch can be used in a nontransferred arc mode to permit melting of ceramics. Refer to Chapter 54, Section 3B, Sixth Edition for principle of operation.

Development is underway for using larger plasma-arc torches for mechanized welding of titanium in gages exceeding 1/8 in. Compared to gas tungsten-arc welds, the welds are half as wide and the quality is as good or better. One-pass welds have been made in 1/2 in. titanium whereas gas tungsten-arc would require five passes.

ELECTRON BEAM WELDING

A primary application for electron beam welding is to obtain ultra-high purity welds in reactive and refractory metals, nickel and stainless steel for the atomic energy and rocket fields.

This process involves heating metals in vacuum and joining them by fusion.

The principle employed to accomplish the heating (directing a concentrated beam of electrons) is not new. The theory is that the kinetic energy of a high velocity beam of electrons can be converted into thermal energy when the electrons strike the metal to be welded. Presently a high vacuum is the only efficient environment for maintaining the focused stream of electrons and preventing arcing between the high-voltage elements.

Some of the advantages of electron beam welding are the results of the high vacuum, and some are a direct result of the process itself. Its advantages include ultra-high purity of the welding atmosphere, small heat-affected zone, close control of the welding heat, high speeds, good joint efficiency and a minimum of shrinkage distortion.

The application of electron beam welding is limited in current aircraft production to a few welds in small components. However, extensive developmental work has been accomplished and application of electron beam welding has increased considerably within the past few years. Preliminary work with the process has indicated that it has decided advantages for welding reactive and refractory materials for high-temperature application. Large chambers are now available, and methods of utilizing sectional or portable chambers have been developed to allow the welding of large structural components.

RESISTANCE WELDING

Most of the resistance welding processes have application in the aircraft industry. Various design criteria dictate the choice of process, with about 50% of the design influenced by cost and 50% by other factors. Quality requirements are established by the applicable military specifications, and nonmilitary aircraft welding is generally performed according to similar requirements.

As aircraft welding applications are often critical, the general quality demands are higher than those in industries where less integrity is required from the weld. To meet these higher requirements, specialized equipment has been evolved.

A resistance welding machine must have the inherent capacity to repeat a prescribed welding cycle in an identical manner for each successive cycle, as well as for each successive change in set-up. This requires control adjustments that are precise and have sufficient resolution. Maintenance procedures must avoid alteration of the machine "calibration."

Spot welding machine operators are generally not as highly specialized, trained or paid as are other welders or welding machine operators. A welding operator test is not required by specification, and, in general, no formal test is required by the employer. Some companies do qualify their spot welding machine operators as a means of reducing the technical surveillance required to maintain product reliability. Equipment used for spot and seam welding heat-treatable aluminum alloys probably has the most critical instrumentation requirements. For spot welding aluminum, the magnitude of current and electrode force with relation to welding time is extremely important. One technique for determining the values involves the use of an inductive pickup and a force pickup attached to the mechanical system.

The current and force systems are frequently used with a two-channel direct writing recorder and an appropriate amplifier to couple the current and force transducers to the writing system. Time is a function of paper speed. A typical portable instrumentation cart shown in Fig. 91.10, is connected to transducers

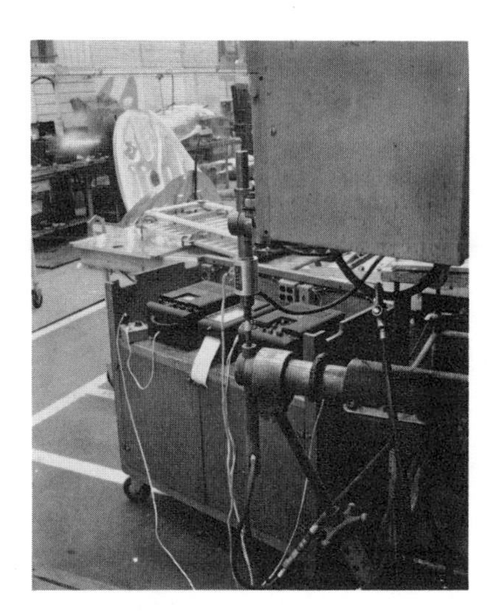

Fig. 91.10.–Instrumentation cart with force and current transducers as used to verify welding machine operational cycle.

in a spot welding machine. The type of trace generated by such a system is shown in Fig. 91.11.

Examination of the trace is made more meaningful if the force trace is calibrated in pounds and the current trace in amperes. However, knowledge of the absolute value of these standard units of force and current is not necessary for the trace to be a valuable tool if care is taken to duplicate the set-up each time the instrumentation is used. Many companies have "standard settings" used to check their equipment on a weekly or monthly basis and they rely on a comparison of traces to indicate any drift in the welding equipment. Others record certain data from the trace in numerical form, as shown in Fig. 91.11.

Similar instrumentation is applied to flash welding to relate platen motion to welding current or voltage.

In addition to oscillographic instrumentation, aircraft companies performing aluminum spot welding control their cleaning process with a surface resistance analyzer. Variations in cleaning seriously affect weld quality and are controlled to bring surface resistance within a standard range.

Other items of supporting equipment include tensile testing machines for on-the-spot strength evaluation. Cutting, polishing and etching equipment is also required to permit rapid macrosection examination.

SPOT, SEAM AND STITCH WELDING

Seam welds are principally used to produce fluid-tight joints. Stitch welds are spot welds spaced two nugget diameters, or less, apart. Aircraft spot welds, by specification, fall into class "A"–critical or structural, and class "B"–general purpose. The strength of "A" and "B" is the same, but more rigid quality control procedures are established for class "A." To these basic classes, some companies have added two additional classes: "C" for tack or substrength welds and "SP" (special purpose) for special structural applications. The "SP" class is generally higher in strength than classes "A" and "B."

Two characteristics of spot welds are used as control criteria, either separately

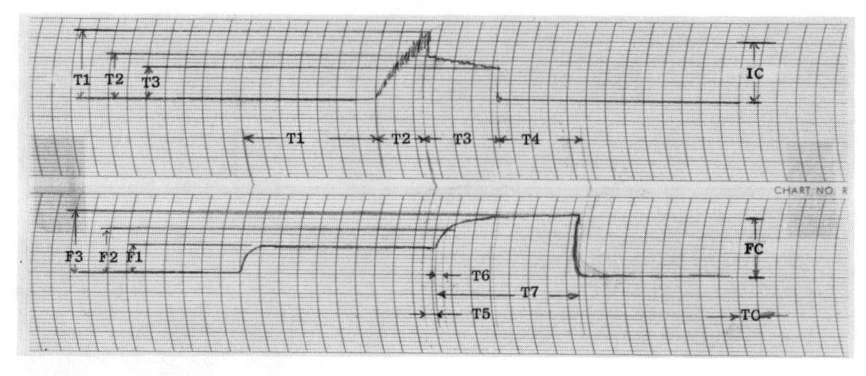

T_1 *(Squeeze Time)*	*175 ms*
T_2 *(Weld Time)*	*66 ms*
T_3 *(Decay Time)*	*83 ms*
T_4 *(Hold Time)*	*75 ms*
T_5 *(Forge Delay)*	*17 ms*
T_6 *(Time to 50% Forge)*	*10 ms*
T_7 *(Forge Time)*	*175 ms*
I_1 *(Peak Current)*	*35000 A*
I_2 *(Start Decay)*	*22000 A*
I_3 *(Finish Decay)*	*16000 A*
F_1 *(Weld Force)*	*1400 lb*
F_2 *(50% of Forge)*	*2250 lb*
F_3 *(Final Force)*	*3000 lb*

$$\text{Rate of rise (forge)} = \frac{(F_2 - F_1)}{T_6} = \frac{850}{0.010} = 85000 \text{ lb/s}$$

Fig. 91.11.–Typical traces and developed data produced by instrumentation system shown in Fig. 91.10. Gain for force system (FC) assumed as 15 lines=3000 lb; current system (IC) is 15 lines=30,000 A and each horizontal division equals 25 milliseconds.

or in combination. The first is strength as read on a tensile tester; the second is size, as shown by macrosection. Both are generally used to establish a setting for class "A" or "B," and additional requirements are added for class "SP." Those companies performing class "B" or more critical work usually maintain quality control charts for each setting used on each machine. Standard quality control techniques are applied to establish action limits and maintain an active control of product reliability.

Similar principles of control are applied to seam or stitch welds with macrosections providing measurable samples.

Figure 91.12 shows a typical application of seam welding in the fabrication of aircraft components.

FLASH WELDING

In aircraft applications, flash welding machines range in size from about 20 kVA to over 1000 kVA. They include an automatic flashing and upsetting cycle. Some include provision for a postheating cycle. Probably the largest volume of similar parts are the various engine control push-pull rods. The largest

Fig. 91.12.–Roll seam welding of stainless steel high-pressure ducts.

cross-section parts are used for landing gear components such as the oleo-strut, braces and retracting links. One of the most highly stressed parts is the catapult hook used for carrier-borne aircraft.

The metal most often used in flash welding applications is heat-treatable low-alloy steel. Some use is made of austenitic stainless steel and titanium alloy, which have somewhat similar welding characteristics. The use of aluminum is gradually increasing.

Since many flash-welded assemblies are structural in nature, military and company specifications relating to the process are quite rigid and provide strong testing requirements. As a result, aircraft flash welding quality has been maintained at a very high level, with nearly 100% joint efficiency realized even at heat-treating levels approaching 300,000 psi ultimate strength.

Figure 91.13 shows a flash-welded catapult hook which is clamped into a tool for removing excess flash. This application is typical of those aircraft components assembled by flash welding.

FOIL BUTT WELDING

Foil butt welding and butt seam welding as such do not have extensive application in aircraft production. However, a modification of these processes using thicker filler strips is finding application for joining titanium and precipitation-hardening stainless steels. In order to produce aircraft quality butt welded joints by resistance welding, strips of about the same thickness as the sheets are required. This results in an equal thickness stackup and makes good weld quality possible. The reinforcing strip is machined off after welding if a flush joint is required. However, good strength advantages can be gained if the buildup can be left on. Applications include honeycomb panel joints and sheet joints where arc welding shrinkage and distortion present a serious problem.

Fig. 91.13.–Catapult hook in place in a tool for removing excess flash.

Reference should be made to Chapter 26, Section 2 for detailed information on foil butt seam welding and butt seam welding.

BRAZING

Brazing has shown a rapid growth trend in its application to the aircraft industry. This growth is expected to continue because of the increasing emphasis on high thrust, lightweight fabricated structures. The use of extremely thin gage sheet material often requires brazing as the fabrication process.

The sources of heat presented in Chapter 60 of Section 3B are all used to varying degrees by the aircraft industry. Aircraft quality requirements necessitate that heating rates, atmospheres, joint fit-up, alloy preplacement and cleanliness be closely controlled. Experience has shown that these requirements can be achieved more consistently with heating methods which lend themselves to automatic rather than to manual operations. Consequently, the major types of heating sources used include furnace heating,, employing various atmospheres, induction heating, salt bath and vacuum furnace heating. Manual oxyacetylene heating is still employed to a limited extent.

Some typical brazing alloys used in the aircraft industry are tabulated in Table 91.8. Brazing alloy selection depends on such factors as service temperature and strength required, base metals joined, corrosion resistance, oxidation resistance, joint erosion and metallurgical characteristics, cost and capillarity and filleting characteristics.

The silver and copper alloys are used mainly when service temperatures are

Table 91.8—Typical brazing alloys used in the aircraft industry

Brazing Alloy	Nominal Composition (%)							Approximate Thermal Points	
	Ag	Cu	Ni	Cr	Si	B	Other	Solidus, F	Liquidus, F
Ag-Ni	..	..	18	..	..	..	Au-82	1740	1740
B Ag-1	45	15	..	..	..	..	Zn-16 Cd-24	1125	1145
B Ag-la	50	15.5	..	..	..	..	Zn-16.5 Cd-18	1160	1175
B Ag-3	50	15.5	3	..	..	..	Zn-15.5 Cd-16	1170	1270
B Ag-8	72	28	..	..	..	..	..	1435	1435
B Ag-13	54	40	1	..	..	..	Zn-5	1340	1575
B Ag-Mn	85	..	..	..	..	..	Mn-15	1760	1780
B Ag-19	92.5	7	..	..	..	..	Li-0.2	1435	1635
B Ni-1	..	..	73	14	4	3.5	Fe-4.5	1790	1900
B Ni-2	..	..	82	7	4.5	3.5	Fe-3	1730	1830
B Ni-3	..	..	91	..	4.5	3.1	..	1800	1900
B Ni-4	..	..	71	19	10	..	..	1975	2075

relatively low (approximately 800 F), as in controls, accessories, wing structures and forward engine sections. The trend toward higher speed and elevated temperatures is resulting in a greater use of higher temperature brazing alloys.

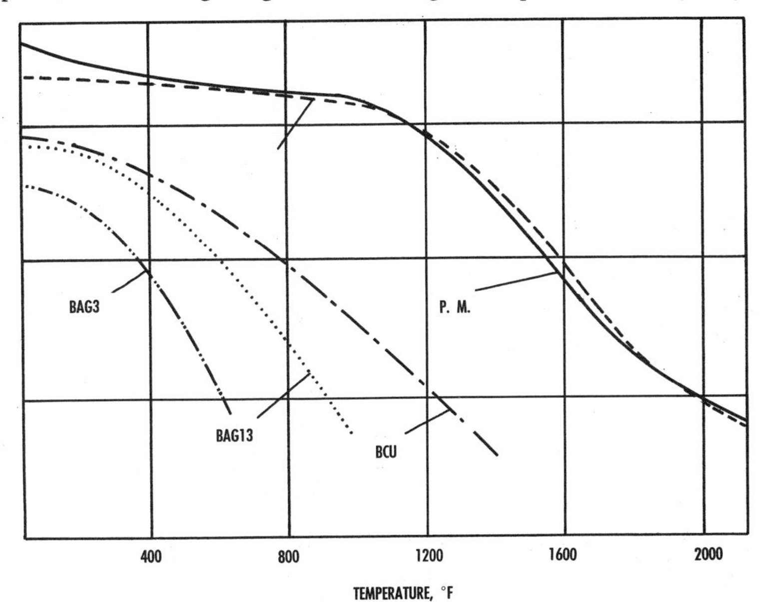

Fig. 91.14.—Comparative strengths of brazing alloys.

Figure 91.14 shows a plot of strength versus temperature for silver-base, copper and a high-temperature nickel-base alloy for a given joint clearance.

INSPECTION

The importance of brazing inspection is recognized throughout the industry since reliability is a requirement. Often visual inspection is satisfactory. Visual inspection can often be implemented by selectively preplacing alloy shims or rings so that positive flow of the alloy through the joint will occur; this can be inspected on the other side of the joint. Radiography is used in many applications, although problems in interpretation may occur. This method of inspection must be used with discretion.

Thermographic inspection is becoming increasingly significant for inspecting honeycomb and other thin material constructions (Fig. 91.15). This process is based on the fact that temperature differentials cause thermographic test fluids to change viscosity and surface tension. Braze-bonded areas act as heat sinks and stay relatively cool, attracting the thermographic liquid and revealing the bond pattern. Lack of bonding areas or poorly brazed joints will heat up, and no pattern inspection will show. The heat source generally used is infrared heat lamps.

COPPER

Copper is used as a brazing alloy on carbon and alloy steels of the 1000, 4000, 6000, 8000, 52100 series and the 300 series corrosion and heat-resistant alloys.

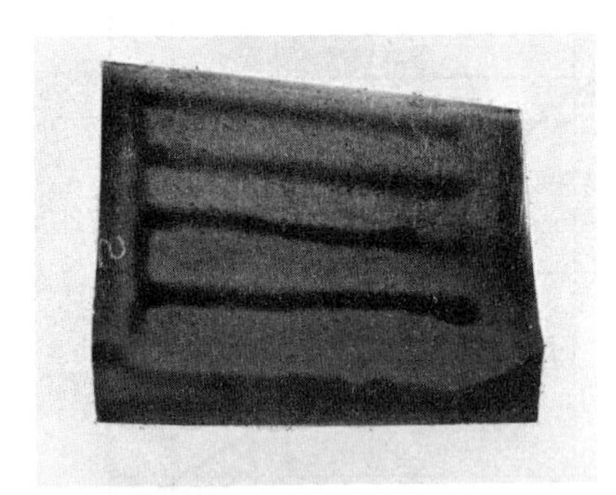

Fig. 91.15.–Thermographic pattern on engine nozzle partition.

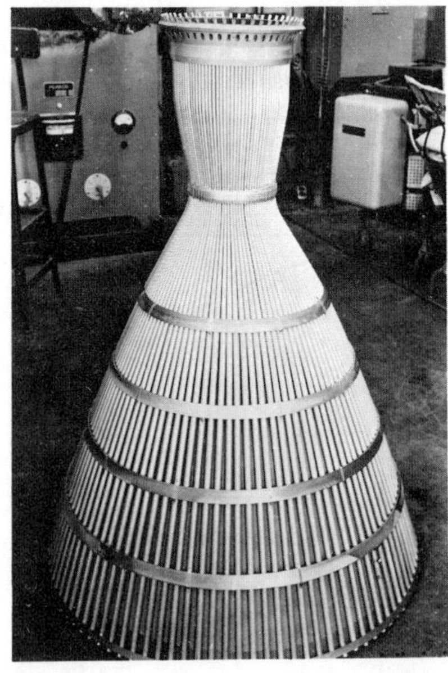

Fig. 91.16.–Rocket nozzle assembly joined with high-temperature brazing alloys.

The main advantages of copper brazing over using a silver-base brazing alloy are that the assembly may be heat-treated after brazing, since the 1981 F melting point of copper is well above the heat-treatment temperatures, and copper has a maximum service temperature of 800 F. The main disadvantage is the close tolerance that must be maintained in mating parts.

SILVER

Silver brazing is used extensively throughout the aircraft industry for applications involving service temperatures of 500 F or below. There are a few higher temperature silver brazing applications. Reference should be made to Chapter 60, Section 3B for detailed information.

Furnace and induction heating are widely used for mass production of silver-brazed components. Often, manual torch oxyacetylene or oxyhydrogen heating is permissible and economical.

Vacuum and blanket heating sources are being evaluated and specified for many components where materials and material configurations preclude the use of other heating sources from a reliability or economic standpoint.

NICKEL-BASE ALLOYS

Nickel-base brazing alloys are used extensively in the aircraft industries where extreme heat and corrosion resistance are required. Although very strong, these filler metals generally tend to be brittle and erosive. Care must be used in selecting and evaluating these alloys for use (See Chapter 60, Section 3B). Constituents of some of these alloys act as fluxes, and the alloys are said to be partially self-fluxing. Even with this behavior, there are super-alloys which must be nickel-plated to obtain proper braze flow consistently.

Furnace heating in a hydrogen atmosphere is used extensively with the Ni-Cr brazing alloys to fabricate complicated assemblies (Fig. 91.16).

Vacuum heating methods are often used where exotic materials preclude the controlled atmosphere furnace heating method. Induction heating sources with and without controlled atmosphere are used in applications where economics are a major consideration.

BRAZED HONEYCOMB STRUCTURE

The brazed honeycomb structure is used because of its stiffness, high strength-to-weight ratio, equidirectional load-carrying and thermal insulation capability, and because it has higher usable temperatures than adhesive-bonded structures.

Brazing of the honeycomb structure is accomplished in sealed and inert gas purged retorts which may be heated in conventional atmosphere or luminous-wall furnaces, electric blankets or by banks of radiant quartz lamps. Brazing is also done in specially designed vacuum furnaces.

Many metals, including corrosion and heat-resistant steels, precipitation-hardening stainless steels, super-alloys and titanium alloys can be brazed into honeycomb structures. The majority of the development and application of brazed honeycomb structure has been with the precipitation-hardening stainless steels. A brazing cycle is normally selected which combines both the brazing and heat-treating operation.

A large variety of brazing alloys is available for use in this application. The specific alloy selected should possess the following characteristics:

1. Adequate mechanical properties and corrosion resistance.
2. Adequate size fillets for the core-to-face sheet bond.
3. The ability to braze satisfactorily at a temperature compatible with the heat treatment of the base metal.
4. Proper thermal conductivity for the application.
5. The capacity to flow uniformly along the core nodes to increase column strength.
6. The capacity to remain in place on curved panels rather than flow to the low points.

CONVENTIONAL ATMOSPHERE FURNACE BRAZING

Successful brazing of honeycomb structures in conventional atmosphere furnaces requires the use of a sealed retort. The retort normally contains a suitable hearth or hearths and is continuously purged with an inert gas. The honeycomb structure is held in contact with the hearth by maintaining a slightly negative pressure within the retort. Because of the mass of the retort and hearth, means for increasing the rate of heating and cooling are continually being investigated. One partial solution to this problem was the development of the gas-fired luminous-wall furnace (Fig. 91.17). An atmosphere control unit which regulates the rate of purge and the pressures in the inner and outer retorts, and which is equipped with a polargraph recorder, a manometer for each retort and a titanium "chip-getter" for purifying the argon gas is generally employed.

Fig. 91.17.–Typical luminous-wall furnace.

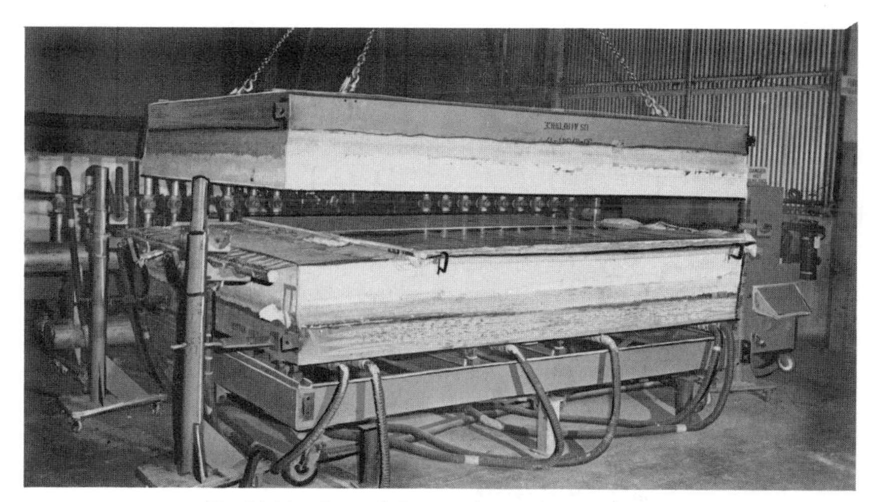

Fig. 91.18.–Typical electric blanket brazing fixture.

ELECTRIC-BLANKET BRAZING

Electric-blanket brazing of honeycomb structures, Fig. 91.18, utilizes cast ceramic platens with integral heating elements and grid cooling passages. The contour of the heating surfaces of the platens is accurately machined to the required contour of the part to be brazed. To obtain uniform heat over the platen surface, the heating elements are covered with a thin di-electric blanket and a 1/4 in. nickel-plated copper plate.

The PH15-7Mo honeycomb structure is enclosed in a thin, stainless or mild steel sealed envelope having suitable thermocouples attached for temperature control. An adequate atmosphere for brazing is obtained by initially evacuating the sealed envelope and subsequently continuously purging the envelope (retort) with inert gas. The complete brazing, cooling and aging thermal treatments are accomplished as one continuous cycle without removing the honeycomb structure from between the platens. The cycle begins with the heating of the part by Inconel tube or strip heating elements to the brazing temperature (1725 F). The part is then cooled to near room temperature by passing air through drilled holes in the cooling passage grid. The air is turned off and, using the same mechanism, liquid nitrogen is introduced to reduce the temperature to –75 F. This assures transformation of the base metal to the desired martensitic structure. The strip heating elements are again used to heat the part to the desired aging treatment.

RADIANT QUARTZ LAMP BRAZING

This method of brazing utilizes banks of electronically controlled radiant quartz lamps to provide the heat for brazing. The panel to be brazed is enclosed in a sealed, thin stainless steel envelope. The interior of the envelope is evacuated and purged with argon gas. To hold the honeycomb structure in place, a slightly negative pressure is maintained within the envelope.

The envelope can be mounted vertically in a picture frame fixture, suspended

Fig. 91.19.–Typical radiant quartz lamp brazing installation.

on a monorail conveyor and arranged to pass through a brazing station and a chill form as illustrated in Fig. 91.19. At the brazing station, banks of radiant quartz lamps, mounted in specially designed reflectors, are located close to the external surfaces of the envelope. This permits a high heat flux density to flow into the brazement and results in an extremely short heat-up cycle. Control of the heat to local areas in the panel being brazed can be achieved (1) by control of banks of lamps or each separate lamp (2) by coating the braze envelope with a pattern of low or high emissivity coating or (3) by interposing fixed or changing mask stencils between the radiant source and the panel. After the fast brazing cycle, the chill platens are indexed to the envelope by pneumatic pressure. The temperature in the chill platens may also be controlled.

Normally, stationary lamps are used, and the work is moved into position. The types of parts brazed in this manner include flat panels with and without edge closure members and heat sinks, wedge panels and single and compound curvature panels.

VACUUM FURNACE BRAZING

This method of brazing honeycomb structures is accomplished in special cold-wall type vacuum furnaces, Fig. 91.20, designed to permit application of the necessary holding pressure during the brazing cycle. The holding pressure is applied either by a differential in pressure between the inside and outside of a retort or by clamping systems acting through the cold wall of the furnace. The assembly is cooled after brazing by the use of a built-in carbon dioxide system.

INDUCTION BRAZING

The aircraft industry employs induction brazing on some of its more sophisticated aircraft. This brazing is being done on 304, 304L, 321 and 347

Fig. 91.20.–Typical vacuum brazing installation.

steel tubing for hydraulic lines. The unions or fittings are the same material in the annealed condition. The 82Au-18Ni braze alloy comes preplaced in the beaded section of each fitting. The brazed joint concept utilizes capillary flow of the braze alloy in two directions to ensure a greater surface area with additional strength and a more reliable seal.

Brazing is accomplished by applying heat to the brazing coil which in turn supplies heat to the outside of the unions. The heat is applied in accordance with schedules using typical time-temperature calibration curves. The braze alloy melts at 1750 F and flows in the longitudinal direction between the outside wall of the tubing and inside wall of the union. The heat cycle can vary between 25 and 90 seconds or longer depending on the wall thicknesses of the tubing and fittings. During this time the maximum temperature must not exceed the melting temperature of the alloy by more than 200 F. During heating, the operation is performed in an atmosphere of dry argon or mixed purge gas. After the connection has cooled, the alloy should have flowed to the edges of the union and tube.

Currently, diameters ranging from 3/16 to 2 1/2 in. are being brazed with wall thicknesses ranging from 0.016 to 0.057 in.

MILITARY SPECIFICATIONS

Materials and processes used in the fabrication of aircraft and missile components are covered by both nongovernment and government specifications. Nongovernment specifications include the Aeronautical Materials Specifications prepared by the Society of Automotive Engineers. Aeronautical Materials

Specifications are frequently used by the manufacturers of aircraft engines. The government specifications include federal specifications used by all departments of the federal government, military specifications used by all services in the Department of Defense and specifications used by only one military service. Each military service publishes indexes and bulletins which list the specifications in current use. The Department of Defense publishes alphabetical and numerical listings of specifications by the military service. When no applicable government specification has been established, special arrangements are made and the procuring agency and the manufacturer agree on specifications.

Methods of computing the design loads for aircraft metal structures are given in Military Handbook MIL-HDBK-5, Metallic Materials and Elements for Flight Vehicle Structures. This handbook also gives data on the allowable design loads for both welded and brazed joints. Additional instructions regarding the use of welding and brazing in aircraft structures are given in design handbooks prepared by the military services.

Military specifications on fusion welding of aircraft and missile components cover processing, quality control procedures, requirements for filler metal and equipment and qualification and certification of welders and welding machine operators.

PROCESS SPECIFICATIONS

Specifications on processing procedures include MIL-W-8611, Welding, Metal Arc and Gas, Steels and Corrosion and Heat Resistant Alloys; MIL-W-8604, Welding of Aluminum Alloys and MIL-W-18326, Fusion Welding of Magnesium Alloys. Specification MIL-W-8611 presents general requirements for fusion welding of low-alloy aircraft steels, stainless steels and heat-resistant alloys. Specification MIL-W-8604 and MIL-W-18326 respectively present similar requirements for welding the aluminum and magnesium alloys. Welding of alloys not covered by these specifications is governed by special requirements and specifications drawn up between the procuring agency and the manufacturer.

QUALITY CONTROL SPECIFICATIONS

General requirements for the inspection of aircraft structures and components are presented in Specification MIL-I-6870, Inspection, Requirements, Nondestructive; for Aircraft Materials and Parts. The following inspection methods, which are described in Chapter 6, Section 1 are used in the inspection of aerospace and missile weldments.

MAGNETIC PARTICLE INSPECTION

Weldments made from magnetic materials such as carbon and low-alloy steels may be examined by magnetic particle inspection. This inspection is performed in accordance with Specifications MIL-I-6868, Inspection, Process, Magnetic Particle, and MIL-STD-410, Qualification of Inspection Personnel, Magnetic Particle and Penetrant.

PENETRANT INSPECTION

This method is used in the inspection of aircraft parts in accordance with Specifications MIL-I-6866, Inspection; Penetrant Method of; and MIL-STD-410, Qualification of Inspection Personnel, Magnetic Particle and Penetrant; Materials, Penetrant.

RADIOGRAPHIC INSPECTION

This inspection method is applicable to any of the alloys used in weldments for aerospace structures and is governed by MIL-STD-453, Inspection, Radiographic, which covers requirements for radiographic inspection and procedures for certification of X-ray laboratories.

OTHER INSPECTION METHODS

Ultrasonic methods of inspection are used in special applications, but no military specifications, applicable in the inspection of welded aerospace parts, have been established. Welds on aircraft parts are always inspected visually to ensure that they present a workmanlike appearance equal to that which would be acceptable for operator qualification tests. Welds in tanks, pressure vessels, exhaust stacks, air ducts and similar parts are frequently inspected by the application of internal pressure. Welds in important structural parts are sometimes proof tested in tension, compression or bending. Requirements for pressure tests and proof testing are usually covered in the performance specifications for the part.

FILLER METAL SPECIFICATIONS

Military specifications on filler metal are listed in bulletins and indexes published by each military service. Reference to these publications should be made to obtain the applicable specification. Military specifications for filler metal are usually based on commercial specifications. One exception is in aircraft and missile applications employing low-alloy steel, in which the strength of the welded joint is increased by heat treatment of the assembly after welding. For these applications, the requirements for the filler metal are modified to obtain the desired response in heat treatment. For example, Specification MIL-E-8697, Electrodes, Welding, Coated, Low-Hydrogen, Heat Treatable Steel, covers requirements for metal-arc welding electrodes suitable for manual welding of the low-alloy steels which may be heat-treated after welding to develop high tensile strengths. This specification includes requirements for all weld metal tensile specimens which are heat-treated in accordance with procedures specified for the base metal.

Requirements for filler metal used in gas welding and inert gas metal-arc welding of low-alloy steels are included in Specification MIL-R-5632, Rods and Wire, Steel, Welding. This specification covers requirements for two classes of filler metals: for welding steels which are not subsequently heat-treated; for welding steels which are heat-treated after welding to obtain high mechanical properties. The major use of the second class of wire is in welding of missile cases which are heat-treated to high yield strengths. Other specifications for low-alloy steel filler metals are included in the process procedures which are approved by the procuring agency. These specifications usually require a lower carbon content to improve weld ductility and extra low sulfur and phosphorus content to ensure freedom from hot cracking. The overall chemistry of the filler metal is sufficiently similar to the base metal to obtain the desired response in heat treatment.

EQUIPMENT SPECIFICATIONS

Military specifications cover welding equipment that is procured by the military services. Compliance to the detail requirements in these specifications is not mandatory for the equipment used by the manufacturer in the fabrication of aircraft parts in accordance with military specifications. The process specifications usually require that the equipment be capable of producing welds which meet the requirements established in the welding procedure certification.

QUALIFICATION SPECIFICATIONS

The welder must produce acceptable qualification test plates or joints before he is certified to do production welding. The qualification tests are described in MIL-T-5021, Tests; Aircraft and Missile Welding Operators' Qualification. For the purpose of these tests, the alloys are divided into six groups, as shown in Table 91.9.

Table 91.9—Aircraft welding operators' certification test groups

Group	Alloy
I	Carbon steel and alloy steels
II	Corrosion-resistant and heat-resistant alloys
III	Copper and nickel alloys
IV	Aluminum alloys
V	Magnesium alloys
VI	Titanium and titanium alloys

In the case of Group I, there is an additional breakdown into Class A and B. Class IA welders are permitted to weld any parts within Group I; Class IB welders are limited to nonstructural welding of Group I alloys. In a similar manner, the welding processes are listed as follows:

a. Gas (oxyacetylene, oxyhydrogen)
b. Metal-arc with covered electrode
c. Submerged arc
d. Gas tungsten-arc
e. Gas metal-arc
f. Atomic hydrogen
g. Carbon arc

The welder is required to pass a qualification test for each group of alloys and for each welding process that will be used in production. A welder who has passed the qualification tests for one alloy in one group may weld all alloys in the same group without taking additional qualification tests, provided the welding process is not changed. The qualification test joints for general aircraft welding are described in MIL-T-5021. Provisions are also made in the specification for the substitution of alternate qualification test specimens for specialized applications which often involve the use of automatic welding equipment. The certified status of the welder is maintained by completing re-examination tests every six months, or by a continuing record of satisfactory workmanship which includes the inspection of welds in production parts.

RESISTANCE WELDING

Military specifications on resistance welding processes include requirements as to materials, qualification tests of equipment, certification of welding schedules used in production and production quality control.

SPOT AND SEAM WELDING

Specification MIL-W-6858 covers general requirements for the resitance spot and seam welding of aluminum, magnesium, nonhardening steels or alloys, nickel alloys, heat-resistant alloys and titanium. The scope of the specification is limited to alloys of these groups in which the spot weld exhibits a tensile pull-out strength not less than 25% of the minimum shear strength of the spot weld.

Welding of materials not meeting this requirement is considered outside of the scope of the specification and is governed by other specifications established or approved by the procuring agency.

Equipment used in spot and seam welding must pass qualification tests as required by the specification. The initial machine qualification tests consist of making a minimum of two sets of test specimens, one representing the maximum thickness and one representing the minimum thickness that will be welded in production. For spot welding, each set consists of not less than 105 consecutive welds. For seam welding, a minimum of 24 in. is required for the length of the welds in each set. The welds must meet the requirements for strength, consistency and soundness required by the specification before the equipment is used in production.

Certification of the welding schedule is required for each thickness combination, within specification tolerance limitations, to be welded in production. The welding schedule includes the cleaning of the material prior to welding, details of machine set-up and the control settings for each machine to be used in production welding. The suitability of the welding schedule is established by testing and examining the welds in accordance with the procedures established in the specification. Specimens used include the tensile shear specimens and metallurgical specimens. The number of specimens required for a certification test varies with the material and class of welding. The maximum number of specimens (20 tension shear and 5 metallurgical specimens) is required for critical structural joints in aluminum alloys or magnesium alloys. For other materials and less critical joints, a smaller number of specimens is required.

Production quality control is accomplished by testing or examining welded specimens at intervals during production. The intervals range from one to two hours, depending on the class of welding that is required. The specimens may be a spot welded lap joint or single spot shear specimens or specimens containing a minimum of 3 in. of seam weld, as applicable, or a simulated section of the production joint. Nugget diameter measurements may be used instead of the tension shear tests. All production parts are visually examined for the presence of external defects. Production welds may be examined for internal defects by metallographic or radiographic examination.

FLASH WELDING

The requirements for flash welding are covered by Specification MIL-W-6873, Welding; Flash, Carbon and Alloy Steel. The scope of this specification is limited to the flash welding of plain carbon and alloy steels having a nominal carbon content of not more than 0.40% and a tensile strength of not more than 240,000 psi. The flash welding of other materials is outside the scope of this specification and is governed by special requirements established or approved by the procuring agency.

The initial qualification of the equipment is included in the certification of the three initial welding schedules for joints representing the greastest, intermediate and least cross-sectional areas to be welded in production. The acceptability of each initial welding schedule is determined by making 15 or more welded specimens which are tested to destruction in tension or bending.

Certification of the welding schedules for other cross-sectional areas is accomplished by welding five specimens which are tested to destruction in tension or bending.

Production quality control includes visual inspection, magnetic particle inspection, destructive testing of selected samples and proof testing. Each welded joint is examined for defects and inspected by magnetic particle inspection. In destructive tests on welded specimens, no joint tested in the "as-welded" condition should fail in the weld. For whole specimens heat-treated after welding, failures in welds are acceptable, provided the stress at failure is not less than 95% of the ultimate strength of the base metal. Welds too large to be tested as a single joint are cut into strips and all strips are tested. The average strength of all the strips must be 95% or more of the ultimate strength of the base metal. No single strip can drop below 80%; however, occurrence of a single test value not lower than 60% is acceptable if the previous history of tests for the particular setting shows that the frequency of occurence of the lower value is not greater than 1%. Selected production samples are proof tested in tension or bending. Whenever possible, the proof load should correspond to at least 67% of the ultimate strength specified for the material.

BRAZING

Specification MIL-B-7883 covers requirements for the brazing of steels, copper, copper alloys and nickel alloys. The brazing methods include torch, furnace, induction, resistance and dip brazing. The filler metal includes copper, copper-base alloys and silver-base alloys. Specification MIL-B-23362 covers the brazing of aluminum and aluminum alloys.

Formal certification of brazing operators and brazing procedures is not required by this specification. Quality control of the brazed joints is based on the inspection of the joints in production parts. Each joint is examined visually for penetration of the filler alloy into the interstice of the joint, and for smoothness of the exposed edges or fillets formed by the filler metal. When the filler metal is applied at one edge, complete penetration of the molten alloy to the opposite edges is required. Selected joints are subjected to radiographic or ultrasonic examination, peel tests and metallurgical examination to determine the presence of voids and other defects. Defects having an aggregate area not exceeding 20% of the faying surface for aluminum and aluminum alloys and 15% for all other alloys are acceptable. The maximum extent of a single defect cannot exceed 20% of the overlap distance of the joint in aluminum and aluminum alloys and 15% in all other alloys.

BONDED HONEYCOMB

General requirements for bonded honeycomb are covered in MIL-HDBK-23, Parts 1 and 2, "Composite Construction for Flight Vehicles." Part 1 of this handbook furnishes information on the materials, fabrication procedures and inspection methods which are applicable to bonded honeycomb structures. Information is also furnished on the durability and repair of these structures. Part 2 contains materials, properties and design criteria for honeycomb cores and completed panels. Reference is made to specifications covering requirements for cores, adhesives and bonding procedures.

BRAZED HONEYCOMB

MIL-HDBK-23 also furnishes information which is applicable to brazed honeycomb constructions. Data are furnished on the core material and cover sheets. Brazing alloys, brazing procedures and inspection methods are described briefly. Except for this document, requirements for the fabrication of brazed honeycomb panels are not specifically covered by military specifications.

BIBLIOGRAPHY

"Vacuum Brazing as a Repair technique for Aero Engine Components," C. J. Baker, *Welding Journal*, 50 (8) 559–566 (1971).

"Brazing Filler Metal Evaluation for an Aircraft Gas Turbine Engine Application," R. P. Schaefer, J. E. Flynn and J. R. Doyle, *Welding Journal*, 50 (9) 394s–400s (1971).

"Vacuum Brazed Stainless Steel for Hot Gas Actuators for Aerospace Components," C. J. Barfield, *Welding Journal*, 49 (7) 559–564 (1970).

"Aluminum Flame Sprayed Coating Process for Reinforced Plastic Aircraft Assemblies," R. W. Whitfield and V. S. Thompson, *Welding Journal*, 47 (1) 31–36 (1968).

"Aircraft Engine Parts Reclaimed by Brazing," *Welding Journal*, 46 (12) 1013–1014 (1967).

"Design and Fabrication of D6AC Steel Weldments for Aircraft Structures," R. E. Key, J. C. Collins and H. I. McHenry, *Welding Journal*, 46 (12) 991–1000 (1967).

"Exothermically Brazed Hydraulic Fittings for Aircraft," N. E. Weare and R. A. Long, *Welding Journal*, 46 (1) 29–30 (1967).

"Arc Welding Magnesium," Dow Chemical Company (1966).

"Brazing, Bonding, Soldering and Special Welding of Magnesium," Dow Chemical Company (1966).

"Resistance Welding Magnesium," Dow Chemical Company (1966).

"B-70 Spurs Wider Use of Sandwich Design," R. Hawthorne, *Space-Aeronautics*, 35 (3), 46–49 (1961).

"New Methods Set Pace of Progress for Sandwich Production," R. Hawthorne, *Space-Aeronautics*, 35 (3) 50–55 (1961).
"Metals and Fabrication Methods for the B-70," W. A. Reinsch, *Metal Progress*, 79 (3), 70–77 (March 1961).
"Sandwich Design Use," *Space-Aeronautics*, 35 (3) (1961).
"Joining of Refractory Metals," *Material Advisory Board No. MAB-171M* (March 20, 1961).
"Design Information on Nickel Base Alloys for Aircraft and Missiles," DMIC *Report 132*, Battelle Memorial Institute (July 20, 1960).

CHAPTER 92

LAUNCH VEHICLES

REVIEWED BY THE WELDING HANDBOOK COMMITTEE

Chapter 92, as it appears here, is essentially a reprint from the Fifth Edition. Although the chapter contents have been reviewed with respect to typography, the general text and references to standard specifications have not been updated. The decision not to present a complete revision was based on current circumstances in the launch vehicle industry (see Preface) and the feeling on the part of the Welding Handbook Committee that the Fifth Edition chapter is still a valuable source of information for anyone seeking useful and relevant data on launch vehicles.

CHAPTER 92

LAUNCH VEHICLES

INTRODUCTION

Many aspects of launch vehicles fabrication are quite similar to fabrication techniques which have been applied in the aircraft and aerospace industries. The technical challenge which is unique to the launch vehicle industry is the fabrication of lightweight or weight-sensitive pressure vessels for aerospace use.

MATERIALS SELECTION FOR PRESSURE VESSELS

Launch vehicle pressure vessels have been fabricated from low-alloy steels, maraging steels, stainless steels, titanium and aluminum alloys, as well as fiber glass-epoxy synthetics which have challenged all of these alloys. From this diversity of materials it is apparent that pressure vessels for launch vehicle application must operate over a considerable range of temperature, stress, corrosion and economic environments.

The higher strength-to-density materials on this list, including fiber glass, titanium-6A1-4V, high-strength low-alloy steels and maraging steels, have been selected for solid propellant motor cases where corrosion is not a particular problem, but where the strength-weight ratio is critical.

For the larger volume, lower stage pressure vessels, both stainless and aluminum alloys have been extensively used; aluminum or titanium is the most likely candidate for any future, first-stage designs. The particular aluminum alloy selected depends somewhat on the size or propellant volume of the stage in question. 2014-T6, with a somewhat advantageous strength-weight ratio over 2219-T87, finds wider acceptance in smaller intermediate stages (10 to 20 ft diam) because of two factors: the strength-weight ratio difference and the ability to be welded conveniently in thicknesses up to approximately 5/8 in. Alloy 2219-T87 is more competitive in the larger stages where thicker materials must be welded and where weight is not quite so critical. Ti-6A1-4V has found further application where corrosion and cleanness requirements are prevalent in the storage of high pressure gas or in the more critical weight of the liquid propellants applications. The high pressure gas "bottles" are used for miscellaneous auxiliary power purposes; the bottles are designed for high stress, putting to use the relatively high strength-weight ratio.

ALUMINUM ALLOYS

APPLICATIONS

Aluminum alloys find primary application in very large volume tankage for liquid propellant rockets such as the several sizes of tankage involved in the Titan and Saturn rocket systems. These range in diameter from 10 to 33 ft. Each of these tanks is fabricated of 2014-T6, 2219-T87 or 5456 alloys with details of 6061-T6. Some experimental work has been done to evaluate the applicability of the newer weldable 7XXX alloys (7039, 7106) for large liquid storage tankage.

METALLURGICAL CHARACTERISTICS

The aluminum alloys of primary interest, the 2XXX series A1-Cu age-hardenable alloys, are welded in the aged condition, T6 or T87. In this condition the heat-affected zone includes, in sequence of low to high temperatures, (1) softening by overaging, approximately 1/8 in. out from the fusion line, (2) slightly harder solution-hardened zone, closer into the weld bead and (3) a zone of copper enrichment in the grain boundaries within the fusion zone itself. The weak link in a transverse tensile fracture is either in the overaged zone, the fusion zone or the weld metal itself, depending on welding conditions, sheet thickness and filler alloy. The fusion zone, where a mixture of grains and liquid grain boundary phase did exist at the moment of peak temperature exposure, is a potential crack zone if restraint and shrinkage provide a load to the structure at this critical period in the welding cycle.

WELDING TECHNIQUES

Aluminum launch vehicle tankage is welded by both the gas metal-arc and gas tungsten-arc welding processes. Gas metal-arc welding is used in the fabrication of Saturn V, third stage, while most other aluminum tankage fabricators have used gas tungsten-arc for the critical pressure vessel butt welds.

Gas metal-arc, if used on materials of between 3/16 and 5/16 in. with single pass, introduces approximately the same or somewhat wider heat effects as dc gas tungsten-arc. However, on thicker joints with joint preparation and multiple passes, gas metal-arc usually provides lesser heat effects. An attribute of lower heat (higher solidification rate) in addition to the lower and narrower heat effect, is a wider distribution of porosity with fewer concentrations.

Direct current-straight polarity gas tungsten-arc provides good penetration with helium shielding gas in square butt welds; approximately 60% can be obtained on plate as thick as 1 in. This degree of penetration is obtained with a "penetrating or buried" arc allowing the jet force from the arc to "drill" a deeply penetrating crater which is filled in as the weld progresses. This technique requires accurate control over arc length, arc voltage, seam tracking and arc current; it has led to the widespread adoption of rather sophisticated and precisely controlled welding power sources and torch heads.

For thicknesses less than approximately 0.200 to 0.220 in., ac may offer some jigging advantage over dc. In this thickness range the importance of enhanced penetration with dc diminishes. Therefore, the less sensitive, unburied arc a-c process provides greater reliability and reproducibility and can tolerate greater departures from a perfect joint alignment; cleanliness of flux is another advantage. These advantages are obtained with a joint strength reduction of only about 2 ksi in 2014-T6.

Fig. 92.1.–Rotation fixture. Performs girth welds in flat position.

Joint Design

The joint itself is square butt with zero root opening and zero mismatch. The tolerances placed on the "zeros" are rather tight. Preparation of these butt joints just prior to welding is quite important. A variety of chemical and solvent cleaning processes is used prior to the introduction of parts to the welding jig. However, most of these cleaning procedures have in common some means of mechanical removal of surface metal from the abutting planes and from the top and bottom surfaces near the joint, immediately before welding. The method of metal removal may be by sanding, filing, scraping or wire brushing. Filing and scraping are preferred since sanding and wire brushing sometimes produce smearing contamination–which may include anodize or chromate protective coatings–into the surface metal rather than removing it. The sharp square corners of the joint are removed during this last mechanical cleaning operation. Failure to knock these corners off will result in a higher incidence of oxide inclusions in the weld.

Welding Position

Welding of aluminum launch vehicle tanks has been done in flat (downhand), horizontal, and vertical positions. Earlier vehicles of 10 ft diam and under took advantage of the wider process latitude afforded by jigging the parts for flat welding. However, larger vehicles make such jigging impractical, and the welding head or fabrication details have been moved as necessary to hold the axis of the tank vertical; horizontal and vertical welding are required. This welding has not been without difficulty, but it has, nevertheless, been successful. Vertical-up welding requires more precise control over gap and welding power parameters to obtain acceptable bead shape, but suffers less porosity than flat welding. Two-pass welds, one from either side, are preferred. Horizontal welding imposes even more exacting process control, and weld porosity will be more prevalent than in flat welding. The sensitivity to porosity imposes more careful joint preparation and cleaning requirements. Careful attention must be given to torch angle and filler wire placement in the puddle to prevent undercut.

Weld Jigging

Weld jigging requirement is a function of part size. The smaller (10 ft diam) tanks, with wall thicknesses less than 0.350 in., are held with tooling strong and stiff enough to hold the details to shape and in line with mating parts. The conventional rotation, stake and dome positioner fixtures are pictured in Figs. 92.1, 92.2 and 92.3.

Fig. 92.2.–Longitudinal weld fixture. Barrel moves in under-frame which serves as weld carriage track and hold-down fingers.

Fig. 92.3.–Dome fixture. Complex system of hydraulic fingers holds dome segments to backup as dome is moved under stationary arc.

Fig. 92.4.–Vertical-up barrel welding fixture. Parts are not in place, allowing visibility of rubber vacuum pads which hold parts during welding.

For larger, more flexible parts, the tanks are held in their most rigid attitude. For example, longitudinal barrel seams are welded, vertical-up with tooling which is only sufficient to maintain alignment between carefully prepared straight line butt joints (Fig. 92.4). Domes, fabricated of gore segments, are welded by a moving torch, the weld changing from vertical-up to flat. The tooling is just sufficient to maintain local joint alignment. The self-jigging or soft tooling concept is epitomized in the girth weld (Fig. 92.5), where the dome merely sits on the barrel, sometimes with through-joint clamps, while the torch

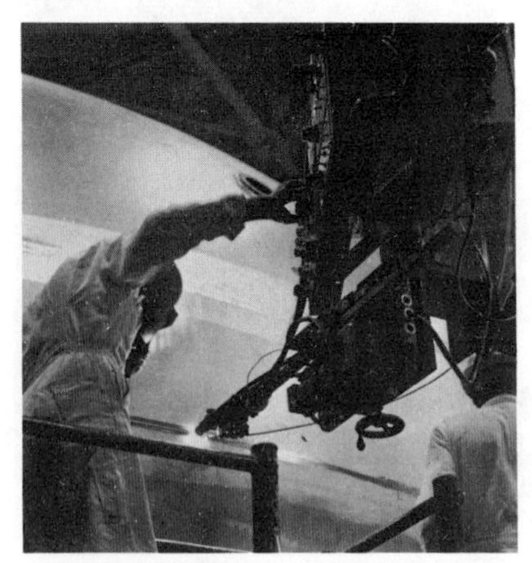

Fig. 92.5.–Horizontal welding of girth welds. Parts are held into alignment by tack welds.

Fig. 92.6.–Through-seam clamps to align horizontal seam as tack welds are performed.

is moved along the joint by a carriage or "skate." The through-joint clamp maintains joint alignment and is removed before tack welding (Fig. 92.6).

The Y-ring is a rather complex part. The three legs of the Y-section terminate in girth welds to the dome, barrel and skirt, or, in some cases, lie in a common bulkhead to a dome. These girth welds are performed as above. However, the Y-ring itself is fabricated of circular arcs; the weld joining these arcs must have the "Y" as its longitudinal section. For Titan, this weld is manual. The larger rings of Saturn are fabricated from three ring segments, 5 1/2 in. thick by 21 in. deep, curved to a 33 ft diam. These segments are joined by the gas metal-arc process requiring some 30 passes on each side to fill a double Vee-joint. The completed ring is then machined to the Y-cross section.

Filler Wire Selection

For alloy 2219, filler wire 2319 is used almost universally. Alloy 2014 is welded with several filler alloys including 2319, 4043 and brazing alloy 716. In general, the tensile strength joint efficiency of the 2000 series alloys is quite low whatever filler wire is selected when postweld heat treatment is not applied. Typical as-welded values are on the order of 70% for 2014-T6 and 65% for 2219-T87. These low joint efficiencies are compensated in the design by chemical mill thinning of all material except that near the joint; thus, the design load of the weld is held down to approximately 40% of the design load of the parent metal. Therefore, relatively soft and weak filler metals such as 4043 can be tolerated. Stronger weld metals, if only marginally stronger, would not eliminate the fabrication step of chem-milling, but would merely alter the thickness ratios. A change in this ratio would result in an insignificant saving, particularly on the lower stages where these alloys find primary application.

MECHANICAL PROPERTIES

The mechanical property requirements, as discussed above, are eased by increasing the cross section of the weld joint itself. (The material away from the weld is thinned by chem-milling.) This allows efficient, or near capability, loading of the 2014-T6 or 2219-T87 aged metal through the as-welded welds.

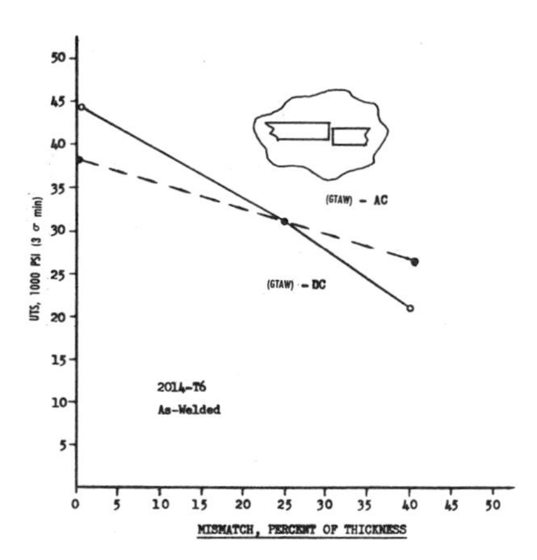

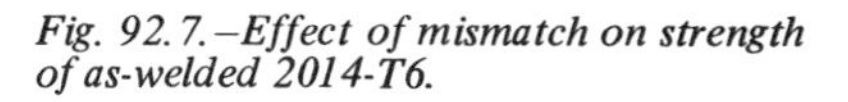
Fig. 92.7.–Effect of mismatch on strength of as-welded 2014-T6.

The welds themselves show typical transverse tensile strengths of about 48 to 50,000 psi and 42 to 45,000 psi for 2014-T6 and 2219-T87, respectively. Small variations from these typical values depend on modifications of welding procedures. For example, two passes—each pass only partially penetrating—can increase strengths by 4 to 5 ksi for 2219-T87 in thicknesses over 1/4 in. Substitution of 716 for 4043 in welding 2014-T6 adds approximately 3 ksi. D-C welds are about 4 ksi stronger than a-c welds in 0.200 in. 2014. Electron beam welds are stronger. These strength-improving procedures have the common attribute of minimizing the width of the heat-affected zone and the time of exposure at given near-peak temperatures.

Shaving the bead flush with the base metal reduces the strength about 3 ksi for 2014-T6. Mismatch, porosity and repair welding reduce tensile strength according to the plots of Figs. 92.7, 92.8 and 92.9. However, there is no evidence that flaws, either spherical or sharp, result in catastrophic subyield-point failure of aluminum pressure vessels.

Design stresses are considerably lower than typical strength in large aluminum launch vehicle pressure vessels because of high probability that some defect or repair will exist at the time of flight.

TYPICAL DEFECTS AND INSPECTION

Defects which are found in aluminum welds include porosity, dross or oxide inclusions, incomplete fusion, inadequate joint penetration and cracking. For the most part, some of the causes of these defects are at least qualitatively understood. Porosity results from any one of several possible sources of hydrogen contamination: shielding gas, base metal, filler wire or aluminum surfaces can introduce hydrogen. Dross or oxide inclusions will result from inclusions in base or filler metal or from aluminum "fines" in the path of the arc which can oxidize before inundation by the puddle. Incomplete fusion results

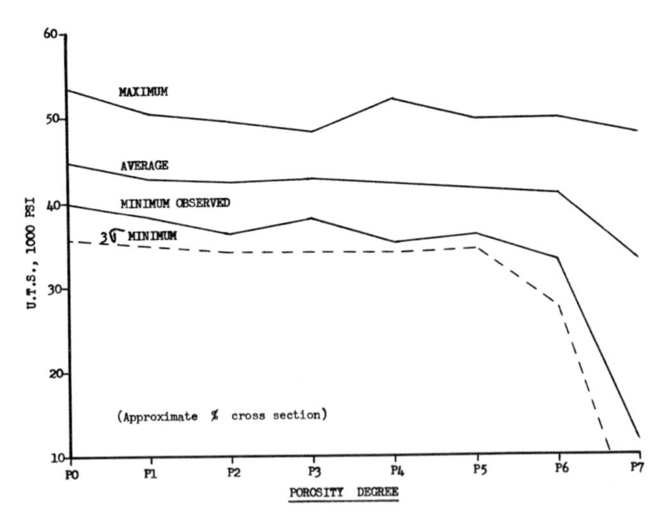

Fig. 92.8.–Effect of porosity on strength of as-welded 2014-T6. PO-P2, % trace; P3–2%; P4–6%; P5–12%; P6–24%; P7–over 24%.

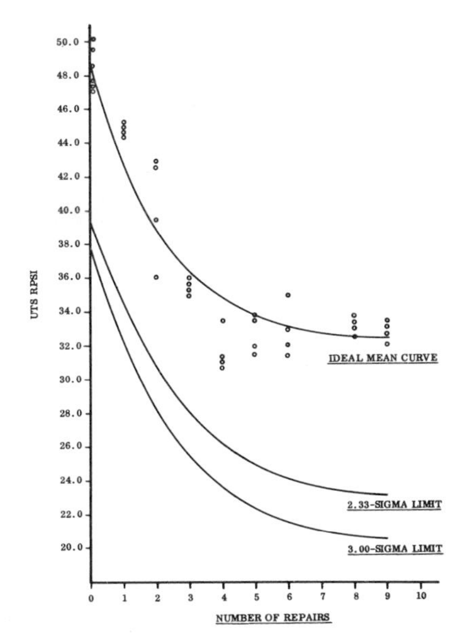

Fig. 92.9.–Weld strength versus number of repairs, 1/4 in. 2014-T6.

from poor wetting of the kerf by the puddle. Inadequate joint penetration results either from failure to melt to the required depth or from failure of the molten root to disrupt the abutting plane oxide layers. Seam tracking errors are a very significant practical cause of these discrepancies, especially for thick plate butt welds partially penetrated from either side.

Cracking, particularly in aluminum-copper heat-treatable alloys (2XXX), may result when the weld is placed in a field of high mechanical restraint and when the welding procedure causes high shrinkage. Practical causes include welding too rapidly or repairing too frequently in the same location.

INSPECTION

These defects are primarily inspected by X-ray, with dye penetrant, visual and ultrasonic as backup. Little difficulty is encountered in finding porosity or dross, particularly if one or both bead surfaces are shaved for clarity of X-ray. However, quantitative interpretation is sometimes difficult.

The "plane defects" can remain undetected unless they extend to a surface where dye penetrant can be utilized. Subsurface cracks or incomplete fusion can be found by X-ray, if favorably oriented, or by ultrasonic methods. Inadequate joint penetration will be subsurface for two-pass two-side welds; typically, it is so tightly closed by weld shrinkage that detection by X-ray may be difficult. The best insurance against this defect is preproduction weld checkout to ensure adequate, or more than adequate, penetration. Ultrasonic inspection offers great promise; detection probability is increased with this method if the defect is not tightly closed.

Another defect that can result from repair welding is an excessively wide heat-affected zone. With the chemically-milled weld land, it is assumed that the thinner portion away from the weld will exhibit the full-aged properties. Multiple repairs in one spot can overage out beyond the land. However, electrical conductivity changes with overaging, affording a convenient means of detection with 2014-T6. Conductivity is measured with an eddy current device to ensure that overaging does not extend beyond the land.

TITANIUM ALLOYS

APPLICATION

Titanium alloys have found application in liquid propellant tanks, high pressure gas storage tanks and solid rocket motor cases. The two alloys which have found widest applicability are the all-α Ti-5Al-2 1/2Sn and the α-β Ti-6Al-4V. Both of these alloys are among the oldest in the titanium family of alloys and much fabrication and property information is available. Reference should be made to Section 4, Chapter 73 for detailed discussion of the metallurgical and welding aspects relating to titanium and its alloys. The α alloy is not age-hardenable and has lower strength potential than the α-β, but it is somewhat easier to weld in thick sections. Ti-5Al-2 1/2Sn is also favored for toughness reasons down to liquid hydrogen temperatures, but at the temperature of liquid oxygen, Ti-6Al-4V has adequate toughness and would be favored because of higher strength. Both alloys are used in an ELI (extra low interstitial) grade for cryogenic tankage.

WELDING TECHNIQUES

Titanium is welded almost exclusively by the gas tungsten-arc process with straight polarity. The shielding gas is usually argon, but may be helium or a mixture of the two if greater penetration is desired for a straight butt weld. Gas metal-arc welding does have application with the thicker walled high pressure gas bottles. Gas metal-arc welding is used with joint preparations and allows for filler metal addition. Plasma-arc welding is also being used for the welding of titanium.

Joint Design

Square butt joints are used only up to approximately 0.125 in. in titanium. The "buried-arc" technique, so effective with aluminum, is not effective with titanium and thus welds in thicker sheets must incorporate joint preparation. Some filler metal is added—usually commercially pure titanium with either of the alloys mentioned. For square butt welds, filler metal is added to provide a weld bead which contains approximately 30% by volume of filler metal and the balance, resolidified base metal. For some Ti-6Al-4V applications, the roof of the initial pass is ground out for refilling to get more filler alloy into the weld bead.

Weld Jigging

Since the titanium tankage produced is within size limitations for convenient jigging, almost all welding has been performed in the flat (downhand) position. Also, forgings have been used for tank cone and cylinder sections to eliminate longitudinal welds. The jigging is quite similar to that used for girth welds in aluminum except for incorporation of a trailing shield on the torch, and gas-filled backup grooves to provide more complete inert gas protection. Welding fixtures have incorporated both inside-out and outside-in welding. The primary advantage for inside-out welding is that local expansion in the heated weld puddle area tends to push the metal out hard against the outside diameter backup bar, thus assuring good contact with the bar in the area of welding. Partial fit-up is more critical with titanium than with aluminum; the weld puddle is more fluid and may rupture if too much bridging is required. A "burn-thru" caused by a ruptured puddle is almost sure to occur if the weld is initiated on a gapped part of the seam. Gaps can be tolerated away from the initiation point; gaps that open at a gentle angle will close by a thermal-associated shrinkage mechanism as the arc approaches. Burn-thru can also result from poor coordination of the initiation of filler wire and puddle melting on arc initiation.

Some smaller spherical pressure vessels are automatically welded in atmosphere-controlled chambers, and much manual welding of clips and brackets is also done in chambers.

Table 92.1—Typical mechanical properties of Ti-6A1-4V butt welds

Specimen	Strength		Elongation
	Ultimate ksi	Yield ksi	% in 2 in.
Solution Treated and Aged Base Metal	164	154	9.8
STA, As-Welded	153	141	2.0
STA, Welded, Stress Relieved	153	141	2.2

MECHANICAL PROPERTIES

Mechanical properties of Ti-6A1-4V, measured transverse to the direction of welding, are as shown in Table 92.1. The tensile strength of approximately 140 ksi compares with 164 ksi in the solution-treated and aged base metal. Thus, thickness buildup at the weld joint is required for an efficient structure. This strength is appreciably higher than that of the commercially pure alloy used as filler wire. In order to ensure such properties, welding and weld repair techniques must not cause excessive dilution by the unalloyed filler wire.

Ductility of titanium welds must receive attention, since any discrepancy in the welding process which allows atmospheric contamination, can cause marked weld embrittlement. Thus, a weld procedure verification test is required with the most critical criterion for acceptability of this test based on the measure of as-welded ductility.

In the case of Ti-6A1-4V, components are sometimes stress-relieved after welding. Stress relief requires approximately four hours at 50 F below the aging temperature, about 1000 F. This treatment does not markedly improve mechanical properties, but does provide insurance against stress corrosion.

TYPICAL DEFECTS AND INSPECTION

Defects which appear in titanium welds include porosity, inadequate joint penetration, burn-thru and interstitial contamination. Cracks may appear in thick (> 1/2 in.) titanium but only rarely in thicknesses of current interest to propellant tankage, 0.180 in. and under. The only defect of this group which is not caused by an obvious deviation from accepted welding procedure is porosity. The cause of porosity in titanium still has not been adequately described in the literature, and its occurrence appears to be left to chance. Careful cleaning procedures and joint preparation appear to reduce the incidence of porosity somewhat. Recent sources have attributed porosity to sulfur and to hydrogen; both contaminants would result from improper cleaning. However, porosity to a modest degree is not too damaging to mechanical properties and can be tolerated if suitable design support data are generated. Spherical tungsten inclusions can be treated as porosity. Interstitial contamination of the weld bead should receive attention because the results, in terms of influence on mechanical properties, can be deleterious; also, nondestructive inspection by the conventional flaw detection techniques is not effective. The practical nondestructive inspection is to observe for surface discoloration indicating oxide or nitride films. The difficulty here is that a particular oxide film may have been formed by exposure of the surface to air at a temperature less than 1000 F, and therefore, is not indicative of bulk contamination. Thus, it is necessary to develop a technique of determining whether a given discoloration pattern is indicative of high or low temperature contamination. In cases of doubt, chemical anaylsis samples are taken from the weld bead to determine levels of bulk interstitial. This sampling, of course, is locally destructive and requires a repair if the bead is found to be acceptable. If bulk contamination is found by analysis, the part is scrapped or possibly salvaged by removal of the contaminated metal.

Additional insurance against interstitial contamination is provided by preparation of a "preproduction weld panel." Immediately before striking the arc on a titanium fabrication, a test panel is welded with gas shielding and power parameters similar to those to be used on the actual production weld. This panel is then evaluated for porosity, mechanical properties (ductility) and interstitial contamination. Only after collection of these data is the production weld initiated. Several ductility tests have been proposed as the basis for acceptance in this routine. The simplest is a bend test. However, there has been some difficulty in interpretation of this test. The nominal requirement is that welds in Ti-6A1-4V pass a bend radius test ten times the thickness without development of cracks. However, the large-grained weld bead of titanium deforms in such a way that the definition of "crack" is difficult to apply. Small fissures open up between grain pairs and can be detected by a low power microscope. Tests have been run to establish a correlation between the development of these small, one-grain-length cracks with such factors as filler wire lot, shielding gas efficiency, welding heat input and percentage of weld metal-to-base metal alloy mixing. None of these attempted correlations has been established with any success. A method has been used of establishing a definitive criterion for observing onset of cracking by placing strain gages on the tensile surface of the specimen which will register sudden increases in strain as the bend is performed. A correlation between cracks detected in this manner and base metal "quality" as indicated by comparison of several base metal lots has been noted.

If, in spite of these preventive measures, a weld bead is obtained which is contaminated, repair is difficult. Since contamination usually occurs over an appreciable length—if not the whole length—of the weld, the local grindout and refill technique successful with pores and other localized defects is impractical. However, some expensive parts have been salvaged by cutting out the whole weld bead and rewelding either the same two parts or a new part to an old with adjustments in geometry to account for the material loss. Such rewelds, even though introducing a double heat-affected zone, have performed satisfactorily.

Repair of burn-thru or an interstitial analysis sampling hole, can be accomplished by careful puddling and filling of the hole. Such repairs may result in excessive dilution of the bead by the weaker commercially pure filler metal, and thus strength requirements must be considered. An alloy (Ti-5A1-2 1/2Sn) can be substituted for pure titanium. Repair of inadequate joint penetration can be accomplished by remelting the bead from either side. Experiments have shown no deleterious effect resulting from multiple passes.

Weld shrinkage may cause distortion, particularly where repairs have been performed or where welding concentration is high—such as on small plumbing outlets. Such shrinkage distortion has been corrected by local planishing of the bead whose lateral shrinkage caused the distortion.

LOW-ALLOY STEELS

APPLICATIONS

High-strength, high-carbon, low-alloy steels have been used extensively in all versions of the aerospace field from landing gears and reentry capsules to large diameter rocket booster cases. These applications employ structural weldments with thicknesses of 1 and 2 in. as well as thin walls of 0.040 and 0.060 in. For motor case application, choices include D6AC, 4330-435V, 4340 and 4330-Si modified. Based on fracture toughness, crack-resistant high-strength designs have more recently employed D6AC steel for solid rocket propellant tankage.

The low-alloy steels are the only metals used for motor case or liquid rocket tankage which require a full postweld heat treatment. Titanium and aluminum are solution-treated and aged before welding and used either as-welded or as-stress relieved. Maraging steels are also welded in a (martensitic) solution-treated condition and require only aging after welding. Low-alloy steels cannot be welded in the quenched condition, ready for postweld tempering, because the martensite structure is too brittle. Some early missile work (Pershing) did use such steels in an as-welded condition after preweld quench and temper, but current motor case designs are too weight-sensitive to allow the increased land thickness (at the weld) necessary to tolerate reduced mechanical properties.

METALLURGICAL CHARACTERISTICS

The entire thermal cycle experienced by the base metal is important in determining the final microstructure and, therefore, the mechanical properties of the heat-affected zone of the weld. The mechanical properties of this zone are important even before final heat treatment because of the tendency of cracks to form under the influence of residual stress or while handling prior to final heat treatment.

The preferred thermal cycle, considering only the time-temperature-transformation behavior of the steels, yields bainite rather than martensite as a product of austenite decomposition on cooling from the welding operation. This thermal cycle, as shown in the time-temperature-transformation curve in Fig. 92.10 (typical of the D6AC or low-alloy group), arrests the cooling at a temperature too high for martensite to form and for a long enough period for bainite to form. This dictates a preheat interpass and hold temperature (30

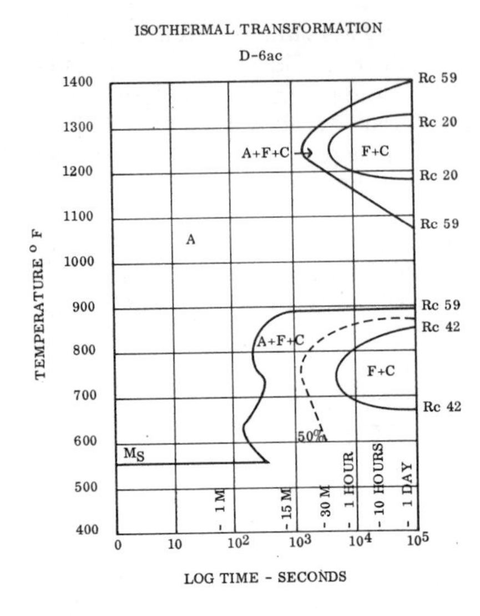

Fig. 92.10.–Time-temperature-transformation characteristics of D6AC steel.

minutes after welding) of 600 to 700 F–just above the M_s temperature for the steel in question. However, this satisfactory metallurgical solution introduces practical welding difficulties with the formation of oxide scale and with welder discomfort. For this reason, a compromise is sometimes made by specifying preheat, interpass and hold temperature minimums of 300 to 450 F. On cooling to these ambient temperatures the weld area microstructure is a mixture of martensite, austenite and, after some time for transformation, bainite. This structure, not highly susceptible to cracking as martensite which does form, will be tempered.

With moderate preheat temperature the remaining austenite will transform to martensite if the part is allowed to cool to room temperature. Therefore, fabricators who use the lower preheat temperatures generally maintain that temperature after welding until a stress relief or a re-austenitization heat treatment is started.

If the postweld treatment is a stress relief, it would be more efficient to hesitate, on heating, at the nose of the bainite knee (e.g., 800 F in Fig. 92.10), since transformation is faster at this temperature than at 1150 ± 50 F, which is generally used for stress relief.

WELDING TECHNIQUES

D6AC steel is welded almost exclusively by the gas tungsten-arc process which has also been qualified for all low-alloy steels employed in highly stressed components. Plasma-arc welding has shown promising properties. Electron beam welding processes (in vacuum and out-of-vacuum) have shown much success under stringent and intricate requirements. Its use for welding large rocket motor cases has been held up by equipment cost, radiation control requirements and lack of versatility. The gas metal-arc process is used in various techniques, and varying success has been achieved in high-strength, high-efficiency weldments. In general, it is accepted for use only after extensive development of procedures and establishment of automatic process controls. Submerged arc welding is not currently finding acceptance in solid rocket motor case fabrication.

The shielding gas for gas tungsten-arc is argon or helium; their mixtures depend on penetration, weld speed and joint configuration. When high preheat temperatures are used, backup gas should be used to prevent excessive scaling.

Joint Design

In welding D6AC steel missile cases, the joint design depends on filler metal composition, backup tooling, wall thickness, location of weld (longitudinal or circumferential, etc.) and fabricator experience. Sound welds can be produced using four different filler metal compositions and many joint configurations, if appropriate heat and filler wire inputs are made during welding. In general, joints are welded from one side using 45 to 70 deg included angles with a 1/16 to 1/8 in. root face about 1/16 to 3/32 in. from the inside surface. The critical root passes consist of one or two gas tungsten-arc fusion passes and several gas tungsten-arc filler passes. D6AC steel requires high preheat (600 F) for welding and a postheat treatment (700 F-four hours) after welding. This procedure prevents any appreciable transformation to martensite. Cracking can occur with misalignment, inadequate joint penetration or poor gas backup or if improper handling disturbs the part before stress relief or heat treatment. If stringer bead passes are used side by side, care must be taken to balance the weld travel speed and the width of the weld to prevent edge microcracking during cooling.

Weld Jigging

The most reliable weldments are made on tooling with "built-in" preheat, hold and postheat fixtures and controls. Welder or welding operator fatigue is decreased if weld tooling provides visual control and physical accessibility for making adjustments. Tooling must be designed to prevent a buildup in residual stresses peculiar to a weld configuration. These high residual tensile welding stresses can cause difficulty in the dendritic structure of the unrefined portions of weld beads, sometimes resulting in intergranular micro- or macrocracking during subsequent thermal cycles. Weld backup grooves also preheated may be copper or ceramic-coated stainless steel with uniform gas distribution to prevent scaling. The backup groove is deep enough to prevent underbead quench on the groove itself under normal heat input control. Incomplete joint penetration if it occurs must be eliminated by a positive removal technique such as underside root grind-out.

MECHANICAL PROPERTIES

The trend in material applications has varied from designs employing higher and higher strengths to present production considerations using a more conservative range of yield strength from 150 to 200 ksi. In rocket motor application a compromise is used between yield strength and fracture toughness. D6AC is most reliable at 200,000 psi strength with plane-strain fracture toughness K_{IC}, minimum properties of not less than 75 in the weld. The toughness of plate and forged material is somewhat higher. The toughness of the heat-affected zones generally is equivalent to the base metal. The 0.2% yield strength is about 190,000 psi, elongation in the weld is 14% and there is a 45%

reduction in area. Higher strengths are obtainable, but such increases are accompanied by decreased toughness. Allowable flaw size is related to toughness (resistance to crack initiation or propagation or both) and inspection capability can become a limiting condition. Therefore, establishment of accept-reject criteria depends upon the ability of an inspection technique to detect defects in very highly stressed components.

TYPICAL DEFECTS AND INSPECTION

Defects that appear in low-alloy steels can be detected by radiography, penetrant, ultrasonic and magnetic suspension or magnetic powder techniques. Inspection procedures and criteria have been established with a fairly ample experience level to substantiate them in most applications. Where complex weld configurations are required, two or more types of inspection may be warranted. Practice generally has established that thorough inspection is required after welding, heat treatment and leak testing.

D6AC when welded with filler similar to base metal chemistry reflects uniform properties and high joint efficiency after heat treatment. Inspection should be accomplished before heat treatment when repairs can be made without possible degradation of the strength of the joint. When it is austenitized, quenched and tempered, the structure is essentially tempered martensite. However, subsequent repair introduces some untempered martensite which must be tempered by local heating.

Probably the most difficult defect to detect is incomplete fusion. Metals that have low fracture toughness and, consequently, a small critical flaw size, require more sensitive inspection methods. Excessive design strength levels or improper designs subjecting welds to unnecessary strains would result in inspection requirements beyond present capabilities.

MARAGING STEELS

APPLICATIONS

Applications for maraging steels in launch vehicles have been limited to large, 156 and 260 in. diam, solid rocket motor chambers. A photograph of a 260 in. maraging steel chamber is shown in Fig. 92.11. Maraging steels are selected for these applications because they are readily formed and welded, require only aging heat treatment at 800 to 950 F after fabrication, have excellent dimensional stability during heat treatment and have better fracture toughness at yield strength levels above 200 ksi than other suitable alloys.

There are several types of maraging steel, but only the 18% nickel alloy has been fabricated into rocket motor chambers. The other types include 25%, 20% and 12% nickel alloys, of which the 12% nickel alloy appears the most promising.

METALLURGICAL CHARACTERISTICS

Welding techniques for maraging steels are influenced by the metallurgical characteristics of these alloys. Therefore, these characteristics will be reviewed briefly. The information in this section will be devoted to 18% nickel maraging

Fig. 92.11.–Two hundred sixty inch diameter rocket motor chamber.

steels. However, much of the information is also applicable to the 12% nickel alloy because of its similar metallurgical and welding characteristics.

Table 92.2 shows typical compositions and tensile properties for three grades of 18% nickel maraging steel. The nickel content of the alloys stabilizes austenite to low temperatures so that the steels transform to martensite regardless of cooling rate. The low-carbon content ensures that the martensite is soft and ductile. Cobalt, molybdenum and titanium additions develop precipitation-hardening mechanisms in the alloy. Variations in amounts of these latter elements are used to adjust the strength of alloy (Table 92.2). Statistical

Table 92.2–Typical composition and tensile properties of 1/2 in. thick 18% nickel maraging steel plate

	Composition, Wt. %					Tensile Properties[2]		
Grade	C[1]	Ni	Co	Mo	Ti	Ultimate Strength ksi	2 % Offset Yield Strength ksi	Elongation in 1 in. %
200	0.03	17.0 19.0	8.0 9.0	3.0 3.5	0.15 0.25	234	225	11
250	0.03	17.0 19.0	7.0 8.5	4.5 5.5	0.30 0.50	269	257	8
300	0.03	18.0 19.0	8.5 9.5	4.5 5.5	0.50 0.80	310	300	9

[1] Maximum value.
[2] Aged at 900 F for 8 hours.

techniques have shown the following relationship between strength and alloying elements:

Yield strength, ksi = 38.1 + 8.8 (% Co) + 22.6 (% Mo) + 87.7 (% Ti).

This relationship shows that titanium has over nine times the strengthening effects of cobalt and three times that of molybdenum.

Maraging steels undergo a number of metallurgical reactions with changing temperature. When heated above 1500 F, 18% nickel maraging steels transform to austenite. Upon cooling, the austenite transforms to ductile martensite which has a tensile strength of about 145 ksi, regardless of cooling rate. When such solution-treated alloys are reheated from 850 to 950 F for three to eight hours, compounds of nickel, molybdenum and titanium precipitation-harden to achieve the tensile properties shown in Table 92.2. If either the solution-treated or aged metal is heated to between 1000 and 1400 F, a small portion of the martensite reverts to austenite. The austenite thus formed contains sufficient nickel to be stable at room temperature and, consequently, does not transform to martensite or respond to age-hardening treatments.

Welds are subject to all of these reactions during welding operations. In a single-pass, the weld fusion zone and high-temperature heat-affected zones undergo thermal cycles which simulate the solution-treated condition; lower temperature areas of the heat-affected zones undergo thermal cycles that cause austenite reversion and, finally, age hardening. However, short heating times can minimize austenite reversion and age hardening.

In multipass welds, the thermal cycles are more complex. Figure 92.12 shows a multipass gas tungsten-arc weld in solution-treated Grade 250, 18% nickel plate, as-welded. The lines on the figure represent areas of equal hardness. Outside of the weld heat-affected zones, the hardness of the solution-treated base metal is R_c 30. Near the weld, but in the heat-affected zones, the hardness increases slightly due to aging because of welding low-temperature thermal cycles. In the weld fusion zone, the lowest hardness is found in the last weld pass, which is essentially in the solution-treated condition and has a hardness of R_c 30. Progressing from top to bottom of the weld fusion zone, the hardness increases in the weld passes that were subjected to low-temperature thermal passes as weld beads were deposited.

When a weld such as that shown in Fig. 92.12 is aged, the hardness of the base metal and weld deposit increases to about R_c 50, variations depending on

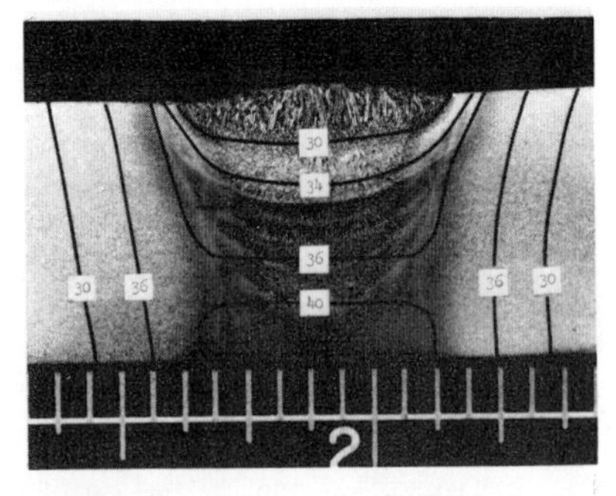

Fig. 92.12.–Hardness variations in as-welded, gas tungsten-arc weldment in Grade 250, 18% nickel maraging steel plate.

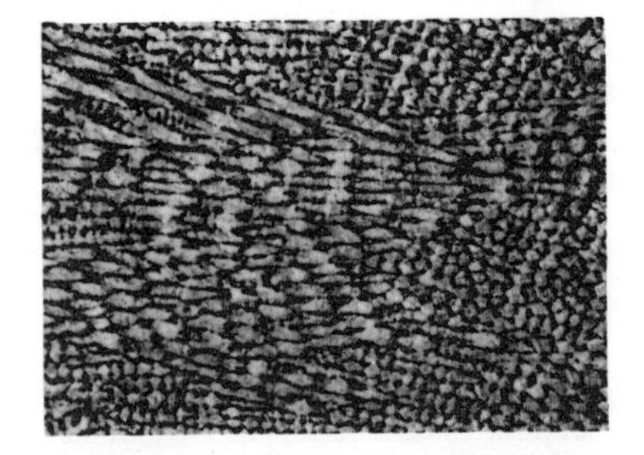

Fig. 92.13.–Typical dendritic microstructures in weld fusion zones in 18% nickel maraging steel plate.

location in the weld. After age-hardening, the zones which had the lowest hardness in the as-welded condition–i.e., fusion zone of last weld pass and unaffected base metal–have slightly higher hardness than zones which were subjected to low-temperature thermal cycles where a slight amount of age-hardening occurred. The slightly lower hardness is attributed to austenite reversion occurring during the welding operations. Through the use of proper welding techniques, however, austenite reversion can be minimized and the tensile properties of the weldment made comparable to those of the base metal.

The microstructures of welded joints reveal additional information regarding the metallurgical and welding characteristics of maraging steels. Weld fusion zones exhibit dendritic structures as shown in Fig. 92.13. Based on phase diagrams (Fe-Ni, Fe-Mo, Fe-Co and Fe-Ti), the dendrites contain a lower alloy content than the average composition of the weld fusion zone; the interdendritic areas contain a higher alloy content than the average composition of the weld metal. In addition, an interdendritic phase is present which has the appearance of retained austenite. This interdendritic phase also contains larger amounts of cobalt, molybdenum, nickel and titanium than the average weld metal composition. Alloy segregation in weld metals is believed to reduce the strengthening effects of the cobalt, molybdenum and titanium additions.

WELDING PROCESSES

Maraging steels are readily adaptable to fabrication with the gas tungsten-arc, gas metal-arc and submerged arc welding processes. All processes can be used without preheat or postheat; interpasss temperatures are kept low. However, care must be maintained with all processes to assure weldments meet aerospace requirements.

The gas tungsten-arc process has proven to be most reliable for rocket motor chamber fabrication. With this process, sufficient inert gas shielding must be supplied to protect the face and root of the weld from oxidation; surface scale on the face of the weld must be removed before deposition of subsequent weld beads. If these precautions are not taken, excessive porosity may be encountered. Also, filler wire surfaces must be clean to minimize weld porosity and to reduce hydrogen which can cause weld cracking.

The major problem in joining maraging steels with the gas metal-arc process, as compared with tungsten-arc, is a greater incidence of porosity. Also, subsequent sections can demonstrate that the fracture toughness of gas metal-arc welds are inferior to those of gas tungsten-arc welds.

Flux composition and moisture are the primary problems which discourage submerged arc welding of maraging steels. Fluxes that minimize silicon pickup in the weld metal are required to prevent weld cracking. Also, fluxes must be handled and stored in a manner which will keep them dry. Otherwise, weld cracking may be encountered through hydrogen pickup–hydrogen contents of even 5 ppm may cause weld cracking.

Welding Parameters

Mechanized welding operations and conventional joint designs and welding parameters are used to fabricate maraging steel rocket motor chambers. Welds in maraging steel rocket motor chambers are prepared in the downhand position to assure maximum reliability in the welding operation. Also, grooved backup bars are used to supply inert gas protection. Single-, double U and Vee joint designs are employed.

Figure 92.14 shows parameters for welding 1/2 in. thick, 18% nickel maraging steel plate with the gas tungsten-arc process. A single-U joint with a 0.06 in. root land and zero in. root opening is used. The plate is not preheated and the interpass temperature is maintained below 200 F to minimize austenite reversion. Small multiple stringer-type weld beads are deposited with low or intermediate heat input to further minimize austenite reversion. Deviations from these procedures within good welding practice are possible without impairing the soundness or properties of weldments.

Welding Fixtures

Conventional welding fixtures are used to fabricate maraging steel rocket motor chambers. These fixtures are designed to maintain alignment between the

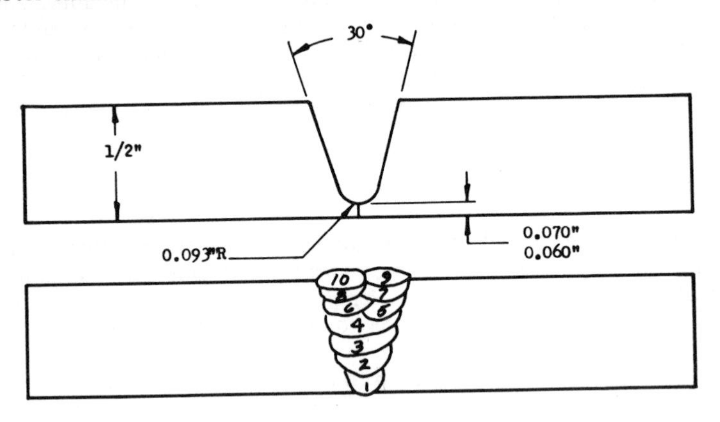

Weld(1) Pass	Amps(2)	Volts	Travel ipm	Wire(3) Feed, ipm	Torch Gas Argon, cfh	Backup Gas cfh
1	150	10	6	18	25	10
2	150	8.5	6	18	↓	↓
3	160	10	6	24		
4	160	↓	6	↓		
5	180		6			
6	↓		6			
7			7			
8			7			
9			7			
10			10			

(1) Electrode was thoriated tungsten with a 1/8 in. diameter.
(2) Direct current, straight polarity.
(3) Filler wire diameter was 1/16 in.

Fig. 92.14.–Parameters for welding 1/2 in. thick maraging steel plate with the gas tungsten-arc welding process.

parts to be welded and to provide a suitable backup for the weld. Examples of fixtures are those used to fabricate the 260 in. diam rocket motor chamber shown in Fig. 92.11.

Hemispherical heads are fabricated for 260 in. diam chambers by welding formed gore sections. The internal fixture for welding the gore sections is shown in Fig. 92.15. The fixture is machined to the inside diameter of the head and is mounted on a welding positioner that can be tilted under the welding head. The gore sections are placed on the internal fixture and held in place with the external fixture shown in Fig. 92.16. The external fixture breaks down into segments which correspond to the gore sections. As each gore section is placed on the internal fixture, the corresponding segment of the external fixture is mounted to hold the gore in place. The segments of the external fixture are connected by tubular flanged sections which are removed to provide access for welding. Grooved mild steel backup bars are fitted from the inside after the gore sections are in place.

When the gore sections are assembled, they are tack welded to maintain alignment. The tack welds are ground so that they will not interfere with the actual welding operation. During welding, the welding positioner is tilted to traverse the part under the welding head which is mounted on a side beam above the part. The tilting speed is set to obtain the desired weld travel speed.

The welding fixtures for the heads cannot exert sufficient force to provide alignment if the gore sections are improperly formed or machined. Therefore, precision forming and machining are required to achieve dimensional tolerances and to minimize mismatch and poor fit-up which are kept below 0.060 in. when welding 0.600 in. thick components.

The cylindrical sections of the chambers are fabricated by roll-forming plate into semicircular cylinders that are joined by longitudinal welds. Two longitudinal welds are made in each cylindrical section in the chamber. The cylindrical sections are subsequently joined by circumferential welds.

An expanding internal fixture is used to provide proper alignment for the circumferential welds. This fixture is fitted with backup bars that span the weld joint (Fig. 92.17). Hydraulic cylinders are incorporated to exert force in expanding the fixture. The force obtained from this fixture is sufficient to deform plastically the cylindrical sections to minimize contour discontinuities in the weld area.

Fig. 92.15.–Internal fixture and positioner for welding 260 in. diam heads.

Fig. 92.16.–External fixture for welding 260 in. diam heads.

Fig. 92.17.–Internal fixture for welding circumferential joints in 260 in. diam rocket motor chamber.

Fig. 92.18.–Two hundred sixty inch diameter chamber on rolls for welding circumferential joint.

A roll and weld technique is used to deposit the circumferential welds. The cylindrical sections are placed on power-driven rolls and aligned on the internal fixture. The joint is tack welded to maintain alignment and the sections are rotated under a welding head to provide weld travel. Figure 92.18 shows the system used to rotate the cylindrical section.

Heat Treatment

Aging-type heat treatments are performed on completed maraging steel rocket motor chambers because the strength of weldments in the as-welded condition is comparable to that of solution-treated material. Two different welding and heat-treating sequences are used for rocket motor chambers:

1. Weld solution-treated material and age the entire chamber.
2. Weld age-hardened material and locally age the weldments.

The 260-in. diam chambers are fabricated from solution-treated material and the entire chamber is age-hardened. To perform this operation, a large maraging furnace is used. The furnace (Fig. 92.19) is assembled in sections which are placed over the top of the chamber. Air, heated in a gas-fired heat exchanger, is circulated through the furnace. The heat exchanger is used so that the products of combustion are not in contact with the chamber during the aging process. The chambers are aged at 900 F for eight hours. The furnace maintains this temperature within a range of 40 F which is adequate for aging maraging steels.

Filler Wire Compositions

Welding filler wire composition is an important variable in welding 18% nickel maraging steels. Through variations in filler wire composition, it is possible to match the tensile strength of the weld metal with all grades of base metal. However, the fracture toughness of the weld metal decreases rapidly as its strength increases. Since fracture toughness is important in rocket motor chamber performance, filler metal compositions are selected to match the

Fig. 92.19.–Maraging furnace for 260 in. diam rocket motor chamber.

minimum tensile strength level of the base metal where they exhibit maximum fracture toughness rather than a higher strength level where fracture toughness is sacrificed.

Filler wire compositions are similar to those of the base metal except for the higher titanium content which is required to achieve the same strength as the base metal. The need for increased titanium content is attributed to the following factors.

1. Freezing segregation which reduces the strengthening afforded by cobalt, molybdenum and titanium.
2. Titanium lost in transferring through the arc by oxidation.

Filler wires for rocket motor chambers are vacuum-, arc- or induction-melted.

Table 92.3–Typical tensile properties of weldments in ½ in. thick 18% nickel maraging steel plate

		Tensile Properties[1]	
Grade	Welding Process	Ultimate Strength (ksi)	Elongation in 1 in. %
250	GTAW	253[2]	4[2]
250	GMAW	267[2]	6[2]
250	SAW	260[2]	4[2]
200	GTAW	213[3]	12[3]
200	GMAW	220[3]	11[3]

[1] Transverse Tensile Tests.
[2] Aged at 900 F for 4 hours after welding.
[3] Aged at 900 F for 8 hours after welding.

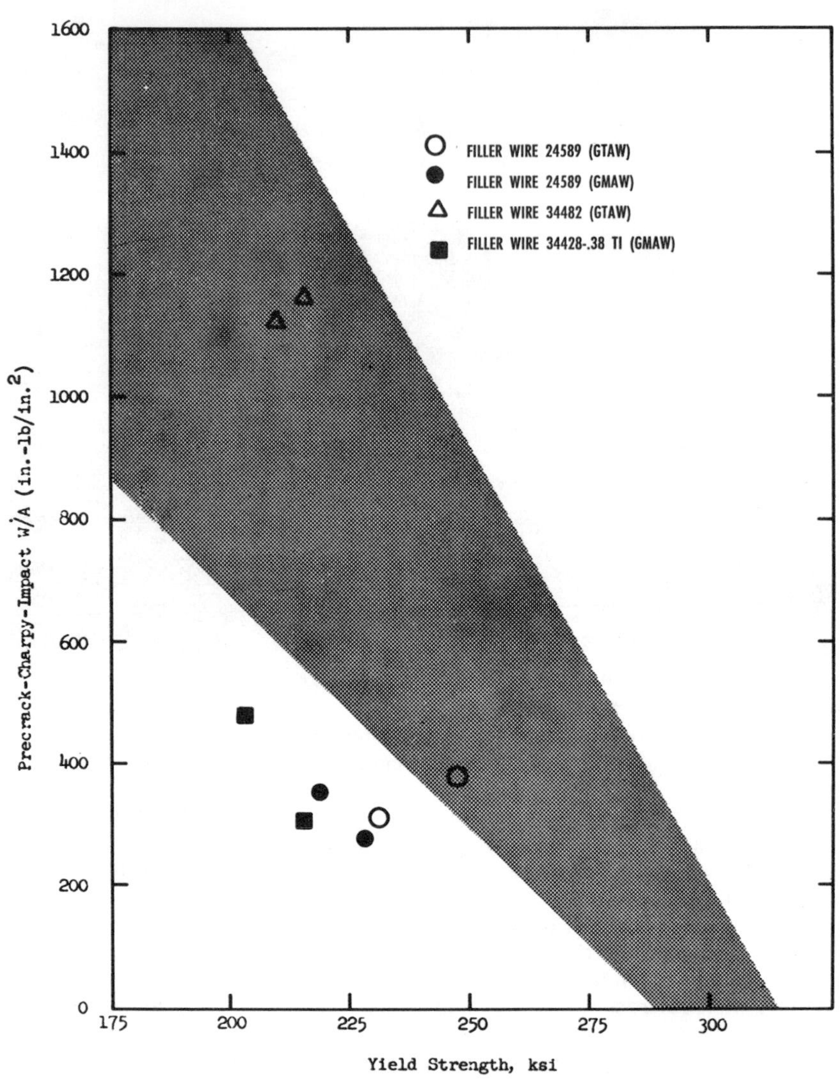

Fig. 92.20.–Relationship between yield strength and precracked Charpy impact fracture toughness of 18% nickel maraging steel plate and weld metals. (Shaded area represents base metal fracture toughness.)

They are fabricated to eliminate seams and other surface defects. They are carefully cleaned to eliminate scale and lubricants; they are also vacuum-annealed and degassed to prevent surface contamination.

MECHANICAL PROPERTIES

Weldments with sufficient tensile strength and fracture toughness to withstand service stresses in the presence of small defects are required to obtain reliable

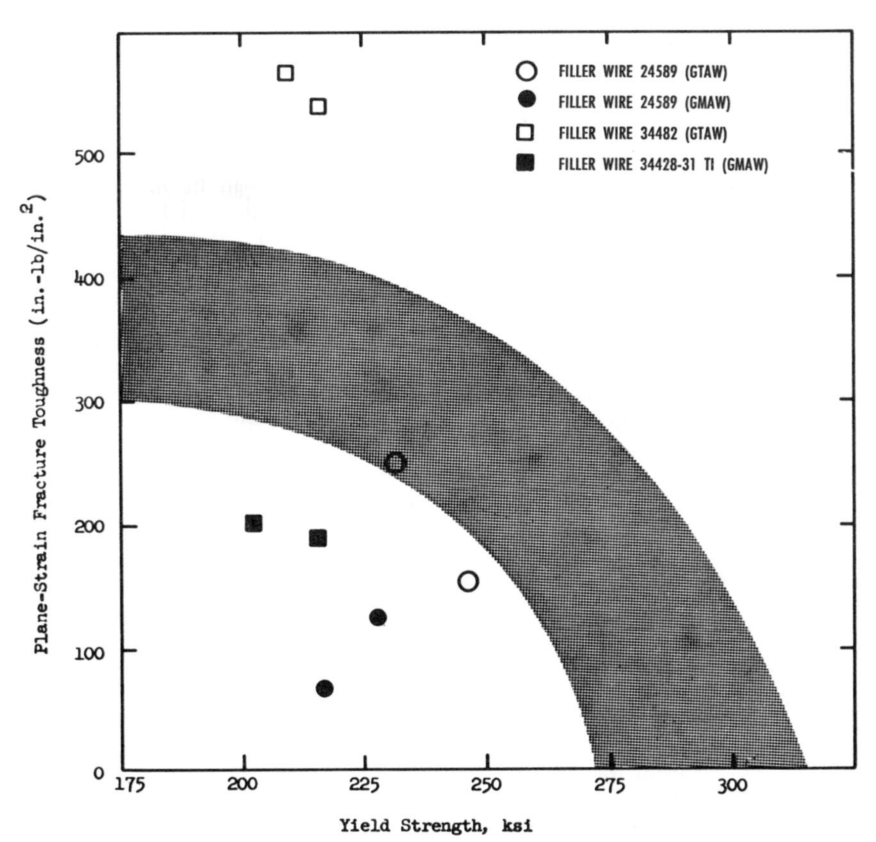

Fig. 92.21.–Relationship between yield strength and plane strain fracture toughness of 18% nickel maraging steel plate and weld metals. (Shaded area represents base metal fracture toughness.)

performance from large diameter rocket motor chambers. Adequate uniaxial tensile properties are readily achieved in various grades of 18% nickel maraging steel weldments made with different welding processes as shown in Table 92.3. Fracture toughness properties, however, require careful consideration.

Figures 92.20 and 92.21 show relationships between the yield strength and fracture toughness of 18% nickel maraging steel base metals and weld metals. The fracture toughness of both decrease with increasing yield strength. Also, the fracture toughness of gas tungsten-arc weld metals exceed those of gas metal-arc weld metals. The use of Grade 200 18% nickel maraging steel and the gas tungsten-arc process for 260 in. diam rocket motor chambers results in maximum fracture toughness in the chambers. For this reason, the filler metal composition for the chambers is selected to exceed the minimum strength rather than the maximum strength of the base metal. By selecting materials and welding procedures which result in maximum fracture toughness, the critical flaw size (which might cause premature failures) is sufficiently large that it can be detected nondestructively.

TYPICAL DEFECTS AND INSPECTION

Porosity is the major defect in maraging steel weldments even when satisfactory welding procedures are used. This defect can be minimized by carefully cleaning the weld filler metals and base metals, and by the use of adequate inert gas shielding for the root and face of gas tungsten-arc and gas metal-arc weldments. However, porosity is always troublesome in large multipass welds.

Under circumstances where hydrogen is introduced in amounts exceeding about 5 ppm, weld cracking also may be encountered. Other defects such as undercut, inadequate joint penetration or incomplete fusion can be minimized by proper adjustment of welding parameters.

Radiographic, ultrasonic and liquid penetrant inspection techniques are suitable for maraging steel rocket motor chambers. Magnetic particle inspection techniques are less than 100% effective because austenite reversion, which occurs adjacent to each weld metal bead, obscures defects that might be present.

BIBLIOGRAPHY

"Strength of Welded Joints in Aluminum Alloy 6061-T6 Tubular Members," R. L. Moore, J. R. Jombock and R. A. Kelsey, *Welding Journal*, 50 (4) 238–247 (1971).

"Effect of Discontinuities on Weld Strength of Aluminum Alloys," F. G. Nelson and M. Holt, *Welding Journal*, 50 (10) 427s–434s (1971).

"Strength Toughness Properties of Welds in Plates of Commercial Titanium Alloys," L. E. Stark, *Welding Journal*, 50 (2) 58s–70s (1971).

"Fatigue Properties of a Welded Low-Alloy Steel," M. R. Baren and R. P. Hurlebaus, *Welding Journal*, 50 (5) 207s–212s (1971).

"Welding of Maraging Steel," F. H. Lang and N. Kenyon, *Welding Research Council Bulletin*, No. 159 (1971).

"Effects of Porosity on High Strength 7039 Aluminum," R. J. Shore and R. B. McCauley, *Welding Journal*, 49 (7) 311s–321s (1970).

"The Weld Heat-Affected Zone of the 18 Ni Maraging Steels," J. J. Pepe and W. F. Savage, *Welding Journal*, 49 (12) 545s–553s (1970).

"Hydrogen Segregation in Ti-6A1-4V Weldments Made with Unalloyed Titanium Filler Metal," D. N. Williams, B. G. Koehl and R. A. Mueller, *Welding Journal*, 49 (11) 497s–504s (1970).

"Effect of Experimental 2219 and 2014 Aluminum Weld Composition Variations," D. L. Cheever, P. A. Kammer, R. E. Monroe and D. C. Martin, *Welding Journal*, 48 (8) 348s–358s (1969).

"Design of Base Metals for Weldability–Maraging Steels," N. Kenyon, *Welding Journal*, 48 (3) 105s–109s (1969).

"Effect of Austenite on the Toughness of Maraging Steel Welds," Norman Kenyon, *Welding Journal*, 48 (5) 193s–198s (1969).

"Maraging Steel as Influenced by Specimen Type and Thermal and Mechanical Treatment, 300 Grade 18-Nickel K_{Ic}," H. W. Maynor, Jr. and C. C. Busch, *Welding Journal* 43 (9) 428s–432s (1964).

CHAPTER 93

CLAD STEEL AND APPLIED LINERS

PREPARED BY A COMMITTEE CONSISTING OF

L. K. KEAY, *Chairman*
Lukens Steel Co.

R. E. LORENTZ, JR.
Combustion Engineering, Inc.

F. B. SNYDER
The Babcock & Wilcox Co.

G. S. SANGDAHL
Chicago Bridge & Iron Co.

W. L. WILCOX
The Babcock & Wilcox Co.

CHAPTER 93

CLAD STEEL AND APPLIED LINERS

DEFINITION AND SCOPE

This chapter deals primarily with the manufacture and fabrication of clad and lined steel plate and forgings used in such areas as the manufacture of pressure vessels and tanks.

Clad metals are composites of two or more metals joined in a continuous manner by a metallurgical bond. In this respect, they differ from liners which are composites having two or more metals attached in an intermittent manner.

Composite metal is manufactured in a variety of shapes, sizes and combinations for many applications. Applications include jewelry, instruments, machine tools, aircraft structures, wire, ships, tubes, refrigeration equipment and coinage.

TYPES OF CLAD STEEL

Clad steel plate and forgings are manufactured either as a raw material, such as by the steel mill, or by the fabricator in his own plant, as by weld overlay. The major methods of producing clad steel are described in the following paragraphs.

ROLL CLADDING

Roll-clad plates and sheets are produced by hot rolling a welded assembly of steel and some other metal (or metals) such as nickel or stainless steel. The internal faying surfaces are cleaned and the periphery of the assembly or "sandwich" is welded both to exclude air and hold the parts together during rolling. In rolling, the high temperature and working create a solid-phase weld between the steel and cladding metal. Stainless steels and other metals are commonly nickel-plated to facilitate the welding together of the surfaces. The types of cladding usually produced by this process are stainless steels, nickel, copper, high-nickel alloys and copper-nickel alloys.

CLADDING BY EXPLOSION WELDING (EXW)

Clad metals fabricated by explosion welding have a continuous metallurgical bond between the two metal surfaces. Joining occurs as the faying surfaces of the metals are thrust together by energy released from an explosive source. The high velocity collision of the two metals produces extensive shear deformation at the welded surfaces and creates a metallic jet which sweeps away surface films and oxides from the areas being joined. Thus ideal metal-to-metal contact is established (See Fig. 93.1).

The weld obtained by explosion cladding may be of three types depending on whether the metallic jet (liquid) escapes completely, remains as a continuous alloy bond zone, or, as is true of most commercial products, is partly expelled, giving a weld obtained partly by solid phase contact and partly by melting (see Fig. 93.2). The bond is strong and complete, enabling the clad object to maintain integrity through forming, welding and heat-treating operations normally used in metal fabrication.

The explosion cladding process is versatile. Large and small plates and forgings have been clad in a wide range of thicknesses. No external heat is required to perform the welding operation although some heat is generated at the weld interface because of the absorption of energy. Because explosion welding is a "cold" process, the components can be heat-treated before cladding. Because of the low diffusion or intermelting, metal combinations difficult or impossible to clad by other methods are usually readily weldable by the explosion process. Possible material exceptions are those in which the base metal(s) has a low impact resistance. Brittle materials will fragment easily under impulsive loading.

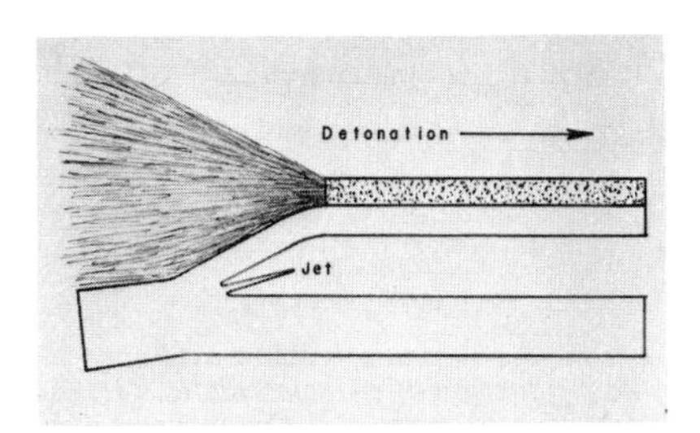

Fig. 93.1.–Metallic jet emanating from collision point during cladding.

Fig. 93.2.–Solidified metallic jet (clean white areas at top right of crests and bottom right in valleys) regularly spaced in this bond zone of Type 410 stainless steel on A387-D low-alloy chromium-molybdenum steel.

FURNACE BRAZE-CLADDING IN VACUUM

A proprietary process of considerable commercial importance, vacuum braze-cladding makes use of the relaxation of plates under the influence of elevated temperature. A brazing alloy compatible with the base metal and the clad metal is contained within a "sandwich" and melts as the two metals come into intimate contact (see Fig. 93.3.).

During the brazing cycle, a vacuum is drawn between the faying surfaces of the sandwich structure to pull the relaxed materials together and eliminate contamination. Thus, the vacuum assures cleanliness, and a strong continuous brazed joint is formed. One, two or more plates may be clad at one time. The process is applicable to pipes, forgings, rings, curved shapes and channel plates, as well as to large flat plates.

PRESS-BRAZE CLADDING

In this process, two or more metal layers are continuously brazed between the platens of a press, utilizing a furnace and special atmospheres or vacuum. Little or no thickness reduction is obtained. The composite may be used as-brazed for items such as tube headers, or may subsequently be rolled.

FURNACE BRAZE CLADDING WITH FLUX OR ATMOSPHERES

Tubesheets and various small parts may be clad by brazing with flux (which usually gives an incomplete joint) or by the use of vacuum or dry hydrogen atmospheres.

WELD OVERLAY CLADDING

Weld overlay cladding is a versatile and widely used method for producing continuously-joined layers of corrosion- or wear-resisting alloys on a base metal, usually steel. It is used by the fabricator for cladding flat or formed surfaces as an alternative to the purchase of clad plate, dished heads or forgings from the metal supplier. It is also used to salvage or extend the life of existing vessels by welding in place. Cladding materials of special compositions that are not readily available in other forms or not producible by other processes can often be applied as cladding by weld overlay.

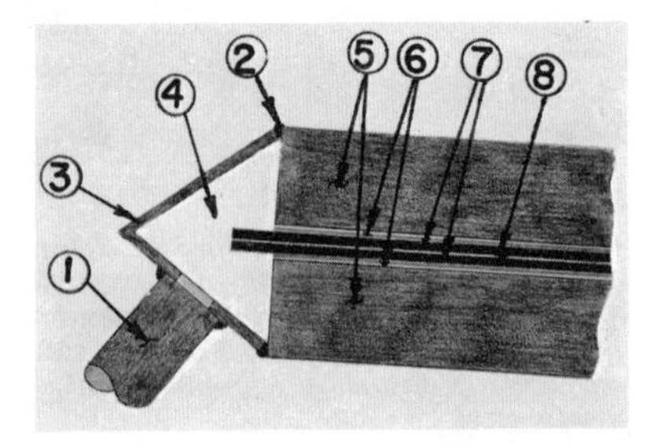

Fig. 93.3.—Vacuum brazing assembly for the production of two clad plates. (1) Vacuum pipe to pump. (2) Channel seal weld. (3) Edge seal bar. (4) Evacuated channel. (5) Base plates. (6) Bonding material. (7) Cladding sheets. (8) Separating material.

WELD OVERLAY CLADDING

CHARACTERISTICS

The three most important characteristics of the weld overlay cladding process are:

1. Dilution (the percentage of base metal in the weld overlay cladding metal)
2. Deposition speed.
3. Deposit thickness.

In the selection of a weld cladding process, the user should consider the ability of the process to limit dilution while maintaining a high deposition rate without having to apply a deposit thickness greater than the application requires. For low-carbon deposits, the base-metal carbon content, dilution and carbon content of the filler metal must be held to low levels to avoid an excessive number of passes. Apparatus and process control become more critical with those processes designed to give minimum dilution. Multilayers or the use of chemically-enriched filler metals, having a surplus of desired alloying elements, are often used to reduce the effect of dilution.

Aside from the effect of dilution, the characteristics of the shielding medium, whether it be gas or flux, alter the transfer of elements from filler metal to deposit and must be considered along with filler metal analysis in establishing welding procedures. The use of chemically-enriched filler metals or the use of alloy or neutral fluxes is the most common method of compensating for loss of elements in transfer.

The three process characteristics listed above are generally related (Fig. 93.4). A family of curves can be drawn for each weld cladding process. Their curvature and position with respect to the coordinates will vary from process to process, but the general relation will remain the same for most processes. For example, for any given process, deposition rates can be raised only by accepting increased dilution with a given travel speed. Figure 93.4 also indicates that low dilution overlays can be obtained at the higher deposition rates if thick deposits (lower travel speed) are economically acceptable. However, under extreme conditions, the deposited bead assumes an undesirable cross-section shape which may present tie-in difficulties or slag inclusions between adjacent beads. Welding conditions must, therefore, be selected to achieve the best balance between suitable bead shape and acceptable dilution, deposition and deposit thickness. Generally, it is more difficult to produce thin deposits (1/8 in. or less) with the very high deposition-rate processes.

Although some processes have been developed specifically for weld cladding, many applications are accomplished using conventional welding processes operated with modifications and procedures that minimize dilution. Of the processed listed in Table 93.1, shielded metal-arc welding (see the pertinent paragraphs in Chapter 22, Section 2 and Chapter 43, Section 3A), plasma-arc (without hot wire) weld cladding (Chapter 44, Section 3A and Chapter 54, Section 3B), gas tungsten-arc welding (Chapter 23, Section 2 and Chapter 44, Section 3A) and gas metal short-circuiting arc welding (Chapter 23, Section 2) are low deposition-rate processes. The use of these processes, therefore, is generally restricted to small assemblies where cost savings associated with lower dilution, thinner deposits, greater equipment flexibility or lower capital investment may override greater productivity. Conversely, the higher deposition rates associated with the remaining processes make them economically more attractive for surfacing very large areas such as the interior of vessels.

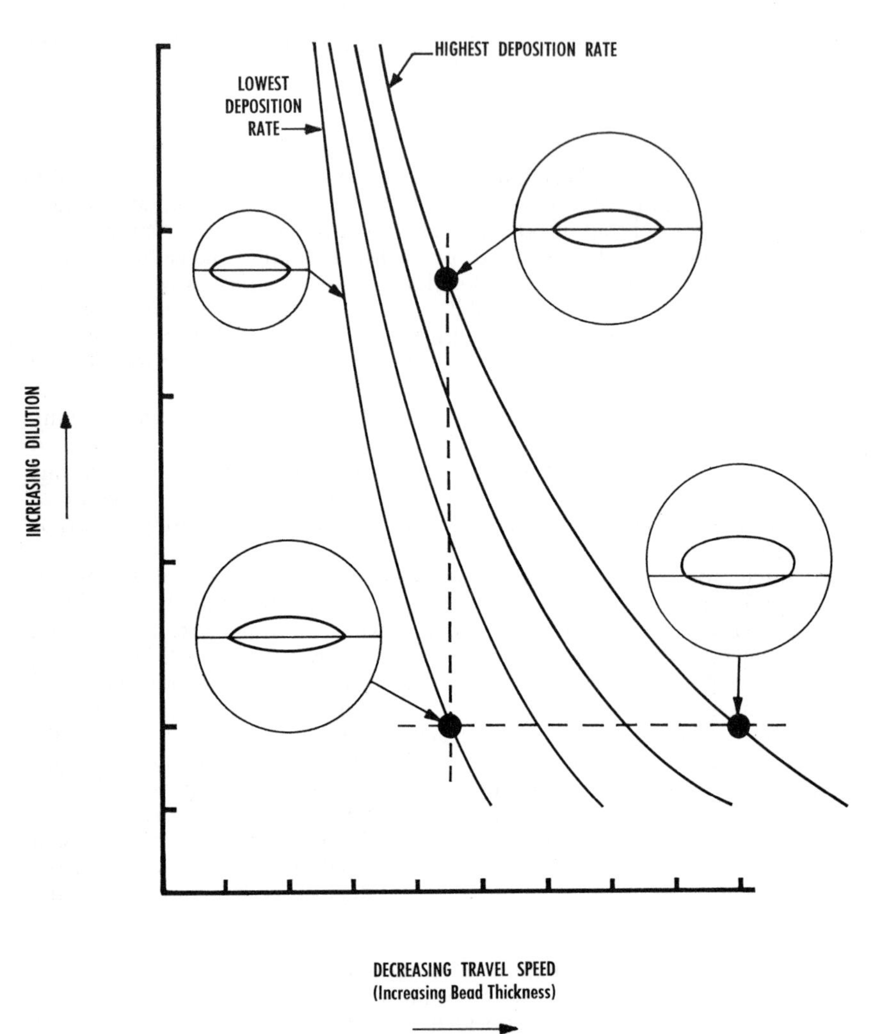

Fig. 93.4.–Typical surfacing process characteristic curves.

SUBMERGED ARC WELDING PROCESSES (SAW)

Single-Wire Technique

A description of the process and apparatus is given in Chapter 24, Section 2. The control system may be either constant voltage or constant current. D-C power operating either reverse (dcrp) or straight (dcsp) polarity is generally

Table 93.1—Weld cladding processes

Basic Welding Process	Process Variation	Mode of Application	Filler Metal Form	General Application Technique
Shielded metal-arc	None	Manual	Covered electrodes	Stringer or weave
Submerged arc	Single wire	Semiautomatic or automatic	Continuous electrode	Stringer or oscillation
	Multiwire	Automatic	Continuous electrode	Stringer or oscillation
	Series arc	Automatic	Continuous electrode	Stringer
	Strip electrode	Automatic	Continuous electrode	Stringer
	Strip electrode with auxiliary cold strip	Automatic	Continuous electrode	Stringer
	Electrode with granular metal	Automatic	Continuous electrode & metal powder	Stringer or oscillation
	Hot wire	Automatic	Continuous electrode & auxiliary wire	Stringer or oscillation
Gas shielded-arc	GTAW	Manual or automatic	Wire or rod	Stringer or oscillation
	GMAW	Semiautomatic or automatic	Continuous electrode	Stringer or oscillation
	GMAW interrupted	Semiautomatic or automatic	Continuous electrode	Stringer or oscillation
	GMAW with auxiliary wire	Automatic	Continuous electrode & auxiliary wire	Oscillation
Plasma-arc	None	Automatic	Powder	Oscillation
	Hot wire	Automatic	Energized wires	Oscillation

recommended. Single-wire stringer beads can reasonably be expected to deposit about 14 lb/h at approximately 15 to 50% dilution using 350 amperes (A) with special techniques.

The oscillating technique described in Chapter 28, Section 2 can typically deposit 25 1b/h with a 25 to 40% dilution level. Beads as wide as 3.5 in. may be made. A variation of the oscillating method employs a small diameter (1/16 in.) wire operating on dcsp power to take advantage of resistance preheat and greater melting rate. This method will deposit about 15 lb/h with 20% dilution; deposits are about 1 in. wide (also see Section 2).

Multiwire Technique

Two or more wires electrically connected in parallel, fed at the same speed and operating in the same weld puddle, with or without oscillation, are sometimes used to increase deposition rate. The deposition rate can be increased considerably by adding more wires while maintaining the same dilution expected with a single wire.

Series Arc Technique

The series arc method also described in Chapter 24 and Chapter 44 will deposit about 25 to 50 lb/h with 10 to 20% dilution. Deposition rates on the high side of this range (35 to 50 lb/h) are obtained by feeding a cold wire into the arc. Beads are approximately 1 1/4 to 1 1/2 in. wide. A-C constant current power is preferred to achieve a uniform penetration pattern.

Strip Process

Using a strip electrode rather than a round wire, the submerged arc welding process is capable of depositing relatively thin, flat weld cladding at deposition rates of 60 to 100 lb/h or more. The technique works best when an auxiliary nonelectrode ("cold") strip is placed beneath the arc to provide additional filler metal. With this refinement the deposition rate is increased and the dilution reduced. Dilution can be held to between 10 and 15% with the cold strip addition, whereas 30 to 40% is typical with a single strip. Strip dimensions usually are 2 to 3 in. by 0.030 to 0.060 in. for the electrode and 1 5/8 in. by 0.050 to 0.060 in. for the cold strip.

The large cross-section area of the strip electrode permits currents as high as 1500 A to be used. Normal conditions call for approximately 1200 A, 32 V and a travel speed of 15 ipm, giving a weld cladding thickness of 3/16 in. Thickness can be varied between 5/32 and 3/8 in. by adjusting the travel speed and the feeding rate and dimensions of the cold strip. Savings are derived not only from higher deposition rates, but also from lower filler metal consumption. Flux consumption is reduced by two-thirds over conventional electrode work, and the low penetration of the process reduces dilution and permits a thinner weld cladding to be deposited.

A strip weld cladding and the equipment used are shown in Fig. 93.5. The cold strip in this case was not powered by its own drive rolls, but was held against the surface to be weld clad and fed into the weld puddle by the rotation of the vessel. With only minor equipment changes, metal powder can be substituted for the cold strip.

DCSP or dcrp is used for the welding current, although ac can be used. In any

Fig. 93.5.–A submerged arc welding process using coil strip electrodes.

case, a power source with a constant potential volt-ampere characteristic is most suitable.

Granular Metal Technique

This technique is a process wherein metal may be composed and deposited by passing an arc between a consumable electrode and metered granular metal.

In the submerged arc version of the process, the granular metal is metered on to the work and covered with flux; this is followed by the consumable electrode and arc which usually oscillate and melt all the granular metal with the electrode to produce the deposit.

As used for weld cladding, an alloy is produced and welded to the base. The amount of granular metal used bears a fixed relation to the electrode used and ranges from 1.5 to 3 times the weight of the wire. Most frequently, the electrode is mild steel and the granular metals supply the alloy and the other elements.

The process features deposition rates up to four times that possible with an electrode alone without increase in current. Penetration is controlled to give about 15% dilution since the arc does not contact the base and the available heat is used to melt metals for deposition. Deposition rates in excess of 100 lb/h may be obtained; special alloys in noncastable or nonworkable forms may be produced.

INERT GAS PROCESSES

Gas Metal-Arc Welding (GMAW)

This process, like submerged arc welding, is basically a high dilution process when operating in the spray-transfer mode, and when used for weld cladding can be expected to deposit 12 to 15 lb/h at dilution levels of about 30 to 50%. The technique employed—with or without oscillation—the gas or gas mixtures used and the weld cladding metal will directly affect dilution to a limited degree. For additional discussion of the process, see Chapter 23, Section 2.

Gas Metal-Arc Welding Auxiliary Wire

This method is similar to gas metal-arc spray transfer except that an auxiliary nonelectrode "cold" wire is fed into the arc in order to increase the deposition rate and reduce dilution. In this way, the deposition rate is increased to 20 to 30 lb/h while maintaining dilution to about 20%. For a discussion of auxiliary or cold wire feeder equipment, refer to Chapter 44, Section 3A.

Plasma-Arc Welding (PAW)

Two wires connected in series and energized by an a-c constant potential power source intersect in the puddle beneath a plasma-arc torch. The wires are heated by resistance to the point that without the torch, a molten bead would form without arcing. The plasma arc is constricted making it possible to produce smooth beads of uniform penetration and low dilution. The process is capable of deposition rates up to 60 lb/h, dilution rates under 10% and cladding thicknesses in one layer from 3/32 to 1/4 in.

WELD OVERLAY CLADDING FOR CORROSION RESISTANCE

Most weld overlay cladding on steel is used primarily for corrosion resistance and employs various types of austenitic stainless steels, nickel and nickel-base alloys such as nickel-chromium-iron and nickel-copper alloys or copper-base alloys such as copper-nickel. For further discussion of weld overlay cladding for corrosion resistance, refer to Chapter 44, Section 3A on surfacing and Section 4 on the individual metal alloys such as the austenitic stainless steels, the nickel-base alloys and the copper-base alloys.

METALLURGY OF WELD OVERLAY CLADDING

The successful deposition of an adequately corrosion-resistant cladding on a carbon or low-alloy steel surface requires a thorough understanding of the metallurgy of the steel base metal and of the cladding metal. The cladding operation involves a heterogeneous (dissimilar metal) weld between the ferritic base metal and the first layer of cladding, and more or less homogeneous welds between that layer and any succeeding layers of cladding.

The same metallurgical considerations which govern the soundness of stainless steel welds are important in cladding. The Schaeffler diagram or other modified diagrams can be used to select proper filler metals and estimate the ferrite content of deposited weld metal as a function of the chemical composition of the weld metal. The diagram in Fig. 93.6 illustrates the structure for various weld deposits on a 0.35% carbon steel (ASTM A 105, Grade 2). It will be remembered that deposits with fully austenitic structures are susceptible to micro-fissuring or hot cracking, and those which contain martensite are brittle. A sound, ductile overlay usually contains 3 to 15% ferrite.

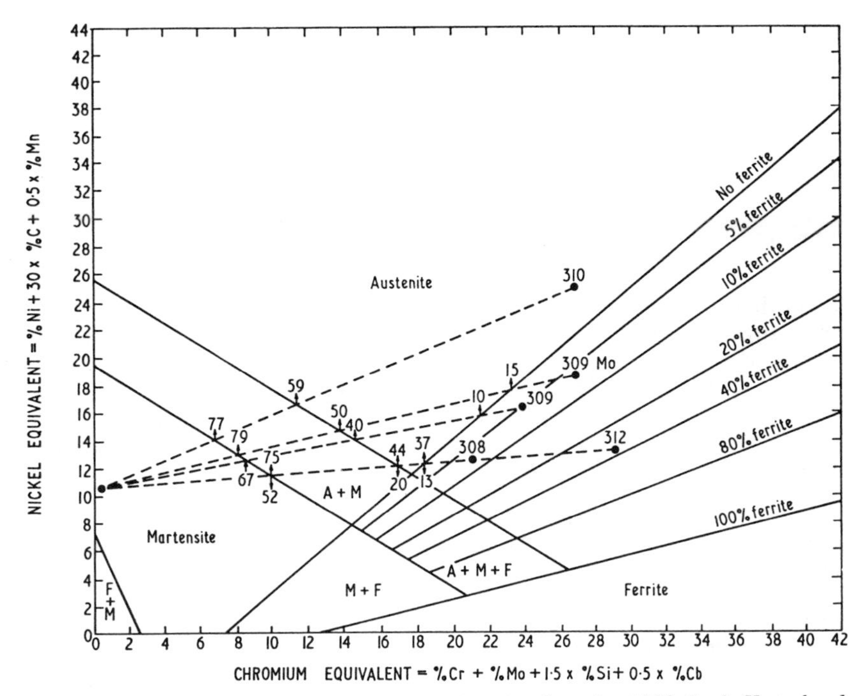

Fig. 93.6.–Schaeffler diagram showing dilution direction lines for A105 Grade II steel and various stainless steel electrodes.

The satisfactory maximum dilution precentages for each electrode composition shown in Fig. 93.6 are represented in Table 93.2. These data show the important influence of dilution on the characteristics of the deposit. The welding process, the welding procedure and the filler metal selection must be tailored to match the base metal being clad. In the diagram, an average composition was assumed for the weld deposit and for the base metal. These each have commercial ranges which must be considered along with the carbon level permitted in the deposit. Successful cladding operations depend on the control of the details involved in the process.

Other filler metals used to clad steels are governed by different criteria. Filler metals such as nickel-base alloys, cupronickel, etc., are essentially single-phase metals. The deposit soundness is dependent upon carbon, manganese or columbium coupled with close control of silicon, phosphorus and sulfur. In the case of the copper-bearing alloys, the iron content of the cladding must be kept low. Specific control levels of these filler metals can be determined by the supplier.

In stainless steel cladding, other metallurgical considerations apply which should be taken into account in process applications. These are the coefficient of expansion difference, the migration of carbon from the base metal to the weld metal and the formation of sigma phase These three conditions are important in considering stress-relief applications, as well as service at temperatures of 850 F or higher.

The coefficient of thermal expansion of austenitic stainless steel is about 1.5

Table 93.2—Dilution percentages for average composition weld metals on A105 Grade II forgings

Material	Austenite Boundary	Austenite & Martensite
E308	13	20
E312	37	44
E309	10	40
E310	–	59
E309 Mo	15	50

times that of carbon steel. Hence, each time the cladding is heated and cooled, shear stresses of yield strength magnitude are imposed on the fusion line of the weld.

Because of the alloy differences between the weld and base metal, diffusion of carbon occurs from the ferritic base metal to the austenitic weld metal. The rate of diffusion is a function of alloy content, temperature, phase relationships, etc. This diffusion occurs at temperatures of 1050 F and higher, the rate increasing with the temperature.

The formation of sigma phase also is associated with stress relief or elevated-temperature service. At an elevated temperature, ferrite may transform into sigma phase. Large percentages of sigma phase reduce corrosion resistance and ductility. It is good practice to subject samples of the cladding to the thermal treatments expected during fabrication to ensure that the resultant ductility is adequate. See Chapter 65, Section 4 for additional information.

HEAT TREATMENT OF CLAD METALS

Depending upon the metals involved, clad steels manufactured by processes utilizing high temperature may need to be heat-treated after bonding. Austenitic stainless clad steels are usually solution-annealed by heating to elevated temperatures and then air cooling, sometimes under fans. Certain types of backing steels will require a subsequent heat treatment such as normalizing, normalizing and tempering or austenitizing, quenching and tempering. Ferritic stainless clad steels are annealed by an austenitizing plus tempering treatment which is compatible with many types of backing steels. Generally, the final heat treatments given a clad plate will be determined by the requirements of the backing-steel specification and the applicable construction code.

In cladding processes such as weld overlay or explosion welding, the base metals normally will have been heat-treated prior to cladding and the only thermal treatment which may be required is a postweld stress-relief operation. If hot forming in fabrication is necessary, however, attention must be given to the final heat treatment of the steel base metal.

FABRICATING CLAD STEEL

FORMING

Generally, integrally-clad steels may be cold-formed to a degree limited by the ductility of the component metals rather than by the strength of the bond.

However, the fabricator planning to cold form a type of cladding with which he is unfamiliar would be well-advised to consult first with the manufacturer of the clad product. The same precaution applies in hot forming.

Clad steels are premium products normally requiring the same cleanliness, purity and smoothness of finish in the final surface as is required for the homogeneous corrosion-resisting metals. Therefore, in all stages of handling and fabrication, it is necessary to avoid scratching, gouging, impressing foreign material and contaminating the surface with spatter from welding or from oxygen cutting.

CUTTING

Clad plate is usually sheared with the cladding side up to avoid scratching the surface. Clad plate may be plasma-arc cut or oxygen cut with powder injection (except for copper clad). Stainless- and nickel alloy-clad plate (except for thicknesses under about 3/8 in.) having more than about 30% cladding may be oxygen cut from the base steel side without flux using large tips and low-oxygen pressures (see Table 93.3).

Table 93.3–Typical machine settings employed for the oxygen cutting of clad steel

Plate Gage, in.	Cutting Speed, ipm	Oxygen Cutting Pressure Range, lb	Approximate Cutting Orifice Diam, in.
3/16	22–25	5–7	0.046
1/4	19–21	6–8	0.055
3/8	16–18	7–10	0.063
1/2	15–17	7–12	0.073
5/8	14–16	8–13	0.082
3/4	13–15	9–14	0.096
1	12–14	10–16	0.111
1 1/2	10–12	12–18	0.128
2	9–11	12–18	0.147
2 1/2	7–9	16–24	0.169

In order to oxygen cut copper-clad steel, it is necessary to remove the copper by grinding or chiseling along the line to be cut. Copper clad may be plasma-arc cut without removing the copper.

CLEANING CLAD PLATE

Cladding surfaces may be cleaned by blasting with clean sand or other iron-free grit or by filling or scrubbing with a solution appropriate to the type of metal. Grease may be removed with detergents and loosely adherent metal by stainless steel wire brushes. Remaining particles of iron or iron oxide can be detected by keeping surfaces exposed to a humid atmosphere for 24 hours. Detailed information on examination and cleaning should be obtained from the manufacturer. ASTM Part 3, Standard A 380 prescribes cleaning methods for stainless steel.

WELDING CLAD STEEL PLATE

The making of joints in clad plate (regardless of the process used to manufacture the cladding) presents a problem in dissimilar metal welding, the complexity of which depends upon the type of cladding. Austenitic stainless clads, for example, are relatively easy to join, while titanium-clad steel requires very special procedures.

It is a cardinal rule in prescribing joint geometry and procedures for welding clad plate that steel weld metal shall not be allowed to melt in cladding metal. The alloys created by such admixture may be harder and less ductile at best and completely brittle and unsound at worst. On the other hand, the stainless steel, nickel, high-nickel alloy and some copper-base electrodes and rods used in industry are tolerant of dilution by steel to an extent that results are predictable and controllable. Table 93.4, 93.5 and 93.6 give shielded metal-arc and bare electrodes and rods used for stainless, nickel-base and copper-base alloy types of clad steel.

Where the clad steel joint is to be a composite weld—carbon-steel or low-alloy filler plus high-alloy or "pure" metal filler—the backing steel is usually welded first, and the joint is designed so that the first pass of carbon-steel weld metal will not penetrate into the cladding. As shown in Fig. 93.7, this may be done by (1) ensuring that a sufficient portion of the root face is backing steel or (2) beveling or stripping back the cladding. Alternatives, shown in Figs. 93.8 and 93.9, are to weld the entire joint with the high-alloy filler. The latter procedure simplifies edge preparation and back gouging operations at the expense of increased consumption of high-alloy weld metal.

Making the weld on the clad side of the joint involves consideration of dilution

Table 93.4—Electrodes and rods for the clad side of nonferrous clad steel

Cladding Type	For Covering Steel with First Pass or Layer		For Completing the Weld	
	Covered Electrode*	Bare Rod or Electrode**	Covered Electrode*	Bare Rod or Electrode**
Nickel & Low-Carbon Nickel	Eni-1	ERNi-3	ENi-1	ERNi-3
Nickel-copper	ENiCu-1, -2, -4	ERNiCu-7	ENiCu-1, -2, -4	ERNiCu-7
Nickel-chromium-iron	ENiCr-1 ENiCrFe-2	ERNiCrFe-5, -6	ENiCrFe-2, -3	ERNiCrFe-5, -6
Copper-Nickel				
70-30	ECuNi	ERNiCu-7	ECuNi	RCuNi***
90-10	ECuNi	ERNiCu-7	ECuNi	RCuNi***
Copper	ENiCu-1, -2, -4 ENi-1 ECuA1-A2	ECu, ERNiCu-7 ERNi-3 ECuA1-A2	NA NA NA	ECu, RCu*** ECu, RCu*** ECu, RCu***

* –See AWS Specification A5.11 or A5.6
** –See AWS Specification A5.14, A5.6 or A5.7
***–Gas Tungsten-arc Welding process

effects, ferrite content (of austenitic stainless steel welds) and other factors just as in weld overlay cladding. Refer to METALLURGY OF WELD OVERLAY CLADDING for a discussion of these factors.

Many specifications for joining clad steel require that the cladding be stripped back as in Fig. 93.7 (center) before welding to allow the entire thickness of the base plate to be welded with similar analysis filler metal. In this case, the clad weld is restored by making a weld overlay using the same techniques and processes employed in the original cladding operation. In practice, special equipment designed for weld overlay cladding will not always be available for making joints in clad steel, thus special techniques may have to be relied upon to minimize dilution. The following measures can be taken:

1. Use small diameter electrodes, and stringer beads directing the arc against the previously deposited bead.
2. Use richer alloy electrodes than that of the cladding.
3. Allow for extra layers of weld and grind away part of the first layer if necessary.
4. In automatic welding, use oscillation.
5. Where possible, keep the arc on the molten puddle as the weld is advanced.

Table 93.5—Electrodes and rods for the clad side of 300 series stainless clad steel

Cladding Type	For Covering Steel with First Pass or Layer		For Completing the Weld	
	Covered Electrode*	Bare Rod or Electrode**	Covered Electrode*	Bare Rod or Electrode**
304	E309	ER309	E308	ER308
304L	E309L[2] E309Cb	ER309L[1, 2]	E308L	ER308L
321	E309Cb	ER309Cb[2]	E347	ER347
347	E309Cb	ER309Cb[2]	E347	ER347
309	E309[1]	ER309[1]	E309	ER309
310	E310[1]	ER310[1]	E310	ER310
316	E309Mo	ER309Mo[2]	E316	ER316
316L	E309MoL[1, 2] E317L[1]	ER309MoL[1, 2] ER317L	E316L	ER316L
317	E309Mo E317[1]	ER309Mo[1, 2] ER317[1]	E317	ER317
317L	E309MoL[1, 2] E317L[1, 2]	ER309MoL[1, 2] ER317L[1, 2]	E317L[2]	ER317L[2]
20 Cb	E320[1]	ER320[1]	E320	ER320

* –Refer to AWS Specification A5.4.
** –Refer to AWS Specification A5.9.

Notes:
1. Two layers of weld deposit should be used. On light gages, it may be necessary to grind off part of the first layer to make room for the second layer.
2. Not a standard classification.

Table 93.6–Electrodes and rods for the clad side of 400 series stainless clad steel

Cladding Type	For Covering Steel with First Pass or Layer: Covered Electrode*	For Covering Steel with First Pass or Layer: Bare Rod or Electrode**	For Completing the Weld: Covered Electrode*	For Completing the Weld: Bare Rod or Electrode**
405	ENiCrFe-2[1] ENiCrFe-3[1] E309[1] E310[1] E430[2]	ERNiCrFe-5[1] ERNiCrFe-6[1,5] ER309[1] ER310[1] ER430[2]	ENiCrFe-2[1] ENiCrFe-3[1] E309[1] E310[1] E430[2,3] E410[2,3]	ERNiCrFe-5[1] ERNiCrFe-6[1,5] ER309[1] ER310[1] ER430[2,3]
410	ENiCrFe-2[1] ENiCrFe-3[1] E309[1] E310[1] E430[2]	ERNiCrFe-5[1] ERNiCrFe-6[1,5] ER309[1] ER310[1] ER430[2]	ENiCrFe-2[1] ENiCrFe-3[1] E309[1] E310[1] E410[2,3] E430[2,3]	ERNiCrFe-5[1,3] ERNiCrFe-6[1,3,5] ER309[1,3] ER310[1,3] ER410[2,3]
410S	ENiCrFe-2[1] ENiCrFe-3[1] E309[1] E310[1] E430[2]	ERNiCrFe-5[1] ERNiCrFe-6[1,5] ER309[1] ER310[1] ER430[2]	ENiCrFe-2[1] ENiCrFe-3[1] E309[1] E310[1] E430[2,3]	ERNiCrFe-5[1,3] ERNiCrFe-6[1,3,5] ER309[1,3] ER310[1,3] ER410[2,3]
429	ENiCrFe-2[1] ENiCrFe-3[1] E309[1] E310[1] E430[2]	ERNiCrFe-5[1] ERNiCrFe-6[1,5] ER309[1] ER310[1] ER430[2]	ENiCrFe-2[1,3] ENiCrFe-3[1,3] E309[1] E310[1] E430[2,3]	ERNiCrFe-5[1] ERNiCrFe-6[1,5] ER309[1] ER310[1] ER430[2,3]
430	ENiCrFe-2[1] ENiCrFe-3[1] E309[1] E310[1] E430[1]	ERNiCrFe-5[1] ERNiCrFe-6[1,5] ER309[1] ER310[1] ER430[1,4]	ENiCrFe-2[1,3] ENiCrFe-3[1,3] E309[1,3] E310[1,3] E430[1,3,4]	ERNiCrFe-5[1,3] ERNiCrFe-6[1,3,5] ER309[1,3] ER310[1,3] ER430[1,3,4]

* –Refer to AWS Specification A5.4 or A5.11.
**–Refer to AWS Specification A5.9 or A5.14.

Notes:
1. Welding on plate colder than 50 F is not advisable.
2. Preheat of 200 F minimum is advisable, particularly on plate over 1/2 in. in thickness.
3. Stress relief at 1100–1350 F (compatible with backing steel) is desirable.
4. Type 430 electrodes and rods may not deposit 16% minimum Cr.
5. Age-hardenable–consult supplier.

When the cladding in composite joints is not stripped back prior to welding, the backing will have narrower, deeper grooves on the clad side after back gouging. This type of joint is advantageous because more layers of the cladding weld metal can be put in, reducing dilution in the final layer. One hazard which should be guarded against when the no-strip-back geometry is used (top of Fig. 93.7) is possible burn-thru by submerged arc welding or other deeply penetrating processes. Careful control or use of barrier layers deposited by low-penetration processes is necessary to avoid pickup of cladding metal by the backing-steel weld.

In making the first pass on the backing-steel side of joints in which the

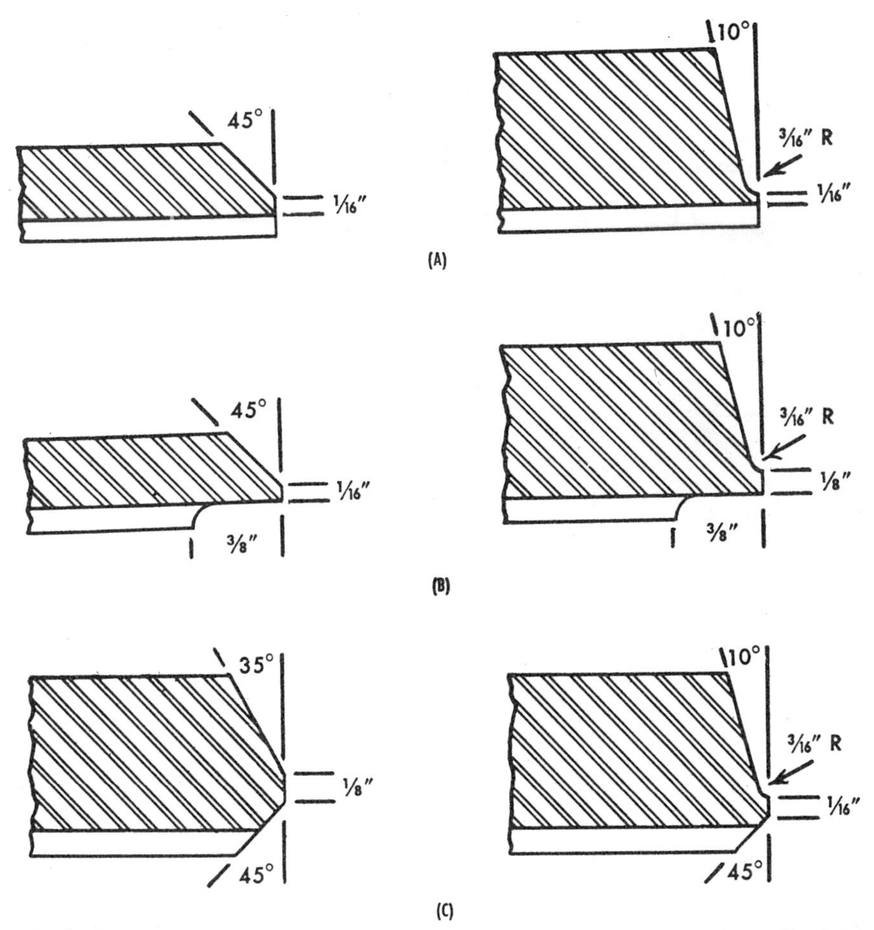

Fig. 93.7.–Plate edge preparation for butt welding clad steels: (A) single-Vee or Vee-joint with a carbon steel lip (B) Stripped back joint (C) Double-bevel joint.

cladding forms part of the root face, the best practice dictates use of low-hydrogen electrodes where the shielded metal-arc welding process is to be used. The low-hydrogen electrodes do not penetrate deeply, and the deposit is less likely to crack if some cladding metal is inadvertently melted. It should be noted that while some dilution from the austenitic stainless steels can be tolerated, even small amounts of dilution by such metals as the copper or copper alloys must be avoided. In some structures employing clad plate, partial penetration welds are adequate. As illustrated in Fig. 93.10, such welds simplify control of dilution. In addition, some engineers consider that an austenitic steel weld penetrating deeply into ferritic steel backing could cause thermal fatigue under conditions of thermal cycling.

Welding Copper Clad Steel

Copper in clad plate or sheet form will usually be one of the deoxidized or oxygen-free weldable grades. The best method of welding copper, the cupro-

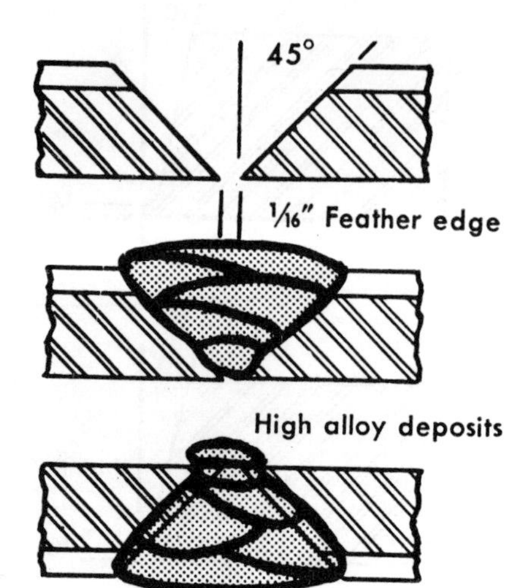

Fig. 93.8.–Full alloy welding of light-gage clad steel using a reserve bevel plate edge preparation.

nickels or bronzes is by using one of the gas-shielded arc welding processes. The joints used are the same as those employed for welding the other types of clad metals (Figs. 93.11 and 93.12). For cladding copper thicknesses over 1/8 in., a preheat temperature of about 300 F will be required for consumable electrode wires smaller than 1/16 in. diam. Above 3/16 in., preheat requirements will increase rapidly to about 800 F, and dilution by the base steel will increase. As a result, it may be necessary to remove a part of the first weld by grinding after the steel is covered so that dilution can be reduced in subsequent layers.

Where the cladding is less than 3/16 in. thick, copper welds may be deposited directly on steel with very low-iron pickup if care is used. The semiautomatic gas metal-arc welding process, in which a backhand motion is used with the arc kept on the cladding and the puddle at all times, will give first-layer iron contents of less than 5%.

It will frequently be advantageous to cover the steel with a barrier layer of nickle-copper or nickel prior to making the copper weld. This procedure is particularly recommended for cupronickel clads, since nickel-copper and nickel are more tolerant of iron dilution than deoxidized copper wire or cupronickel wires (Table 93.4). Furthermore, barrier layers prevent copper penetration of steel grain boundaries which would cause cracking.

Blind Joints

Where joints must be made entirely from the backing-steel side, it is necessary to use a weld metal for the backing-steel portion which will be compatible with the cladding weld and the steel. For stainless steel clads, an alternative requiring great caution, is to make a low-carbon, low-hydrogen type ferritic deposit directly on the stainless weld. Small diameter ingot iron electrodes and the GMAW short-circuiting arc process have been used to make this transition layer between the high-alloy weld and a carbon or low-alloy steel deposit. (See the pertinent bibliographical entry.)

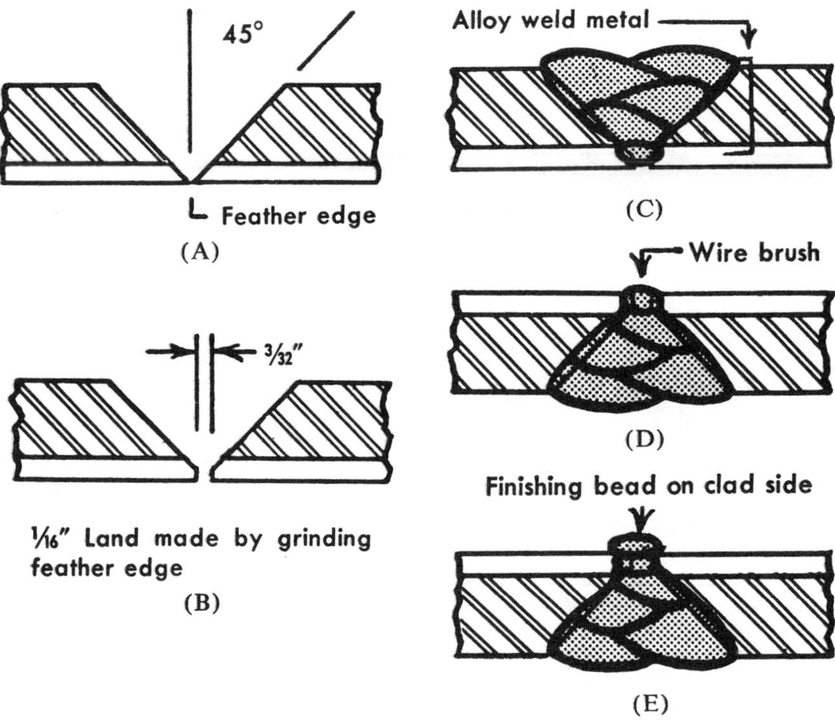

Fig. 93.9.–Standard procedure for full alloy welding of light-gage clad steel.

CODES AND SPECIFICATIONS FOR CLAD STEEL

The American Society for Testing and Materials and the American Society of Mechanical Engineers issued these standards for clad steel plate: A (SA) 263–chromium stainless types, A (SA) 264–chromium-nickel stainless types, A (SA) 265–nickel and nickel-alloy types and B 452–copper and copper-alloy types.

The ASME Pressure Vessel Code, Section VIII, Division I and Section III give rules for clad steel construction, including quality assurance procedures. NAVSHIPS 250-1500-1 and MIL-STD 271 give rules for fabrication, inspection and nondestructive testing.

SHEET OR STRIP LINING

DESCRIPTION AND MATERIALS

This type of construction is usually used for pressure vessels and tanks in the process industries such as the chemical, petrochemical, oil refining, paper and

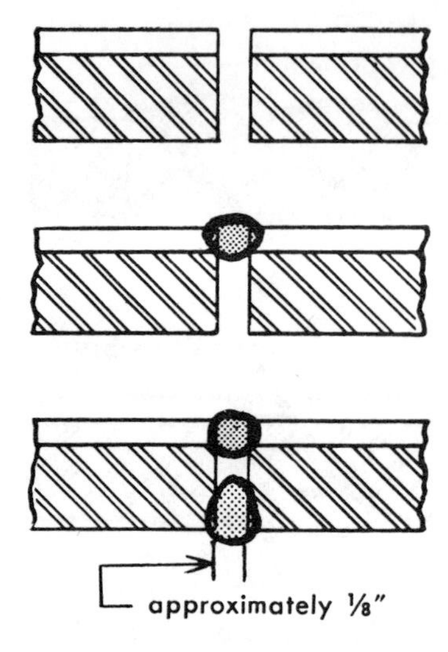

Fig. 93.10.–Full alloy butt welding of light-gage clad steel using a square edge preparation.

mining industries. Design of lined vessels, including the base steel specification, is commonly based upon ASME, API or other codes.

The metal used for liners includes a wide variety of stainless and nonferrous metals in sheet or strip form. The dimensions of liner sheets depend largely upon the method of attachment. Widths from 24 to 48 in. are commonly used for sheets which are to be attached by resistance welding. Lengths may vary between 3 and 12 ft. The dimension is also determined by the type of alloy used and the sizes in which the liner sheets are normally available from mills. The thickness also depends upon the service life required and the type of metal used. In general, thickness varies from 1/16 to 3/16 in.; the most commonly used thicknesses are from 5/64 to 7/64 in. Liner metal thinner than 1/16 in. is seldom used, except for such expensive metals as silver, platinum, tantalum and titanium. The lighter gages are harder to weld and are therefore seldom used except when the greater thicknesses are prohibitive in cost.

FILLER METAL

Electrodes for joining liners to steel are selected with due consideration of dilution and other factors. (Refer to the section on welding clad steel plate under FABRICATING CLAD STEEL.)

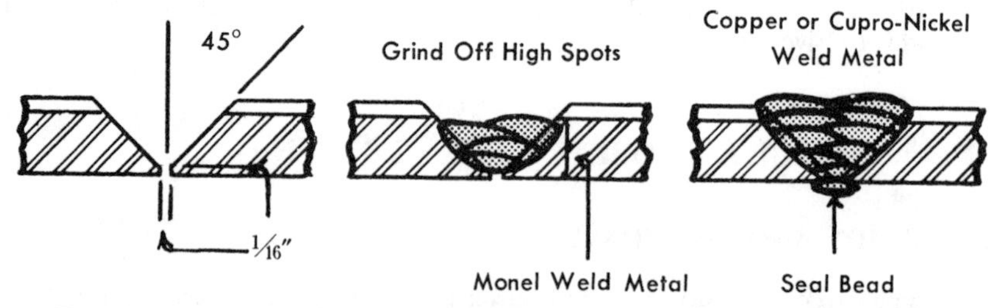

Fig. 93.11.–Welding copper or cupronickel clad plate in gages from 3/16 to 3/8 in.

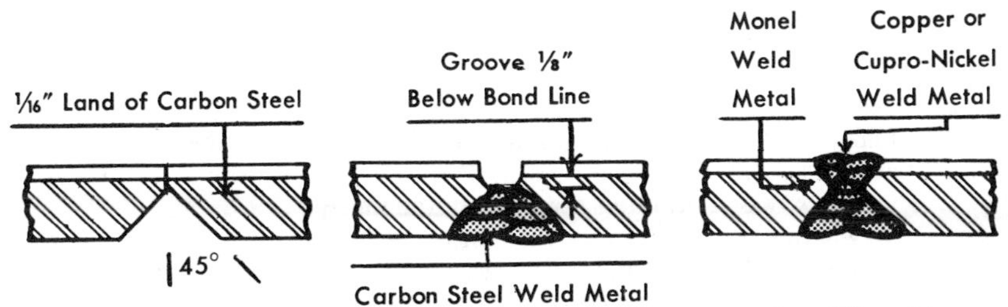

Fig. 93.12.–Welding copper or cupronickel clad plate in gages from 3/8 to 1 in.

LINER DESIGN

A wide liner design variation is possible when spacing attachment welds. At one extreme would be the vessel with no attachment welds at all. The inner layer of the corrosion-resistant metal would be placed inside the vessel in as large a sheet as can be obtained commercially, and the edges of the sheets would be attached to the vessel with no intermediate welding. The other extreme would be nearly 100% attachment by overlapping welds such as could be accomplished with some resistance welding processes. The important factors determining the spacing of attachment welds are service conditions, thickness of lining, differences in thermal expansion rates between the liner and the base plate and economy of construction.

Impairment in corrosion resistance is generally restricted to the weld zone. Dilution of the weld metal or weld nugget by pickup from the base metal may result in a composition with reduced corrosion resistance. In arc welding, the effect of dilution may be overcome by using electrodes of higher alloy content. Where this is not possible, the effect may be minimized by applying the weld metal in two or more beads, as shown in Fig. 93.13b. In resistance welding, dilution may be kept from the surface of the lining by careful control of the welding operation. The corrosion resistance of many lining metals may also be impaired by cracking in the weld or in the heat-affected zone. The incidence of cracking may be reduced by specifying a wider spacing between attachment welds, if temperature and the difference in thermal expansion of the liner and base plate are the cause of the cracks.

Austenitic steel and many nonferrous alloys have higher rates of thermal expansion than the mild steel base plate metal. The points of attachment of the alloy liner will conform to the expansion of the thicker base plate. The liner can do this by compressing, buckling or bulging. Compression is promoted by close spacing of attachment welds. It begins as elastic compression, but, for any considerable rise in temperature, transforms into plastic flow. Buckling occurs with wide attachment spacing. The liner is elastic at the wider spacings, and, at intermediate spacings, buckling may be accompanied by plastic flow. Repeated buckling during service is likely to result in fatigue cracking in the liner, its attachment weld or heat-affected zone. Generally, the attachment spacing must be closer where wide fluctuations in service temperatures or pressures are encountered. Often the spacing must be as close as 3 to 4 in. However, no specific rules for spacing can be detailed since each case must be given special consideration.

PRE-LINING FLAT SURFACES

Special resistance-welding equipment is used for the spot or seam welding of flat surfaces. A typical spot-welding machine is shown in Fig. 93.14. This machine makes two spot welds simultaneously with electrodes connected in series. Spacing is from 3 to 12 in. apart. The resistance-seam welder, shown in Fig. 93.15, makes two continuous seams simultaneously.

Arc-seam welding, sometimes called "plow welding," is usually done with automatic submerged arc or automatic gas metal-arc welding equipment. The weld is controlled to penetrate through the liner to the base steel.

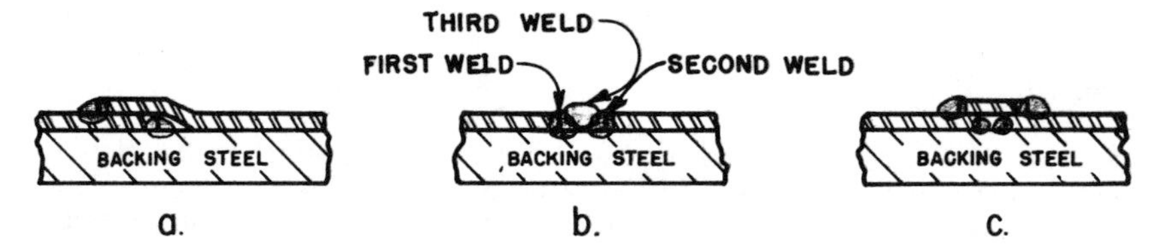

Fig. 93.13.–Typical joints used for connecting liners.

ATTACHMENT OF LINERS

Narrow strips may be attached only at the joints. Larger liner panels are attached between joints by spot-, seam or plug welds. Plug welds are made by filling holes in the liner against the base steel. Spot welds may be made by the GTAW and GMAW, as well as by resistance welding processes. Liners up to 1/16 in. thick need no edge preparation. Thicker linings may be pre-beveled by chipping or grinding in place. Seam welding by the submerged arc welding process is illustrated in Fig. 93.16.

METHOD OF STRIP LINING

Almost any weldable metal available in sheet or strip form can be applied as a strip lining. The length of the strips may vary from 24 to 40 in. for field-applied liners and from 24 to 120 in. for shop-applied liners. The widths of the strips depend to a great extent upon service temperatures and may vary between 2 and 6 in. or more. The narrower widths are used for the higher temperatures.

Strip linings may be applied to vessels that have been in service, as well as to new vessels. Field attachment is usually performed manually. Welding procedures for strip linings are described in the preceding section and shown in Fig. 93.12.

In field lining, it is extremely important that the liner be pressed against the shell by some mechanical means to ensure full contact. In some cases, it is necessary, for small diameters, to roll the liner plates in the shop to the proper curvature before applying in the field. If possible, on field-erected vessels, all oxidation and scale should be removed from the base metal, either by wire brushing or sandblasting. Sandblasting is the most satisfactory method to reveal any corrosion pits or defects in the base plate. To minimize the amount of welding, the joints between the strips should be kept as narrow as possible.

Fig. 93.14.–Typical spot welding machine used for applying liners.

However, they should be far enough apart so that the two strips can be attached to the base metal and then covered with a final bead (Fig. 93.13b).

INSPECTION OF SHEET AND STRIP LINING

The method of testing liners applied in the field should be agreed upon between the purchaser and the fabricator.

Pressure Testing

A pressure test behind the liner, using air, nitrogen or other fluid, is easy to make and is widely used. Pressure is applied to the liner through a hole drilled in the liner or the base plate. If the hole is in the liner, other test methods must be used on the weld that closes up the hole. A soap solution is commonly used to detect air leaking through the liner or the welds. The test pressure has to be low enough so that the air will bubble through the soap solution rather than blow it away.

The hydrostatic test required by the ASME Boiler and Pressure Vessel Code is sometimes used as a test of the liner. One hydrostatic testing method employs oil which is steamed out of the vessel at the conclusion of the test. In this test, back seepage or bleeding of oil from the back of the liner shows up visibly. Another method that has been used is to provide weep holes through the base metal behind each panel, and then test with water. If water is evident at the weep holes during the hydrostatic test, the leaks can be located with an air and soap solution test applied from the other side.

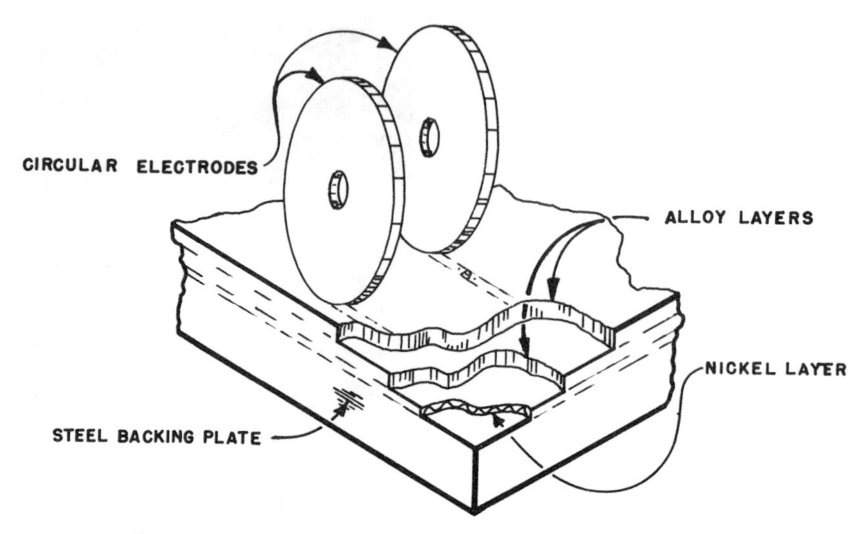

Fig. 93.15.–Multiple alloy liner sheets beings attached by seam welding.

Fig. 93.16.–Strip lining being applied to a vessel.

The pressure test appears to be simple to use and reasonably certain to detect any leaks that may be present. In practice, it may fall short of this. The difficulty results from the application of pressure behind the liner. This reverse pressure alters the configuration of the liner and may close up leaks during testing that show up under operating conditions. For the same reason, it fails to locate leaks that show up under the hydrostatic test.

After the hydrostatic test, care should be taken that all moisture is removed from behind the lining before the test plugs are closed. This can be accomplished by heating the lining with a torch or, in the case of small or medium-sized pressure vessels, inserting them in a stress-relieving furnace and heating to a temperature above 212 F.

Other Inspection Methods

Leakage of the liner will occur principally in the welds and heat-affected zones. Magnetic particle, dye-penetrant or fluorescent-penetrant inspection is used to examine inner welds. Radiography is specified by the ASME Boiler and Pressure Vessel Code. The halogen sniffer test is also used.

BIBLIOGRAPHY

"Submerged Arc Welding and Surfacing," L. Van Dyke and G. Wittstick, *Welding Journal*, 51 (5) 317–325 (1972).

"Plasma-Arc Hot-Wire Surfacing," E. C. Garrabrant and R. S. Zuchowski, *Welding Journal*, 48 (5) 385–395 (1969).

"Corrosion-Resistant Weld Overlays by the Dual Strip Process," R. D. Thomas, *British Welding Journal*, 13 (May, 1966).

"Cladding and Overlay Welding with Strip Electrodes," H. C. Campbell and W. C. Johnson, *Welding Journal*, 45 (5) 399–409 (1966).

"Nickel Cladding with Strip Electrodes," G. Almquist and N. Egeman, *Welding Journal*, 45 (5) 275–283 (1966).

"Stainless Weld Overlays–Engineering the Variables," H. N. Farmer, Jr., *Fusion Facts*, 21, 112–118 (Winter, 1965–66).

"Multi-Cathode Gas Tungsten-Arc Welding," J. E. Anderson and D. M. Yenni, *Welding Journal*, 44 (7) 327s–331s (1965).

"Ultrasonic Inspection of Welds," B. Otrofsky, *Welding Journal*, 44 (3) 97s–105s (1965).

"Weld Deposited Cladding of Pressure Vessels," R. D. Wylie, J. McDonald, Jr. and A. L. Lowenberg, *British Welding Journal*, 378–393 (August 1965).

"Bonding of Metals with Explosives," A. H. Holtzman and G. R. Cowan, *Welding Research Council Bulletin*, No. 104, (April 1965).

"Overlay Welding for Mammoth Reactor," A. Ujiie, S. Sato, J. Nagata, T. Nagaoka and J. Kobayashi, Technical Review, Kobe Technical Institute, Mitsubishi Heavy Industries, Ltd. (May, 1965).

"Strip Cladding Speeds Stainless Overlay," James E. Norcross, *Welding Engineer* (October, 1965).

"Strip Electrodes Solve Cladding Problem," Goran Alquist and Nils Egeman, *Welding Engineer* (October, 1965).

"Properties of Tubular Stainless Steel Electrodes," H. N. Farmer, Jr., *Welding Journal*, 43 (8) 667–678 (1964).

"Clad Plate Production by Explosion Bonding," A. Pocalyko and C. P. Williams, *Welding Journal*, 43 (10) 854–861 (1964).

"New Development In Braze-Cladding Research," R. C. Bertossa, *Western Machinery and Steel World* (1963).

"Bulk Process Welding," R. F. Arnoldy, *Welding Journal*, 42 (11) 885–891 (1963).

"Plasma Arc Weld Surfacing," R. S. Zuchowski and R. P. Culbertson, *Welding Journal*, 41 (6) 548–555 (1962).

"Repair of Kraft Digesters by Welded Overlay," W. L. Wilcox and H. C. Campbell, *Welding Journal*, 40 (8) 839–844 (1961).

"Use of Inconel Deposited Weld Metal for Nuclear Component Parts," R. W. Minga and

W. H. Richardson, *Welding Journal*, 40 (7) 726–735 (1961).
"New Techniques for Cladding with the Gas Shielded Process," R. D. Engel, *Welding Journal*, 39 (12) 1222–1229 (1960).
"Interpretative Report on Welding of Nickel Clad and Stainless Clad Steel Plate," W. H. Funk, *Welding Research Council Bulletin,* No. 61, Welding Research Council of the Engineering Foundation (June, 1960).
"Suggested Procedures for Flame Cutting Clad Steels," Lukens Steel Co. (1959).
"Development of Titanium-Clad Steel Using a Vacuum-Brazing Process," D. Canonico and H. Schwartzbart, *Welding Journal*, 38 (2) 71s–77s (1959).
"Applications of Explosion Bonding," J. J. Douglass, E. I. Dupont De Nemours & Co., Explosives Dept., Metal Cladding Section, Wilmington, Delaware 19898.
"Unfired Pressure Vessels," ASME Boiler and Pressure Vessel Code, Section VIII; "Material Specifications," Section II (1959).
"Peripheral Welding of Internally-Clad Steel for Nuclear Reactor Application," W. Leonard and J. C. Thompson, Jr., *Welding Journal*, 36 (3) 243–251 (1957).
"Bonded Fluxes for Submerged-Arc Welding of Alloy Steels," H. C. Campbell and W. C. Johnson, *Welding Journal*, 36 (11) 1078–1084 (1957).
"The Inert-Gas Metal-Arc Overlay Process," C. B. Felmley, *Welding Journal,* 34 (6) 542–550 (1955).
"Stainless Clad Steels," H. Thielsch, *Welding Journal,* 31 (3) 142s–159s (1952).
"Stainless-Steel Applied Liners," H. Thielsch, *Welding Journal,* 31 (7) 321s–337s (1952).

CHAPTER 94

FILLER METALS

PREPARED BY A COMMITTEE CONSISTING OF:

G. R. CRAWMER, *Chairman*
General Electric Co.

H. S. AVERY
Abex Corporation

H. C. CAMPBELL
Arcos Corporation

L. J. CHRISTENSEN
Chicago Bridge & Iron Co.

P. B. DICKERSON
Alcoa Process Development Lab.

J. HINKEL
Lincoln Electric Co.

R. B. HITCHCOCK
E. I. Dupont DeNumours & Co.

P. A. KAMMER
The Mckay Co.

A. I. KAYSER
Weapons Engineering Standardization Office

L. F. LOCKWOOD
Dow Chemical Co.

A. E. WIEHE
Hobart Brothers Co.

CHAPTER 94

FILLER METALS

INTRODUCTION

The development, acceptance and classification of filler metals are taking place continuously. This chapter attempts to compile in synopsis form what is known about the many filler metals in use. Major emphasis has been placed on those widely accepted filler metals which are classified under an American Welding Society Specification. However, other filler metals and specifications are mentioned where applicable.

Although each of the following sections differs in format to suit the characteristics of the filler metal discussed, each section provides basic application information for the Handbook user under the general topics of selection, ordering of materials, storage and handling, and use.

In providing this information, there is naturally some overlapping as relates to the base metal weldability and welding process sections of the Handbook and AWS specifications. In fact, portions of some AWS specifications may be found in this chapter verbatim. However, the chapter is not intended to simply reproduce these sections and specifications, but rather to provide the Handbook user with one convenient reference on filler metals.

The particular procedure used to deposit a filler metal may vary the deposit properties over a rather wide range. Thus, all values given in the chapter should be treated as representative rather than absolute values. The actual properties of any particular weld made with a given filler metal may be predicted only by tests and strict adherence to proven fabrication procedures.

USE OF FILLER METALS

The use of welding and brazing filler metals liberates fumes and gases the extent of which depends on the process and procedure used to deposit the filler metal. Thus, adequate ventilation when using filler metals has always been necessary. To alert the user to this necessity, filler metals supplied under AWS specifications are now required to bear a cautionary or warning label, as applicable, on the package. The appearance of the precautionary label should not be interpreted as an indication of any change in the health hazard from welding or brazing fumes. Rather, its purpose is to promote safe welding practices.

PACKAGING OF FILLER METALS

Some filler metals last indefinitely in storage while others have rather distinct shelf lives. When unusual environmental or extra long time storage conditions occur, the manufacturer can, when requested by the purchaser, use special packaging methods to minimize deterioration such as described in MIL-W-1043C.

FILLER METAL SPECIFICATIONS AND IDENTIFICATION

Essentially all filler metals discussed in this chapter are classified according to one of the following AWS specifications:

Mild Steel Covered Arc-Welding Electrodes, A5.1
Iron and Steel Gas Welding Rods, A5.2
Aluminum and Aluminum Alloy Arc-Weiding Electrodes, A5.3
Corrosion-Resisting Chromium and Chromium-Nickel Steel Covered Welding Electrodes, A5.4
Low-Alloy Steel Covered Arc-Welding Electrodes, A5.5
Copper and Copper-Alloy Arc-Welding Electrodes, A5.6
Copper and Copper-Alloy Welding Rods, A5.7
Brazing Filler Metal, A5.8
Corrosion-Resisting Chromium and Chromium-Nickel Steel Welding Rods and Bare Electrodes, A5.9
Aluminum and Aluminum Alloy Welding Rods and Bare Electrodes, A5.10
Nickel and Nickel-Alloy Covered Welding Electrodes, A5.11
Surfacing Welding Rods and Electrodes, A5.13
Nickel and Nickel-Alloy Bare Welding Rods and Electrodes, A5.14
Welding Rods and Covered Electrodes for Welding Cast Iron, A5.15
Titanium and Titanium Alloy Bare Welding Rods and Electrodes, A5.16
Bare Mild Steel Electrodes and Fluxes for Submerged-Arc Welding, A5.17
Mild Steel Electrodes for Gas Metal-Arc Welding, A5.18
Magnesium-Alloy Welding Rods and Bare Electrodes, A5.19
Mild Steel Electrodes for Flux-Cored Arc Welding, A5.20
Composite Surfacing Welding Rods and Electrodes, A5.21

It is the purpose of each of these specifications to establish a filler metal classification based on the mechanical properties and/or chemical composition of each filler metal. These specifications also set forth the conditions under which the material should be tested. For some materials, radiographic standards of acceptability are given, and usability tests are also required by some of these specifications. Brand name listings for the materials covered by these specifications are shown in AWS 5.0, Filler Metal Comparison Charts.

It is important to read the paragraphs on manufacture and acceptability in all AWS filler metal specifications which state, "The electrodes may be made by any method yielding a product conforming to the requirements of this specification. At the option and expense of the purchaser any or all of the tests required by this specification may be used as a basis for acceptance of electrodes."

The American Welding Society specifications require that all covered electrodes be identified by imprinting the AWS classification number on the coating as shown in Fig. 94.1. Coils of wire and straight welding or brazing rods are required to be identified by marking the classification number on the containers. Color coding of electrodes is no longer required by AWS; however, some manufacturers still follow this practice and many users continue to employ this as a means of identifying electrodes.

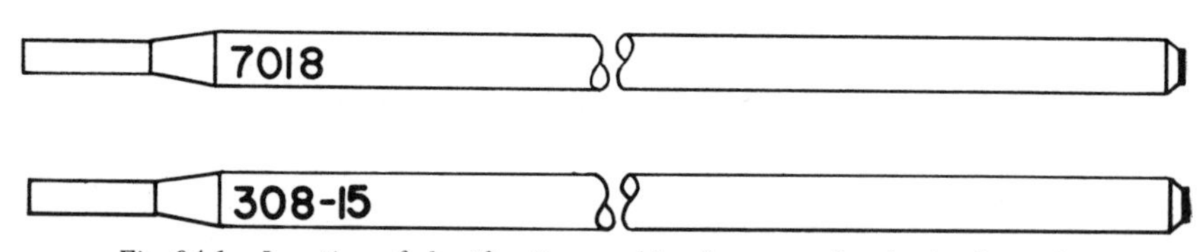

Fig. 94.1.–Location of classification marking for covered and grip electrodes.

AWS CLASSIFICATION SYSTEM

The AWS numerical system provides a means of identifying all electrodes, filler metals, rods, etc. Within the classification system, a clue to the identity of any product can be found in the prefix letter or the classification number.

E, R, B and RG Prefix Letters

In their simplest form, these prefix letters signify the following:

1. E indicates an arc-welding electrode.
2. R indicates a welding rod as distinguished from electrodes or brazing filler metals.
3. B indicates a brazing filler metal.
4. RG indicates a welding rod to be used expressly with gas.

Typical E-prefix products are E6010, AWS A5.1; E8018, A5.5; and E308-16, AWS A5.4.

The letter R at the beginning of each number indicates a welding rod, as distinguished from electrodes and brazing filler metals. An RB at the beginning indicates that the filler metal is suitable as a welding rod and as a brazing filler metal. RG indicates a welding rod to be used in gas welding.

The following products illustrate the use of these prefixes: RCuSn-A, AWS A5.6; RBCuZn-A, AWS A5.7 and RG 65, AWS A5.2.

The B prefix indicates a filler metal for brazing. As indicated above, RB indicates that the filler metal is suitable as a welding rod and as a brazing filler metal. Examples include BAlSi-2, AWS A5.8 and RBCuZn-A, AWS A5.7.

The ER Prefix

The letter E at the beginning of each number indicates an electrode, and the R indicates a welding rod. In the case of filler metals which may be used in submerged arc welding, gas metal-arc welding, atomic hydrogen welding and gas tungsten-arc welding, both letters are used. Examples include ER308, AWS A5.9 and ERTi-1, AWS A5.16.

One exception to the above is the specification for Aluminum and Aluminum Alloy Arc-Welding Electrodes, AWS A5.3, which does not follow the system used in the other filler metal specifications. This classification consists of Al-2 or Al-43 which indicates a covered electrode whose chemical composition is equivalent to Aluminum Association alloy designations 1100 and 4043 respectively.

MILD STEEL AND LOW-ALLOY STEEL FILLER METALS

GENERAL

Despite differing trade names, filler metals which meet the requirements of any given AWS classification may have major characteristics that are similar if not identical. Certain minor variations continue to exist among brands due to differences in production facilities and the usual variety of opinion that exists in any field concerning specific operating characteristics. The more important characteristics of each classification are discussed later.

In weldments to be heat-treated, depositing metal having a heat-treated response comparable to that of the base metal would be preferred. Some alloy steels have compositions that possess special properties, such as corrosion resistance. When welding these steels, an attempt should be made to obtain deposited weld metal having similar properties. Deposited weld metal of the same composition as the base metal will usually meet such special requirements; however, some alloying elements in the steel cannot be effectively transferred across the arc. It then becomes necessary to substitute another element that can be satisfactorily deposited.

Because of the great variety of compositions available, it is very important to make a careful selection of low-alloy steel electrodes where properties other than mechanical properties are required. Since the characteristics of various alloy steels are generally known, it is possible to predict the characteristics of weld metal of a given composition e.g., alloys containing molybdenum have increased creep-resistant properties, alloys containing chromium have improved oxidation resistance and alloys containing nickel have improved low-temperature notch toughness.

Many weldments made with low-alloy steel filler metals are stress-relieved[1]

[1] It is preferable to refer to this process as postweld heat treatment rather than stress relief.

before being placed in service. Therefore, the specifications often define stress-relieved (at a specific temperature) mechanical properties of the weld metal. The as-welded deposits will have a higher strength with lower ductility than the stress-relieved deposits. Recommendations for preheat, interpass temperatures and postweld heat treatments for weldments using mild steel and low-alloy steel filler metals are given in Chapters 61 and 63, Section 4.

Impact properties of the weld deposit are dependent both on the chemistry of the electrode and the heat input (which influences the cooling rate). Where specific impact requirements must be met, it is important to duplicate the conditions anticipated on the production weld to get reliable service results.

COVERED ARC-WELDING ELECTRODES

Two separate specifications, Mild Steel Covered Arc-Welding Electrodes (AWS A5.1) and Low-Alloy Steel Covered Arc-Welding Electrodes (AWS A5.5) cover these electrodes. The former specifications cover electrodes of the E60XX and E70XX series while the latter deals with alloyed electrodes of the E70XX through E120XX series of classifications. Many general considerations for each specification are similar.

The electrodes described in the specifications are grouped according to (1) operating characteristics (2) type of covering and (3) properties of the deposited metal. Operating characteristics and types of covering and the mechanical and chemical requirements are described in detail in the specifications.

The classification system used in the specifications follows the established pattern. The letter E designates an electrode; the first two digits (or three digits of a five-digit number), 70, for example, designate the nominal minimum tensile strength of the deposited metal in 1000 psi. The third digit (or fourth digit of a five-digit number) indicates the position in which the electrode is capable of making satisfactory welds. Thus the "1", as in E7010, means that the electrode is usable in all positions (flat, vertical, overhead and horizontal). The "2", as in E7020, indicates that the electrode is suitable for depositing welds in the flat position and for making horizontal fillet welds. The last digit of the classification indicates the type of covering on the electrode as prescribed in the specifications. In addition, a letter suffix such as, Al, designates the chemical composition of the deposited weld metal for low-alloy steel electrodes. Thus, a complete classification of a low-alloy steel electrode would be E7010-A1, E8016-C2, etc.

Mild Steel Covered Electrodes

Typical operating characteristics, current ranges and mechanical properties for mild steel covered electrodes are shown in Tables 94.1 and 94.2.

E6010 Electrodes.–Classification E6010 electrodes are designed to produce the best possible mechanical properties consistent with good usability characteristics in all welding positions. They are characterized by a deeply penetrating, forceful, spray-type arc and readily removable, thin, friable slag which may not seem to completely cover the deposit. Fillet welds are usually relatively flat in profile and have a rather coarse, unevenly-spaced ripple. The coverings are high in cellulose, usually exceeding 30% by weight. The other materials generally used in the covering include titanium dioxide, various types of magnesium or aluminum silicates, metallic deoxidizers such as ferromanganese, and liquid

Table 94.1—Operating characteristics of mild steel and low-alloy steel electrodes

Classification	Current and Polarity	Welding Positions
EXX10	DCRP (electrode positive)	All
EXX11	AC or DCRP (electrode positive)	All
E6012	AC or DCSP (electrode negative)	All
EXX13	AC or DSCP (electrode negative)	All
EXX14	DC either polarity	All
EXX15	DCRP (electrode positive)	All
EXX16	AC or DCRP (electrode positive)	All
EXX18	AC or DCRP (electrode positive)	All
EXX20	DCSP (electrode negative) or AC for H fillets; DC either polarity or AC for flat position welding	H-fillets and flat
EXX24	AC or DC either polarity	H-fillets and flat
EXX27	DCSP (electrode negative) or AC for horizontal fillet welds and DC either polarity or AC for flat position	H-fillets and flat
EXX28	AC or DCRP (electrode positive)	H-fillets and flat

	Type of Covering	Penetration	Surface Appearance	Slag
EXX10	High-cellulose sodium	Deep	Flat, wavy	Thin
EXX11	High-cellulose potassium	Deep	Flat, wavy	Thin
E6012	High-titania sodium	Medium	Convex, rippled	Heavy
EXX13	High-titania potassium	Shallow	Flat or concave, slight ripple	Medium
EXX14	Iron powder, titania	Medium	Smooth, fine ripples	Medium
EXX15	Low-hydrogen sodium	Medium	Flat, wavy	Medium
EXX16	Low-hydrogen potassium	Medium	Flat, wavy	Medium
EXX18	Iron powder, low hydrogen	Medium	Convex, smooth even ripple	Medium
EXX20	High iron oxide	Medium	Flat or concave, smooth	Heavy
EXX24	Iron powder, titania	Shallow	Convex, smooth fine ripple	Medium
EXX27	Iron powder, iron oxide	Medium	Slightly concave, smooth, even ripple	Heavy
EXX28	Iron powder, low hydrogen	Shallow	Flat to concave, smooth, fine ripple	Heavy

sodium silicate. Because of the covering composition, these electrodes are classified as the high-cellulose sodium type.

These electrodes (or E6011) are recommended for all-position work where the quality of the deposit is of greatest importance, particularly on multiple-pass applications in the vertical and overhead positions, where radiographic requirements must be met.

The majority of applications for these electrodes is on mild steel. However, they have been used to advantage on galvanized plate and on some low-alloy steels. Typical applications include shipbuilding, structures such as buildings and bridges, storage tanks, piping and pressure vessel fittings. Since the applications are so widespread, a discussion of each is impractical. Sizes up to 3/16 in. are easily used in all positions.

Generally speaking, the maximum currents that can be used with the larger

Table 94.2–Current range for mild steel electrodes (amperes)

Electrode Diameter, in.	E6010 and E6011	E6012	E6013	E6020	E6027
1/16	.....	20 to 40	20 to 40	.....	.....
5/64	.....	25 to 60	25 to 60	.....	.....
3/32	40 to 80	35 to 85	45 to 90	.....	.....
1/8	75 to 125	80 to 140	80 to 130	100 to 150	125 to 185
5/32	110 to 170	110 to 190	105 to 180	130 to 190	160 to 240
3/16	140 to 215	140 to 240	150 to 230	175 to 250	210 to 300
7/32	170 to 250	200 to 320	210 to 300	225 to 310	250 to 350
1/4	210 to 320	250 to 400	250 to 350	275 to 375	300 to 420
5/16	275 to 425	300 to 500	320 to 430	340 to 450	375 to 475
	E7014	E7015 and E7016	E7018	E7024 and E7028	
1/16	.....	.....	.....	.....	
5/64	.....	.....	.....	.....	
3/32	80 to 125	65 to 110	70 to 100	100 to 145*	
1/8	110 to 160	100 to 150	115 to 165	140 to 190	
5/32	150 to 210	140 to 200	150 to 220	180 to 250	
3/16	200 to 275	180 to 255	200 to 275	230 to 305	
7/32	260 to 340	240 to 320	260 to 340	275 to 365	
1/4	330 to 415	300 to 390	315 to 400	335 to 430	
5/16	390 to 500	375 to 475	375 to 470	400 to 525	

*These values do not apply to the E7028 classification.

sizes of these electrodes are somewhat limited as compared to other classifications due to the high spatter loss that occurs with high currents.

E6011 Electrodes.–Classification E6011 electrodes are designed to duplicate the usability characteristics and mechanical properties of the E6010 classification using ac as the source of power. Although also usable with dcrp, a sacrifice in usability characteristics will be noted when compared to the E6010 electrodes. Penetration, arc action, slag and fillet-weld appearance are very similar to that of the E6010 classification of electrodes.

The coverings are also high in cellulose and are designated as the high-cellulose potassium type. In addition to the other ingredients usually found in the E6010 electrode coverings, small quantities of calcium and potassium are usually present.

Like the E6010 electrodes, sizes larger than 3/16 in. are not generally used in all positions. The current and voltage ranges usually recommended are identical to those of the E6010 electrode. As in the case of E6010 electrodes, high currents result in high spatter loss. The ductility, tensile strength and yield strength are higher than obtained with E6010 electrodes.

E6012 Electrodes.–Classification E6012 electrodes are characterized by medium penetration, a rather quiet type of arc, slight spatter and dense slag completely covering the deposit. The coverings are high in titania, usually exceeding 35% by weight, which is why these coverings are usually referred to as the titania or rutile type. In addition, the coverings usually contain various siliceous materials such as feldspar and clay, small amounts of cellulose, ferro-manganese and sodium silicate as a binder. Also, small amounts of calcium compounds may be used to produce satisfactory arc characteristics on straight polarity. Whereas single-pass welds may meet radiographic requirements, multipass welds fall far short.

Generally speaking, fillet welds tend to be convex in profile, having a smooth, even ripple in the horizontal position and a widely spaced convex ripple in the vertical position which becomes smoother and more uniform as the size of the weld is increased. Ordinarily, a larger size fillet weld must be made in the vertical and overhead positions than with electrodes of the E6010 and E6011 classifications of the same diameter, if good fusion and profile are to be obtained.

Although E6012 electrodes are considered as all-position electrodes, far greater tonnages of them are used in the flat and horizontal positions than in the vertical and overhead positions. They are especially recommended for single-pass, high-speed, high-current, horizontal fillet welds. Ease of handling, good fillet weld profile and ability to withstand high current and to bridge gaps under conditions of poor fit-up make them very well-suited to this type of work. When used for vertical and overhead welding, the electrode size used is frequently one size smaller than would be used if an E6010 or E6011 electrode were used.

These electrodes have been used to advantage on many low-alloy steels, particularly of the higher carbon varieties. This is probably due to the fact that the penetration obtained, although adequate, is by no means that obtained with the E6010 or E6011 classifications. As a result, the pickup of alloying ingredients is not as great, which undoubtedly has a beneficial effect from the standpoint of cracking.

The weld metal deposited by these electrodes is lower in ductility and higher in yield strength than weld metal from either the E6010 or E6011 classification of electrodes.

E6013 Electrodes.–Classification E6013 electrodes, although very similar to the E6012 electrodes, possess some worthwhile differences. Slag removal is somewhat better and the arc can be established and maintained more readily, particularly in the case of the small diameters (1/16, 5/64 and 3/32 in.), thus permitting satisfactory operation with lower open-circuit voltage. Originally, these electrodes were designed specifically for light sheet metal work. However, the larger diameters are being used on many applications previously welded with classification E6012 electrodes. Even less penetration is obtained than with the E6012 classification of electrodes.

The coverings used are very similar to those employed with E6012 electrodes, containing rutile, siliceous materials, cellulose, ferro-manganese and liquid silicate binders. An important difference is that easily ionized materials are incorporated into the coverings which permit the establishment and maintenance of an arc with ac at low welding currents and low open-circuit voltages.

E6013 electrodes are similar to the E6012 electrodes in operation and appearance of deposit. The arc action tends to be quieter, the bead surface smoother and with a finer ripple. The different brands of E6013 electrodes have varied characteristics. Some of these electrodes are usually recommended for sheet metal applications where their ability to weld satisfactorily in the vertical position from the top down is an advantage. Others with a more fluid slag are in part replacing E6012 electrodes for horizontal fillet welds and other general purpose welding. Rather than the convex contour characteristic of the classification E6012 electrodes, these electrodes produce a flat fillet weld similar to that of the E6020 electrode classification. They are also readily usable in making groove welds because of the concave bead shape and easily-removed slag. In addition, the weld metal is definitely freer of slag and oxide inclusions than E6012 weld metal and the radiographic quality is better. In fact, the radiographic quality of welds made with the smaller diameter E6013 electrodes often meets the Grade I requirement of this specification.

Ordinarily, the high welding currents possible with the classification E6012 electrodes cannot be used. When welding in the vertical and overhead positions, however, the current and voltage will be very similar.

E7014 Electrodes.–Classification E7014 electrodes have coverings similar to E6012 and E6013 electrodes with the addition of iron powder. The amount of covering and the percentage of iron powder in the covering are usually less than those of E7024 electrodes.

The characteristics of the E7014 electrodes are a compromise between the E6013 and E7024 electrodes. The deposition rate is higher than that of the E6013 electrodes, but is not as high as that obtained with E7024 electrodes. The amount and character of the slag permit the E7014 electrodes to be used in all-position welding. They are more versatile than the E7024 electrodes, but not as versatile as the E6012 and E6013 electrodes.

The E7014 electrodes are suitable for welding mild and low-alloy steels. Typical weld beads are smooth with fine ripples. Penetration is approximately the same as that obtained with E6012 electrodes, which is advantageous when welding over gaps due to poor fit-up. The profile of fillet welds tends to be flat to slightly convex. The slag is easily removed and the welds are self-cleaning in many cases. The usable currents of E7014 electrodes are higher than for E6012 and E6013 electrodes.

E7015 Electrodes.–Classification E7015 electrodes are commonly known as low-hydrogen electrodes and are used with dcrp. They have coverings high in limestone and other ingredients low in hydrogen content which form a basic slag. They were developed for welding hardenable steels, in which electrodes other than low-hydrogen electrodes produce a phoenomenon known as "underbead cracking." These underbead cracks occur in the base metal usually just below the weld metal and are caused by the hydrogen absorbed from the arc atmosphere. The elimination of hydrogen with its consequent underbead cracking permits the welding of "difficult to weld steels" with less preheat than required for non-low-hydrogen electrodes, thus making for better welding conditions. Although these cracks do not occur in mild steels, they may occur whenever a non-low-hydrogen electrode is used on high-tensile steels.

Another use for the E7015 classification of electrodes is the welding of high sulfur steels. The non-low-hydrogen electrode deposits on these steels (which contain 0.10 to 0.25% sulfur) are badly honey-combed. The E7015 classification of electrodes can be used to weld these steels with less difficulty.

Their arc is moderately penetrating. Their slag is heavy, friable and easily removed, and the deposited weld metal lies in a flat bead or may even be slightly convex.

E7015 electrodes are used in all positions up to and including 5/32 in. diam. The larger sizes are useful for fillet welds in the horizontal and flat positions.

Generally speaking, the currents used with these electrodes are higher than those recommended for E6010 electrodes of the same diameter. It is recommended that the shortest possible arc be maintained for all positions of welding in order to get the best results. The reduced tendency for underbead cracking and the quality of as-welded deposits of these electrodes should materially reduce the necessity of preheat and postheat of weldments, thus making for better welding conditions.

These electrodes were originally developed for the welding of hardenable steels and, in addition to their use on alloy steels, high-carbon steels and high-sulfur steels, they have been found useful on malleable iron, spring steels and the mild steel side of clad plates. These electrodes are commonly used for making small welds on heavy sections since they are less susceptible to cracking than non-low-hydrogen electrodes. They are also used extensively in the welding of steels which will subsequently be enameled and in all those steels which contain selenium.

E7016 Electrodes.—Classification E7016 electrodes have all the characteristics of the E7015 electrodes. The core wire and coverings are very similar except for the use of a certain amount of potassium silicate or other potassium salts in the coverings of the E7016 electrodes to facilitate their use on ac. All that has been said of the E7015 electrodes applies equally well to the E7016 electrodes.

E7018 Electrodes.—Classification E7018 electrodes have coverings containing high percentages of iron powder in combination with low-hydrogen ingredients similar to those commonly used in E7015 and E7016 electrodes. As a rule, the coverings on these electrodes are slightly thicker than those of the E7015 and E7016 electrodes. The iron powder in the coverings usually amounts to between 25 and 40% of the covering weight.

These low-hydrogen electrodes are usable with both ac and dcrp. They are designed for the same applications as the E7015 electrodes. As is common with all low-hydrogen electrodes, a short arc should be maintained at all times.

In addition to their use for mild steel, the E7018 electrodes are well-suited for fillet welds in high-strength, high-carbon or alloy steels. The fillet welds made in the horizontal and flat positions are slightly convex in profile, with a smooth and finely rippled surface. The electrodes are characterized by a smooth, quiet arc, very low spatter and low penetration; they can be used at relatively high lineal speeds.

Low-Hydrogen Covered Electrodes

The absolute moisture in the covering is reduced to less than 0.2% in some grades by baking at high temperatures. The electrodes are then packaged in moisture-proof containers. The coverings of these electrodes, however, are somewhat hygroscopic, and it is important to keep in mind that the low-hydrogen type electrodes will re-absorb a considerable quantity of moisture when exposed to air of high humidity.

Furthermore, the coverings of the low-hydrogen electrodes may absorb moisture during storage or prolonged exposure to high humidity "on the job." Electrode users should be careful to maintain conditions to keep the moisture content of the electrode covering low. If it becomes necessary to lower the moisture content before production use, some electrodes may be rebaked at a temperature up to 800 F for a period of one to three hours. The electrode manufacturer should be requested to indicate the proper length of time and temperature for this purpose. Table 94.3 shows some suggestions for storage and rebaking of mild steel covered electrodes.

The mechanical and impact properties of deposits made from conventional cellulose, titania and iron oxide covered electrodes are generally improved by stress relieving, but the deposits of low-hydrogen electrodes are changed only slightly. The low-temperature impact properties of these electrodes are exceptionally good and are among their outstanding qualities.

E6020 Electrodes.—Classification E6020 electrodes, having a high iron-oxide, sodium-type covering, are designed to produce flat or slightly concave, horizontal fillet welds with either ac or dcsp. They will produce satisfactory results on fillet and groove welds in the flat position with ac or dc, either polarity. These electrodes are characterized by a spray-type arc and a heavy slag,

Table 94.3–Typical storage and rebake conditions for mild steel covered arc-welding electrodes

AWS Classification	Storage Conditions		
	Normal Room	Holding Ovens	Rebake[a]
E6010, E6011	80 F ± 20 F 20 to 60% relative humidity	Baking is not usually recommended. Suppliers should be consulted for storage and rebake conditions	
E6012, E6013 E6020, E6027 E7014, E7024	80 F ± 20 F 50% max relative humidity	20 to 40 F above ambient temperature	275 ± 25 F, 1 hr at temperature
E7018, E7028	80 F ± 20 F 50% max relative humidity	50 to 250 F above ambient temperature	750 ± 50 F, 1 hr at temperature
E7015, E7016	80 F ± 20 F 50% max relative humidity	50 to 250 F above ambient temperature	750 ± 50 F, 1 hr at temperature

[a]Because of inherent differences in manufacture, the suppliers of these electrodes should be consulted for the exact rebake conditions.

well honey-combed on the under side, which completely covers the deposit and can be readily removed. Due to the heavy slag, welders must be careful not to trap slag between weld passes. When normal welding currents and techniques are employed, penetration is considered to be medium. However, E6020 electrodes are capable of operating satisfactorily at high currents which results in deep penetration, particularly when a technique is employed such as in deep fillet welding. Of all the classifications of covered electrodes produced, this classification is generally considered to be the best for this specialized procedure.

The E6020 electrodes are essentially mineral-covered having high percentages of iron oxide, manganese compounds and silica in their covering along with sufficient deoxidizers to give a deposit of the desired composition. The slag coverage is so extensive and the slag-metal reaction of such a nature that the electrodes do not normally depend on gaseous protection. The coverings are such that an iron oxide, manganese oxide, silica slag is usually produced. Other material such as aluminum, magnesium or sodium may be present to modify the slag. Ferro-manganese is used as the main deoxidizer; sodium silicate is used as the binder. Quantities of basic oxide, acid silica and silicates, and deoxidizers must be carefully controlled to produce satisfactory operation and good weld metal.

Fillet welds tend to have a flat or concave profile and a smooth, even ripple. In many cases the surface of the deposit is dimpled. The more restricted the opening in which the metal is deposited, the greater tendency toward dimples. Dimples are to be expected in practically all cases on the first few passes of deep groove welds. As the weld nears completion, this tendency decreases. It is often noted that the use of ac tends to promote this dimpled condition to a greater extent than dc. No undesirable mechanical or physical defects are associated with this surface condition.

Generally speaking, these electrodes are recommended for horizontal fillet welds and flat-position welds where radiographic requirements must be met. High deposition rates can be obtained on such welds in heavy plate. Usually these electrodes are not used on thin sections due to increased warpage resulting from the use of the higher currents commonly employed.

When making horizontal fillet welds using the conventional technique, current values nearer the lower end of the range indicated should be employed if undercutting is to be held to a minimum. If the deep fillet welding technique is employed, the higher currents are used. Applications include pressure vessels, heavy machine bases and structural parts where thickness of section permits.

E7024 Electrodes.–Classification E7024 electrodes have coverings containing high percentages of iron powder in combination with ingredients similar to those commonly used in E6012 and E6013 electrodes. As a rule, the coverings on E7024 electrodes are very heavy, usually amounting to about 50 of the covered electrode weight.

The E7024 electrodes are well-suited for making fillet welds in mild steel. The welds are slightly convex in profile, with a very smooth surface and extremely fine ripples. These electrodes are characterized by a smooth, quiet arc, very low spatter and low penetration, and they can be used at high lineal speeds.

This classification operates with ac or dc, either polarity. The E7024 electrodes, although most generally used on mild steel, also produce satisfactory welds on many low-alloy, medium- and high-carbon steels.

E6027 Electrodes.–Classification E6027 electrodes have coverings containing high percentages of iron powder in combination with ingredients similar to those commonly found in E6020 electrodes. As a rule, the coverings on E6027 electrodes are very heavy, usually amounting to about 50% of the covered electrode weight.

The E6027 electrodes are designed to produce satisfactory fillet or groove welds in the flat position with ac or dc, either polarity, and will produce flat or slightly concave, horizontal fillet welds with either ac or dcsp.

This classification has a spray-type metal transfer and deposits metal at high lineal speeds. Penetration is medium and spatter loss is very low. The coverings on these electrodes produce a heavy slag, honey-combed on the under side, which covers the weld deposit and is quite friable and easy to remove.

Welds produced with E6027 electrodes have a flat to slightly concave profile with a smooth, fine, even ripple and with good metal wash up the sides of the joint. The weld metal is apt to vary in radiographic quality and be somewhat inferior to that from E6020 electrodes. High currents can be used since a considerable portion of the electrical energy passing through the electrodes is used to melt the covering and the iron powder contained therein. These electrodes are well-suited to the welding of fairly heavy sections.

E7028 Electrodes.–Classification E7028 electrodes are, in many respects, very much like the E7018 electrodes. However, there are differences between the two classifications which are described in the following paragraphs.

E7018 electrodes are all-position electrodes, whereas E7028 electrodes are suitable for horizontal-fillet and flat-position welding only.

The coverings of the E7028 electrodes are much thicker than that of the E7018 electrodes; the coverings of E7028 electrodes represent approximately 50% of the weight of the electrodes. The iron content of the E7028 coverings is higher than that of the E7018 coverings. As a consequence, on horizontal-fillet and flat-position welding, E7028 electrodes give a higher deposition rate than the E7018 electrodes for any given size of electrode. The iron powder content of the E7028 coverings represents about 50% of the weight of the coverings.

E7028 electrodes have a spray-type transfer; E7018 electrodes have a globular-type transfer. The ratio of the weight of the weld metal to the weight of the core wire consumed is about 1.05 min for the E7018 electrodes and about

1.30 min for the E7028 electrodes. Apart from the above differences, all that has been said about the E7018 electrodes applies to the E7028 electrodes.

Low-Alloy Steel Covered Electrodes

The electrode coverings used for low-alloy steel covered electrodes are generally similar to those used for mild steel with a corresponding covering classification number. For instance, the covering of an E9016 low-alloy steel electrode is similar to that for an E7016 mild steel electrode (although the E9016 covering may contain small amounts of alloying ingredients). Therefore, the descriptions of operating characteristics and current ranges contained in the preceding paragraphs are applicable generally to the low-alloy steel electrodes.

Most of the low-alloy steel electrode classifications are based on the low-hydrogen type (also known as the basic type) coverings due to the reduced possibility of underbead cracking when using these coverings. The as-welded mechanical properties of the weld metal deposited with low-hydrogen electrodes are superior to those of conventional electrodes such as E6010 and E6020. Tensile strengths of up to 120,000 psi or better with elongations of over 20% may be obtained in the as-welded condition of some of the electrodes.

The characteristics of the basic covering make it possible to add alloying elements such as carbon, manganese, silicon, chromium, nickel, molybdenum, vanadium, etc., to produce various ranges of weld metal analysis and strength. Heat-treatable deposits with an extremely wide range of mechanical properties may be produced from these electrodes.

The increased strength of the weld deposit is obtained by alloying. Thus the E8018 through E12018 electrodes contain alloying elements added in the core wire or more often in the covering since it is more economical to add the alloying elements in the covering than it is to use alloy core wire. Refer to AWS Specification A5.5-69, Low-Alloy Steel Covered Arc-Welding Electrodes, for some typical chemical analyses and mechanical properties of the EXX16 and EXX18 electrodes.

As discussed previously, the chemical composition is designated by a letter and number suffix such as A1, in the designation E7010-A1. The chemical compositions and mechanical properties are specified in AWS A5.5-69. Welding metallurgy, preheat, interpass temperatures and postweld heat treatments for low-alloy steels are discussed in Chapter 63 of this Handbook. In most cases, the electrodes are designed for stress-relieving operations and not for quench and temper properties.

The tolerance for hydrogen in low-alloy steels is usually less than that found for mild steels. Thus, control of moisture in the electrode coverings through proper storage and handling of electrodes can become important for some alloy systems. Refer to the pertinent sections of this chapter and Chapter 63 of Section 4 for storage, handling and moisture control recommendations.

BARE MILD STEEL ELECTRODES AND FLUXES FOR SUBMERGED ARC WELDING

Classification

The AWS specification A5.17 covers the requirements for mild steel electrodes and fluxes for submerged arc welding of mild and low-alloy steels. Electrodes are classified on the basis of their chemical composition as manufactured.

Fluxes are classified on the basis of the mechanical properties of a weld deposit made with a particular electrode. The classification designation given to a flux consists of a prefix F followed by a two-digit number representative of the tensile strength and impact property requirements for test welds made in accordance with the specification. This is then followed by a set of letters and numbers corresponding to the electrode classification used to classify the flux.

The classification system used for submerged arc welding electrodes follows as closely as possible the standard pattern used in other AWS filler metal specifications. However, the inherent nature of the electrodes being classified has necessitated specific changes which allow for more accurate electrode classification.

As an example, in the classification EL8K, the prefix E designates an electrode as in other specifications. The letter L indicates that this is an electrode which has a comparatively low-manganese content (0.60% max). This letter may also be M or H indicating a medium (1.25% max) or high (2.25% max) manganese content electrode. The digit or digits in the designation (8 above) indicate the nominal carbon content of the electrode. The letter K which appears in some designations indicates that the electrode is made from a heat of steel which has been silicon-killed.

The electrodes are classified only on the basis of their as-manufactured chemical analysis which is given in A5.17. Fluxes are classified on the basis of the mechanical properties of a weld deposit made using the flux in combination with any of the electrodes classified in the specification. As examples, consider the classifications F60-EH14, F71-EM12K and F62-EL8K. The prefix F designates a flux. The numbers 6 or 7, which immediately follow this prefix, designate the required minimum tensile strength in 10,000 psi of a weld made using the flux in combination with a specific electrode classification. The second digit after the prefix–0,1,2,3 or 4–designates the required minimum impact strength (0–no impact requirements, 1–20 ft-lb at 0 F, 2–20 ft-lb at -20 F, 3–20 ft-lb at -40 F and 4–20 ft-lb at -60 F) of a weld made using the flux in combination with a specific electrode classification. The suffix which is included in the designation (EH14, EM12K, etc.) after the hyphen, indicates the electrode classification with which the flux will meet the specified mechanical property requirements when tested as prescribed in the specification.

Use

Welding parameters, positions and welding techniques for submerged arc welding are covered in detail in Chapter 24 of this Handbook.

Storage of Electrodes and Fluxes

Under normal conditions, the standard packaging of most fluxes will protect the products for a considerable length of time: approximately one year. However, under humid or wet conditions some fluxes tend to be hygroscopic, and moist flux can adversely affect both the welding operation and the weld deposit. Flux which has picked up moisture may usually be reconditioned by baking, although flux which is wet to the touch must be replaced. Welding should only be conducted with completely dry flux. The manufacturer should be consulted for specific recommendations for reconditioning.

Solid electrodes usually have a long shelf life if not allowed to become excessively dirty or rusty. Extremely dirty or rusty electrodes should be replaced since they will cause poor feeding and/or poor weld deposits. Most of the submerged arc electrodes have a copper flash to inhibit rusting.

MILD STEEL ELECTRODES FOR GAS METAL-ARC WELDING

The AWS specification relating to these electrodes is A5.18-69. This specification deals with the various grades of solid gas metal-arc welding electrodes including one electrode with an emissive covering. The electrodes are grouped according to Charpy V-notch impact requirements.

Electrodes E70S-2, E70S-6, E70S-1B and E70U-1 are in one group and require 20 ft-lb at -20 F when tested as prescribed in the specification. The E70S-3 electrode requires 20 ft-lb at 0 F. The remaining electrodes do not have impact requirements.

All of the electrodes are continuous solid wire with only minor differences; they are principally in manganese and deoxidizers such as silicon, titanium, zirconium and aluminum. The higher strength alloy E70S-1B contains 1/2% molybdenum. In cases where weldments must be heat-treated, depositing a weld metal having a heat-treated response comparable to that of the base metal would be preferred.

Because of the great variety of compositions of plate available, it is important to make a careful selection of the solid gas metal-arc welding electrode to be used. Manganese and silicon in the solid-wire electrodes produce an attractive welding operation and, when automatic wires are utilized by experienced welders, welds of excellent bead contour, high efficiency, low spatter, good penetration and good mechanical properties are produced. These electrodes are primarily designed to be run with shielding gases such as carbon dioxide, argon-carbon dioxide mixtures or argon and oxygen. It is possible to vary the composition of the weld metal by the use of the various gases. The mechanical properties given in A5.18 are made with carbon dioxide gas.

The gas metal-arc welding electrodes are generally run with dcrp or electrode positive. The welding characteristics and composition of the weld deposit can be varied considerably by both voltage and amperage. In any given requirement where gas metal-arc welding electrodes must be used, it is mandatory that close control of arc voltage and amperage be exercized in order to achieve the desired chemical composition of the deposited weld metal. Deposited weld metal composition must be carefully watched because it has a definite effect on both impact and tensile properties. A favorable range of carbon, manganese and silicon must be maintained at all times if the desirable impact values are to be achieved. (Typical welding conditions are shown in Table 94.4.) Refer to Chapter 23, Section 2 for technical details of gas metal-arc welding.

Description and Intended Use of Electrodes

E70S-1 Electrodes.–Electrodes of the E70S-1 classification contain the lowest silicon content of the solid electrode classifications and are tested with an argon-oxygen mixture as a shielding gas in order to meet the requirements of the specification. Commercially, these electrodes may be used with carbon dioxide shielding gas when weld quality requirements are not critical and lower cost welding materials are desired.

E70S-2 Electrodes.–This classification covers multiple-deoxidized steel electrodes which contain a nominal combined total of 0.20% zirconium, titanium

Table 94.4–Current ranges for gas metal-arc welding electrodes

Size in in. (mm.)	E70S-1B, E70S-3, E70S-5, E70S-G,	E70S-2 E70S-4 E70S-6 E70S-GE	E70U-1 E70S-1	
	Amperes (A)	Volts (V)	A	V
0.030 (0.762)	30–130	15–19		
0.035 (0.889)	65–215	15–24		
0.045 (1.143)	125–320	20–29	200–320	25–30
0.052 (1.321)	130–400	23–31	230–400	26–31
0.062 (1.575)	150–475	25–32	250–475	27–32

and aluminum in addition to the silicon and manganese content. These electrodes are capable of producing sound welds in semi-killed and rimmed steels as well as in killed steels of various carbon levels. Because of the added deoxidants, these electrodes can be used for welding steels which have a rusty or dirty surface, with a possible sacrifice of weld quality depending on the degree of surface contamination. These electrodes can be used with a shielding gas of argon-oxygen mixtures, carbon dioxide or argon-carbon dioxide mixtures. They are preferred for out-of-position welding by the short-circuiting type of transfer due to the ease of operation.

E70S-3 Electrodes.–The E70S-3 electrode classification is similar to the E70S-1 classification except that the E70S-3 electrodes have a higher silicon content. They will meet the requirements of the specification with either carbon dioxide or argon-oxygen as a shielding gas. These electrodes are used primarily on single-pass welds, although they can be used on multi-pass welds, especially when welding killed or semi-killed steel. They can be used for out-of-position welding with small diameter electrodes using the short-circuiting type transfer with Ar-CO_2 mixtures or CO_2 shielding gases.

When classifying E70S-3 electrodes, it should be noted that the use of CO_2 shielding gas in conjunction with excessively high heat inputs for the preparation of the test assemblies may result in a failure to meet the minimum specified tensile and yield strength.

E70S-4 Electrodes.–Electrodes of this classification contain a slightly higher silicon content than those of the E70S-3 classification and produce a weld deposit of higher tensile strength. The primary use of these electrodes is for CO_2 shielded welding applications where a slightly longer arc or conditions, or both, require more deoxidation than electrodes of the E70S-1 and E70S-3 classifications provide.

E70S-5 Electrodes.–This classification covers electrodes which contain aluminum in addition to manganese and silicon as deoxidizers. These electrodes can be used when welding rimmed, killed or semi-killed steels with CO_2 shielding gas and high welding currents. The relatively large amount of aluminum assures the deposition of well-deoxidized and sound weld metal. Because of the aluminum, these electrodes are not used for the short-circuiting type transfer. They can be used for welding steels which have a rusty or dirty surface, with a possible sacrifice of weld quality, depending on the degree of surface contamination.

E70S-6 Electrodes.–The E70S-6 electrodes will produce welds which will meet the highest impact property requirements of the specification when used with CO_2 shielding gas. They may be used with high currents when welding rimmed steels and may also be used to weld sheet metals when smooth weld beads are desired. They can be used for out-of-position welding with the short-circuiting type transfer. They may also be used on steels which have moderate amounts of rust and mill scale on their surfaces. The quality of the weld will depend on the degree of surface impurities.

E70S-G Electrodes.—The E70S-G electrode classification includes those solid electrodes which are not included in the preceding classes. The electrode supplier should be consulted for the characteristics and intended use of these electrodes. The E70S-G electrodes are not required to meet chemical or impact requirements; however, they must be capable of meeting all other requirements of the specification.

E70S-1B Electrodes.—Electrodes of this classification contain a properly balanced chemical content with adequate deoxidizers to control porosity during welding with CO_2 as the shielding gas. This electrode will give dense radiographic quality welds in both ordinary and difficult to weld low-carbon and low-alloy steels. The quick freeze weld puddle characteristic permits out-of-position welds. When porosity is a problem due to dirty or rusty surfaces, or a high-sulphur steel content, these electrodes will provide for consistently good weld quality. These electrodes may be employed for single and multi-pass welding of mild steel or low-alloy steel.

E70S-GB Electrodes.—This classification includes those electrodes to which alloy additions have been made for deoxidation and usability improvement which are not covered by the E70S-1B classification. The electrode supplier should be consulted for the characteristics and intended use of these electrodes. The E70S-GB electrodes are not required to meet chemical or impact requirements; however, they must meet all other requirements of the specification.

E70U-1 Electrodes.—The E70U-1 electrode classification covers emissive-covered electrodes that have a specially treated surface which enables the electrode to be used with dcsp, with reasonable arc stability. These electrodes are designed to give a spatter-free, spray-type weld metal transfer using argon as a shielding gas and dcsp. These electrodes may be used in the flat and horizontal positions for fillet or multi-pass welds on rimmed steels. Use of dcsp with these electrodes permits higher deposition rates than those obtained when used with dcrp.

Electrode Surface Finish

Most of the gas metal-arc welding electrodes on the market are sold with a light copper flash coating applied to the surface of the wire. This copper coating adds to the appearance and storage life of the wire and improves the electrical conductivity between the solid wire and the welding gun contact tip. Copper coatings occasionally show a tendency to flake off causing interruption of feeding during automatic welding. If this happens, the manufacturer should be notified. Electrodes generally are slightly lubricated to promote manufacture and feeding during welding. Some electrodes are not copper coated and depend on a surface lubricating film for adequate corrosion protection. The necessary surface film of lubricant which is applied to the wire must be of such a nature that it lubricates but does not tend to build up in the gun head and must also have no harmful effect on usability.

Storage

Solid electrodes usually have a long shelf life if not allowed to become excessively dirty or rusty. Dirty or rusty electrodes should not be utilized for quality welding.

Electrodes for Flux Cored Arc Welding

Flux cored electrodes consist of a metallic outer sheath or tube enclosing a core of fluxing materials and, in many cases, gas- or vapor-forming materials, deoxidizers and alloying metals. During welding, these materials protect the molten metal from the atmosphere, deoxidize and/or denitrify the weld metal and improve its mechanical properties. Two main variations exist: (1) gas-shielded electrodes that require additional gas-shielding (usually CO_2) around the arc and weld puddle and (2) self-shielded electrodes that generate their own complete shielding. Within these variations are found electrodes designed to join many different types of base metals including mild steels, low-alloy steels and stainless steels.

Filler Metal Specifications

At the present time, only the mild steel electrodes are covered by an AWS filler metal specification (see Mild Steel Electrodes for Flux-Cored Arc Welding, AWS A5.20-69). Specifications are being prepared to cover flux cored electrodes for welding low-alloy steels and to cover electrodes for welding chromium and chromium-nickel corrosion-resisting steels. These specifications should be available in 1973. Many flux cored electrodes of these types, however, are presently available and are in commercial use.

Two military specifications covering flux cored electrodes exist at the present time. Military Specification MIL-E-24403 (SHIPS) is a specification covering the general requirements for flux cored arc welding electrodes designed to operate with the flux cored arc welding process using gas shielding. Military Specification MIL-E-24403/1 (SHIPS) is a detailed specification covering mild steel flux cored arc welding electrodes designed for use with carbon dioxide shielding gas; it is the first of a probable series of detailed specifications covering various types of flux cored electrodes.

The Structural Welding Code, AWS D1.1-72, is not a filler metal specification per se, but it does establish weld metal mechanical property requirements for several grades of flux cored electrodes used in producing weld metals with tensile strengths from 80,000 to 110,000 psi. Flux cored electrodes are included in the ASME Boiler and Pressure Vessel Code, Section IX as the F6 classification and some of the electrodes have been qualified to NAVSHIPS 250-1500-1, American National Standards for Pressure Piping and American Bureau of Shipping.

Mild Steel Flux Cored Electrodes

AWS Specification A5.20 prescribes requirements for mild steel electrodes for flux cored arc welding of mild and low-alloy steels. The electrodes are classified on the basis of whether or not carbon dioxide is required as a separate shielding gas, the type of current, their usability for either single- or multiple-pass applications and the chemical composition and as-welded mechanical properties of deposited weld metal. The specification should be referred to for a description of these electrode classifications and for details on the standard test procedures which are used to obtain the properties of these electrodes. It should be recognized that the properties may vary widely depending on electrode size; amperage, voltage and shielding gas used; plate thickness and composition; joint geometry; admixture with the deposited metal; etc.

The designation system used in AWS A5.20 follows as closely as possible the standard pattern used in other AWS filler metal specifications. The inherent nature of flux cored electrodes does, however, necessitate specific changes to facilitate electrode classification E70T-1. The prefix E designates an electrode as, i.e. in other specifications. The number 60 or 70 refers to the minimum as-welded tensile strength in 1,000 psi. The letter T indicates a tubular electrode. The suffix 1 indicates a particular classification based on deposited weld metal chemical analysis, shielding gas used and usability of the electrode for single- or multiple-pass applications.

E60T-7 Electrodes.—Electrodes of this classification are used without externally applied gas shielding and may be used for single- and multiple-pass applications. The weld deposits have a low sensitivity to cracking.

E60T-8 Electrodes.—Electrodes of this classification are used without externally applied gas shielding and may be used for single- and multiple-pass applications in the flat and horizontal positions. Due to low penetration and to other properties, the weld deposits have a low sensitivity to cracking. The weld deposits have good notch toughness properties at 0 F.

E70T-1 Electrodes.—Electrodes of this classification are designed to be used with carbon dioxide shielding gas for single- and multiple-pass welding in the flat position and for horizontal fillets. Some electrodes in this classification require that joints be relatively clean and free of oil, excessive oxide and scale in order that welds of radiographic quality can be obtained. A quiet arc, high deposition rate, low spatter loss, flat-to-slightly convex bead configuration and easily controllable and removable slag are characteristics of these electrodes. Weld deposits made with these electrodes have good impact properties.

E70T-2 Electrodes.—Electrodes of this classification are used with carbon dioxide shielding gas and are designed primarily for single-pass welding in the flat position and for horizontal fillets. Multiple-pass welds with some E70T-2 electrodes require an appreciable amount of admixture of the base and filler metals. However, in some of the more recently developed electrodes, where manganese is used as the deoxidizing ingredient, this admixture of base and filler metal is not required. Electrodes of this high-manganese type result in good mechanical properties in both single- and multiple-pass applications. These electrodes can be used for welding metal which has heavier mill scale, rust or other foreign materials on its surface than can be tolerated by some electrodes of the E70T-1 classification and can still produce welds of radiographic quality. The arc characteristics and deposition rates are similar to those of the E70T-1 electrodes.

E70T-3 Electrodes.—Electrodes of this classification are used without externally applied gas shielding and are intended primarily for depositing single-pass, high-speed welds in the flat and horizontal positions on light plate and gage thickness base metals. They should not be used on heavy sections or for multiple-pass applications.

E70T-4 Electrodes.—Electrodes of this classification are used without externally applied gas shielding and may be used for single- and multiple-pass applications in the flat and horizontal positions. Due to low penetration and other properties, the weld deposits have a low sensitivity to cracking.

E70T-5 Electrodes.—This classification covers electrodes primarily designed for flat fillet or groove welds with or without externally applied shielding gas. Horizontal fillet welds can be made satisfactorily, but at lower deposition rates than that obtainable in flat groove welds. They are designed for use with or without externally applied gas shielding in the event of drafts or malfunctions that would interrupt gas flow. Welds made using CO_2 shielding gas are of higher quality than those made with no shielding gas. The E70T-5 electrodes can be used in single-pass applications with minimum surface preparation, as described for the E70T-2 electrodes, without impairing their use for multiple-pass work. These electrodes have a globular transfer, low penetration, slightly convex bead configuration and a thin, easily removable slag. Weld deposits will meet the highest impact requirements of the specification.

Military Specification MIL-E-24403/1 (SHIPS) covers one mild steel flux cored electrode, MIL-70T-5, which is based on the AWS E70T-5 classification. The requirements for the MIL-70T-5 electrode are for carbon dioxide shielding gas only and are more detailed and slightly more stringent than those for the E70T-5 electrode. Both stress-relieved and as-welded property requirements are included along with tighter ranges on the chemical composition.

E70T-6 Electrodes.—Electrodes of this classification are similar to those of the E70T-5 classification, but are designed for use without an externally applied shielding gas.

E70T-G Electrodes.—This classification includes those composite electrodes which are not included in the preceding classes. The electrode supplier should be consulted for the characteristics and intended use of these electrodes. They may be used with or without gas shielding and may be used for multiple-pass work or limited to single-pass applications. The E70T-G electrodes are not required to meet chemical, radiographic, bend test or impact requirements; however, they are required to meet tension test requirements (obtained from either an all-weld-metal tension or transverse tension test specimen) and all other requirements of the specification.

Low-Alloy Flux Cored Electrodes

As described previously, low-alloy flux cored electrodes are not presently classified by an AWS filler metal specification, but electrodes may be obtained which generally meet the chemical and mechanical property requirements of AWS Specification A5.5, Low-Alloy Steel Covered Arc-Welding Electrodes. This specification can be used as a guide until a separate specification is available. Alloys representative of those available as flux cored electrodes are shown in Table 94.5.

In many instances, the electrode manufacturer will use the chemical suffix (A1, B2, B3, etc.) from the low-alloy covered electrode specification to describe his low-alloy flux cored electrodes. Most of the low-alloy flux cored electrodes available today are based on the flux system of either the E70T-1 or E70T-5 classifications. The former results in better weldability while the latter results in weld metals with greater notch toughness. The manufacturer's literature should be consulted for further details.

As mentioned previously, AWS D1.1-72 has established four grades of electrodes on the basis of certain weld metal mechanical properties. Most E70T-1 mild steel electrodes will meet the requirements of the E80T grade. Low-alloy electrodes are available which meet the requirements of each of the

Table 94.5–Representative non-classified low-alloy steel flux cored electrodes

Type of Steel	Nominal Alloy Composition	Chemistry Suffix From AWS A5.4	ASME Code, Section IX Weld Metal Analysis Number
Carbon-molybdenum	0.5% Mo	A1	A2
Chromium-molybdenum	1.25% Cr-0.5% Mo	B2	A3
	2.25% Cr-1.0% Mo	B3	A4
Nickel	1% Ni	C3	A3
	2.5% Ni	C1	A3
	3.25% Ni	C2	A3
Manganese-molybdenum	1.75% Mn-0.5% Mo	D1	–
	2.0% Mn-0.5% Mo	D2	–
Manganese-molybdenum Nickel-chromium	1.8% Mn-0.50% Mo 2.5% Ni-0.30% Cr	M	A3

four grades. Refer to AWS D1.1-72 for the appropriate table listing these requirements.

Sizes and Forms of Flux Cored Electrodes

Flux cored electrodes are available in diameters ranging from 0.045 in. to 5/32 in. The most widely used sizes are 1/16 in. for out-of-position semiautomatic welding, 3/32 in. for flat- and horizontal-position semiautomatic welding and 7/64 and 1/8 in. for flat-position machine welding.

Flux cored electrodes are available in coils, with or without support, on spools and in drums. AWS specification A5.20 should be consulted for details.

Welding with Flux Cored Electrodes

Flux cored electrodes are used for both semiautomatic (hand-held) and machine welding. All of the electrodes are capable of flat-position welds. Some small diameter flux cored electrodes can be used to make vertical and overhead welds: a few are specifically designed for out-of-position welding.

Deposition rates and efficiencies of flux cored electrodes may vary between and within the AWS classifications from one manufacturer to another. The penetration characteristics of different flux cored electrode types will also vary. Welding current, electrode stickout, polarity, core composition and gas or no-gas shielding will have a strong influence on penetration. Although most of the initial gas-shielded electrodes were intended for CO_2 gas shielding, some of the electrodes now available are intended for use with other gases or gas mixtures such as 75% Ar-25% CO_2.

Chapter 58, Section 3B of the WELDING HANDBOOK should be consulted for details on equipment and other welding data.

Special Characteristics of Flux Cored Electrodes–Storage

Under normal conditions, the standard packaging of most flux cored electrodes will protect them from moisture pickup for at least one year. If the packages become wet or if the electrode has rust on its surface, harmful moisture contamination may result. Excessive moisture in most mild steel and stainless steel flux cored electrodes will result in weld porosity. In higher strength deposits, excessive moisture may produce hydrogen-caused (delayed) cracking.

Electrodes which have picked up harmful amounts of moisture and which show little or no signs of rust may usually be reconditioned by baking. The electrode manufacturer should be consulted for a specific recommendation. Electrodes which are extremely rusty cannot usually be reconditioned because the rust will often cause poor feeding even if the moisture is baked off.

Opened packages of most flux cored electrodes may be left in dry shop atmospheres for periods of two to three days as long as they are protected from excessive contamination by dust and dirt. If shop conditions are humid, the safest practice is to store opened packages in heated or dehumidified storage areas. Since the necessary precautions may vary widely between different types of electrodes and between the same type produced by different manufacturers, the respective manufacturer should be consulted for his recommendations.

Health and Safety

Flux cored electrodes are normally used at higher welding currents and therefore higher deposition rates than many other welding electrodes (such as covered electrodes). For this reason, the concentrations of airborne substances evolved during welding may reach correspondingly higher levels. Recommended safety and ventilation practices should be adhered to. See Chapter 9, Section 1 of the WELDING HANDBOOK and ANSI Standard Z49.1, Safety in Welding and Cutting.

GAS WELDING RODS

The AWS specification relating to gas welding rods is AWS A5.2. These rods or wires are bare steel rods having no coverings. The welding operation is determined solely by the composition of the rods and the welding flame used. The various classes of gas welding rods are briefly described in the following paragraphs.

Class RG65 welding rods are used for the oxyacetylene welding of carbon and low-alloy steels which exhibit strengths in the range of 65,000 to 75,000 psi. They are used on sheet, plate, tubes and pipes. When a base-metal alloy analysis is used for some specific property such as creep resistance or corrosion resistance, then the filler metal analysis should match the base-metal alloy analysis. Class RG65 welding rods are of low-alloy steel analysis.

Class RG60 welding rods are used for the oxyacetylene welding of carbon steels in the strength range of 50,000 to 65,000 psi and for welding wrought iron. They may also be used for such low-alloy steels as fall in this range. These are the general-purpose gas welding rods, of medium strength and good ductility, that are most commonly used for the welding of carbon-steel pipes for power plants, process piping and other conditions of severe service.

Class RG45 welding rods are of a simple low-carbon steel analysis. Most rods of this class are of the following nominal composition: carbon, 0.07% max; manganese, 0.25% max; phosphorus and sulfur, each 0.04% max; silicon, 0.08% max. These welding rods are general-purpose rods and may be used to join wrought iron.

Iron and steel gas welding rods are designed to yield deposited metal of a desired composition. Changes take place during welding. For example, recovery of such elements as carbon and chromium depends largely upon the presence of deoxidizers such as manganese and silicon.

Carbon in general is a strengthening element but causes a decrease in ductility. In oxyacetylene welding, carbon may be controlled to a certain extent by adjustment of the welding flame. The use of an oxidizing flame lowers the carbon in the weld and may cause a porous deposit. A reducing flame introduces carbon into the filler metal. In addition, carbon is introduced into a narrow surface zone in the heated base metal which lowers the melting point and results in less base-metal melting. The recovery of carbon in the range up to 0.20% under a neutral flame is usually complete, provided deoxidizers are present.

Manganese and silicon act as deoxidizers in the molten filler metal. Silicon is the prime deoxidizer and serves to protect carbon and other readily oxidizable additions. Manganese is added to increase the activity of the silicon and to aid in the formation of a fluid slag. In the presence of manganese, silicon becomes a more powerful deoxidizer due to the formation of a stable, highly fluid manganese silicate. The manganese silicate floats to the surface of the melted metal and prevents appreciable oxidation by the atmosphere surrounding the weld. Manganese-to-silicon ratios vary from 4:1 to 1.5:1, depending on the intended use of the welding rod.

Phosphorus and sulfur in the welding rod and base metal are retained in the weld metal without much change, although there is generally some elimination of sulfur. Specifications set a maximum sulfur content of 0.040% for iron and steel gas welding rods. Sulfur is particularly harmful and is one of the main causes of porosity in weld metal due to the evolution of gas which is trapped when the metal solidifies.

Chromium and vanadium are carbide formers which strengthen the weld metal, providing welds with high-tensile strength and good ductility. Molybdenum and nickel, both used as strengthening additions, do not appreciably change the characteristics of the molten puddle when present in small amounts. Molybdenum helps to prevent hot cracking, while nickel in amounts up to 1% causes a more fluid puddle.

Excessive heating during the welding operation adversely influences the impact strength of gas welds. Low-impact strength may be corrected by depositing the weld metal in a number of small stringer beads instead of wide weave beads.

CORROSION-RESISTANT CHROMIUM AND CHROMIUM-NICKEL STEEL FILLER METALS

Arc-welding electrodes and welding rods, which deposit ferrous weld metal containing more than 4% chromium and less than 50% nickel, are considered to be corrosion-resistant chromium and chromium-nickel steel filler metals.[2] Because of the corrosion- and heat-resistant properties of these alloys, they are frequently referred to as the stainless electrode series. There is an important distinction between the heat and corrosion-resistant-cast alloys, based on carbon content. The higher carbon castings of the A.C.I. "H" series have greater hot strength, creep resistance and life expectancy under stress at red heats. The

[2] Refer to AWS Specifications A5.4 and A5.9.

lower carbon A.C.I. "C" series castings, wrought products and most electrode wires have better corrosion resistance. The two series are not interchangeable. If low-carbon welds are made on heat-resistant castings, the welds may prove faulty if critically stressed in high-temperature service.

This difference in hot strength can also lead to fissuring of low-carbon welds on high-carbon castings under restrained cooling. Sometimes the hot strength of the weld can be increased with columbium. In other cases, carbon is included in the electrode covering to compensate. Sound engineering usually requires that the carbon content of the weld match that of the base metal for high-temperature service.

Oxidation resistance is chiefly a function of the chromium content. Care must be taken to completely remove the fluoride-containing slag from shielded metal-arc welding as it may adversely affect the protective oxide at higher temperatures.

STAINLESS ELECTRODES

The tests required by the specifications[3] classify the electrodes according to composition and welding position. By standardizing weld metal compositions determined by chemical analysis, uniformity of corrosion resistance is ensured. Interchangeability of electrode brands having the same classification number can be achieved because the electrodes meet usability standards for welding position and pass all-weld-metal tension tests for quality.

STAINLESS STEEL WELDING RODS AND BARE ELECTRODES

The oxyacetylene, atomic hydrogen, inert-gas and submerged arc welding processes are all applicable to joining chromium-nickel and straight chromium steels, and frequently require added filler metal.[4] The welding rod is supplied in wire form varying from 3/64 to 3/16 in. diam, straightened and cut to 36-in. lengths, and as coiled wire in diameters from 1/16 to 1/4 in. For the consumable-electrode processes (submerged arc and gas metal-arc), wire is supplied on spools, ranging from 0.030 in. to 1/4 in. diam. For submerged arc welding, standard coils of wire weigh 25, 50, 65 or 150 lb. Welding electrodes wound on spools are usually 25 lb net weight, and on rims are usually 25 or 60 lb.

FILLER METAL CLASSIFICATIONS

The composition of wire used for welding rods often differs from that used for electrode core wire because all of the alloying elements must be present in the wire, whereas, in an electrode, many of the elements are in the covering. If unusual losses of elements occur during welding, compensation for these losses must be made by increasing the alloy content of the wire.

For submerged arc welding of materials of similar composition, the core wire

---

[3] AWS A5.4.
[4] AWS Specifications A5.9.

may be of comparable composition. In this case, however, the core wire usually contains somewhat more chromium to compensate for oxidation losses. Submerged arc welding fluxes that are chromium-reinforced are also available. When welding dissimilar metals such as stainless steel to mild steel or stainless clad steel, oxidation losses and dilution losses must be considered in selecting the filler metal alloy as well as the flux. Chromium oxidation losses have been estimated to be about 10% of the chromium content of the filler metal alloy. This percentage assumes equal chromium oxidation in the fused base metal and in the filler metal. For estimating dilution effects, the amount of fused base metal should be determined.

WELD METAL COMPOSITIONS

Over the past fifty years, the development of stainless steels has resulted in many different compositions, most of which are readily weldable. Chapters 64 and 65 in Section 4 indicate the variety of stainless steel base metals which are presently standardized by steel suppliers. For most of the 300, 400 and 500 series steels, there are comparable compositions of filler metal. The 200 series steels (low-nickel, high-manganese are welded using the corresponding 300-type electrodes and rods. The electrodes covered by specifications are classified according to composition.

E308, ER308, E308L and ER308L Electrodes and Rods

This is the most widely used series of classifications, commonly designated the 19-9 electrodes. The familiar 18-8 chromium-nickel alloy has many modifications which are welded with the E308 electrodes. By specifying that the weld metal shall have 0.08% maximum carbon and suitable minimum chromium and nickel contents, the E308 electrodes can be used for all of the medium carbon AISI wrought stainless steels—numbered Types 301 to 308 inclusive with the exception of Type 304L—and for the ACI Casting Types CF-8, CF-29 and HF. However, E308 deposits have lower creep strength than HF castings, which usually carry about 0.30% carbon. E347 welds are nearer to HF-30 in hot strength. The 304L plate may be welded with either E308L electrodes or E347 stabilized electrodes. An E308-type electrode, because of its higher carbon content, should not be used to weld 304L plate, as this would permit weld decay under heat-treating and service conditions. The problems involved in welding Type 304L alloy are discussed in Chapter 65.

E309, ER309, E309Cb and E309Mo Electrodes and Rods

For more rigid service requirements where higher chromium and nickel contents are required, the so-called 25-12 electrodes are applicable. The basic composition corresponds to that of wrought stainless steel of Type 309 and the CG-12, CH-10, CH-20 and HH castings. HH castings, when balanced to be free of ferrite, are usually stronger in the temperature range from 1200 to 2100 F than E309 welds. Columbium can partially compensate for this, but it confers greater vulnerability to oxidation. Because of its higher alloy content, the 25-12 electrode is frequently employed in welding 18-8 clad steels so that fusion with the mild steel backing will not dilute the weld-metal composition below that of the alloy cladding.

Occasionally, for specific applications, columbium is added to the steel or casting to stabilize the alloy, in which case the E309Cb electrode should be used. The E309Cb modification is also used on Type 347 clad steel to offset dilution. The E309Mo electrode is principally used in welding the 18-8 Mo clad steels, Type 316.

E310, ER310, E310Cb and E310Mo Electrodes and Rods

The so-called 25-20 electrodes are most properly used more for special applications than for welding the corresponding wrought Type 310, cast CK-20 and HK steels. For heat resistance, the ordinary E310 electrodes do not match the high hot strength of HK castings. At 1400 F and up, HK castings with 0.30% carbon have about twice the creep strength of a 0.08 to 0.15% carbon weld deposit, and the difference is even greater with high-carbon castings. It is advisable to obtain modified electrodes that will deposit 26% Cr–20% Ni metal with a carbon content that matches the base metal. These are available with 0.25% maximum, 0.25 to 0.35% and 0.35 to 0.45% carbon.

The E310 electrodes produce a strong and ductile weld when used for welding carbon steels or low-alloy steels. For this reason, E310 electrodes have been widely used for welding armor plate, particularly naval armor in heavy sections. The welding of low-alloy hardenable steels with this type of electrode is simplified by the electrode minimizing the need for preheat or postheat treatment, even for very hardenable steels. These electrodes are also used for joining stainless to mild steel and stainless clad steels. For molybdenum or columbium-bearing stainless clad steels, E310Mo or E310Cb electrodes are made by introducing the required molybdenum or columbium through the electrode covering.

E312 and ER312 Electrodes and Rods

The nominal composition of this weld metal is 29% chromium, 9% nickel. These electrodes were originally designed to weld cast alloys of similar composition. More recently, they have been found valuable in welding dissimilar metal compositions, one component of which is high in nickel. This alloy gives a two-phase weld deposit with substantial percentages of ferrite in an austenitic matrix. Even with considerable dilution by austenite-forming elements such as nickel, the microstructure remains two-phase and thus highly resistant to weld-metal cracks and fissures. The 29% Cr–9% Ni composition is very susceptible to development of the sigma phase (FeCr) after exposure in the 1300 to 1700 F range. The ferrite of a weld deposit is likely to convert to sigma at 1600 F. Sigma precipitation causes marked embrittlement, evidenced at ordinary temperatures.

E16-8-2 Electrodes

Electrodes of the 16-8-2 type contain approximately 16% chromium, 8% nickel and 1 to 2% molybdenum. This composition gives a weld deposit of austenite with up to 5% ferrite present, with good heat ductility properties. The welds are notably free from fissuring and crater cracking, even under conditions of high restraint. The electrodes are used primarily for the welding of Type 316, Type 317 and Type 347 components employed in high-pressure, high-temperature piping systems. In particular, the heat-affected zones of the base metal seem less subject to tearing when welded with E16-8-2 electrodes than with E347. The weld metal has excellent mechanical properties in the as-welded or solution-treated condition. Because of the relatively low chromium and molybdenum content it does not readily transform to the brittle sigma phase, which is a failing of E312 electrodes.

The specified composition range for E16-8-2 electrodes is carefully balanced to achieve the desired weld-metal properties. Corrosion tests indicate that E16-8-2 is similar to Type E316 in most media. If a severely corroding medium is involved, pretests are advisable to determine whether the application may require a columbium-stabilized electrode.

ER321 Bare Electrodes and Rods

Bare electrodes and welding rods of 19-9Ti type are available for inert gas-shielded arc welding of Type 321 steel. Covered electrodes are not practical because of the poor recovery of titanium from such electrodes; for this reason, even the ER321 filler metal is not suitable for the submerged arc welding process. The titanium addition is necessary as a carbide-forming element, to stabilize the alloy against intergranular corrosion in the same way that columbium stabilizes the 347 alloy.

E316, ER316, E316L, ER316L, E317, ER317, E318 and ER318 Electrodes and Rods

The introduction of molybdenum to ordinary 18-8 stainless steel makes the alloy more resistant to the corrosive attack of many organic acids, brine and sulfurous and sulfuric acids. Two molybdenum alloys are standard, one containing about 2% molybdenum and the other between 3 and 4%. For this reason, two electrodes are specified to match the compositions of the corresponding wrought alloys, Types 316 and 317. The E316 electrodes are also applicable for welding the cast CF-8M and CF-12M alloys. Stabilized Type 316 alloys containing columbium (now designated Type 318), which are welded with E318 electrodes with columbium added through the covering, are available. Stabilization is also accomplished by reducing the carbon content to below 0.03%. For welding this type of steel, Type 316L electrodes and rods are usually recommended.

In the 18-8 steels, the introduction of molybdenum improves the corrosion resistance as mentioned, but it also gives the alloy improved creep properties at elevated temperatures. Occasionally E316 electrodes are applicable for certain welds involving the so-called super-alloys used in gas turbines. These applications, however, generally require special electrode compositions containing tungsten, columbium, molybdenum and cobalt.

E320 and ER320 Electrodes and Rods

The introduction of copper in a stainless steel imparts resistance to a wide range of chemicals, including sulfuric and sulfurous acids and their salts. To balance the alloy when it contains 3 to 4% copper, the composition is held at a nominal 20% chromium, 34% nickel, 2.5% molybdenum, with columbium added for stabilization against intergranular corrosion. These filler metals are used to weld wrought alloys and castings of similar composition, such as No. 20, No. 20Cb3 and CN-7M. Modified electrodes and rods without columbium are available for repairing castings which do not contain columbium, but such welds require solution-annealing for stabilization.

E330 Electrodes

The iron-base alloy with 35% nickel content is the Type 330 alloy, containing 15 Cr-35 Ni. This alloy is more frequently supplied in cast form, designated HT, than in the wrought form, Type 330. It is popular for furnace parts and other heat- and scale-resistant applications. The use of E330 electrodes is usually limited to the welding of similar compositions. Nearly a quarter of the heat-resistant castings made at the present time are of the HT type, containing about 0.50% carbon. Modified E330 electrodes with extra carbon in the covering to permit matching the weld to the base metal composition are available. They should be used for high-temperature applications.

E347, ER347 and ER348 Electrodes and Rods

The introduction of columbium or titanium for the stabilization of austenitic alloys against intergranular corrosion is discussed in Chapters 65 and 66 of Section 4. E347 electrodes match the composition of the columbium-stabilized stainless steels, designated as Type 347 in wrought form and CF-8C in cast form. They are also used for welding the titanium-stabilized steel, Type 321, since titanium-bearing covered electrodes are not made because of the difficulty of satisfactorily transferring titanium across the arc.

The ideal ER347 rod for gas tungsten-arc welding is not the same as that for gas welding. For example, MIL-R-5031 specifies a low-columbium grade (type 5A) for gas tungsten-arc welding to minimize hot-cracking problems, and a high Cb grade (type 5) for gas welding to offset transfer loss. Type E347 electrodes may also be used to weld 304L stainless plate.

For nuclear energy applications, a restriction of holding tantalum to 0.10% maximum is necessary because of the high neutron-capturing capacity of tantalum. In bare electrodes and welding rods, a special grade ER348 has been created to cover this composition. In covered electrodes, there is no separate classification for the low tantalum grade, but many E347 brands will meet the restriction. Electrodes with tantalum-restricted coverings are made with low-tantalum ferro-columbium.

E349 and ER349 Electrodes and Rods

A filler metal commonly called 19-9WMo is used to weld steels of similar composition: 19-9WMo, 19-9DL and 19-9DX. The combination of columbium, molybdenum and tungsten with the chromium and nickel gives good high-temperature rupture strength. The chemical composition of the weld metal results in an appreciable amount of ferrite, increasing the crack resistance of the weld metal.

E410, ER410 and ER420 Electrodes and Rods

The 12% chromium steel is the lowest alloy steel within the commonly accepted stainless steel group. It is a hardenable steel, popular for many corrosion and elevated temperature applications, such as in power plants and petroleum refineries. The E410 electrodes, when used to weld Types 405, 410 and 414 wrought steels and Type CA-15 cast steel, require preheat and postheat treatment for most engineering purposes.

Frequently, electrodes of this classification are used for surfacing mild steel parts to resist corrosion, erosion and abrasion, occurring in valve seats and other valve parts. By specifying higher carbon than in E410 and ER410, this filler metal becomes applicable for many surfacing applications requiring corrosion resistance provided by 12% chromium along with somewhat higher hardnesses for increased resistance to abrasion. The ER420 composition fills this need.

E430 and ER430 Electrodes and Rods

Type 430 wrought steel is most frequently employed for its corrosion resistance to oxidizing mineral acids such as nitric acid. The chromium content is usually between 15 and 17%, and with its low-carbon content, the alloy consists of martensite and nonhardenable ferrite. As the chromium increases, the nonhardenable ferrite constituent increases, and the strength, ductility and impact properties suffer while corrosion resistance improves. A satisfactory balance for many applications is achieved at about 16% chromium; consequently, this is one of the most widely used of the straight chromium steels. Because of its partially hardenable nature, preheat and postheat treatments must be applied when welding with E430 and ER430 electrodes. Electrodes having higher chromium content, up to 28%, are available, but because of their limited application, they have not been included in the specifications.

E502 and ER502 Electrodes and Rods

Although not a stainless steel in the usual sense, the 5% chromium-molybdenum steel is widely used in refinery pipelines for its corrosion resistance to hot and crude oils and its strength at elevated temperatures. It is an extremely hardenable steel, and therefore preheat and postheat treatments are required when welding with E502 and ER502 electrodes.

E505 Electrodes

Electrodes of the 505 type are commonly referred to as the 8 to 10% chrome type, or 9 Cr-Mo because of the 1% molybdenum. They are used for welding base metal of the same composition, which is usually furnished in pipe or tube form as well as castings. Type 505 is an air-hardening material. It must be welded employing both preheat and stress relief for satisfactory results.

E7Cr Electrodes

Electrodes of the 7 chrome classification are principally used to weld base metal of similar composition: 6 to 8% chromium with about 0.5% molybdenum. The 7 chrome base metal is usually furnished in tube and pipe forms or castings. This is an air-hardening material which requires both preheat and stress relief for satisfactory welding.

Stainless Steel Flux Cored Electrodes

True flux cored electrodes (i.e., electrodes containing significant flux or slag formers) for welding stainless steels are a comparatively recent development. As yet, they are not covered by an AWS filler metal specification. Electrodes of both the gas-shielded and the self-shielded types are available. The gas-shielded types may use carbon dioxide, argon, argon/oxygen or other gas mixtures. Until a separate filler metal specification is available, the chemical requirements of either AWS A5.4, Corrosion-Resisting Chromium and Chromium-Nickel Steel Covered Welding Electrodes or of AWS A5.9, Corrosion-Resisting Chromium and Chromium-Nickel Steel Welding Rods and Bare Electrodes and the mechanical property requirements of A5.4 may be used in selecting these flux cored electrodes. Many deposit analyses are available including 308L, 309L, 316L, 347, 410, 430 and 502.

Table 94.6—Usable positions and types of current for chromium and chromium-nickel electrodes

	Direct Current		Alternating Current	
Usability Designation	Flat and Horizontal Fillet	Vertical and Overhead	Flat and Horizontal Fillet	Vertical and Overhead
−15	All sizes	5/32 in. and smaller	Not recommended	Not recommended
−16	All sizes	5/32 in. and smaller	All sizes	5/32 in. and smaller

USABILITY CHARACTERISTICS AND TYPES OF COVERINGS

The specifications for covered stainless electrodes recognize the various types of coverings by requiring a fillet weld test in all of the usable positions and with each usable type of current. Two usability classifications, as shown in Table 94.6, are thus defined, each designated by a two-digit number following the composition designation (e.g., E308-15).

The terms "lime," "lime-titania" and "titania" are frequently used to designate various types of coverings for stainless electrodes. In general, a lime-type covering is one whose mineral ingredients include chiefly limestone and fluorspar with minor amounts (up to about 8%) of titanium dioxide. Coverings containing more than 20% titanium dioxide are usually considered to be the titania type, and those containing between 8 and 20% are considered to be the lime-titania type.

Lime Type

The basic lime-type covering is usually applicable for electrodes operating on dcrp only. In the flat position, the bead ripples are less uniformly spaced and the fillet weld contours are more convex than with the titania type. In the vertical position, this type of electrode is usually preferred because smooth, uniform welds can be made easily. Fully austenitic chromium-nickel welds from lime-type electrodes may show considerably less tendency to crack or fissure than those from titania-type electrodes. The -15 classification is usually applicable to lime-type covered electrodes.

Titania Type

For a-c welding, the covering always contains some titania, usually sufficient to classify it as a titania type. In addition, readily ionizing elements such as potassium are added for arc stability. Such electrodes are designated a-c/d c electrodes, and although adjusted to make them usable with ac, they are more frequently used with dcrp. They are used in preference to the lime-type electrodes for flat-position welding, particularly for smooth concave fillet welds or butt welds requiring finish grinding, because of the ease of uniform metal deposit and the minimum of reinforcement. If they qualify as an a-c/d-c electrode, they may be designated -16. Titania-type electrodes are also available for dcrp only (-15).

Lime-Titania Type

This covering type is more often applied to the straight chromium or chromium-molybdenum steels than to the chromium-nickel steels. Depending on the remaining ingredients, electrodes with this type of covering may be used with dc only or on both ac and dc.

Tables of usable currents and voltages are not included here because they may vary widely for different brands conforming to the same classification.

MECHANICAL PROPERTIES

The specifications indicate minimum tensile strength and elongation for each electrode. All-weld-metal tension specimens are specified because the properties of the weld metal thus determined are not influenced by the characteristics of the base metal. The properties of the chromium-nickel alloys are usually tested in the as-welded condition, and the straight chromium or chromium-molybdenum steels are usually tested after suitable postheat treatments. Refer to AWS Specifications A5.4 and A5.9 for the typical mechanical properties of chromium and chromium-nickel weld metal (all-weld-metal specimens).

CORROSION RESISTANCE

In the annealed condition, stainless steel weld metal is usually comparable to wrought or cast stainless steel from the standpoint of corrosion resistance. In many applications, however, welds are placed in corrosion-resisting service in the as-welded or stress-relieved (or stabilized) condition. Certain corrosive media may attack the welds more rapidly than the base metal, so that unless service experience is already available, it is usually desirable to conduct some type of corrosion test on the welds. The commonly used tests are described in Chapter 65, Section 4. Typical corrosion test data on all-weld-metal specimens subjected to the Huey test (boiling 65% nitric acid) are given in Table 94.7.

FILLER METALS FOR CAST IRON

Cast iron is welded by the oxyacetylene, carbon-arc and shielded metal-arc welding processes. The filler metals available are suitable for welding gray cast iron, malleable iron and some alloy cast irons. These filler metals are identified by the usual letters E, R and B. The chemical symbols for the principle elements in each composition appear in the classification designation. Subclassification is provided by letters (A,B,C, etc.) and digits (1,2). A suffix CI is added to the nickel-base filler metals to avoid confusion with filler metals used to weld nickel-base alloys.

Table 94.7—Typical corrosion rates of corrosion-resistant weld metal in boiling 65% nitric acid

	Penetration per Month, in.[1]			
	E308	E316	E347	E430
As-welded	0.0013	0.0015	0.0010	.....
Sensitized[2]	0.0151	0.0200	0.0051	.....
Quench anneal[3]	0.0006	0.0008	0.0007	.....
Subcritical anneal[4]	.....	.....	.....	0.0045

[1] Average of five 48-hour periods.
[2] 1200 F for 4 hours, air-cooled.
[3] 1950 F for 15 min, water-quenched.
[4] 1450 F for 4 hours, furnace-cooled 100 F per hour to 1000 F, air-cooled.

ECI, RCI, RCI-A and RCI-B CAST IRON WELDING RODS AND ELECTRODES

The ECI heavily covered electrode and the RCI oxyacetylene rod are ordinary gray cast-iron compositions which produce machinable weld deposits under proper cooling conditions. Suggestions for preheating and welding procedure are given in the appendix to the A5.15 Specification.

The RCI-A welding rod is an alloy-cast-iron composition with molybdenum and nickel. It is used to obtain greater tensile strength and finer grain structure (the molten weld metal is fluid, permitting rapid welding).

The RCI-B welding rod is a nodular iron (ductile cast iron) used to weld high-strength gray iron and malleable, nodular iron castings with the oxyacetylene process. Most of the graphitic carbon in the weld is in the form of nodules, giving good ductility, easy machining and good color match to the base metal.

RBCuZn-A, RCuZn-B, RCuZn-C, RBCuZn-D, ECuSn-Z, RCuSn-Z, ECuSn-C AND ECuA1-A2 COPPER-BASE WELDING RODS AND ELECTRODES

Cast iron can be braze-welded with copper-base filler metals due to the low residual stresses produced by these fillers which yield plastically during cooling. These weld deposits are rich in the soft and ductile alpha phase and stretch harmlessly to accommodate the contraction of the cast iron. On cooling from 500 F to ambient, the strength of the deposit increases markedly with little change in ductility.

The copper-zinc welding rods (60Cu-40Zn) are used where color match is not required, and where there will be no electrolytic corrosion or high-temperature exposure. These rods are deposited by oxyacetylene welding only.

RBCuZn-A is a naval brass containing 1% tin for improved strength and corrosion resistance. RCuZn-B is a manganese-bronze with added strength, hardness and corrosion resistance while RCuZn-C is a low-fuming bronze with silicon added to control the oxidation of the zinc. RBCuZn-D is a nickel-bronze which is low-fuming because of a silicon addition. This deposit is silver-white in color, hence a better match for cast iron than the yellow alloys.

The copper-tin phosphor-bronze compositions are available as covered electrodes for shielded metal-arc welding and as bare welding rods. These compositions permit rapid welding and low-heat input, thus minimizing distortion and cracking. ECuSn-A and RCuSn-A contain 5% tin; ECuSn-C contains 8% tin and thus has greater hardness and higher strength.

The aluminum-bronze electrodes, ECuA1-A2 (bare and covered), are a copper-aluminum composition with a relatively low melting point and high deposition rate (thus permitting low-amperage, rapid welding to minimize distortion and to avoid formation of white cast iron in the fusion zone). The yield and ultimate strengths of ECuA1-A2 deposits are almost double those of ECuSn deposits, making these electrodes suitable for welding higher strength cast irons.

EST MILD STEEL ELECTRODES

This classification is a covered, all-position electrode especially designed for welding cast iron. It can be used at low amperage to minimize dilution. However, because it is a steel electrode, it undergoes greater shrinkage than cast iron, thus developing high stresses in the cold deposit. Its use is largely confined to repair of small pits and cracks, and repair of small castings which will require no machining.

ENi-CI, ENiFe-CI, ENiCu-A AND ENiCu-B NICKEL-BASE ELECTRODES

Several high-nickel alloy filler metals, developed primarily for the joining and repair of gray iron castings, are also used to weld ductile iron, malleable iron and other iron-base casting compositions. The 60% nickel-iron type is preferred for welding ductile iron, and for welding castings containing 0.20% or more of phosphorus and other more difficult to weld compositions. The flux coverings of these electrodes contain a large amount of carbon which is transferred to the weld deposit as well-dispersed particles of graphite. Good machinability is provided by these particles.

The ENi-CI electrode is a flux-covered high-nickel composition (at least 85% Ni). The ENiFe-CI electrode is a flux-covered nickel-iron with 40 to 60% Ni. The ENiCu electrodes are modified nickel-copper compositions: ENiCu-A with 50-60% Ni and ENiCu-B with 60-70% Ni.

ALUMINUM AND ALUMINUM-ALLOY FILLER METALS

Aluminum is the second most widely fabricated metal. It is extensively welded by both gas metal-arc and gas tungsten-arc welding processes. Bare filler rods and electrodes used with these processes are covered by AWS Specification A5.10. Covered electrodes are seldom used for welding aluminum because of the relatively poor operability and the necessity for complete flux removal after welding. AWS Specification A5.3 prescribes requirements for shielded metal-arc welding electrodes used in welding aluminum alloys.

Procedures used in welding aluminum and aluminum alloys, as well as data concerning the performance of these welds, are described in Chapter 69, Section 4.

FILLER METAL CLASSIFICATIONS

Aluminum alloys are classified by a four-digit system based upon the major alloying element. Filler metal alloys fall into four major groupings: the 1XXX series (pure aluminum), the 2XXX series (Al-Cu), the 4XXX series (Al-Si) and the 5XXX series (Al-Mg). All groups are used with the gas metal-arc and gas tungsten-arc welding processes. Only the 1XXX and 4XXX series are used with the gas metal-arc and gas welding methods. The four-digit numbers are prefixed with the letter E to indicate suitability as an electrode, R as a welding rod or ER for both products, in accordance with the standard AWS classification system.

The chemical composition of wrought and cast aluminum filler metal alloys is described in AWS Specification A5.10. The ASME F-Number grouping for those filler alloys included in the ASME Unfired Boiler and Pressure Vessel Code and typical mechanical properties of welds with each filler alloy are listed in Table 94.8.

ER1100 and ER1260.—Commercially pure aluminum alloy 1100 can be used to weld all 1XXX series base aluminum alloys, and 3003 and 5005 alloys. It provides adequate tensile strength for butt welds, along with very high weld ductility, good electrical conductivity and good corrosion resistance. For some chemical applications requiring a high-purity base aluminum alloy, a filler metal alloy of equal or higher purity is used; filler metal alloy 1260 is suitable for most of these applications.

ER2319.—This filler metal alloy is used with 2219 and 2014 wrought aluminum alloys, as well as Al-Cu casting alloys. This filler metal alloy is heat-treatable and provides highest strength and ductility with the Al-Cu alloys after postweld heat treatment.

ER4043 and ER4047.—These filler metal alloys can be used more universally than any others and with all arc and gas welding processes, although ER4047 is used most commonly as a brazing filler rod. They may be used with all 1XXX, 3XXX and 6XXX series alloys, as well as 2219, 2014, 5005, 5050, 5052, 7005 and 7039 wrought alloys. They are also used for welding Al-Si and Al-Si-Mg casting alloys or any combinations of these cast and wrought aluminum alloys.

ER4043 and ER4047 exhibit low sensitivity to cracking during welding, moderate strength and good corrosion resistance. The relatively high-silicon content of these fillers produces lower weld ductility than the 1XXX, 2XXX and 5XXX series filler metal alloys. Although neither ER4043 nor ER4047 is a heat-treatable alloy, the weld will respond and increase in strength when the weld metal is diluted with the 2XXX or 6XXX series base alloys, and the solution is heat-treated and aged.

ER4145.—This alloy exhibits the least sensitivity to weld cracking when welding the 2XXX series wrought alloys and the Al-Cu and Al-Si-Cu cast aluminum alloys. It is excellent for repair operations on these base alloys and will respond to heat treatment. ER4145 can be used with all the welding processes and all the other base alloys as the other 4XXX series fillers, but it exhibits slightly lower ductility than ER4043 or ER4047.

ER5XXX series.—Welds with these filler metal alloys possess the highest as-welded strength of any aluminum fillers and the highest ductility with the exception of the pure aluminum fillers. The ER5XXX series is used with the 5XXX and 6XXX series base alloys, 7005 and 7039. With the increase in the magnesium content of these filler metal alloys, weld metal strength increases and sensitivity to cracking decreases. Alloys ER5654, ER5356, ER5183 and ER5556

Table 94.8–Mechanical properties of common aluminum weldments

Filler Alloy					Elongation		
AWS Classification	F-No. Grouping ASME Section IX	Base Metal	Typical Ultimate Tensile Strength (psi x 10^3)	Typical Yield Strength (psi x 10^3) 0.2% Offset in 2 in.	Free Bend (%)	Tensile Elongation (% in 2 in.)	Minimum Shear Strength of Fillet Weld (psi x 10^3)
ER1100	F21	3003	16	7	58	24	7.5
ER1260	F21	1060	10	5	63	29	--
ER2319	--	2219	35	26	20	3	16
ER2319	--	2219	50[2]	38[2]		7[2]	22[2]
ER4043	F23	6061	27	18	16	8	11.5
ER5039	--	7039	48	30	30	10	--
ER5039	--	7039	50[2]	30[2]		12[2]	--
ER5183	F25	5083	43	22	34	16	19
ER5356	F22	5086	39	19	38	17	17
ER5554	F22	5454	35	16	40	17	17
ER5556	F25	5456	46	23	28	14	20
ER5654	F22	5154	33	18	39	17	12

[1] Actuated mechanical properties are greatly affected by welding parameters.

[2] Solution heat-treated and aged after welding.

are not recommended for applications of sustained elevated temperatures in excess of 150 F (66 C), but exhibit excellent characteristics for cryogenic applications.

ER5554.–This alloy is suitable for elevated temperature operations and is used with the 5052 and 5454 base alloys for this purpose. Weld metal tensile strength of 31,000 psi results.

ER5356.–This is the most universally used of the 5XXX series filler, since it combines a weld metal tensile strength of 38,000 psi with high weld metal ductility.

ER5183 and ER5556.–These alloys produce weld metal tensile strengths of 40,000 psi and 42,000 psi, respectively, with high weld ductility.

ER5654.–This filler has close controls of impurities to permit its use for hydrogen peroxide storage applications with 5254 and 5652 base alloys. It is also used with the 5154 alloy and produces a weld metal tensile strength of 30,000 psi with a weld metal elongation of 40%.

ER5039.–This is an Al-Mg-Zn filler metal alloy primarily designed for use with the 7005 and 7039 base alloys to provide highest strength with these alloys following postweld thermal treatments.

Casting filler metal alloys R-C4A, R-CN42A, R-SC51A and R-SG70A are used for foundry repair of castings of the same composition as the filler alloys.

SELECTION OF FILLER METAL

Selection of a filler metal is governed by the alloys welded and the service requirements of the particular application involved. If the strength of the welds is of primary importance, several aluminum fillers may be suitable to develop the minimum tensile strength across butt-welded joints. The selection of the filler may then depend upon another factor such as the highest shear strength for fillet welds or best weld ductility.

Most aluminum base and filler metal alloy combinations are satisfactory for general atmospheric conditions; however, for specific corrosive environments, fillers more nearly matching the chemistry of the base metal should be used for best performance. Chemistry match of specific elements in the filler metal and base alloys is also important to obtain optimum color match of the weld when the part is subjected to anodic treatments. For instance, alloys containing silicon turn gray or black when anodized, while alloys with chromium assume a yellow or gold color. Thus, 4043 or other 4XXX series fillers would give a poor color match with 6063 base metal but a good match with Al-Si casting alloys.

Refer to Chapter 69, Section 4 for the relative ratings of various filler alloys with the commonly welded wrought and cast aluminum alloys for six desired welding or service requirements. Where no rating is given, this combination is not recommended. In some instances, it may be desirable to conduct tests to evaluate the most suitable filler choice for a specific application. If the parts are heat-treated after welding, the choice of filler may be more restricted or different from the information given in Chapter 69.

The aluminum filler metal alloys available today have been carefully balanced chemically to minimize cracking problems, and it is rare to have cracks in any areas other than weld termination points. Since aluminum weld metal shrinks during solidification to a greater degree than ferrous metals, weld crater cracking can be encountered unless techniques are employed to fill the weld crater or to taper the molten weld pool width to a small size prior to extinguishing the arc.

Susceptibility to weld cracking can be influenced by the degree of dilution between the base metal and filler metal. Particularly when groove welding the 2XXX, 6XXX and 7XXX series heat-treatable aluminum alloys, it is desirable to minimize the percentage of the base-metal alloy dilution into the weld. Thus joints in these alloys are often beveled, as illustrated in Fig. 94.2 to increase the percentage of filler metal alloy in the final weld bead and reduce weld-cracking tendencies (see Chapter 69, Section 4).

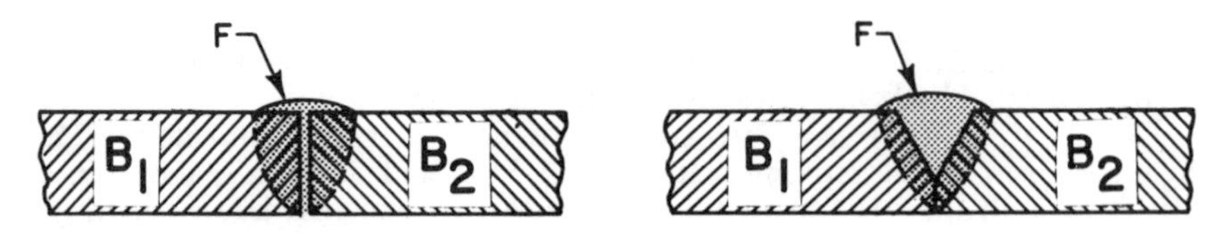

Fig. 94.2.–Square groove weld (left) and single-Vee groove weld (right), showing how joint preparation influences amount of dilution. F is filler metal; B_1 *and* B_2 *indicate base metals.*

Cracking (known as "hot-shortness") in aluminum welds is due to the low-strength or ductility of the weld metal composition or the metallurgical structure of the base metal adjacent to the weld at elevated temperatures. A filler metal alloy with a solidus-melting temperature similar to or below that of the base metal being joined greatly reduces the tendency for intergranular cracking in the base alloy adjacent to the weld. Such a filler metal alloy minimizes the stresses imposed by the solidification shrinkage of the weld metal until low-melting constituents of the base alloy have solidified and developed sufficient strength to resist the stress. Under high-restraint conditions, the pure aluminum filler metal alloys may not possess sufficient weld metal strength when cooling to withstand contraction stresses.

Weld metal compositions known to be sensitive to cracking should be avoided. Combining magnesium and copper in an aluminum weld is not desirable, so a 5XXX series filler metal alloy should be avoided when welding 2XXX series base alloys. Conversely, the 2319 filler metal alloy would not be used with a 5XXX series base alloy. In the case of Al-Si and Al-Mg alloy weld metal compositions, sensitivity to cracking is greatest in the range of 0.5 to 2% of the alloying element. Cracking tendencies decrease when the weld metal composition is below or above this range. For example, 5052 alloy (2.5% Mg), welded with ER5554 (2.75% Mg) filler metal alloy, would exhibit a greater susceptibility to weld cracking than if the weld were made with ER5356 (5% Mg) aluminum filler metal alloy. Figure 94.3 illustrates the weldability of various A1-Cu-Si and A1-Mg-Si alloys based upon the final weld metal chemistry.

MANUFACTURE, PACKAGING AND STORAGE

The surface of aluminum filler rods and electrodes must be free of moisture and lubricants since these become a source of hydrogen which is the major cause of porosity in aluminum welds. Thick or dark oxide films are not desirable since they will slow up and interfere with coalescence of the weld. The 5XXX series filler metal alloys are most sensitive to moisture absorption in their surface oxide, if packaged or stored improperly, and thus should be protected from rapid changes in temperature and humidity during shipment and storage.

All filler metal wire supplies should be stored in a dry, heated place and kept

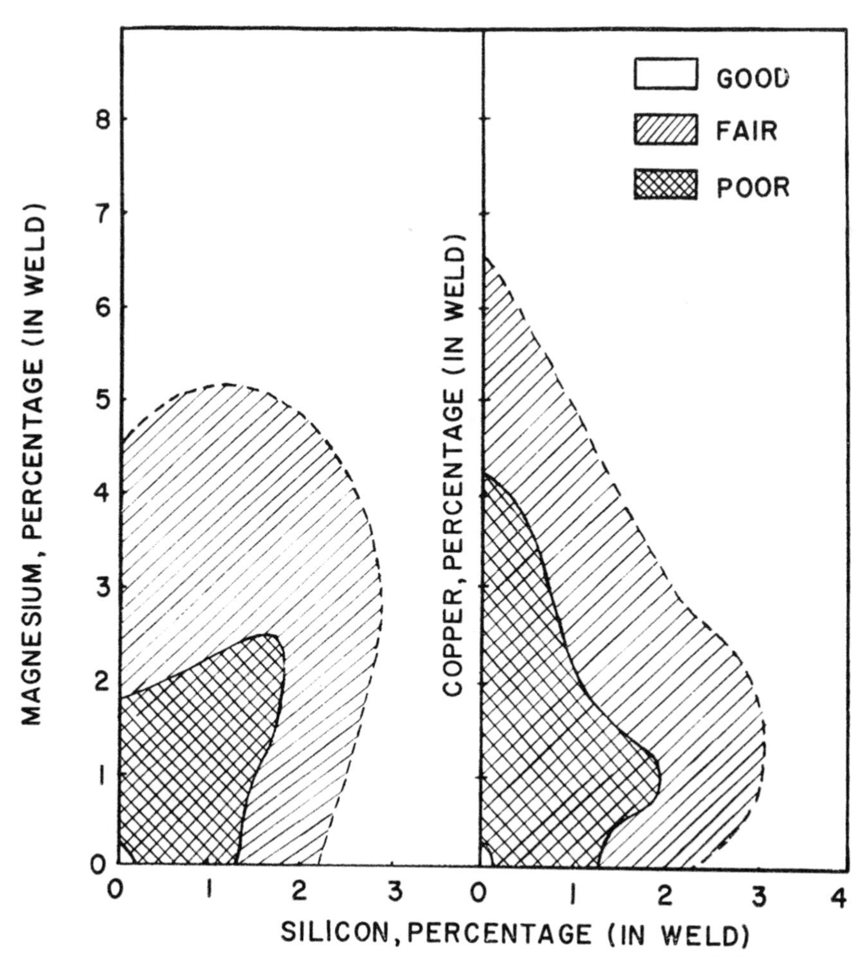

Fig. 94.3.–Diagrams indicating the relative weldability of Al-Mg-Si and Al-Cu-Si alloys.

covered. Electrode spools on the machine should be covered to avoid contamination. If an electrode spool is to be unused for an extended period, it should be returned to its carton and resealed tightly or placed in a desiccating or low-humidity cabinet. Original electrode or rod containers should not be opened until needed for use.

The flux used with covered aluminum rods or electrodes is hygroscopic and must be kept dry for proper performance. Thus these rods or electrodes must be well-protected from moisture during shipment and storage.

ALUMINUM FILLER AVAILABILITY

Standard sizes of bare aluminum welding electrodes on expendable spools are 0.030, 3/64, 1/16, 3/32 and 1/8 in. These are available on 10, 12 1/2 or 15-lb spools. Wire diameters up to 1/16 in. are available on 1-lb spools.

Bare welding rods in straight lengths and coils are supplied in 1/16, 3/32, 1/8, 5/32, 3/16 and 1/4 in. diam. The standard length of straight length aluminum rods is 36 in. and the most common package contains 5 lb. Other lengths and sizes of packages are available and described in AWS A5.10-69.

Covered aluminum electrodes are either 14 in. long, in diameters of 3/32, 1/8, 5/32, 3/16 and 1/4 in., or 18 in. long in diameters of 5/16 and 3/8 in. Standard packages contain 1, 5 and 10 lb net weight.

NICKEL AND HIGH-NICKEL-ALLOY FILLER METALS

GENERAL

The basic requirements for these filler metals are covered by AWS Specification A5.11 and AWS Specification A5.14.

Reference to the above specifications will indicate the wide range of alloy compositions which are predominately nickel and will also provide necessary details of standard dimensions. Military specifications MIL-E-22200/3 and MIL-E-21562 cover the same filler metals as the AWS specifications but interchangeability is not always possible for all alloy designations due to different test requirements.

SELECTION AND USE

The nominal composition of the base metal to be welded should be checked prior to final selection of the filler metal alloy. These analyses should match closely, but frequently, the filler metal will have percentage amounts of other elements which have been added to satisfy usability requirements e.g., to control porosity or hot-cracking tendencies. Two of the most important guidelines for the welder to follow in using the filler metals discussed in the following paragraphs are (1) to operate at low heat inputs and (2) minimize puddling or wide weaving.

Table 94.9—Nickel filler metals

AWS DESIGNATION	USE
ENi–1	Covered electrode for welding nickel to itself and to steel. It may also be used for surfacing steel.
RNi–2	Bare rods used without flux for oxyacetylene welding of nickel.
ERNi-3	Bare wire or rods for gas-shielded arc welding of nickel to itself and to steel. These filler metals may also be used for surfacing steel.

NICKEL

ENi-1, RNi-2 and ERNi-3 comprise those filler metals for welding wrought and weldable grades of cast nickel which, when properly deposited, will develop suitable soundness, strength, ductility and corrosion resistance for the applications for which the base metal was selected. Uses of these filler metals are shown in Table 94.9. As in all welding applications, one of the keys to a successful weld is the term "properly deposited." In the case of the nickel materials, this means special care in cleaning, minimization of puddling and recognition by the welder that nickel weld metal is somewhat more sluggish than carbon steel or stainless steel.

NICKEL-COPPER

The commercially available filler metal alloys that meet the nickel-copper designation are the several grades of Monel[5]; these are listed in Table 94.10.

Table 94.10—Typical nickel-copper filler metals

AWS DESIGNATION	USE
ENi Cu–1	Covered electrode, used for welding Monel to steel and for surfacing steel, as well as the steel on the clad side of Monel clad steel.
ENi Cu–2	Same as above but meets more rigid bend requirements.
ENi Cu–4	Covered electrode used where Cb is undesirable.
RNi Cu–5	Bare wire for oxyacetylene welding of Monel with flux.
ERNi Cu–7	Bare wire or rod for gas-shielded arc and submerged arc welding of Monel to itself and for surfacing steel and the alloy side of clad steel.

Like nickel, the nickel-copper filler metals used with the gas-shielded arc welding processes (R and ER) contain elements to prevent porosity or hot cracking or both, and will, therefore, be of different chemistry than the base metal they are joining. On occasion, this difference may show up in different degrees of corrosion resistance.

NICKEL-CHROMIUM-IRON

This class of filler metals is usually referred to as the Inconel[5] series, and the major ones are listed in Table 94.11. As suggested previously for nickel and nickel-copper, the use of specific nickel-chromium-iron filler metals should be checked chemically for the base metal to be welded.

[5] Registered trade mark of the International Nickel Co., Inc.

Table 94.11—Nickel-chromium-iron filler metals

AWS DESIGNATION	USE
ENiCrFe–1	Covered electrode, e.g., Inconel 132, for welding Inconel 600.
ENiCr–1	Covered electrode, e.g., Chromend* 20/80, for welding high-nickel alloys to carbon steel.
ENiCrFe–3	Covered electrode, e.g., Inconel 182, for more rigid use than ENiCrFe–1; it is used widely for high-temperature applications involving cast base alloys.
RNiCr–4	Bare rod, e.g., Inconel 42, for oxyacetylene welding Inconel 600.
ERNiCrFe–5	Bare rod and wire, e.g., Inconel 62, for gas-shielded arc welding Inconel 600.
ERNiCr–3	Bare wire and rod, e.g., Inconel 82, for gas-shielded arc welding Inconel 600 and related alloys, e.g., cast alloys used for high temperatures.
ERNiCrFe–6	Bare wire and rod for gas-shielded arc welding high-nickel alloys to carbon and austenitic stainless steels.
ERNiCrFe–7	Bare wire and rod for gas-shielded arc welding Inconel X–750 and 722 which are age-hardenable alloys.

*Registered trademark of the Arcos Corp.

DISSIMILAR METALS

Frequent applications of dissimilar metals are found where it is required that a nickel-base alloy must be joined to steel, e.g., monel to steel, type 304 stainless steel to nickel, etc. Table 94.12 indicates a few typical filler metals widely used for these applications including overlay welding on carbon steel.

Table 94.12—Typical nickel-base filler metals for welding dissimilar metals

AWS DESIGNATION	TYPE
ENiCrFe–2	Covered electrode, e.g., Inco weld A[1].
ENiMo–3	Covered electrode, e.g., Hastelloy W[2].
ERNiCrFe–6	Bare wire and rod for gas-shielded arc welding, e.g., Inconel 92[1].
ERNiMo–6	Bare wire and rod for gas-shielded arc welding, e.g., Hastelloy[2].

[1] Registered trademark of International Nickel Co., Inc.
[2] Registered trademark of Stellite Div., Cabot Corp.

Table 94.13—Nickel-molybdenum filler metals

AWS DESIGNATION	USE
ENiMo–1	Covered electrode Hastelloy B for welding the alloy to itself, to steel and to other metals.
ENiMoCr–1	Covered electrode Hastelloy C for welding the alloy to itself.
ENiMo–3	Covered electrode Hastelloy W for welding the alloy to itself, to steel and to other metals.
ERNiMo–4	Bare filler rod or wire for gas-shielded arc welding Hastelloy B as above.
ERNiMo–5	Bare filler rod or wire for gas-shielded arc welding Hastelloy C as above.
ERNiMo–6	Bare filler rod or wire for gas-shielded arc welding Hastelloy W.

NICKEL-MOLYBDENUM

The Hastelloy[6] type filler metal alloys are shown in Table 94.13. There are other Hastelloys for which there are not yet AWS specifications. The alloy manufacturer or filler metal supplier should be contacted for specific details.

COPPER AND COPPER-ALLOY FILLER METALS

CLASSIFICATION

Basic requirements for these filler metals are covered by AWS Specification A5.6-69, Copper and Copper-Alloy Arc-Welding Electrodes (shielded metal-arc, gas metal-arc and submerged arc welding) and AWS Specification A5.7-69, Copper and Copper-Alloy Welding Rods (oxyacetylene and gas tungsten-arc welding). The following classifications for copper-base alloys are commonly accepted categories; the use of trade names where they appear is meant to relate the subject to terms frequently used.

COPPER

The following grades of copper are commonly used industrially in welding applications: (1) CDA[7] 110–electrolytic tough pitch (2) CDA 102–oxygen free (3) CDA 122–phosphorus deoxidized.

---

[6] Registered trademark of Stellite Div., Cabot Corp.

[7] Copper Development Association

The ECu and RCu filler metals are used to weld the above coppers, and each is of the deoxidized and strengthened type with small additions of silicon, tin, phosphorus and manganese. ECu material is used for gas metal-arc welding the heavier thicknesses of copper while the RCu material is used primarily with gas tungsten-arc welding, although RCu may be used equally well in the acetylene process with flux. While the 110 alloy is weldable, the as-welded strength is not as good as that for the 102 or 122 alloys. Maximum weld properties are obtained with the 122 copper.

Due to the high thermal conductivity of the coppers and the high fluidity when melted, great care must be exercised by the welder to prevent incomplete fusion defects. This is best prevented by use of preheat, careful observation of the joint ahead of the weld pool and extensive practice. Preheat temperatures as high as 1000F are common. The ECu alloys are used with dcrp. There are no covered electrodes of the Cu analysis which will work efficiently.

COPPER-SILICON

Alloys in this category are almost always referred to as silicon bronzes, e.g., Everdur[8], Herculoy[9], Olympic[10], etc. CDA 655 (high-silicon-bronze A) is a basic reference. ECuSi and RCuSi-A are filler metals for gas-metal arc and gas tungsten-arc welding respectively. A covered electrode, ECuSi, is available, and the RCuSi-A filler metal may be applied with oxyacetylene using a flux. However, neither of these latter two filler metals are widely used. No preheat is required and the best practice, regardless of the welding process, is to use stringer beads and low heat. The silicon-bronze alloy is "hot short" and restrained joints should be avoided.

COPPER-ALUMINUM

These alloys are referred to as aluminum bronzes. Aluminum is the primary alloying element and the most common alloys are CDA 614 with 7% A1-2% Fe and CDA 954 with a cast alloy composed of approximately 10% A1-4% Fe.

The iron-free electrode ECuA1-A1, used with the gas metal-arc welding process, generally serves as an overlay material. RCuA1-A2 is used only in the gas tungsten-arc welding process with either ac or dcsp for welding copper-aluminum-iron alloys and overlay applications.

The ECuA1-A2 electrodes (Ampco-trode 10[11]) may be covered for shielded metal-arc welding, bare for gas metal-arc welding or stranded for gas metal-arc welding. These are quite versatile electrodes and are applied to aluminum-bronze, silicon-bronze and manganese-bronze, as well as to weld overlays on most metals for corrosion and wear resistance.

RCuA1-B (high iron) electrodes are used with the gas tungsten-arc welding process for applications similar to those for ECuA1-A2 filler metal. ECuA1-B is generally used for surfacing applications.

---

[8] American Brass Company
[9] Revere Copper & Brass, Inc.
[10] Chase Brass & Copper Co., Inc.
[11] Ampco Metal, Inc.

COPPER-NICKEL

CDA 706 (90% Cu-10% Ni) and CDA 715 (70% Cu-30% Ni) are the major cupronickel alloys. Both are welded with ECuNi which may be a covered electrode for shielded metal-arc welding or bare for gas metal-arc welding. Both are also welded with RCuNi which may be used with oxyacetylene (flux required) or gas tungsten-arc welding.

COPPER-TIN

Copper-tin alloys CDA 510 (95% Cu-5% P) and CDA 521 (92% Cu-8% P) are usually referred to as phosphor-bronze A and C respectively. Filler metals for each are the covered electrodes ECuSn-A and ECu-Sn-C which are used in the shielded metal-arc welding process and the bare RCuSn-A which is used with the gas tungsten-arc welding process in joining the 510 alloy.

Since the alloys are hot-short, it is necessary to adjust welding procedures to compensate for this property. Best results require that small weld pools, stringer beads, low-heat input and relatively unrestrained joints be used. The phosphor-bronze filler metals may also be used for overlay on steel to provide bearing surfaces.

COPPER-ZINC

The copper-zinc alloys cover a multitude of brasses and bronzes of which relatively few are considered weldable. Due to the zinc volatilization problem (toxic fumes and zinc depletion from pool), there are no arc-welding filler metals in this category. The following briefly describes the bare filler metals in the copper-zinc classification and their applications.

RBCuZn-A is frequently called Naval Brass and is used primarily with the oxyacetylene process with flux for joining brass (welding), surfacing copper, bronze, steel, etc., for corrosion and wear resistance or for braze-welding steel and the cast irons to themselves.

RCuZn-B is similar but the addition of 0.15% Si (maximum) provides a low-fuming characteristic when the oxyacetylene process is used. Nickel is also present to assist in the iron distribution in the deposit. Properly used with flux and a slightly oxidizing flame, the weld pool will have a slight silicon-oxide film to minimize zinc vaporization (zinc oxide fumes). The alloy is used to braze weld copper and steel and to weld brass.

RCuZn-C is similar to the -B alloy, but does not contain nickel. It is the most widely used low-fuming bronze alloy for the oxyacetylene welding of brass and the braze welding or bronze welding of steel, copper, etc.

RBCuZn-D is a high-nickel alloy and is referred to as a nickel-bronze. Its use is primarily for braze welding steel where it provides the maximum strength of all of the braze-welding alloys.

MAGNESIUM AND MAGNESIUM-ALLOY FILLER METALS

The most commonly used joining methods for magnesium which employ filler metal are the gas-shielded arc welding processes. Rather small quantities of filler metal are also consumed in gas welding and brazing.

ELECTRODE CLASSIFICATION

Magnesium welding rods and electrodes are classified on the basis of filler metal chemistry. The classification used in AWS A5.19 and MIL-R-6944 is based on the standard nomenclature established in ASTM Recommended Practices B275, Codification of Light Metals and Alloys, Cast and Wrought. Information concerning the composition of the several magnesium filler metals in current use is given in Chapter 70, Section 4. With the exception of the AZ91C composition, the filler metals are covered in AWS A5.19 and MIL-R-6944.

SELECTION OF FILLER METAL

The selection of filler metal composition depends primarily on the magnesium alloy to be welded. Experience has shown that certain filler metals are suitable for welding specific base metals or combinations of base metals; AWS Specification A5.19 contains a guide to the choice of filler metals based on this experience. The filler metals listed there give the best combination of weld strength and weldability obtainable in these alloys.

The most commonly used filler metal for the wrought magnesium alloys not containing thorium or rare earth is the AZ61A composition. The higher cost AZ91C, AZ92A and AZ101A compositions, although satisfactory for the wrought alloys, are used primarily for castings because of their superior resistance to weld cracking on these more highly alloyed base metals. The EZ33A composition is used to weld the thorium or rare-earth-containing wrought or cast alloys.

WELDABILITY

The weldability of most magnesium alloys is good to excellent if the proper filler metal is employed. The use of a filler metal with a lower melting point and a wider freezing range than the base metal is advantageous as it improves weldability and minimizes weld cracking. For details concerning the metallurgical characteristics and weldability of magnesium, see Chapter 70, Section 4.

WELD STRENGTH

The strength of welds in most magnesium alloys usually approaches that of the base metal because of the fine-grain size obtained in the weld deposit. Welded joints will frequently fail in the heat-affected zone rather than in the weld itself because the strength of the deposited weld metal exceeds that of the heat-affected base metal. See Chapter 70, Section 4 for a listing of the typical all-weld-metal deposit strengths of the several magnesium fillers. The transverse tensile properties of welds in all commonly welded magnesium base metals will be found in Chapter 70, Section 4.

WELDING CONSIDERATIONS

Wrought magnesium alloys pose no unusual welding problems but those containing more than about 1.5% aluminum do require a postweld stress relief to prevent stress-corrosion cracking in service. Most sand and permanent mold castings are also readily welded; however, they generally require preheating to avoid weld cracking unless the weld is made in a relatively unrestrained area such as an external boss or flange. In general, cast magnesium alloys also require postweld heat treatment to achieve maximum strength in the weld area. Where postweld heat treatment is not necessary or desired to achieve maximum properties, a stress relief must still be applied to the Mg-Al-Zn alloys after welding to prevent stress-corrosion cracking.

Details of joint design, weld precleaning, weld preheat and postheat, weld stress relief and gas-shielded arc welding schedules, including filler metal consumption information, will be found in Chapter 70, Section 4.

MAGNESIUM FILLER METAL AVAILABILITY AND SIZE SELECTION

Filler metals for welding magnesium are available on expendable spools containing 3/4 or 10 lb of wire and in straight length rods. The spooled electrode is offered in standard diameter sizes of 0.040, 3/64, 1/16, 3/32 and 1/8 in. Uncoated straight length rods are 36 in. long and are furnished in diameters of 1/16, 3/32, 1/8, 5/32 and 1/4 in. The diameter of the rod to be selected for gas or gas-shielded arc welding is dependent on the type of weld operation to be performed, e.g., surface buildup or hole filling on a casting, butt weld, fillet weld, etc. With the gas metal-arc process, it is also necessary to take into account the type of metal transfer and the amperage level to be employed in making the desired weld.

See the appropriate table in Chapter 70, Section 4 indicating the range of base metal thickness which can be welded by various filler metal sizes as well as the relative cost of each filler metal diameter. In spray-type arcs, the lowest weld cost is achieved by using the largest diameter filler metal suitable for the job. Even though a small (0.040 in.) diameter filler wire is satisfactory for spray welding all applicable thicknesses, its use would not be economical on the heavier base metal thicknesses because the cost of the small diameter electrode greatly exceeds that of the larger diameter electrode which is also satisfactory.

In the short-circuiting mode of operation, there is no option of substituting filler metal size because operation outside the indicated limits results in the loss of proper transfer of metal from the electrode. Refer to Chapter 70, Section 4 for an illustration of the approximate amperage ranges which exist for the various types of metal transfer as a function of the electrode diameter.

STORAGE AND CARE OF FILLER METAL

Filler metal cleanliness and freedom from surface oxidation are important in gas-shielded arc welding. Oil or other organic material, as well as heavy oxide coatings on the surface of the filler metal, will interfere with coalescence of the weld puddle and cause porosity or other weld defects or both. For this reason, filler metal is manufactured and packaged in such a manner as to prevent contamination.

Proper storage of welding rods and electrodes in the user's plant is essential if the filler metal is to remain free of contamination until used. Packages of filler metal should not be left outdoors or in unheated buildings because the greater variations in temperature and humidity increase the possibility of moisture condensation. Properly protected filler metal can be stored for long periods of time without adverse effects on its performance.

Packages of filler metal should remain sealed until ready for use. Once removed from the container, a spooled electrode should be kept covered, even during use, to prevent surface contamination by dust, moisture or other air-borne foreign material. After welding is completed, the electrode should be returned to its original container for storage. Welding rods should also be kept covered until ready for use. Storage conditions here need not be as rigorous as for spooled electrodes because good welding practice always includes a stainless steel wool rub of the rod just prior to use to remove surface oxidation or other contamination.

TITANIUM AND TITANIUM-ALLOY FILLER METALS

The two main criteria influencing the selection of titanium or titanium-alloy filler metals are the interstitial impurity level and alloy type. The level of interstitial impurities (carbon, oxygen, hydrogen and nitrogen) is of major concern for two reasons. First, small amounts drastically affect the mechanical properties by raising yield strengths and decreasing ductility and toughness. Relatively small amounts of these impurities in the filler metal can be the cause of severe embrittlement in the welded metal. Second, because of titanium's high chemical reactivity at elevated temperatures, these impurities are readily picked up during welding. Oxygen and nitrogen will be readily absorbed from the atmosphere by hot titanium, even at temperatures appreciably below its melting point.

The presence of oxygen and nitrogen in quantities as low as 0.5% will embrittle a weld beyond the point of usefulness. Accordingly, titanium must be shielded from the normal atmosphere during the welding process by blanketing the weld zone with an inert atmosphere of argon or helium. If the shielding is expected to be less than ideal, and ductility and toughness of the weld are important, the impurity level of the filler metal must necessarily be low to accommodate the expected pickup during welding.

Molten titanium is highly reactive with most filler metals, including all the common refractories. This requires that the filler metal, as well as the weld areas, be free from dust, grease and contact with ceramic blocks or other foreign materials during welding. Similarly, covered arc-welding electrodes and other fluxing compounds cannot be used since they cause contamination and embrittlement. Filler metal is limited to bare wire used in conjuction with gas-shielded arc welding processes.

CHEMISTRY AND SELECTION OF FILLER ALLOY

The selection of the correct type of filler depends upon the application, the type of base metal and the expected pickup of impurities during welding. Metallurgically, there are three types or classes of alloys: (1) alpha alloys, including commercially pure titanium (2) beta alloys (3) the two-phase alpha-beta alloys. A discussion of the metallurgy and weldability of these alloys is presented in Chapter 73, Section 4. Titanium cannot be fusion welded to any other alloy systems.

Titanium and titanium-alloy welding rods and electrodes are classified on the basis of chemistry—that is, the composition of the pure metal or alloy used to manufacture the wire or the rod. The composition of the titanium and titanium-alloy filler metals in current use is covered in AWS A5.16 and MIL-R-81588. There are slight differences in some of the values when both documents are compared. AWS A5.16 does not classify the rods and electrodes according to metallurgical types whereas the Government document does. The ER prefix used in AWS A5.16 indicates suitability as an electrode or welding rod in accordance with the standard AWS classification system.

The listed values for the interstitial elements (carbon, oxygen, hydrogen and nitrogen), as well as iron, are the maximum acceptable limits. Excessive iron can reduce the ductility and impact strength of titanium due to formation of a brittle intermetallic compound and solid-solution hardening. As noted in the specifications, these levels of impurities are considered to result in usable filler metal when using refined welding procedures and nearly perfect shielding. Many fabricators of aerospace and high-pressure cryogenic equipment specify the extra low interstitial grades (ELI) where ductility and impact properties are critical.

By filler metal selection, it is possible to control the alloy content of the weld metal. In general, a particular alloy is welded with a filler of the same composition or one somewhat lower in alloy content. A lower alloy content is usually chosen when ductility and toughness are more important than strength. Thus, the commercially pure and alpha filler metals are sometimes employed to weld alpha-beta alloys. Use of these filler metals lowers the beta content of the weld metal which generally results in improved weld ductility and toughness.

Unalloyed Fillers

Unalloyed titanium filler metal (ERTi-1 to ERTi-4) is used to weld unalloyed titanium, alpha alloys and alpha-beta alloys. The higher the number of the ERTi filler, the higher will be the strength with correspondingly lower ductility.

When welding unalloyed titanium, a filler grade is usually chosen on the basis of yield strength. For example, if welding conditions are such that significant pickup of interstitials is expected, a filler of lower yield strength is chosen since the increase in interstitial content in the deposited weld metal will raise the yield strength and lower the ductility. A high-strength base metal should be welded with a high-strength filler only under ideal conditions.

When unalloyed filler is used to weld alpha-beta alloys, the joint efficiency may be reduced to about 90%, but improved weld-joint ductility and weld metal toughness are usually obtained.

Alpha Fillers

The alpha alloy filler metals (ERTi-5A1-2.5Sn, ERTi-8A1-1Mo-1V, and ERTi-6A1-2Cb-1Ta-1Mo) are used to weld base metals of the same composition as the filler metals. They may also be used to weld alpha-beta alloys, but have not been used extensively for this purpose. The ERTi-3A1-2.5V filler metal is also used to weld base metal tubing of the same composition. This filler metal is not as readily weldable as the alpha metals. The ERTi-6A1-2Cb-1Ta-1Mo filler metal has replaced the older ERTi-8A1-2Cb-1Ta filler metal because aluminum must be limited to 7% for maximum weld toughness.

Alpha-beta Filler

Alpha-beta alloy filler metal (ERTi-6Al-4V) is used to join the alpha-beta alloy base. Weldability of other alpha-beta base metals is either limited or not recommended.

Beta Filler

Beta-alloy filler metal (ERTi-13V-11Cr-3Al) is used to join the beta-alloy base metal. It is not used to weld other base metals as better all-around properties are usually obtained with a lower beta content in the weld metal.

Dissimilar Combinations

The alpha and alpha-beta alloys are welded to each other either by using a filler that matches the composition of the base metal with the lowest beta content or by using the unalloyed ERTi-1 through ERTi-4 filler metals. The beta alloy may be difficult to weld to the other alloys because dilution in the weld joint may result in weldments having very low ductility.

The ductility and impact strength of titanium will be greatly reduced if alloyed with other structural metals such as steel and aluminum. For this reason, titanium has never been satisfactorily fusion welded directly to other common structural metals. Experimental methods for process equipment advanced the use of compatible metals such as silver and vanadium as interlayers between steel and titantium.

WELDING OPERATIONS

Typical settings for welding titanium with and without filler are listed in Table 94.14. These settings are for semiautomatic or automatic arc-welding equipment with backup bar, trailing shield and holddown shoes. Settings for manual welding are quite similar when backup bars, trailing shields and holddown shoes or chill bars are used. The arc gas flow may be increased 60 to 100% when a trailing shield is not used. To weld at a slower speed, the amperage should be reduced proportionately. To reduce melt-thru, when no backup or chill bar is being used during either manual or automatic welding operations, the current should be reduced about 20%.

Actual settings may vary widely even when welding similar gage material with different set-ups. To ensure that the correct filler metal is being used, the welding speed is adequate, the amperage is proper and the weld penetration is complete, it is best to first weld test specimens for determining optimum settings for the application.

FILLER METAL AVAILABILITY

Filler metals for welding titanium and its alloys are available in straight lengths, in coils without support or on expendable spools. Standard straight lengths of 36 in. (914 mm) are available in 5-(2.27 kg), 10-(4.53 kg), 25-(11.34 kg) and 50-(22.68 kg) pound (lb) quantities. The spooled wire can be obtained in the 1-lb (0.45 kg), the 10-lb (4.54 kg) and the 20-lb (9.07 kg) size. Spooled wire in the 1-lb (0.45 kg) size is offered in standard diameter sizes: 0.030 in. (0.76 mm), 0.035 in. (0.89 mm), 0.045 in. (1.13 mm); 1/16 in. (2.39 mm) and 1/8 in. (3.18 mm) are available in the heavier weight spools, in coils of 25 lb (11.34 kg) and 50 lb (22.68 kg), and in the standard straight lengths.

Table 94.14—Typical weld settings for "open air" machine welding

	Tungsten Arc Without Filler			Tungsten Arc With Filler			Consumable Electrode			
Gauge, in.	0.030	0.060	0.090	0.060	0.090	0.125	0.125	0.250	0.500	0.625
Electrode diam, in.	1/16	1/16	1/16-3/32	1/16	1/16-3/32	3/32-1/8	1/16	1/16	1/16	1/16
Filler wire diam, in.	–	–	–	1/16	1/16	1/16	–	–	–	–
Wire feed rate, ipm	–	–	–	22	22	20	200-225	300-320	375-400	400-425
Voltage (V)	10	10	12	10	12	12	20	30	40	45
Amperes (A)	25-30	90-100	190-200	120-130	200-210	220-230	250-260	300-320	340-360	350-370
Nozzle id, in.	9/16-5/8	9/16-5/8	5/8-3/4	9/16-5/8	5/8-3/4	5/8-3/4	3/4-1	3/4-1	3/4-1	3/4-1
Primary shield, cfh	15A	15A	20A	15A	20A	20A	50A+15H	50A+15H	50A+15H	50A+15H
Trailing shield, cfh	20A	30A	50A	40A	50A	50A	50A	50A	60A	60A
Back-up shield, cfh	4H	4H	5H	5H	6H	6H	30H	50H	60H	60H
Back-up material	Copper or Steel			Copper or Steel			Cu	Cu	Cu	Cu
Back-up groove, in.	1/4 x 1/16	1/4 x 1/16	3/8 x 1/16-3/16 x 1/16	1/4 x 1/16	3/8 x 1/16-3/16 x 1/16	3/8 x 1/16-3/16 x 1/16	3/8 x 1/16-3/16 x 1/16	1/2 x 1/8-1/4 x 1/16	5/8 x 1/8-1/4 x 1/16	5/8 x 1/8-1/4 x 1/16
Electrode travel, ipm	10	10	10	12	12	10	15	15	15	15
Power supply	DCSP	DCSP	DCSP	DCSP	DCSP	DCSP	DCRP	DCRP	DCRP	DCRP

MANUFACTURE, PACKAGING AND STORAGE

The quality requirements of titanium filler metals are at present established by various specifications, the major ones being AWS A5.16 and MIL-R-81588. To meet these specifications, rods and electrodes must be uniform in composition and color, and they must be smooth and free from injurious defects. Rods must be straight. Spooled wire must be free from scratches, nicks, waves, sharp bends or kinks so that the wire is free to unwind without restrictions caused by overlapping, wedging, tangling and bumping.

To prevent moisture absorption and contamination between manufacture and use, most titanium welding wire, especially for aerospace use, is packaged in vapor-barrier containers. Premium quality filler metals are inner-packaged and placed in an envelope or a bag or in a compartment of a multiple-compartment envelope. Envelopes must be free of oil vapors or other substances detrimental to welding operations or the filler metal.

Envelope material used in packaging should have a water transmission rate lower than 0.05 g per 645.2 sq cm (100 sq in.) per 24 h at 37.8 C (100 F) and 90% relative humidity. The material should also have a sufficient strength to withstand normal handling. To prevent puncturing the envelope, the ends of wire may be capped. For further protection against oxidation and corrosion of wire, envelopes are purged with an inert gas. This gas is maintained as the atmosphere in the envelope. Desiccants may also be placed in the envelope for additional protection prior to heat sealing.

SURFACING FILLER METALS

GENERAL

Since surfacing can be defined as the deposition of filler metal on a metal surface to obtain desired properties or dimensions, any of the filler metals described in this chapter can be used, at times, as surfacing materials. To focus on a more specialized field (hard facing), certain alloys intended to resist wear and a few multiple-purpose types will be described briefly. They are covered in more detail in Chapter 44, Section 3A of this Handbook, which also includes information about surfacing processes and procedures. Many special surfacing alloys are covered by AWS Specifications A5.13 and A5.21, but many more are not included because they have not achieved adequate industrial standardization (they are either proprietary alloys or they have limited availability).

The issuance of two specifications is explained by a difference in manufacturing. For solid drawn or cast rods and electrodes used bare, the grade can be defined by chemical analysis of the bare rod. For covered electrodes, analyses must be made on the deposited metal because the covering may have made some contribution. Such compositions are defined in AWS A5.13.

Filler metals can also be formulated on the basis of the melt that results from welding with a tube filled with powder or granules. Such tube products can be either bare or covered and are defined in AWS A5.21.

Deposit analyses specified for composite filler metals are substantially the same as those given in A5.13 except that only the RFe5-A, RFe5-B and RFeCr-Al classifications are defined for composite rods, and only the EFe5-A,

EFe5-B, EFe5-C, EFeMn-A, EFeMn-B and EFeCr-Al classifications are defined for composite electrodes.

Additionally, the AWS 5.21 Specification defines composite tungsten-carbide welding rods and electrodes as consisting of tungsten-carbide granules in a mild steel tube or sheath. Classification of these products is based on the mesh size of the tungsten-carbide granules preferred by "RWC" for rods and "EWC" for electrodes. Specified mesh size and weight per cent of tungsten-carbide granules for rods are given in AWS A5.21. Specifications for electrodes are similar but cover only 12/30, 20/30, 30/40, 40 and 40/120 mesh sizes.

DESIRED CHARACTERISTICS

A combination of properties including hardness, abrasion resistance, corrosion resistance, impact resistance and heat resistance must be considered when selecting filler metal for surfacing applications. Hardness requirements may be in the hot or "red hardness" range as well as at normal temperatures. Abrasion resistance is sometimes, but not always, related to hardness and depends upon both the type of wear and the individual constituents present in the surfacing metal. Corrosion resistance depends on the service conditions, on the soundness of the weld deposit and on its composition, with the stipulation that the original filler metal may be altered by dilution from the base metal.

Impact resistance (somewhat different from that indicated by conventional notched-bar tests) depends on yield strength to resist plastic flow under battering blows and on toughness to resist spalling and cracking under deformation. Oxidation resistance, which is needed in high-temperature applications, depends chiefly upon chromium content. Metal-to-metal wear applications may involve seizing and galling, which are welding phenomena that are inhibited by high-elastic strength and by surface films. Soft copper-base alloys–leaded bronzes–and those that develop tenacious oxide films–aluminum bronzes–may thus be used for friction surfaces.

Toughness and abrasion resistance tend to be incompatible properties in an alloy (with the notable exception of austenitic manganese steel) and usually force a compromise selection. The harder alloys are prone to cracking and many overlays of these are fissured. Overlay cracking from the thermal stresses of welding is usually undesirable, but is sometimes acceptable if vertical cracks relieve stresses and prevent lateral spalling of the deposit.

CLASSIFICATION OF ALLOYS

The magnitude and variety of the surfacing field have resulted in a great variety of products, making selection of the best filler metal difficult. Careful analysis of service conditions–matching them against weld deposit properties, supplemented by established, reliable and valid service test data–provides the best method of selecting the proper alloy.

Classification can be based on many factors including hardness, composition, service conditions and abrasion resistance in a specific test. For this discussion, the most useful method is a combination of chemical composition and structure of the as-deposited metal, since most surfacing deposits are used in this condition. A basic distinction exists between ferrous and nonferrous base alloys.

An older and unsophisticated classification based on total alloy content (excluding iron) should be avoided. It implies increased merit with increased alloy percentage; there is no need for this oversimplification and it can be very misleading. The alloying elements should be named and percentages given whenever filler metals are to be characterized by composition.

MAJOR FERROUS BASE CLASSIFICATIONS

Martensitic Steels

Carbon content is usually in the range of 0.10 to 0.50%, but may be as high as 1.50%. Other elements, selected for their hardenability contribution, are used in moderate amounts to promote martensite formation as the weld deposit cools. Molybdenum and nickel (rarely above 3%), and chromium (up to about 15%) are the favored alloys; manganese and silicon are usually present, chiefly for deoxidation.

The carbon in these steels is the major property-influencing element. These steels are inexpensive and relatively tough; the lower carbon alloys are tougher and more crack-resistant than the higher carbon types. They are capable of being built up to form thick, crack-free deposits. The deposits have high strength and some ductility. Abrasion resistance is moderate, and, within the group, has a tendency to increase with carbon content and with hardness. The lower hardness deposits may be machinable with tools while grinding is advisable with higher hardnesses. This classification represents the largest volume usage, on a weight basis, of surfacing filler metal.

The moderate price, good welding behavior and broad range of properties of these steels make them popular for many of the bulk-surfacing uses, such as the buildup of shafts, rollers and other machined surfaces; simple moderate abrasion applications involving impact and buildup of badly worn surfaces before finishing with a more expensive and more highly alloyed deposit.

High-speed steels are basically martensitic steels with tungsten, molybdenum and vanadium additives to improve hardness at high temperatures to about 1200 F. They are similar to the molybdenum-type high-speed tool steels and are used in applications where a high-speed tool steel deposit is desired. The air-hardening deposits need no heat treatment.

Pearlitic steels are similar to the martensitic steels but they contain less alloy. Because of this lower alloy content, they form pearlite (a structure softer than martensite) on cooling. Pearlite steels are useful as build-up overlays.

The low-alloy steels, which represent the largest volume use of surfacing filler metal, may be either martensitic or pearlitic in weld deposits. Mixed structures are common. As the potent "hardenability" elements are increased (particularly chromium, the least expensive of these), there is a tendency to raise the amount of retained austenite in the deposit. In such medium-alloy steels, the raw, untempered martensite that forms as the weld cools is relatively brittle and prone to cracking. The austenite is softer and tougher. Some grades, which may be termed "semiaustenitic", are formulated to exploit this toughening while depending on martensite for hardness.

Hard surfacing deposits are seldom heat-treated, the deposit properties being dependent on composition and weld cooling rate. A notable exception is the buildup or repair of special tool steels such as dies, etc. Here the filler metal is usually closely matched to the base metal, and the full tool steel techniques of heat treatment and handling are invoked to achieve a high-integrity product.

Austenitic Steels

The two major types of austenitic steels are those based on high manganese (related to Hadfield manganese steel) and the Cr-Ni-Fe stainless steels. Both are tough, crack-resistant and capable of crack-free deposition in thick, multiple-layer deposits. They are relatively soft as deposited (150 to 230 BHN), but they rapidly work-harden from deformation or under impact.

The stainless grades, notably types 308, 309, 310 and 312, are used for corrosion-resistant overlays and for joining or build-up purposes. They are rarely employed for wear resistance. As a separation layer between carbon or low-alloy steels and manganese or surfacing overlays, they contribute enough alloying elements to minimize or prevent the formation of brittle martensite in the dilution zone. The 309 and 310 grades are good, heat-resistant alloys and serve for surface protection against oxidation up to about 2000 F. Modified 308 types are used for railway trackwork rebuilding (with additives of 0.3 to 0.6% carbon and 4% manganese) and for various hot-wear applications (with higher carbon and possibly molybdenum).

The manganese steel types usually depend on 12 to 15% manganese to ensure the austenitic structure, 0.5 to 0.9% carbon (the base 13% manganese steel may contain between 1.0 to 1.4% carbon) and nickel (2.75 to 5.0%) or molybdenum (0.6 to 1.4%) to enhance toughness. The molybdenum type has higher yield strength and flow resistance. These grades have been widely used to rebuild battered-down railway trackwork, provide metal-to-metal wear resistance coupled with impact and protect the surface or replace worn areas where abrasion is associated with severe impact. Surfacing of power shovel dippers is a common and typical application. However, these grades are seldom appropriate for hot wear because temperatures above 800 F may embrittle the high-manganese austenite.

More complex grades have been developed to provide higher yield strength, more resistance to reheating and less tendency for a vulnerable dilution zone when welded against carbon steels. The following combinations may be encountered: 15% manganese with 15% chromium and 1% nickel; 4% nickel, 4% chromium, 14% manganese and related types with up to 2% vanadium. The 4Ni-4Cr type has displaced the simpler 4Ni-1Mo type because of its superior yield strength and resistance to the effects of welding heat. This complex grade is also considered superior for joining cast and wrought manganese steel where carbon and phosphorus respectively should not exceed 0.9 and 0.035% in order to avoid weld metal fissuring.

The welding of carbon steel to the 13% manganese type should be done cautiously; a low-manganese fusion zone may develop martensite and become quite brittle unless the lowered manganese is compensated with other equally effective alloying elements. The 309 and 310 stainless grades, as well as the complex manganese types, are capable of this.

Arc welding is almost universally used in this area because the heat of gas welding tends to embrittle manganese steel base metal. The manganese steel electrodes are very popular in the mining, mineral processing and earthmoving industries, primarily for surfacing and rebuilding austenitic manganese steel castings.

Irons

The iron-base alloys with high carbon are termed irons because they have the characteristics of cast irons. They have a moderate-to-high alloy content to confer air-hardening properties or to create special hard carbides in the surfacing deposits. Although the carbon range is from 2.0 to 5.5%, 3.5 to 5% is more typical. The irons resist abrasion better than the two previously described alloy types and are preferred up to the point where they lack toughness to withstand the associated impact. They are usually limited to one- or two-layer overlays and, even so, tension cracks are common, especially if large areas are covered.

The highest alloy group is the high-chromium irons with about 24 to 33% chromium. This high chromium, combined with the high carbon, produces hard Cr_7C_3 type carbides in the structure. Frequently, 4 to 8% manganese or 2 to 5% nickel is added to promote an austenitic matrix. Also, tungsten, molybdenum or vanadium may be added to increase hot hardness and abrasion resistance.

Though resistance to low-stress abrasion (for example, a plowshare working in sandy soil) is high for all of this group, the irons with an austenitic matrix are inferior for high-stress grinding abrasion (a ball mill). The high-chromium irons that undergo a martensitic transformation are good for both types of abrasion, especially when heat-treated. High chromium confers oxidation resistance, and in some applications, these alloys resist hot wear quite well, as in rolling mill or piercing mill guides. However, they are inferior to the Cr-Co-W type alloys in hot hardness above 1100 F.

Lower alloy irons, typically with 15% chromium and molybdenum or nickel, have an austenitic matrix and are very popular for their general abrasion resistance (they have more crack resistance than the martensitic irons) and their reasonable price. They have excellent resistance to low-stress abrasion in proportion to the content of hard carbide; since low-stress abrasion usually dominates situations where impact is low or absent, these lower alloys are applicable.

Martensitic irons comprise a group whose alloy content is low enough to permit at least partial transformation to hard martensite as the deposit cools from the welding heat. However, the alloying must be sufficient to prevent pearlite transformation: balance must be maintained for the intended purpose. Chromium, nickel, molybdenum and tungsten are the customary control elements. The usual matrix is a complex carbide-containing islands of martensite with some retained austenite.

The high-chromium martensitic irons are an exception where the austenite-martensite aggregate is the matrix. The presence of considerable martensite confers excellent resistance to high-stress abrasion (recognizable by the way the imposed stress fractures the abrasive), very high compressive strength and, consequently, high resistance to light impact. At the same time, resistance to low-stress abrasion is outstanding and metal-to-metal wear resistance may be high. With some grades, hot hardness is good up through 1000 to 1200 F. Disadvantages are the sensitivity to composition variables and the martensite cracking tendency. Though compressive strength is high, tensile strength is low. Most cracking is from tension resulting from either thermal stresses or the deformation of a soft base under the harder overlay. Proper support of the martensitic irons as overlays is important.

MAJOR NONFERROUS-BASE CLASSIFICATIONS

Composite Tungsten-Carbide Types

This surfacing filler metal is supplied in the form of mild steel tubes filled with crushed and sized granules of cast tungsten carbide, usually in the proportions of

60% carbide and 40% tube, by weight. The carbide, a mixture of WC and W_2C, is very hard, surprisingly tough and very abrasion-resistant. Deposits containing large, undissolved amounts of it have more resistance to all types of abrasion than any other welded overlay. Various grades supply different granule sizes in the filler, usually designated by the screen mesh size of the particles. Differential wear of the deposits will make them rough in proportion to the size of the particles. A grade with 8/12 granules (finer than 8 mesh to the inch and coarser than 12 mesh) might be used for nonskid surfacing of horseshoes: the much finer 40/120 grade is better on plowshares. The 20/30 and 30/40 sizes are popular for general use.

As the heat of welding melts the steel tube, the molten iron dissolves some of the tungsten carbide to form a matrix of high-tungsten steel or iron. The main function of this abrasion-resistant matrix is to anchor and support the undissolved granules. This is uniformly achieved with oxyacetylene welding; but arc welding, especially of the fine-granule electrodes, may dissolve so much of the carbide that abrasion resistance is impaired, though hot hardness is increased. Hot wear applications are unusual while hardness is high (to approximately 1100 F). Use above 1200 F is limited by softening of the matrix and oxidation of the carbide.

Cobalt-base Surfacing Metals

These alloys usually contain 26 to 33% chromium, 3 to 14% tungsten and 0.7 to 3.0% carbon. Three grades are available with hardness, abrasion resistance and crack-sensitivity increasing as the carbon and tungsten contents increase.

These alloys have high oxidation, corrosion and heat resistance. The 1% carbon grade is outstanding for exhaust valves in internal combustion engines. High hardness and creep resistance are retained at temperatures above 1000 F, and the alloys are useful for some types of service at temperatures as high as 1800 F. The cobalt base, combined with the chromium, provides good corrosion resistance in many applications; resistance to metal-to-metal wear is also very good.

Nickel-base Surfacing Metals

This is a broad group with many different categories and varying heat and corrosion resistances. The most common nickel-base surfacing alloys are the series containing 0.3 to 1.0% carbon, 8 to 18% chromium, 2.0 to 4.5% boron and 1.2 to 5.5% each of silicon and iron. There are three alloys in this group, with hardness and abrasion resistance increasing with the carbon, boron, silicon and iron contents. They may be spray-coated. The general comments made on the cobalt-base filler metals also apply to this group, although heat strength and resistance to high-stress abrasion are lower.

Many of the nickel-base filler metals which have been described in this chapter are also used for surfacing applications. Among these are nickel-base alloys with additions of copper, chromium, molybdenum, chromium-molybdenum and chromium-molybdenum-tungsten.

Copper-base Surfacing Metals

Various alloys of copper with aluminum, silicon, tin and zinc are used for corrosion as well as wear applications. Some of these alloys have been described previously in this chapter.

Alloys with aluminum contain 9 to 15% aluminum and up to 5% iron. They are the hardest of the copper surfacing alloys, reaching 380 Brinell hardness, and they are used extensively to minimize metal-to-metal wear. Alloys with tin and zinc are also used for bearing surfaces.

SELECTION OF SURFACING METALS

In general, the loss of metal from a part will involve one or more of the following: sliding or rolling friction, shock and impact, heat, abrasion and corrosion. Other factors, such as smoothness of the deposit desired or the ability of the surfaced part to form an efficient tool, also affect the choice. Previous discussion of each classification indicates the relative merits of the various groups. For specific applications, selection should be based on careful analysis of all pertinent factors, including previous service performance, if available from a clearly comparable situation. Reliable service validation, like the acceptance of the 1% carbon-chromium-cobalt-tungsten alloy for exhaust valves, is practical for many applications. However, careful analysis is better than casual service tests which are frequently misleading because of uncontrolled variables. Well-controlled and statistically valid field tests are expensive and time-consuming and justifiable for only very important applications.

Since most surfacing is a compromise between abrasion resistance and impact resistance as opposing properties, the initial selection is usually the most abrasion-resistant alloy that will give an acceptable deposit and will probably withstand the expected impact. If service causes impact failure, a tougher but less wear-resistant alloy is then substituted. If wear is rapid but toughness is clearly adequate, a step in the other direction can be made.

Alloy cost should not weigh heavily, since welding labor cost is usually more critical. Expensive alloys may more than pay their way with longer life when they are properly selected.

BRAZING FILLER METALS

Brazing filler metals, simply speaking, are metals that are added when making a braze. They have melting points above 800 F but below those of the metal being brazed and have properties suitable for making joints by capillary attraction between closely-fitted sufaces.

CHARACTERISTIC PROPERTIES

Although a wide variety of filler metals is used for brazing, all must satisfy common requirements and have certain characteristic properties:

1. The ability to wet the surfaces of the metals being joined and form a strong bond.
2. A melting point or melting range compatible with the metals being joined and sufficient fluidity at brazing temperature to flow and distribute into properly prepared joints by capillary action.
3. A composition of sufficient homogeneity and stability to minimize separation of constituents (liquation) under normal brazing conditions.
4. The ability to form joints possessing suitable mechanical and physical properties for the intended application.

Melting Behavior of Brazing Filler Metals

Refer to AWS Specification A5.8, Brazing Filler Metal, for the chemical composition ranges of AWS-classified brazing filler metals and the liquidus, solidus and brazing temperature ranges for the various fillers. The pure brazing metals and some brazing alloys melt at single temperatures or have a narrow melting range; such alloys melt and flow freely. Filler alloys with wide melting ranges are partially liquid and partially solid within this range and consequently flow rather sluggishly. The highest temperature at which the metal is completely solid, i.e., the temperature at which melting starts, is termed the solidus. The lowest temperature at which the metal is completely liquid, i.e., the temperature at which freezing starts, is termed the liquidus. The solidus and liquidus for a particular metal are definite and fixed. Other terms, such as flow point and brazing temperature, are not precise and vary with individual interpretation.

When brazing with some brazing filler metals (particularly those with a wide temperature range between solidus and liquidus), the several constituents of the filler metals tend to separate during the melting process. The lower melting constituent will flow, leaving behind a "skull" of the high-melting constituent. This occurrence, called liquation, is undesirable since the unmelted skull does not contribute significantly to the actual brazed joint: it does not flow into the joint. However, where fit-up is not too good, a filler metal with a wide temperature range will usually fill the joint more easily. A more detailed discussion of the behavior of filler metals during brazing is given in Chapter 60, Section 3B.

Brazing Filler Metal Classifications

The American Welding Society classification system for brazing filler metals is based on chemical composition rather than on mechanical property requirements. The mechanical properties of a brazed joint depend, among other things, on the base metal and filler metal used. Therefore, a classification system based on mechanical properties would be misleading since it would only apply if the brazing filler metal were used on a given base metal. If a user of brazing filler metal desires to determine the mechanical properties of a given base metal and filler metal combination, it is recommended that tests be conducted using AWS C3.2, Standard Method of Evaluating the Strength of Brazed Joints.

Brazing filler metals are standardized into seven fairly well-defined groups of classifications. These include the aluminum-silicon, the copper-phosphorus, the silver, the precious metals, the copper and copper zinc, the magnesium and the nickel filler metals. These seven basic groups of brazing filler metals are identified by the principal element or elements in their chemical composition. In a typical example, such as BCuP-2, the B identifies it as a brazing filler metal (like a prefix E identifies electrodes and R designates welding rods in other AWS specifications). The RB in RBCuZn-A and -D indicates that the filler metal is suitable as a welding rod and as a brazing filler metal. CuP is for copper-phosphorus, the two principal elements in this particular brazing filler metal. Similarly, in other brazing filler metals, Si is for silicon, Ag for silver, etc., using standard chemical symbols. The numeral or letter suffix denotes one particular chemical analysis within a group (1 is one analysis within the same group, 2 another, etc.)

OPERATING CHARACTERISTICS AND USABILITY

BAlSi (Aluminum-Silicon) Classification

Brazing filler metals of the BAlSi classification are used for joining the following grades of aluminum and aluminum alloys: 1060, EC, 1100, 3003, 3004, 3005, 5005, 5050, 6053, 6061, 6063, 6951, 7005 and cast alloys A612 and C612. All classified filler metals are suitable for the furnace and dip brazing processes. The BAlSi-3 and -4 filler metals are suitable for the manual and automatic torch brazing processes. These filler metals are also used to a limited extent with the other brazing processes. They are generally used with lap or tee joints rather than butt joints.

Suitable joint clearances for lap lengths less than 1/4 in. would be 0.002 in. to 0.004 in. for dip brazing and 0.004 in. to 0.008 in. for torch, furnace and automatic oxygen brazing. Longer lap lengths may require clearances up to 0.025 in. Fluxing is essential for all processes. After brazing, the brazed parts must be cleaned thoroughly. Immersion in boiling water generally will remove the majority of the residue; following this, the parts are usually immersed in a concentrated commercial nitric acid or other suitable acid solution and then rinsed thoroughly.

BAlSi-2 brazing filler metal is available as a sheet and as a standard cladding on one or both sides of a brazing sheet having a core of either 3003 or 6951 aluminum alloy. It is used for furnace and dip brazing only.

Both BAlSi-3 and BAlSi-4 are general-purpose brazing filler metals. BAlSi-3 is used with all brazing processes, with some casting alloys and where controlled flow is desired. BAlSi-4 is also used with all brazing processes for complete flow and for good corrosion resistance.

BAlSi-5 brazing filler metal is available as a sheet and as a standard cladding on one or both sides of a brazing sheet having a core of 6951 aluminum alloy. BAlSi-5 is used for furnace and dip brazing at lower temperatures than BAlSi-2. The core alloy employed in brazing sheet with this filler metal cladding can be heat-treated and aged after brazing.

BCuP (Copper-Phosphorus) Classification

Brazing filler metals of the BCuP classification are used primarily for joining copper and copper alloys with some limited use on silver, tungsten and molybdenum. These filler metals should not be used on ferrous or nickel-base alloys. While these filler metals are used in the brazing of cupronickels, caution should be exercised in their use for brazing alloys containing greater than 30% nickel. They are suitable for all brazing processes. Lap joints are recommended, but butt joints may be used if requirements are less stringent.

These filler metals have self-fluxing properties when used on copper; however, a flux is recommended when used on all other metals, including the copper alloys. Corrosion resistance is satisfactory except when the joint is in contact with sulfurous atmospheres at elevated temperatures.

These filler metals also have a tendency to liquate if heated slowly. It is important, therefore, especially if the filler metal is preplaced, to heat as rapidly as possible. After brazing, the color is light gray. Immersion in 10% sulfuric acid will restore the copper color.

It will be noted that the brazing temperature ranges for these filler metals begin below the liquidus. It is a basic principle in brazing that the lowest temperature within the recommended range be used, consistent with the joint clearance and the time needed to complete the braze. This in turn will depend on the materials joined, the brazing process employed and the joint design used.

BCuP-1 brazing filler metal is used primarily for preplacing in the joint and is particularly suited for resistance brazing and some furnace-brazing applications. This filler metal is somewhat more ductile than the other BCuP filler metals containing more phosphorus. It is also less fluid at brazing temperature. Joint clearances should be from 0.002 to 0.005 in.

BCuP-2 and -4 brazing filler metals are extremely fluid at brazing temperatures and will penetrate joints with very little clearance. Best results are obtained with clearances of 0.001 to 0.003 in.

BCuP-3 and -5 brazing filler metals may be used where very close fits cannot be held. Joint clearances of 0.001 to 0.005 in. are recommended.

BAg (Silver) Classification

Brazing filler metals of the BAg classification are used for joining most ferrous and nonferrous metals except aluminum and magnesium. These filler metals have good brazing properties and are suitable for preplacement in the joint or for manual feeding into the joint. All methods of heating may be used. Lap joints are generally used; however, butt joints may be used if requirements are less stringent. Joint clearances of 0.002 to 0.005 in. are recommended for proper capillary action. Flux is generally required; however, on most metals, if brazing is done in a suitable atmosphere with BAg-8, BAg-8a, BAg-18 or BAg-19, flux may not be necessary.

BAg-1 brazing filler metal has the lowest flow temperature of the BAg filler metals. It also flows most freely into capillary joints. Its narrow melting range is suitable for rapid or slow methods of heating. BAg-1 is recommended for first consideration on applications requiring a low-temperature, free-flowing BAg filler metal. BAg-1 is more economical (5% less Ag) than BAg-1a. BAg-1a brazing filler metal has properties similar to BAg-1. Either composition may be used where low-temperature, free-flowing filler metals are desired.

BAg-2 brazing filler metal is, like BAg-1, free-flowing and suited for general-purpose work. Its broader melting range is helpful where clearances are not uniform, but, unless heating is rapid, care must be taken that the lower melting constituents do not separate out by liquation. BAg-2a brazing filler metal is similar to BAg-2, but is more economical than BAg-2 since it contains 5% less silver.

BAg-3 brazing filler metal is a modification of BAg-1a. It has good corrosion resistance in marine and caustic media and when used on stainless steel will inhibit interface corrosion. Its nickel content improves its wetting ability on tungsten-carbide tool tips and its largest use is to braze carbide tool assemblies. Melting range and low fluidity make BAg-3 suitable for building fillets or bridging large gaps. BAg-4 brazing filler metal is, like BAg-3, used extensively for carbide tip brazing. It is freer flowing than BAg-3. BAg-5 and -6 brazing filler metals are used particularly for brazing in the electrical industry. They are also used in the dairy and food industries where the use of cadmium-containing alloys might be prohibited. BAg-6 has a broad melting range and is not as free-flowing as BAg-1 and -2; it is a good filler metal for bridging gaps or forming fillets.

BAg-7 brazing filler metal is low-melting with good flow and wetting properties. It is used (1) where lower zinc and cadmium content is helpful in low heating cycles (2) for food equipment where cadmium must be avoided (3) to minimize stress corrosion of nickel or nickel-base alloys at low brazing temperatures and (4) where white color will improve appearance.

BAg-8 brazing filler metal is suitable for use in controlled atmosphere furnace brazing or vacuum brazing without the use of a paste flux. It is usually used on copper or copper alloys. When molten, BAg-8 is very fluid and may flow out over the work surfaces during some furnace brazing applications. It can also be used on stainless steel, nickel-base alloys and on carbon steel, although its wetting action on these metals is slow.

BAg-8a brazing filler metal is used for pure, dry atmosphere brazing and is of particular advantage when brazing precipitation-hardening and other stainless steels in the 1400 to 1600 F range. The lithium content serves to promote wetting and to increase the flow of the filler metal on difficult to braze metals and alloys. Lithium is particularly helpful on base metals containing minor amounts of titanium or aluminum.

BAg-13 brazing filler metal is used for service temperatures up to 700 F. Its low-zinc content makes it suitable for furnace brazing. BAg-13a brazing filler metal is similar to BAg-13 except that it contains no zinc. This is an advantage where zinc volatilization is objectionable in furnace operations.

BAg-18 brazing filler metal is similar to BAg-8 in its applications. Its tin content helps promote wetting on stainless steel, nickel-base alloys and carbon steel. It has a lower liquidus than BAg-8 and is used in step brazing applications where fluxless brazing is important.

BAg-19 brazing filler metal is used for the same applications as BAg-8a. BAg-19 is usually used where the precipitation-hardening heat treatment is combined with the brazing operation while BAg-8a is used where the higher brazing temperatures are not required.

BAu (Precious Metals) Classification

Brazing filler metals of the BAu classification are used for the brazing of iron, nickel and cobalt-base metals where resistance to oxidation or corrosion is required. Because of their low rate of interaction with the base metal, they are commonly used on thin base metals. These filler metals are usually used with induction, furnace or resistance heating in a reducing atmosphere or in a vacuum and with no flux. For other applications, a borax-boric acid flux is used.

BAu-1, -2 and -3 brazing filler metals, when used for different joints in the same assembly, permit variation in brazing temperature so that step brazing can be used.

BAu-4 brazing filler metal is used to braze a wide range of high-temperature iron- and nickel-base alloys.

BCu (Copper) Classification

Brazing filler metals of the BCu classification are used for joining various ferrous and nonferrous metals. They can also be used with various brazing processes. Lap and butt joints are commonly used.

BCu-1 brazing filler metal is used for joining ferrous metals, nickel-base and copper-nickel alloys. It is free-flowing and is often used in furnace brazing with a hydrogen or disassociated ammonia atmosphere and generally no flux. However, on metals that have constituents with difficult-to-reduce oxides (chromium, manganese, silicon, titanium, vanadium, aluminum and zinc), a flux may be required. BCu-1a brazing filler metal is a powder form of BCu-1, and its application and use are similar to BCu-1.

BCu-2 brazing filler metal is supplied as a copper-oxide suspension in an organic vehicle. Its applications are similar to BCu-1 and -1a.

RBCuZn (Copper-Zinc) Classification

Brazing filler metals of the RBCuZn classification are used for the same materials and under the same circumstances as the BCu classifications. However, with the RBCuZn filler metals, overheating must be particularly guarded against since voids may be formed in the joint by entrapped zinc vapors. The corrosion resistance of these filler metals is generally inadequate for joining copper, silicon-bronze, copper-nickel or stainless steel.

RBCuZn-A brazing filler metal is used on steels, copper, copper alloys, nickel, nickel-base alloys and stainless steel where corrosion resistance is not of importance. It is used with the torch, furnace and induction processes. Fluxing is generally required and a borax-boric acid flux is commonly used. Joint clearances from 0.002 to 0.005 in. are suitable.

RBCuZn-D brazing filler metal (called white brass) is used with steel, nickel and nickel-base alloys. It can be used with all brazing processes.

BMg (Magnesium) Classification

Brazing filler metal BMg-1 is used for joining AZ10A, K1A and M1A magnesium base metals while BMg-2a is used for joining these alloys and the AZ31B and ZE10A compositions.

BMg-1 and -2a brazing filler metals are suitable for the torch, dip or furnace brazing processes. The BMg-2a classification, because of its lower melting range, is usually preferred in most brazing applications. Heating must be closely controlled with all filler metals to prevent melting of the base metal. A flux is used with all processes. Joint clearances of 0.004 to 0.010 in. are best for most applications. Corrosion resistance is good if the flux is completely removed after brazing.

BNi (Nickel) Classification

Brazing filler metals of the BNi classification are generally used for their corrosion- and heat-resistant properties. Base metal alloys most commonly brazed with the BNi filler metals are AISI 300 series stainless steels, AISI 400 series stainless steels and nickel- and cobalt-base alloys. Other base metals such as carbon steel, low-alloy steel and copper are also brazed with the BNi filler metals when specific properties are desired.

The BNi filler metals retain their heat resistance at service temperatures up to 1800 F, depending on the specific filler metal. They are also satisfactorily used for room temperature applications and where the service temperatures are equal to the temperatures of liquid oxygen, helium or nitrogen.

The best quality can be obtained by brazing in an atmosphere which is reducing to both the base metal and the brazing filler metal. A vacuum (below 5 μ of mercury), a pure, dry hydrogen (-60 deg FDP—degree Fahrenheit dew point or below) atmosphere or a pure, dry argon (-80 deg FDP or below) atmosphere are examples which are commonly used.

With the application of a suitable flux, any brazing process (torch, furnace, induction, etc.) can be used to braze an assembly with BNi filler metals. The BNi filler metals are particularly suited to vacuum systems and vacuum tube applications because of their very low vapor pressure. The limiting element is chromium in those alloys in which it is employed. No silver, copper, zinc or other high vapor pressure element is employed. It should be noted that when phosphorus is combined with some other elements, these compounds have very low vapor pressures and can be readily used in a vacuum brazing atmosphere of 0.1 micrometer of mercury at 1950 F without removal of the phosphorus.

BNi-1 brazing filler metal is widely used for high-strength and heat-resistant joints. It is used for joining turbine blades, jet engine parts, highly stressed sheet-metal structures and other highly stressed components.

BNi-2 brazing filler metal has properties and uses which are similar to BNi-1, except that brazing can be conducted at lower temperatures with better flow and diffusion.

BNi-3 brazing filler metal is a good heat-resistant metal for lower temperature brazing of highly stressed parts. It is used for applications similar to BNi-1 and will flow well in marginal atmospheres, or where joints of close tolerances or large areas are encountered.

BNi-4 brazing filler metal is similar to BNi-3. It is used for forming larger, more ductile fillets. It can be used to form joints where fairly large gaps are present.

BNi-5 brazing filler metal is used for high-strength and oxidation-resistant joints for elevated temperature service. Its applications are similar to BNi-1 except that it can be used in certain nuclear applications where boron cannot be tolerated.

BNi-6 brazing filler metal is extremely free-flowing and exhibits a minimum amount of erosion with most nickel- and iron-base metals. It is good for use in marginal atmospheres and for brazing low-chromium steels in exothermic atmospheres.

BNi-7 brazing filler metal is used for the brazing of honeycomb structures, thin-walled tube assemblies and other structures which are used at high temperatures. Erosion can be controlled because of low solubility with iron- and nickel-base alloys. It produces strong, leak-proof joints with heat-resistant base metals at relatively low brazing temperatures. It is recommended for nuclear applications where boron cannot be used. The best results are obtained when it is used in the furnace brazing process. Ductility is improved by increasing time at brazing temperature.

UNCLASSIFIED FILLER METALS

In addition to those filler metals classified in AWS A5.8, there are available a number of special-purpose filler metals based on relatively uncommon metals which are not yet sufficiently standardized to be included in the specification. As the newer filler metals come into more common usage and become sufficiently standardized, they will be classified to ease selection.

BRAZING FILLER METAL FORMS

Most of the filler metals are ductile and can be rolled or drawn to wire or strip in various standard forms. The ductile as well as non-ductile filler metals can also be supplied as powders which are made either by mechanical disintegration (filing, milling, etc.) of the cast filler metals or by atomization of the molten filler metals. A listing of the standard forms and sizes of classified filler metals is given in AWS A5.8.

FLUXES AND ATMOSPHERES

Proper fluxes or protective atmospheres are important adjuncts to brazing. For details concerning fluxes and atmospheres, refer to the individual base metal chapters in Section 4.

VENTILATION DURING BRAZING

Three major factors in brazing which govern the amount of contamination to which brazers may be exposed are (1) Dimensions of the space in which brazing is to be done (with special regard to height of ceiling) (2) number of brazers (3) possible evolution of hazardous fumes, gases or dusts according to the base metals and filler metals being used. ANSI Standard Z49.1, Safety in Welding and Cutting, discusses the ventilation that is required during brazing and should be referred to for details. Attention is particularly drawn to Section 8–Health Protection and Ventilation.

Brazing filler metals of classifications BAg-1, BAg-1a, BAg-2, BAg-2a and BAg-3 contain cadmium. If these brazing filler metals are improperly used and subjected to overheating, cadmium oxide fumes can be generated. Cadmium oxide fumes are particularily hazardous, and inhalation of these fumes in excess of standards established can be fatal. Therefore, particular precautions should be exercised when brazing with these cadmium-containing brazing filler metals.

BIBLIOGRAPHY

"Surfacing by Welding for Wear Resistance," *Composite Engineering Laminates*, The MIT Press (1969).

"Registration Record of Aluminum Association Alloy Designations and Chemical Composition Limits for Wrought Aluminum Alloys," The Aluminum Association (May 1, 1965).

Filler Metals for Joining, O. T. Barnett, Reinhold Publishing Corp., New York (1959).

"The Welding of Titanium and Titanium Alloys," G. E. Faulkner and C. B. Voldrich, DMIC Report 122 (Dec. 31, 1959).

"Development, Properties and Usability of Low-Hydrogen Electrodes," D. C. Smith, *Welding Journal*, 38 (9), 377s-392s (1959).

"Elevated-Temperature Properties of Modified Type 347 Weld Metals," T. J. Moore, *Welding Journal*, 38 (12), 457s-474s (1959).

"Filler Wire is the Key to Better Titanium Welds," J. H. Johnson and E. F. Funk, *Welding Engineer* 43 (6), 45, 46 and 48 (1958).

"Metallurgical Considerations in the Argon-Arc Welding of Aluminum," J. E. Tomlinson and J. G. Young, British Welding Research Association, London (1957).

"Effect of Moisture in the Coatings of Low-Hydrogen Iron-Powder Electrodes," D. C. Smith, W. G. Rinehart and K. P. Johannes, *Welding Journal*, 35 (7), 313s-322s (1956).

"Welding Type 347 Stainless Steel for 1100°F Turbine Operation," R. M. Curran and A. W. Rankin, *Welding Journal*, 34 (3), 205-213 (1955).

Welding for Engineers, Udin, Funk and Wulff, John Wiley & Sons, New York (1954).

"The Incidence of Cracking in Welding Type 347 Steels," L. K. Poole, *Welding Journal*, 32 (8), 403s-412s (1953).

"Weldability of Steels," Second Edition, Welding Research Council (1971).

"Welded Joints Between Dissimilar Metals in High Temperature Service," R. W. Emerson and W. R. Hutchinson, *Welding Journal*, 31 (3), 126s-141s (1952).
"Hard Facing for Impact," H. S. Avery, *Welding Journal*, 31 (2), 116-145 (1952).
"Hard Facing Alloys of the Chromium Carbide Type," H. S. Avery and H. J. Chapin, *Welding Journal*, 31 (10), 917-930 (1952).
"Selecting Hard Facing Metals to Resist Impact, Heat, Friction, Abrasion," H. S. Avery, *Product Engineering* (March, 1952).
"Hard Facing Alloys for Steel Mill Use," H. S. Avery, *Iron and Steel Engineer* (September, 1951).
"Some Characteristics of Composite Tungsten Carbide Weld Deposits," H. S. Avery, *Welding Journal*, 30 (2), 144-160 (1951).
"Physical and Welding Metallurgy of Chromium Stainless Steels," H. Thielsch, *Welding Journal*, 30 (5), 209s-250s (1951).
"Corrosion of Molybdenum-Bearing Stainless Steel Weld Metals," A. L. Schaeffler and R. D. Thomas, Jr., *Welding Journal*, 29 (1), 13s-31s (1950).
Aluminum Welding, Reynolds Metals Co., Richmond, Virginia.
Welding Kaiser Aluminum, Kaiser Aluminum and Chemical Corp., Oakland, California.
Welding Alcoa Aluminum, Aluminum Company of America, New Kensington, Pennsylvania.
Brazing Manual, American Welding Society, Miami, Florida (1963).
ANSI Standard Z49.1, Safety in Welding and Cutting, American Welding Society, Miami, Florida.

INDEX

D

G

Q

R

X

Z